Functional Analysis in China

Mathematics and Its Applications

Managing Editor:

M. HAZEWINKEL

Centre for Mathematics and Computer Science, Amsterdam, The Netherlands

Volume 356

Functional Analysis in China

edited by

Bingren Li
*Institute of Mathematics,
Academia Sinica,
Beijing, China*

Shengwang Wang
*Department of Mathematics,
Nanjing University,
Nanjing, China*

Shaozong Yan
*Department of Mathematics,
Fudan University,
Shanghai, China*

and

Chung-Chun Yang
*Department of Mathematics,
The Hong Kong University of Science and Technology,
Clear Water Bay,
Kowloon, Hong Kong*

SPRINGER-SCIENCE+BUSINESS MEDIA, B.V.

A C.I.P. Catalogue record for this book is available from the Library of Congress

ISBN 978-94-010-6567-2   ISBN 978-94-009-0185-8 (eBook)
DOI 10.1007/978-94-009-0185-8

Printed on acid-free paper

TABLE OF CONTENTS

Part Two: Research Papers

Preface

Functional analysis was originally established as a mathematical branch in the 1930s. After almost 60 years of continuous development and with the borrowing of various methods and techniques from analysis, algebra, geometry, and topology, it has become a very important research area itself and an essential research tool in almost every mathematical discipline.

It was the late Professor Chao-chih Kwan who first advocated the study of functional analysis in China during the 1950s, conducting seminars at the Institute of Mathematics, Academia Sinica and offering courses at the Mathematics and Mechanics Department of Peking University. In the late 1940s, Professor Kwan went to Poincare Institute of Paris University and studied under Professor M. Fréchet, one of the founders of the general topology and functional analysis. Through the efforts of Professor Kwan and various other colleagues of several generations, functional analysis has become one of the main mathematical branches in China, with many outstanding contributions and achievements.

Here, due to the limitation of space, we shall mention only two of China's most important and influential functional analysts with their contributions. The first is the late Professor Yuan-Yung Tseng. He was educated at the University of Chicago under Professor E.H. Moore working on the theory of generalized inverse matrix and received his Ph.D. in 1932. While in Chicago, he even attended the courses given by Von Neumann and M. H. Stone. His first important contribution was the derivation of the so-called "Resolution of the identity of self-adjoint operators". In the 1940s, he introduced the concept of the "generalized inverse" for unbounded operators in Hilbert space, made some contributions to the study of the biorthonormal system of Hilbert space and generalized the concept of Hilbert space to inner product space. Professor Dao-Xing Xia is another leading Chinese functional analyst. He studied under Professor I. M. Gelfand. In the early 1960s, he established the singular integral models of hypernormal and semihypernormal operators on Hilbert space. These models have opened up a new direction in the study of this type of operators and have made a significant impact on the mathematical community world-wide. Another important contribution is his systematic studies on the measure and integral theory in infinitely dimensional spaces. This is a very complicated topic in functional analysis.

This book contains thirty survey and research papers by renowned experts and talented young analysts from leading Universities or Institutions. The topics of these papers deal with mainly the following five aspects in functional analysis.

1. Nonlinear part, which includes critical point theory, fixed point theory, topological degree theory and nonsmooth analysis as well as their applications to the theory of nonlinear integral equations and nonlinear partial differential equations.

2. Linear operator theory, which includes hypernormal and semihypernormal operator theory, operator theory in indefinite inner product spaces, Toeplitz operators, completely irreducible operators, almost isometric operators, multivariate operator theory and local spectral theory.

3. Theory of operator algebras, which includes real Von Neumann algebras, liminal (AF) algebras, C^*-crossed products, automorphism groups, completely positive maps and non-commutative tori.

4. Applications that include the solvability of some partial differential equations. Also included are the theory of semigroups, integrated semigroups and C-semigroups and their applications to abstract Cauchy problems and linear systems.

5. Special spaces that contain Banach spaces and topological vector spaces, including locally convex Riesz spaces, Orlicz spaces, Ba-spaces and their applications.

We would like to take this opportunity to thank the contributors for their cooperation and efforts in presenting their contributions. Last but not least, we want to thank Kluwer and its staff for their endorsement of the project and the help they offered throughout the preparation of the book.

| Chung-Chun Yang | 杨重骏 | Shengwang Wang | 王声望 |
| Shaozong Yan | 严绍宗 | Bingren Li | 李炳仁 |

Contributors

Shangquan Bu 步尚全
>Department of Applied Mathematics, Tsinghua University, Beijing 100084

Guangfu Cao 曹广福
>Department of Mathematics, Jilin University,
>Changchun, Jilin 130023

Shihsen Chang (Shisheng Zhang) 张石生
>Department of Mathematics, Sichuan University,
>Chengdu, Sichuan 610064

Shutao Chen 陈述涛
>Department of Mathematics, Harbin Normal University,
>Harbin, Heilongjian 150001

Xiaoman Chen 陈晓漫
>Institute of Mathematics, Fudan University,
>Shanghai 200433

Guanggui Ding 定光桂
>Department of Mathematics, Nankai University,
>Tianjin 300071

Yanheng Ding 丁彦恒
>Institute of Mathematics, Academia Sinica,
>Beijing 100080

Xianling Fan 范先令
>Department of Mathematics, Lanzhou University,
>Lanzhou, Gansu 730001

Dexing Feng 冯德兴
>Institute of Systems Science, Academia Sinica,

Beijing 100080

Dajun Guo　郭大钧
Department of Mathematics, Shandong University,
Jinan, Shandong 250100

Tiexin Guo　郭铁信
Department of Mathematics, Xiamen University,
Xiamen, Fujian 361005

Yuzan He　何育赞
Institute of Mathematics, Academia Sinica,
Beijing 100080

Zhaobo Huang　黄昭波
Institute of Mathematics, Fudan University,
Shanghai 200433

Youqing Ji　纪友清
Department of Mathematics, Jilin University,
Changchun, Jilin 130023

Chunlan Jiang　蒋春澜
Department of Mathematics, Jilin University,
Changchun, Jilin 130023

Banghe Li　李邦河
Institute of Systems Science, Academia Sinica,
Beijing 100080

Bingren Li　李炳仁
Institute of Mathematics, Academia Sinica,
Beijing 100080

Shaokuan Li　李绍宽

Chinese Textile University, Shanghai 200051

Shujie Li 李树杰
Institute of Mathematics, Academia Sinica, Beijing 100080

Yaqing Li 李雅卿
Institute of Systems Science, Academia Sinica,
Beijing 100080

Jin Liang 梁进
Kunming Institute of Technology, Kunming,
Yunnan 650093

Qing Lin 林青
Institute of Mathematics, Academia Sinica,
Beijing 100080

Xi Lin 林熙
Jimei Navigation College, Xiamen,
Fujian 361021

Peizhu Luo 罗佩珠
Institute of Applied Mathematics, Academia Sinica,
Beijing 100080

Shanli Sun 孙善利
Institute of Mathematics, Jilin University,
Changchun, Jilin 130023

Shunhua Sun 孙顺华
Department of Mathematics, Sichuan University,
Chengdu, Sichuan 610064

Yusheng Tong 童裕孙
Institute of Mathematics, Fudan University,

Shanghai 200433

Haiyan Wang　王海燕
　　　Department of Mathematics,
　　　Southeast University, Nanjing, Jiangsu 210096

Shengwang Wang　王声望
　　　Department of Mathematics, Nanjing University,
　　　Nanjing, Jiangsu 210008

Tingfu Wang　王廷辅
　　　Harbin University of Science and Technology,
　　　Harbin, Heilongjian 150001

Zhongyao Wang　王宗尧
　　　Department of Mathematics, East China
　　　University of Science and Technology, Shanghai 200237

Yau-chuen Wong　黄友川
　　　Department of Mathematics, The Chinese University of Hong Kong,
　　　Shatin, N. T., Hong Kong

Congxin Wu　吴从炘
　　　Department of Mathematics,
　　　Harbin Institute of Technology,
　　　Harbin, Heilongjiang 150001

Liangsen Wu　吴良森
　　　Department of Mathematics, East China
　　　Normal University, Shanghai 200062

Tijun Xiao　肖体俊
　　　Department of Mathematics, Yunnan Teachers' University,
　　　Kunming, Yunnan 6500092

Shaozong Yan 严绍宗
>Institute of Mathematics, Fudan University,
>Shanghai 200433

Zhaoyong You 游兆永
>Department of Mathematics, Xi'an Jiaotong University,
>Xi'an, Shaanxi 710049

Dianzhou Zhang 张奠宙
>Department of Mathematics, East China Normal Univeristy,
>Shanghai 200062

Junfeng Zhao 赵俊峰
>Department of Mathematics, Wuhan University,
>Wuhan, Hubei 430072

Kehe Zhu 朱克和
>Department of Mathematics, State University of New York,
>Albany, NY 12222, USA

Linhu Zhu 朱林户
>Department of Applied Mathematics, Xi'an Jiaotong University,
>Xi'an, Shaanxi 710049

Some Problems and Results in the Study of Nonlinear Analysis

Shihsen Chang (Shisheng Zhang)

Department of Mathematics, Sichuan University, Chengdu

This paper is presented as a survey of the latest and new results on some topics in the study of nonlinear analysis, which were obtained by the author and some Chinese mathematicians.

This paper consists of seven sections.

Section 1 is devoted to recalling some definitions, notations and results which will be needed later. Section 2 is devoted to generalizing the famous Fan-Browder theorem. Sections 3 and 4 are devoted to studying the existence problems of solutions for generalized quasi-variational inequalities (in short $GQVI$) and three kinds of complementarity problems. Section 5 presents some more general topological types of KKM theorems, minimax inequality theorems and Fan's matching theorem. Section 6 is devoted to studying the economic equilibrium theorems of Shafer-Sonnschein version.

As applications, in Section 7 we use the results mentioned above to study the optimization problem, nonlinear programming problem, saddle point problem and continuous selection problem.

1. Preliminaries

Let X and Y be nonempty sets. We shall denote by $\mathcal{F}(X)$ the family of all nonempty finite subsets of X. Let $F : X \to 2^Y$ be a multifunction. For $A \subset X$ and $y \in Y$, let

$$F(A) = \cup\{F(x), x \in A\}, \quad F^{-1}(y) = \{x \in X : y \in F(x)\}.$$

Throughout this paper we denote by $(X, \{\Gamma_A\})$ the H-space.

For the definitions and properties of H-space, H-convex set and H-compact set, refer to Bardaro-Ceppitelli [2, 3] and Horvath [39].

1991 Mathematics Subject Classification: 49J40, 90C33, 49J35, 54D05, 54H25, 90A14, 90C30, 65K10

Supported by the National Natural Science Foundation of China

Definition 1.1 [60]. Let X and Y be two topological spaces and $F : X \to 2^Y$ be a multifunction. F is said to be transfer open-valued (resp. transfer closed-valued), if for any $x \in X$, $y \in F(x)$ (resp. $y \notin F(x)$) there exists an $x' \in X$ such that $y \in int(F(x'))$ (resp. $y \notin \overline{F(x')}$). (Throughout this paper we use $int(A)$ and $\bar{A}$ to denote the interior and closure of the set A respectively).

Remark. It is easy to see that a closed-valued (resp. open-valued) mapping is transfer closed-valued (resp. transfer open-valued). But the converse is not true.

Proposition 1.1. Let X be a nonempty set, Y be a topological space and $G' : X \to 2^Y$ be a multifunction.
  (1) G is transfer closed-valued if and only if

$$\bigcap_{x \in X} G(x) = \bigcap_{x \in X} \overline{G(x)}.$$

  (2) G is transfer open-valued if and only if

$$\bigcup_{x \in X} G(x) = \bigcup_{x \in X} int(G(x)).$$

  (3) If X is also a topologival space, $G(x)$ is nonempty for each $x \in X$, and G^{-1} is transfer open-valued, then

$$X = \bigcup_{y \in Y} int(G^{-1}(y)).$$

  (4) In addition, if X is a compact topological space, G is nonempty H-convex-valued and G^{-1} is transfer open-valued, then there exists a continuous selection of G, i.e., there exists a continuous function $f : X \to Y$ such that

$$f(x) \in G(x) \text{ for all } x \in X.$$

2. A generalization of Fan-Browder's fixed point theorem

In 1961, using his own generalization of the KKM theorem, Ky Fan [32] (see also Ky Fan [33-35]) established an important and very basic "geometric" lemma for multifunctions. Later Browder [8] obtained this result in the form of a fixed point theorem. Since then there have appeared numerous generalization of the Fan-Browder fixed point theorem and their applications in various fields as coincidence and fixed point theory, minimax theory, variational inequalities, nonlinear analysis, convex analysis, game theory, mathematical economics and many others.

In this section, we give a new generalization.

Theorem 2.1. Let $\{(X_\alpha, \{\Gamma_{A_\alpha}\}) : \alpha \in I\}$ be a family of H-spaces, I a finite index set, $X = \prod_{\alpha \in I} X_\alpha$ and $\{T_\alpha : \alpha \in I\}$ be a family of multifunctions, where $T_\alpha : X \to 2^{X_\alpha}$ for all $\alpha \in I$. Suppose that

    (i) For any $x \in X$, and for any $\alpha \in I$, $T_\alpha(x)$ is a nonempty H-convex set;

    (ii) For each $\alpha \in I$, $T_\alpha^{-1} : X_\alpha \to 2^X$ is transfer open-valued.

If there exist a compact subset L of X and an H-compact subset K of X such that for each weakly H-convex subset D of X with $K \subset D \subset X$ the following holds :

$$\bigcap_{y \in D} (\overline{\{x \in X : y_\beta \notin T_\beta(x) \text{ for some } \beta \in I\}} \cap D) \subset L, \tag{2.1}$$

where y_β is the projection of y onto X_β, then there exists an $x_* \in X$ such that

$$x_* \in \prod_{\alpha \in I} T_\alpha(x_*), \text{ i.e, } x_{*_\alpha} \in T_\alpha(x_*) \text{ for all } \alpha \in I,$$

where x_{*_α} is the projection of x_* onto X_α for each $\alpha \in I$.

It should be pointed out that, if $\{(X_\alpha, \{\Gamma_{A_\alpha}\}) : \alpha \in I\}$ is a family of compact H-spaces, then the condition (2.1) in Theorem 2.2 holds automatically, so we have the following

Theorem 2.2. Let $(X_\alpha, \{\Gamma_{A_\alpha}\})$ be a family of compact H-spaces, $X = \prod_{\alpha \in I} X_\alpha$ and $\{T_\alpha : \alpha \in I\}$ be a family of multifunctions, where $T_\alpha : X \to 2^{X_\alpha}$ for all $\alpha \in I$. Suppose further that

    (i) For any $x \in X$, and any $\alpha \in I$, $T_\alpha(x)$ is a nonempty H-convex set;

    (ii) For any $\alpha \in I$, $T_\alpha^{-1} : X_\alpha \to 2^X$ is transfer open-valued.

Then there exists an $x_* \in X$ such that $x_* \in \prod_{\alpha \in I} T_\alpha(x_*)$.

3. Generalized quasi-variational inequality problem

The quasi-variational inequality (QVI) was first introduced and studied in Bensoussan-Lions [4, 5], which arises in some problems of random impulse controls, and it has many applications in control and optimization theory, economics and transportation equilibrium, contact problems in elasticity, fluid flow through porous media, game theory and mathematical programming. In recent years, various extensions of QVI problems have been proposed and analysed.

In 1987, Parida-Sen [49] considered the following $GQVI$ problem:

Let $S \subset \mathbb{R}^n$ and $C \subset \mathbb{R}^p$ be two nonempty subsets, $T : S \to 2^C$ be a multifunction and $M : S \times C \to \mathbb{R}^n$, $\eta : S \times S \to \mathbb{R}^n$ be single-valued functions. To find $x \in S$, $y \in T(x)$ such that

$$(M(x, y), \eta(u, x)) \geq 0 \text{ for all } u \in S.$$

Recently, Chang-Shu and Zhang [10, 11, 15], Fu [37], Kum [44], Yao-Guo [64-66] considered the following abstract $GQVI$ problem under various conditions:

Let E, F be two locally convex Hausdorff topological vector spaces, $X \subset E, Y \subset F$ be two nonempty subsets, $S : X \to 2^X, T : X \to 2^Y$ be two multifunctions and $\varphi : X \times Y \times X \to \mathbb{R}^n$ be a continuous function with $\varphi(x, y, x) \geq 0$ for all $x \in X$, $y \in Y$. To find $x \in S(x)$, $y \in T(x)$ such that

$$\varphi(x, y, u) \geq 0 \text{ for all } u \in S(x). \tag{3.1}$$

In this section, we shall use the results presented in Section 2 to study the existence of solutions for $GQVI$ (3.1).

Theorem 3.1. Let $(X, \{\Gamma_A\})$ and $(Y, \{\Gamma_B\})$ be two H-spaces. Suppose that

(i) $T : X \to 2^Y$ is a multifunction with nonempty H-convex values and $T^{-1} : Y \to 2^X$ is transfer open-valued;

(ii) $S : X \to 2^X$ is a multifunction with nonempty compact H-convex values;

(iii) $\varphi : X \times Y \times X \to \mathbb{R}$ is a function such that

(a) $\varphi(x, y, x) \geq 0$ for all $x \in X$ and $y \in T(x)$;

(b) $z \mapsto \varphi(x, y, z)$ is l.s.c.;

(c) For any $(x, y) \in X \times Y$, $\{z \in X : \varphi(x, y, z) = \min_{y \in S(x)} \varphi(x, y, u)\}$ is H-convex;

(d) The multifunction

$$z \mapsto \left\{ (x, y) \in S^{-1}(z) \times Y : \varphi(x, y, z) = \min_{u \in S(x)} \varphi(x, y, u) \right\}$$

is transfer open-valued.

(iv) There exist a compact subset L of $X \times Y$ and an H-compact subset $K \subset X \times Y$ such that for each weakly H-convex subset D of $X \times Y$ with $K \subset D \subset X \times Y$, the following holds:

$$\bigcap_{(u,v) \in D} \overline{(\{(x, y) \in X \times Y : u \notin F_1(x, y) \text{ or } v \notin T(x)\} \cap D)} \subset L,$$

where $F_1 : X \times Y \to 2^X$ is a multifunction defined by

$$F_1(x, y) = \left\{ z \in S(x) : \varphi(x, y, z) = \min_{u \in S(x)} \varphi(x, y, u) \right\} \text{ for each } (x, y) \in X \times Y.$$

Then there exist $\bar{x} \in S(\bar{x})$, $\bar{y} \in T(\bar{x})$ such that

$$\varphi(\bar{x}, \bar{y}, x) \geq 0 \text{ for all } x \in S(\bar{x}).$$

By using Proposition 1.1 and Theorem 2.2, we can prove the following theorems.

Theorem 3.2[21]. Let $(X, \{\Gamma_A\})$ be a compact Hausdorff H-space, and $(Y, \{\Gamma_B\})$ be a Hausdorff H-space. Suppose that

(i) $T : X \to 2^Y$ is a multifunction with nonempty H-convex values and $T^{-1} : Y \to 2^X$ is transfer open-valued;

(ii) $S : X \to 2^X$ is a continuous multifunction with nonempty compact H-convex values and $S^{-1}(x)$ is open for any $x \in X$;

(iii) $\varphi : X \times Y \times X \to \mathbb{R}$ is a continuous function satisfying

(a) $\varphi(x, y, x) \geq 0$ for all $x \in X$ and all $y \in T(x)$;

(b) The function $z \mapsto \varphi(x, y, z)$ is H-quasi-convex.

Then there exist $\bar{x} \in S(\bar{x})$, $\bar{y} \in T(\bar{x})$ such that

$$\varphi(\bar{x}, \bar{y}, x) \geq 0 \text{ for all } x \in S(\bar{x}).$$

Theorem 3.3 [21]. Let $(X, \{\Gamma_A\})$ be a compact Hausdorff H-space, and $(Y, \{\Gamma_B\})$ be a Hausdorff H-space. Suppose further that

(i) $T : X \to 2^Y$ is a multifunction with nonempty H-convex values and $T^{-1} : Y \to 2^X$ is transfer open-valued;

(ii) $S : X \to 2^X$ is an l.s.c. multifunction with nonempty H-convex values and $S^{-1} : X \to 2^X$ is closed-valued.

(iii) $\varphi : X \times Y \times X \to \mathbb{R}$ is a continuous function satisfying

(a) $\varphi(x, y, x) \geq 0$ for all $x \in X$ and all $y \in T(x)$;

(b) For each $(x, y) \in X \times Y$, $\{z \in X : \varphi(x, y, z) = \min_{u \in S(x)} \varphi(x, y, u)\}$ is H-convex;

(iv) For any continuous function $f : X \to Y$, there exists a finite subset $A \subset X$ such that for any $x \in X$, there exists a $z \in A$ satisfying $x \in S^{-1}(z)$ and

$$\varphi(x, f(x), z) = \min_{u \in S(x)} \varphi(x, f(x), u).$$

Then there exists $\bar{x} \in S(\bar{x})$, $\bar{y} \in T(\bar{x})$ such that

$$\varphi(\bar{x}, \bar{y}, x) > 0 \text{ for all } x \in S(\bar{x}).$$

Theorem 3.4 [11]. Let E be a reflexive Banach space, F be a Fréchet space, $X \subset E$ and $Y \subset F$ be two nonempty closed convex sets. Suppose that $T : X \to 2^Y$ is a multifunction with nonempty compact convex values and that it is upper semi-continuous with respect to the weak topology on X and the topolopy on Y. Let $M : X \times Y \to E^*$ be a continuous mapping with respect to the weak topology on X, the topology on Y and the norm topology on E^*. Let $\eta : X \times X \to E$ be a weakly continuous mapping (i.e., a continuous mapping with respect to the topology on X and the weak topology on E) satisfying the following conditions:

(i) $\eta(x, x) = 0$ for all $x \in X$;

(ii) The function $u \mapsto (M(x, y), \eta(u, x))$ is convex and the multifunction $G : X \to 2^{E^*}$ defined by

$$G(x) = \{M(x, y) : y \in T(x)\}$$

is η-monotone;

(iii) There exist $\bar{u} \in X$ and $\bar{v} \in T(\bar{u})$ such that

$$\lim_{\|x\| \to \infty, x \in X} (M(\bar{u}, \bar{v}), \eta(x, \bar{u})) > 0.$$

Then there exist $\bar{x} \in X$, $\bar{y} \in T(\bar{x})$ such that

$$(M(\bar{x}, \bar{y}), \eta(x, \bar{x})) \geq 0 \text{ for all } x \in X.$$

By using Theorem 2.2, we can prove the following theorem, which generalizes Theorem 3.1 and Theorem 3.6 in Yao-Guo [66].

Theorem 3.5. Let X be a nonempty closed convex subset of $\mathbb{R}^n$ and $f : X \to \mathbb{R}^n$. Suppose that

(i) For each $y \in X$, the mapping

$$y \mapsto \{x \in X : (f(x), x - y) \leq 0\}$$

is transfer closed-valued;

(ii) There exists a nonempty bounded subset D of X such that for each $x \in X \backslash D$, there exists a $y \in D$ such that $(f(x), x - y) > 0$.

Then there exists an $\bar{x} \in X$ such that

$$(f(\bar{x}), u - \bar{x}) \geq 0 \text{ for all } u \in X.$$

4. Complementarity problems in Banach spaces

Complementarity theory was introduced and studied by Lemke, Cottle and Dantzig in the early 1960's. Recently, complementarity problems have been extended and generalized in various directions to encompass a large class of problems arising in control, optimization, economics and transportation equilibrium, contact problems in elasticity and fluid flow through porous media (see, for example Isac, Thera [41, 42], Yao [65], Noor [47, 48], Chang, Huang [10, 13], Chen and Yang [26, 62].

In this section, we shall present some results for three kinds of complementarity problems in Banach spaces.

Definition 4.1. Let E be a Banach space, E^* the dual of E, K a nonempty closed convex cone in E, K^* the dual of K, i.e.,

$$K^* = \{x^* \in E^* : (x^*, x) \geq 0 \quad \forall x \in K\},$$

and $T : K \to E^*$ be a mapping.

(1) The so-called complementarity problem is to find an $x_0 \in K$ such that

$$Tx_0 \in K^* \text{ and } (Tx_0, x_0) = 0;$$

(2) In addition, if $S : K \to E^*$ is a linear mapping and $g : K \to K$ is a nonlinear mapping, the so-called implicit complementarity problem is to find an $x_0 \in K$ such that

$$Tg(x_0) + Sg(x_0) \in K^* \text{ and } (Tg(x_0) + Sg(x_0), g(x_0)) = 0;$$

(3) If $C \subset E$ is a subset, and $T : C \to 2^{E^*}$, $g : C \to E$, and $S : C \to E^*$ are three mappings, the so -called generalized implicit complementarity problem (in short, $GICP$) is to find $\bar{x} \in C$ and $\bar{y} \in T(\bar{x})$ such that

$$g(\bar{x}) \in K, \quad S(\bar{x}) + \bar{y} \in K^* \text{ and } (\psi(\bar{x}) + \bar{y}, g(\bar{x})) = 0.$$

The existence of solutions for these kinds of complementarity problems has been considered by many authors. Resently Chang-Shu [11] and Fu [37] obtained some results under suitable conditions.

Theorem 4.1. Let E be a Banach space, K be a locally weakly compact closed convex cone of E, and $T : K \to E^*$, $i = 1, 2$, be two mappings satisfying the following conditions:

(i) For any sequence $\{x_n\} \subset K$, if $x_n @ > w >> \bar{x}$, then $Tx_n @ > w^* >> T\bar{x}$, where $T = T_1 - T_2$;

(ii) For any given $y \in K$, the function $x \mapsto (T_1 x, y - x)$ is weakly u.s.c. and $x \mapsto (T_2 x, y - x)$ is weakly l.s.c.;

(iii) T_1 is a positively homogeneous mapping of p-order and

$$(T_1 x, x) > 0 \text{ for all } x \in K, \quad x \neq 0;$$

(iv)

$$\limsup_{\substack{\|x\| \to \infty \\ x \in K}} \frac{(T_2 x, x)}{\|x\|^{p+1}} \leq 0.$$

Then there exists an $x_0 \in K$ such that

$$Tx_0 \in K^* \text{ and } (Tx_0, x_0) = 0.$$

Remark. Theorem 4.1 improves the main result in Thera [59].

Theorem 4.2 (Fu [37]). Let E be a Banach space, K be a locally compact convex cone of E, and $T : K \to 2^{E^*}$ be u.s.c. with nonempty compact convex values. Let $g : K \to K$ and $S : P \to E^*$ be continuous. Assume that

(i) For each fixed $y \in K^*$, $(y, g(u))$ is convex in u;

(ii) There exist $\bar{u} \in K$, $r > 0$, $\|\bar{u}\| < r$ such that

$$(S(x) + \bar{y}, g(\bar{u}) - g(x)) \leq 0 \text{ for all } x \in K, \quad \|x\| = r, \quad y \in T(x);$$

(iii) g is surjective.

Then the $GICP$ has a solution.

Remark. We would like to point out that how to obtain more general and more better existence theorems for these kinds of complementarity problems remains unsolved.

5. Topological type of KKM theorem and minimax theorem

The celebrated Knaster-Kuratowski-Mazurkiewicz theorem (KKM theorem for short) was first generalized to the infinite dimensional cases by Ky Fan [32], and since then this theorem has become an important theoretical foundation of the KKM technique in dealing with nonlinear problems. Recently Horvath [39], Bardaro-Ceppitelli [2, 3], Shioji [54], Park [50, 51], Liu [46], Ding [29], Ding-Tan [30] and many other authors generalized the KKM theorem by replacing the convexity assumptions with the mere topological property, i.e., contractibility. In Chang-Zhang [12] and Chang-Ma [14], the authors extended the concept of KKM mapping to generalized KKM mapping, obtained a general version of the KKM theorem, Fan's minimax inequality and established a necessary and sufficient conditions that the family of sets has the finite intersection property.

In this section, we shall give some more general topological versions of KKM theorem, Fan's minimax and matching theorems.

Definition 5.1. Let X be a Hausdorff topological space.

(1) X is called an interval space [56], if there exists a mapping $[\cdot, \cdot] : X \times X \to C(X)$ (the family of all connected subsets of X) such that for any $x, y \in X$,

$$x, y \in [x, y] = [y, x];$$

(2) $(X, \{C_A\})$ is called a W-space, if $\{C_A\}$ is a family of nonempty connected subsets of X indexed by a finite subset A of X such that $A \subset C_A$.

Remark. It is easy to see that the Hausdorff topological vector space, convex space, contractible space and connected space are all special cases of interval space and W-space. In addition, if $(X, \{\Gamma_A\})$ is an H-space with $A \subset \Gamma_A$ for all $A \in \mathcal{F}(X)$, then it is also a special case of interval space and W-space.

We first give the following topological version of KKM theorem, which generalizes the corresponding results of Ky Fan [32, 36], Bardaro-Ceppitelli [2, 3], Park [51], Bielawski [6] and Horvath [39].

Theorem 5.1 [17]. Let $(X, \{C_A\})$ be a W-space, Y be a topological space and $F : X \to 2^Y$ be a mapping satisfying the following conditions:

(i) $F(x)$ is nonempty open (closed) for all $x \in X$;

(ii) For any $A \in \mathcal{F}(X)$, $\bigcap_{x \in A} F(x)$ is connected;

(iii) $F^{-1}(y)$ is open for each $y \in Y$;

(iv) For any $x_1, x_2 \in X$, $F(C_{\{x_1, x_2\}}) \subset F(x_1) \cup F(x_2)$.

Then (1) The family $\{F(x) : x \in X\}$ of sets $F(x)$ has the finite intersection property;

(2) If for each $x \in X$, $F(x)$ is closed and there exists an $x_0 \in X$ such that $F(x_0)$ is compact, then $\bigcap_{x \in X} F(x) \neq \emptyset$.

Next we give a topological type of minimax theorem, which contains the main results of Lin-Quan [45], Geraghty and Lin [38], and Wu [67] as its special cases.

Theorem 5.2 [18]. Let $(X, \{C_A\})$ be a W-space, Y be a Hausdorff topological space, f and g be two functions from $X \times Y$ to $\mathbb{R}$ satisfying the following conditions:

(i) $y \mapsto g(x, y)$ and $x \mapsto g(x, y)$ are u.s.c. and $y \mapsto f(x, y)$ is l.s.c..

(ii) $f(x, y) \leq g(x, y)$ for all $(x, y) \in X \times Y$;

(iii) (a) for any $A \in \mathcal{F}(X)$ and $r \in \mathbb{R}$, the set $\{y \in Y : g(x, y) < r, \quad \forall x \in A\}$ is connected;

      (b) For any $\{x_1, x_2\} \subset X$, there exists a connected subset $C_{\{x_1, x_2\}} \subset X$ with $\{x_1, x_2\} \subset C_{\{x_1, x_2\}}$ such that

$$g(x, y) \geq \min\{g(x_1, y), g(x_2, y)\} \text{ for all } x \in C_{\{x_1, x_2\}} \text{ and } y \in Y;$$

(iv) Y is compact.

Then

$$\sup_{x \in X} \inf_{y \in Y} g(x, y) \geq \inf_{y \in Y} \sup_{x \in X} f(x, y).$$

Recently, in [19, 20], the authors studied the topological type of KKM theorem and minimax theorem in interval spaces and obtained the following results.

Theorem 5.3 [20]. Let X be a strongly Dedekind complete Hausdorff interval space, Y an interval space and $F : X \to 2^Y$ a multifunction with nonempty compact values. If the following conditions are satisfied:

(i) For any $A \in \mathcal{F}(X)$, $\bigcap_{x \in A} F(x)$ is connected;

(ii) For any $x, y \in X$, $F(u) \subset F(x) \cup F(y)$ for all $u \in [x, y]$;

(iii) For any $y \in Y$, $F^{-1}(y)$ is closed,

then $\bigcap_{x \in X} F(x) \neq \emptyset$.

Theorem 5.4 [19]. Let X be an interval space, Y be a topological space and Z be a complete dense totally ordered space. If $f : X \times Y \to Z$ satisfies the following conditions:

(i) $f(x, \cdot)$ is u.s.c. for all $x \in X$;

(ii) For any $A \in \mathcal{F}(X)$ and for all $z \in Z$, the set

$$\bigcap_{x \in A} \{y \in Y : f(x, y) > z\}$$

is connected;

(iii) $f(\cdot, y)$ is quasi-convex for all $y \in Y$ and is l.s.c. on any connected set of X;

(iv) There exist $x_0 \in X$, $z_0 < \inf_{x \in X} \sup_{y \in Y} f(x, y)$ and a compact subset $L \subset Y$ such that

$$f(z_0, y) < z_0 \text{ for all } y \in Y \backslash L,$$

then

$$\sup_{y \in Y} \inf_{x \in X} f(x, y) = \inf_{x \in X} \sup_{y \in Y} f(x, y).$$

Remark. Theorem 5.4 contains the main results in Cheng-Lin [28] as a special case, and so it also contains the main results in Brezis et al. [7], Komornik [43] and Geraghty-Lin [38] as special cases. Some of its corollaries also generalize and improve the corresponding results in Lin-Quan [45], Sion [55] and Stachó [56].

6. Economic equilibrium theorem of Shafer-Sonnenschein version

Definition 6.1. An abstract economy (or generalized game) $\Gamma = (X_i, A_i, B_i, P_i)_{i \in I}$ is defined as a family or ordered quadruples (X_i, A_i, B_i, P_i), where I is a finite or an infinite set of agents, X_i is a nonempty topological space (a choice set), $A_i, B_i : \prod_{k \in I} X_k \to 2^{X_i}$ are constraint correspondences, and $P_i : \prod_{k \in I} X_k \to 2^{X_i}$ is a preference correspondence. An equilibrium for Γ is a point $\bar{x} \in X = \prod_{i \in I} X_i$ such that for each $i \in I$, $\bar{x}_i \in \overline{B}_i(\bar{x})$ and $P_i(\bar{x}) \cap A_i(\bar{x}) = \emptyset$. When $A_i = B_i$ for each $i \in I$ and X_i is a topological vector space, $i \in I$, then the definitions of an abstract economy and an equilibrium coincide with the standard definition of Shafer-Sonnenschein [53].

Definition 6.2. Let X be an interval space and B a subset of X.

(1) B is said to be W-convex, if for any $x, y \in B$ we have $[x, y] \subset B$;

(2) The intersection of all W-convex and closed subsets in X containing B is called the W-closed convex hull of B. We denote it by W-$\overline{co}B$.

By using Theorem 5.3 we can prove the following economic equilibrium theorems of Shafer-Sonnenschein version.

Theorem 6.1 [20]. Let $\Gamma = (X_i, A_i, B_i, P_i)_{i \in I}$ be an abstract economy satisfying the following conditions:

(i) For any $i \in I$, X_i is a Hausdorff interval space and their product interval space $X := \prod_{i \in I} X_i$ is strongly Dedekind complete;

(ii) For any $i \in I$, D_i is a nonempty compact W-convex subset of X_i; the constraint correspondences A_i, $B_i : X \to 2^{D_i}$ and the preference correspondence $P_i : X \to 2^{D_i}$ satisfy the following conditions:

(a) For all $x \in X$, $\emptyset \neq A_i(x) \subset B_i(x)$ is W-convex closed;

(b) The set $M_i := \{x \in X : A_i(x) \cap P_i(x) \neq \emptyset\}$ is open, $B_i^{-1}(y_i)$ and $(W$-$co(A_i \cap P_i))^{-1}(y_i)$ are both closed for all $y_i \in D_i$;

(c) For any $y \in D := \prod_{i \in I} D_i$, $y_i \notin W$-$\overline{co}(A_i \cap P_i)(y)$;

(iii) For any $y \in D$, for any $x_1, x_2 \in X$, and for any $x \in (x_1, x_2) := [x_1, x_2] \backslash \{x_1, x_2\}$, there exists an $i_0 \in I$ such that $A_{i_0}(x) \cap P_{i_0}(x) \neq \emptyset$ and $y_{i_0} \notin W$-$\overline{co}(A_{i_0} \cap P_{i_0})(x)$.

Then Γ has an equilibrium $\bar{x}$ in D, i.e., for any $i \in I$

$$\bar{x}_i \in B_i(\bar{x}) \quad \text{and} \quad A_i(\bar{x}) \cap P_i(\bar{x}) = \emptyset.$$

Theorem 6.2 [20]. Let $\Gamma = (X_i, A_i, B_i, P_i)_{i \in I}$ be an abstract economy satisfying the following conditions:

(i) For each $i \in I$, X_i is a Hausdorff interval space, D_i is a nonempty compact subset of X_i, and $X := \prod_{i \in I} X_i$ is the product interval space;

(ii) For each $i \in I$ the constraint correspondence A_i, $B_i : X \to 2^{D_i}$ and the preference correspondence $P_i : X \to 2^{D_i}$ satisfy the following conditions:

(a) For each $x \in X$, $\emptyset \neq A_i(x) \subset B_i(x)$ and $B_i(x)$ is W-convex closed;

(b) $M_i = \{x \in X : (A_i \cap P_i)(x) \neq \emptyset\}$ is open;

(c) For all $x \in D := \prod_{i \in I} D_i$, $x_i \notin W$-$\overline{co}(A_i \cap P_i)(x)$;

(d) B_i and W-$\overline{co}(A_i \cap P_i) : X \to 2^{D_i}$ both are upper semi-continuous;

(iii) For any $y \in D$, for any $x_1, x_2 \in X$, and for any $x \in (x_1, x_2)$ there exists an $i_0 \in I$ such that $A_{i_0}(x) \cap P_{i_0}(x) \neq \emptyset$ and $y_{i_0} \notin W\text{-}\overline{co}(A_{i_0} \cap P_{i_0})(x)$.
Then Γ has an equilibrium $\bar{x}$ in D.

7. Applications

(I) Minimization problem

Theorem 7.1. Let X be a closed convex subset of $\mathbb{R}^n$, and let f be a Gâteaux differentiable function from an open convex set Ω containing X into $\mathbb{R}$. Suppose that f is pseudoconvex on Ω, i.e., for every $x, u \in \Omega$ such that $x \neq u$, if $(\nabla f(x), u - x) \geq 0$ implies $f(u) \geq f(x)$, and the Gâteaux differential ∇f satisfies the following conditions;
  (i) For any $y \in X$, the mapping

$$y \mapsto \{x \in X : (\nabla f(x), x - y) \leq 0\}$$

is transfer closed.
  (ii) There exists an $x_0 \in X$ such that

$$\lim_{\|x\| \to \infty, x \in X} (\nabla f(x), x - x_0) > 0. \tag{7.1}$$

Then there exists an $\bar{x} \in X$ such that

$$f(\bar{x}) = \min_{x \in X} f(x).$$

Remark. This theorem generalizes Theorem 7.3 in Yao-Guo [66].

(II) Nonlinear programming and Saddle point problem

Let E be a reflexive Banach space, F a Fréchet space, $X \subset E$ and $Y \subset F$ be two nonempty subsets, and $L : X \times Y \to \mathbb{R}$ be a function.

Definition 7.1. The first kind of nonlinear programming problem is to find $(\bar{x}, \bar{y}) \in U$ such that

$$L(\bar{x}, \bar{y}) = \min_{(x,y) \in U} L(x, y), \tag{P_1}$$

where

$$U = \{(x, y) \in X \times Y : L(x, y) = \max_{v \in Y} L(x, y)\}.$$

The second kind of nonlinear programming problem is to find $(\bar{x}, \bar{y}) \in W$ such that

$$L(\bar{x}, \bar{y}) = \max_{(x,y) \in W} L(x, y), \qquad (P_2)$$

where

$$W = \{(x, y) \in X \times Y : L(x, y) = \min_{u \in X} L(u, y)\}.$$

Definition 7.2. Let E be a Banach space, $\Omega \subset E$ a nonempty open subset, $\psi : \Omega \to \mathbb{R}$ be a Gâteaux differentiable functional and $\eta : \Omega \times \Omega \to E$ be a function such that
  (i) $\eta(x, x) = 0$ for $x \in \Omega$;
  (ii) $\psi(x) - \psi(u) \geq (\nabla\psi(u), \eta(x, u))$ for all $x, u \in \Omega$,
where $\nabla\psi(u)$ is the Gâteaux derivative of ψ at $u \in \Omega$. Then ψ is said to be η-convex.

By using Theorem 3.4 we can prove the following

Theorem 7.2 [11]. Let E be a reflexive Banach space, F a Fréchet space, $X \subset E$ a nonempty closed convex set, $Y \subset F$ a nonempty compact convex set and Ω an open subset. Let $L : \Omega \times Y \to \mathbb{R}$ be a function and $\eta : \Omega \times \Omega \to E$ be a weakly continuous function satisfying the following conditions:
  (i) The function $x \mapsto L(x, y)$ is η-convex and weakly continuous;
  (ii) The function $Y \mapsto L(x, y)$ is concave and continuous;
  (iii) $\nabla_x L(u, v) : X \times Y \to E^*$ is continuous with respect to the weak topology on X, the topology on Y and the norm topology on E^*.
  If there exist $\bar{u} \in X$ and $\bar{v} \in Y$ such that

$$L(\bar{u}, \bar{v}) = \max_{v \in Y} L(\bar{u}, v) \text{ and } \lim \frac{\|x\| \to \infty}{x \in X} (\nabla_x L(\bar{u}, \bar{v}), \eta(x, \bar{u})) > 0,$$

then the first and second nonlinear programming problems have solutions.

(III) Saddle point problem

Definition 7.3. Let X and Y be topological spaces and $f : X \times Y \to \mathbb{R}$ be a mapping.
  (1) For a given $\epsilon > 0$, a point $(x_\epsilon, y_\epsilon) \in X \times Y$ is said to be an ϵ-saddle point of f, if

$$f(x, y_\epsilon) - \epsilon < f(x_\epsilon, y_\epsilon) < f(x_\epsilon, y) + \epsilon \text{ for all } x \in X, y \in Y.$$

  (2) $f : X \times Y \to [-\infty, +\infty]$ is said to be r-transfer lower (upper, respectively) semicontinuous in y [60], if for any $(x, y) \in X \times Y$ with $f(x, y) > r(< r,$ respectively), there exist a point $x' \in X$ and a neighborhood $N(y)$ of y such that $f(x', z) > r(< r,$ respectively) for any $z \in N(y)$.

By using Theorem 2.1 we can establish an existence theorem for ϵ-saddle points in non-compact H-space which is a generalization of Theorem 1 in Tan-Yu-Yuan [57].

Theorem 7.3. Let $(X, \{\Gamma_A\})$, $(Y, \{\Gamma_B\})$ be H-spaces, and $f : X \times X \to \mathbb{R}$ be a function satisfying the following conditions : for any fixed $\epsilon > 0$

(i) For each $(x, y) \in X \times Y$

$$\inf_{v \in Y} f(x, v) > -\infty \quad \text{and} \quad \sup_{u \in X} f(u, y) < +\infty;$$

(ii) For each fixed $y \in Y$, $x \mapsto f(x, y)$ is H-quasi concave and $x \mapsto f(x, y) - \inf_{v \in Y} f(x, v)$ is ϵ-transfer u.s.c. in x;

(iii) For each fixed $x \in X$, $y \mapsto f(x, y)$ is H-quasi-convex and $y \mapsto f(x, y) - \sup_{u \in X} f(u, y)$ is ϵ-transfer l.s.c. in y;

(iv) There exist a compact subset L_1 of X, an H-compact subset K_1 of X, a compact subset L_2 of Y and an H-compact subset K_2 of Y such that for each weakly H-convex subset D_1 of K with $K_1 \subset D_1 \subset X$ and for each weakly H-convex subset D_2 of Y with $K_2 \subset D_2 \subset Y$ the following holds

$$\bigcap_{(w,z) \in D_1 \times D_2} \left\{ (x,y) \in X \times Y \Big| \begin{array}{l} f(w,y) - \sup_{u \in X} f(u,y) \le -\epsilon \text{ or} \\ f(x,z) - \inf_{v \in Y} f(x,v) \ge \epsilon \end{array} \right\} \cap (D_1 \cap D_2) \subset L_1 \times L_2.$$

Then f has an ϵ-saddle point $(x_\epsilon, y_\epsilon) \in X \times Y$.

(IV) Continuous Selection Problem

By using Proposition 1.1, we can prove the following continuous selection theorem which is a generalization of Yannelis-Prabhakar's selection theorem [63], and it is also equivalent to the Ding-Kim-Tan's selection theorem [31].

Theorem 7.4. Let X be a nonempty paracompact Hausdorff topological space and Y be a nonempty convex subset of a topological vector space. Suppose that $S, T : X \to 2^Y$ are two multifunctions satisfying

(i) For each $x \in X$, $coS(x) \subset T(x)$ and $S(x) \ne \emptyset$;

(ii) $S^{-1} : Y \to 2^X$ is transfer open-valued.

Then T has a continuous selection.

References

[1] J. P. Aubin and I. Ekeland, Applied Nonlinear Analysis, John Wiley and Sons, (1984).

[2] C. Bardaro and R. Ceppitelli, Some further generalizations of Knaster-Kuratowski-Mazurkiewicz theorem to minimax inequalities, J. Math. Anal. Appl., 132(1988), 484–490.

[3] C. Bardaro and R. Ceppitelli, Applications of the generalized Knaster- Kuratowski-Mazurkiewicz theorem to variational inequalities, J. Math. Anal. Appl. 137 (1989), 46–58.

[4] A. Bensoussan and J. L. Lions, Nouville formulation de problems decontrole impulsionnel et applications, C. R. Acad. Sci. Paris, 276(1973), 1189–1192.

[5] A. Bensoussan and J. L. Lions, Controle impulsionnel et inequations quasi-variationnelles stationnaries, C. R. Acad. Sci. Paris, 276(1973), 1279–1284.

[6] R. Bielawski, Simplicial convexity and its applications, J. Math. Anal. Appl., 127 (1987), 155–171.

[7] H. Brezis, L. Nirenberg and G. Stampacchia, A remark on Ky Fan's minimax principle, Boll. Un. Mat. Ital. 6(1972), 293–300.

[8] F. E. Browder, The fixed point theory of multivalued mappings in topological vector space, Math. Ann., 177(1968), 283–301.

[9] D. Chan and J. S. Pang, The generalized quasi-variational inequality problem, Math. Opera. Res., 7:2(1982), 211–222.

[10] S. S. Chang, Variational Inequality and Complementarity Problem Theory with Applications, Shanghai Scientific and Technological Literature Publishing House, Shanghai, (1991).

[11] S. S. Chang and Yong-Lu Shu, Variational inequalities for multivalued mappings with applications to nonlinear programming and saddle point problems, Acta Math. Appl. Sinica, 14:1 (1991), 32–39.

[12] S. S. Chang and Y. Zhang, Generalized KKM theorem and variational inequalities, J. Math. Anal. Appl., 159(1991), 208–222.

[13] S. S. Chang and N. J. Huang, Generalized strongly nonlinear quasi-complementarity problems in Hilbert spaces, J. Math. Anal. Appl., 158(1991), 194–202.

[14] S. S. Chang and Y. H. Ma, Generalized KKM theorem on H-space with applications, J. Math. Anal. Appl., 163 (1992), 406–420.

[15] S. S. Chang and C. J. Zhang, On a class of generalized variational inequalities and quasi-variational inequalities, J. Math. Anal. Appl., 179(1993), 250–259.

[16] S. S. Chang and L. Yang, Section theorem on H-space with applications, J. Math. Anal. Appl., 179(1993), 214–231.

[17] S. S. Chang, Y. J. Cho, X. Wu and Y. Zhang, The topological versions of KKM theorem and Fan's matching theorem with applications, Topological Methods in Nonlinear Anal., 1(1993), 231–245.

[18] S. S. Chang, X. Wu and S. W. Xiang, A topological KKM theorem and minimax theorems, J. Math. Anal. Appl., 182(1994), 756–767.

[19] S. S. Chang, X. Wu and D. C. Wang, Further generalizations of minimax inequalities for mixed concave-convex functions and applications J. Math. Anal. Appl., 186(1994), 402–413.

[20] S. S. Chang, X. Wu and Z. F. Shen, Economic equilibrium theorem of Safer-Sonnenschein version and nonempty intersection theorems in interval spaces, J. Math. Anal. Appl., 189(1995), 297–309.

[21] S. S. Chang, B. S. Lee, X. Wu and G. M. Lee, On the generalized quasi-variational inequality problems, Set-Valued Analysis (to appear).

[22] G. Y. Chen, Existence of solutions for a vector variational inequality: an extension of the Hartman-Stampacchia theorem, J. Optim. Th. Appl. 74(3)(1992), 445–456.

[23] G. Y. Chen and G. M. Cheng, Vector variational inequality and vector optimization, Lect. Notes in Econ. & Math. Syst. 285, 408–416. Springer-Verlag, (1987).

[24] G. Y. Chen and B. D. Craven, Approximate dual and approximate vector variational inequality for multiobjective optimization, J. Austral. Math. Soc. (Series A) 47(1989), 418–423.

[25] G. Y. Chen and B. D. Craven, A vector variational inequality and optimization over an efficient set, Zeitscrift für Operations Research 3(1990), 1–12.

[26] G. Y. Chen and X. Q. Yang, The vector complementarity problem and its equivalence with the weak minimal element in ordered sets, J. Math. Anal. Appl. 153 (1990), 136–158.

[27] G. Y. Chen, A generalized section theorem and a minimax inequality for a vector-valued mapping, Optimization, 22(1991), 745–754.

[28] C. Z. Cheng and Y. H. Liu, On the generalization of minimax theorem for the type of fixed point, Acta Math. Sinica, 34(1991), 502–507.

[29] X. P. Ding, New generalization of an $H\text{-}KKM$ type theorem and their applications, Bull. Austral. Math. Soc., 48(1993), 451–464.

[30] X. P. Ding and K. K. Tan, Generalizations of KKM theorem and applications to best approximations and fixed point theorems, SEA. Bull. Math., 17(1993), 139–150.

[31] X. P. Ding, W. K. Kim and K. K. Tan, A selection theorem and its applications, Bull. Austral. Math. Soc., 46(1992), 205–212.

[32] K. Fan, A generalization of Tychonoff's fixed point theorem, Math. Ann., 142(1961), 305–310.

[33] K. Fan, A minimax inequality and application, Inequalities (ed. O. Shisha), V. 3, pp. 103–113, Academic Press, New York, (1972).

[34] K. Fan, Sur un théorème minimax, C. R. Acad. Sci. Groupe I, 259(1964), 3925–3928.

[35] K. Fan, Applications of a theorem concerning sets with convex section, Math. Annalen, 163(1966), 189–203.

[36] K. Fan, Some properties of convex sets related to fixed point theorem, Math. Ann., 266(1984), 519–537.

[37] J. Y. Fu, Implicit variational inequalities for multivalued mappings, J. Math. Anal. Appl., 189(1995), 801–814.

[38] M. A. Geraghty and B. L. Lin, Topological minimax theorem, Proc. Amer, Math. Soc., 91(1984), 377–380.

[39] C. Horvath, Some results on multi-valued mappings and inequalities without convexity, in Nonlinear and Convex Analysis (B. L. Lin and S. Simons. eds), Marcel Dekker, (1987), 96–106.

[40] Y. S. Huang, Fixed point theorems with an application in generalized games, J. Math. Anal. Appl., 186(1994), 634–642.

[41] G. Isac, The numerical range theory and boundedness of solutions of the complementarity problem, J. Math. Anal. Appl., 143(1989), 235–251.

[42] G. Isac and M. Thera, A variational principle application to the nonlinear complementarity problem, Lecture Notes in Pure and Appl. Math., V. 107(1987), Dekker, New York, 127–145.

[43] V. Komornk, Minimax theorems for upper semi-continuous function, Acta Math., Acad. Sci. Hungar., 40(1982), 159–163.

[44] S. H. Kum, A generalization of generalized quasi-variational inequalities, J. Math. Anal. Appl., 182(1994), 158–164.

[45] B. L. Lin and X. C. Quan, A noncompact topological minimax theorem, J. Math. Anal. Appl., 16 (1991), 587–590.

[46] F. C. Liu, On a form of KKM principle and SupInfSup inequalities of von Neumann and of Ky Fan type, J. Math. Anal. Appl., 155(1991), 420–436.

[47] M. A. Noor, The quasi-complementarity problem, J. Math. Anal. Appl., 130(1988), 344–353.

[48] M. A. Noor, Fixed point approach for complementarity problem, J. Math. Anal. Appl., 133(1988), 437–448.

[49] J. Parida and A. Sen, A variational-like inequality for multifunctions with applications, J. Math. Anal. Appl., 124(1987), 73–78.

[50] S. Park, A unified approach to generalizations of the KKM-type theorems related to acyclic maps, Numer. Funct. Anal. and Optimiz., 15(1994), 105–119.

[51] S. Park, Generalizations of Ky Fan's matching theorem and their applications, J. Math. Anal. Appl., 141(1989), 164–176.

[52] S. Park, On minimax inequalities on spaces having certain contractible subsets, Bull. Austral. Math. Soc. 47(1993), 25–40.

[53] W. Shafer and H. Sonnenschein, Equilibrium in abstract economics without ordered preferences, J. Math. Ecomom, 2(1975), 345–348.

[54] N. Shioji, A further generalization of the KKM theorem, Proc. Amer. Math. Soc., 111(1991), 187–195.

[55] M. Sion, On general minimax theorems, Pacific J. Math. 8 (1958), 171–176.

[56] L. L. Stachó, Minimax theorems beyond topological vector spaces, Acta Sci. Math., 42(1990), 157–164.

[57] K. K. Tan, J. Yu and X. Z. Yuan, Note on ϵ-saddle point theorems, Acta Math. Hungrica, 65(1994), 395–401.

[58] E. Tarafdar, Fixed point theorems in H-spaces and equilibrium points of abstract economics, J. Austral. Math. Soc. (Ser. A), 53(1992), 252–260.

[59] M. Thera, Existence results for the nonlinear complementarity problem and applications to nonlinear analysis, J. Math. Anal. Appl., 154(1991), 572–584.

[60] G. Q. Tian, Generalizations of $FKKM$ theorem and the Ky Fan minimax inequality with applications to maximal elements, price equilibrium and complementarity, J. Math. Anal. Appl., 170(1992), 457–471.

[61] X. Q. Yang, Vector variational inequality and its duality, Nonlinear Anal. T. M. A., 21(1993), 869–877.

[62] X. Q. Yang, Vector complementarity and minimal element problem, J. of Optim. Theory and Appl., 77(1993), 483–495.

[63] N. C. Yannelis and N. D. Prabhakar, Existence of maximal elements and equilibra in linear topological spaces, J. Math. Econom., 12(1983), 233–245.

[64] J. C. Yao, The generalized quasi-variational inequality problem with applications, J. Math. Anal. Appl., 158(1991), 139–160.

[65] J. C. Yao, On the generalized complementarity problem, J. Austral. Math. Ser. B, 35(1994), 420–428.

[66] J. C. Yao and J. S. Guo, Variational and generalized variational inequalities with discontinuous mappings, J. Math. Anal. Appl., 182(1994), 371–392.

[67] W. T. Wu, A remark on the fundamental theorem in the theory of games, Sci. Record, New Ser., 3(1959), 229-233.

Topics on the Approximation Problem of Almost Isometric Operators by Isometric Operators

Guanggui Ding

Department of Mathematics, Nankai University, Tianjin

Introduction

Let E, E_1 be two Banach spaces and $0 \leq \epsilon < 1$. We say a mapping $T \in \mathbb{B}(E, E_1)$ is "ϵ−isometric" or "ϵ−almost isometric" if

$$(1 - \epsilon)\|x\| \ \leq \ \|Tx\| \ \leq \ (1 + \epsilon)\|x\| \tag{1}$$

for all $x \in E$. We say T is "isometric" if we can take $\epsilon = 0$.

We have known a lot of conclusions about isometric theory. In 1966 E. Michael and A. Pelczynski[1] proved that, for every $T \in B(l^\infty_{(m)}, l^\infty_{(n)})$ with $\|x\| \leq \|Tx\| \leq (1+\eta)\|x\|$ for all $x \in l^\infty_{(m)}$, where $0 < \eta < 1$, there exists an isometric embedding $V \in \mathbb{B}(l^\infty_{(m)}, l^\infty_{(n)})$ such that $\|T - V\| \leq \eta$. Enlightened by this result, we put forward the "isometric approximation problem" (IAP for short) on the relation between the isometric and almost isometric operators such as:

Problem (A). Is there a function $\delta(\epsilon)$ defined on $(0, 1)$ such that $\lim\limits_{\epsilon \to 0} \delta(\epsilon) = 0$, and for each ϵ−isometric $T \in B(E, E_1)$ there exists an isometric $V \in B(E, E_1)$ with $\|T - V\| \leq \delta(\epsilon)$?

Moreover, we are also interested in the varieties of Problem (A). Let's generalize the definition of almost isometry. Let P be a nonempty subset of E. A mapping $T \in B(E, E_1)$ is called an ϵ−isometric or ϵ−almost isometric operator on P if (1) holds for all $x \in P$; we say T is isometric on P if we can take $\epsilon = 0$.

Problem (A_1). Let P be a subset of E. Is there a function $\delta(\epsilon)$ defined on $(0, 1)$ such that $\lim\limits_{\epsilon \to 0} \delta(\epsilon) = 0$ and for each ϵ−isometric $T\colon E \to E_1$ on P there exists an isometric $V\colon E \to E_1$ on P with $\|T - V\| < \delta(\epsilon)$?

1991 Mathematics Subject Classification: 46B03, 46B99, 47A58, 47B99

In this paper, we assume P is the "positive cone" of a Banach lattice.

Problem (B). Let $D \subset B(E, E_1)$. Is there a function $\delta(\epsilon)$ defined on $(0,1)$ such that $\lim\limits_{\epsilon \to 0} \delta(\epsilon) = 0$ and for each ϵ-isometric $T \in D$ there exists an isometric $V \in D$ with $\|T - V\| \leq \delta(\epsilon)$?

In particular, we have

Problem (B_1). The same as the above Problem (B) except for that $\mathbb{D}$ is the set of "surjective" operators in $\mathbb{B}(E, E_1)$.

Problem (B_2). The same as the Problem (B) except for that $\mathbb{D}$ is the set of "positive" operators in $\mathbb{B}(E, E_1)$ when both E and E_1 are Banach lattices.

We want to say that it is important to study these Problems, owing to the following observations:

(1) The problems depend closely on the characterizations of spaces E and E_1. Through the study, we will at least have a deeper knowledge of the difference between the usual Banach spaces.

(2) Our study will enable us to keep a tight hold on the isometric theory of Banach spaces.

(3) In our study, we would like to use the influence of the richer topic "complemented subspaces".

(4) We find our study to be in contact with the topic "Problem of density of norm attaining operators".

(5) Obviously, our study is also closely linked with the representation theory of bounded linear operators from E to E_1.

We say a pair (X, Y) of Banach spaces to be "proper" if there exists an ϵ-isometric $T \in B(X, Y)$ for every $0 < \epsilon < 1$. Clearly one can not expect any pair of Banach spaces to be proper, but Huang Senzhong[2] proved that for any 2-dimensional Banach space X there exists a Banach space Y satisfying that pair (X, Y) is proper. He also showed that the answer to Problem (A) is negative for this space $B(X, Y)$. From this result it seems impossible for us to find any general method in the research for the IAP.

1. Some simple conclusions

We can easily get the following conclusions:

Theorem 1.1. Let E, E_1 be Banach spaces.

(a) If E is non-separable and E_1 is separable, or E is non-uniformly convex and E_1 is uniformly convex, then the pair (E, E_1) is not proper.

(b) If E is non-strictly convex and E_1 is strictly convex, or E is non-reflexive and E_1 is reflexive, then there is no any isometry in $\mathbb{B}(E, E_1)$.

(c) If E is isometric to a subspace of a Banach space $\hat{E}$, then if there exists no isometry or almost isometry in $B(E, E_1)$, so is not in $\mathbb{B}(\hat{E}, E_1)$. In particular, if E is separable, then so is not in $B(C[a, b], E_1)$.

Theorem 1.2. (a)[3] The pairs (l^p, c) and (c_0, l^1) (hence $\mathbb{B}(c, l^1)$ and $\mathbb{B}(C[0, 1], l^1)$) are not proper.

(b)[4] There is no isometry in the spaces $\mathbb{B}(C[0, 1], L^p[0, 1])$ and $\mathbb{B}(C[0, 1], l^p)$ $(p \geq 1$, including $p = 1!)$.

Theorem 1.3. If E, E_1 are finitely dimensional normed spaces, then the answers to Problems (A) and (A$_1$) for the space $B(E, E_1)$ are affirmative.

Using the polar decomposition theorem we can prove

Theorem 1.4. If H is a Herbert space, then the answers Problems (A) and (A$_1$) for the space $B(H)$ are affirmative.

Since the collection of inverse maps of $B(E, E_1)$ is open in $B(E, E_1)$, so we have

Theorem 1.5. For $\mathbb{B}(E, E_1)$, if the answer to Problem (A) is affirmative, then so is the answer to Problem (B$_1$).

2. About $B(E, E_1)$, where E is of finite dimension and E_1 is of infinite dimension

The IAP is quite difficult even if E is finitely dimensional. There are still some open questions about this type.

The following result was given by Wang Sichun:

Theorem 2.1.[5] (a) For the real $\mathbb{B}(l^\infty_{(2)} \to L^1(\mu))$, the answer to Problem (A) is affirmative, but there is no isometry in the complex $\mathbb{B}(l^\infty_{(2)} \to L^1(\mu))$;

(b) In real $\mathbb{B}(l^\infty_{(3)} \to L^1(\mu))$, there is no ϵ−isometry $(0 \leq \epsilon \leq \frac{1}{37})$.

(c) For real $\mathbb{B}(l^1_{(n)} \to L^1(\mu))$, the answer to Problem (A) is affirmative.

G. Schechtman[6] has proved that the pair $(E^p_{(m)}, l^1)$ is proper for every m—dimensional $(0 < m < \infty)$ subspace $E^p_{(m)}$ of L^p when $1 < p < 2$. For general real two-dimensional (B)-space, Wang Risheng obtained the following result.

Theorem 2.2.[7] For any real 2-dimensional Banach spaces $E_{(2)}$ the pair $(E_{(2)}, l^1)$ is proper.

Since there is no isometry in $\mathbb{B}(l^p_{(2)}, l^1)$ $(2 \leq p < \infty)$[8], we get the following conclusion immediately:

Theorem 2.3. For real $\mathbb{B}(l^p_{(2)}, l^1)$ $(2 \leq p < \infty)$, the answer to Problem (A) is negative.

However, any two-dimensional (B)-space $E_{(2)}$ can be embedded isometrically into $L^1[0, 1]$[8], so we may ask:

Question 1. What about Problem (A) for $\mathbb{B}(E_{(2)}, L^1[0, 1])$?

By imitating the technique which Johnson used to construct the universal space[9], Hang Senzhong obtained a very interesting result.

Theorem 2.4.[2] There exists a separable (B)-space E such that for any n-dimensional (B)-space $E_{(n)}$ $(\infty > n \geq 2)$, the pair $(E_{(n)}, E)$ is proper, but the answer to Problem (A) for $B(E_{(n)}, E)$ is negative.

3. About $B(L^p(\mu), L^p(n))(0 < p < \infty)$

The author acquired the following results in Mittag-Leffler Institute in Sweden in 1980 , but the paper was published in China in 1985.

Theorem 3.1.[10] Let $T \in \mathbb{B}(L^1[0, 1])$ be an ϵ—isometry on the positive cone $(L^1)^+$. If $T^+ \chi_{[0,1]} \wedge T^- \chi_{[0,1]} = 0$, then there exists a $V \in \mathbb{B}(L^1)$ such that $\|V\| = 1$, $\|V(|f|)\| = \|f\|$ for all $f \in L^1$ and $\|T - V\| \leq \frac{3\epsilon + \epsilon^2}{1 + \epsilon}$.

The technique used in the proof of the theorem is the so-called "tree method". A sequence $\{\Delta_{ni} : n = 0, 1, 2, \ldots, i = 1, 2, \ldots, 2^n\}$ of non-void measurable sets of a measure space (Ω, Σ, μ) is called a "tree" if:
  (i) $\Delta_{01} = \Omega$,

(ii) $\mu(\Delta_{ni} \cap \Delta_{nj}) = 0$ when $i \neq j$,

and

(iii) $\Delta_{ni} = \Delta_{n+1,2i-1} \cup \Delta_{n+1,2i}$.

In this section we take $\Delta_{ni} = \left(\frac{i-1}{2^n}, \frac{i}{2^n}\right)$ (we omit the end-points). Obviously each linear operator on $L^1[0,1]$ is completely determined by the tree $\{\Delta_{ni}\}$.

By means of the tree method we can also construct an isometry $V \in B(L^1)$ on the positive cone $(L^1)^+$ such that $\|S - V\| \geq \frac{1}{2}$ for any isometric $S \in \mathbb{B}(L^1)$, but we have:

Theorem 3.2.[10] Let $T \in \mathbb{B}(L^1)$ be an ϵ-isometry on the positive cone of L^1. If T maps arbitrarily two disjoint elements of $\{\Delta_{ni}\}$ (Here we write A rather than 1_A for any measurable set A) to disjoint elements and $\|T\Delta_{n+1,2i-1}\| = \|T\Delta_{n+1,2i}\|$, $n = 0, 1, \ldots, i = 1, 2, \ldots, 2^n$, then there is an isometric $S \in \mathbb{B}(L^1)$ on the whole L^1 with $\|S - T\| \leq \frac{3\epsilon + \epsilon^2}{1 + \epsilon}$.

Now let us consider the space $\mathbb{B}(L^p(\mu), L^p(\nu))$ $(1 \leq p < \infty)$ The next two theorems were given by Huang Senzhong[12].

Theorem 3.3. (a) If $p \in [2, \infty)$, $T \in \mathbb{B}(L^p(\mu), L^p(\nu))$ is ϵ-isometric on the positive cone of $L^p(\mu)$, then T is an ϵ_1-isometry on the whole space, where $\epsilon_1 \to 0$ as $\epsilon \to 0$.

(b) If $p \in (1, \infty)$, $T \in B(L^p(\mu), L^p(\nu))$ is a "positive" ϵ-isometric on the positive cone of $L^p(\mu)$, then T is ϵ_2-isometric on the whole space $L^p(\mu)$, where $\epsilon_2 \to 0$ as $\epsilon \to 0$.

(c) If $p \in (1, 2)$, $T \in B(L^p(\mu), L^p(\nu))$ is a "surjective" ϵ-isometric on the positive cone of $L^p(\mu)$, then T is an ϵ_3-isometry on the whole space $L^p(\mu)$, where $\epsilon_3 \to 0$ as $\epsilon \to 0$.

The Problem (A) on $\mathbb{B}(L^p(\mu), L^p(\nu))$ has been solved positively by D. E. Alspach[13], so we have the following result.

Theorem 3.4. Assume the hypothesis of any one of (a), (b), and (c) of Theorem 3.3 to hold. Then there exists an isometry $V \in \mathbb{B}(L^p(\mu), L^p(\nu))$ such that $\|T - V\| \leq \delta(\epsilon)$, where $\delta(\epsilon) \to 0$ as $\epsilon \to 0$.

Obviously many questions can be asked from Theorem 3.3, for example:

Question 2. What about Problem (A1) for $\mathbb{B}(L^p(\mu), L^p(\nu))$ $(p \in [1, 2))$?

Huang Senzhong even studied the IAP about the p-normed spaces L^p where $0 < p < 1$, he proved that each L^p-space E is order isomorphic and isometric to $\left(\sum_{\alpha \in A} L^p[0,1]^{m_\alpha}\right)_{l^p}$, where m_α's $(\alpha \in A)$ are cardinals, therefore he extended Alspach's theorem as follows:

Theorem 3.5.[14] Alspach's result is still true when $0 < p < 1$.

    Huang also proved that $L^p(\mu)$ is isometric to $L^p(\nu)$ if the Mazur distance between L^p (μ) and $L^p(\nu))$ is less than $3 - 2^{p/2}$.

    But about Problem (B2) on $B(L^p(\mu), L^p(\nu))$ we know nothing, so we ask the following

Question 3. What about Problem (B2) for $B(L^p(\mu), L^p(\nu))$ $(0 < p < \infty$)?

4. About $B(C(K), C(S))$ and $B(L^\infty(\Omega_1), L^\infty(\Omega_2))$

    Let K be a compact metric space, S a compact T_2–space. Y. Benyamini solved positively Problem (A) on the space $\mathbb{B}(C(K), C(S))$, he also constructed two compact T_2–spaces K_0 and S_0 (K_0 is not assumed to be metrizable) such that the pair $(C(K_0, C(S_0))$ is proper, but there is no isometry in $\mathbb{B}(C(K_0), C(S_0))$ [16,17].

    The author proved the following results.

Theorem 4.1.[18] For the space $B(l^\infty, L^\infty(\Omega, \Sigma, \mu))$, where (Ω, Σ, μ) is a measure space, the answer to Problem (A) is affirmative.

Corollary. For the spaces $B(c_0)$, $B(c)$ and $B(c_0, c)$ the answer to Problem (A) is affirmative.

Theorem 4.2.[18] Let $(\Omega_1, \Sigma_1, \mu_1)$, $(\Omega_2, \Sigma_2, \mu_2)$ be measure spaces. If $T : L^\infty(\Omega_1, \Sigma_1, \mu_1)$ $\to L^\infty(\Omega_2, \Sigma_2, \mu_2)$ is ϵ-isometric on the positive cone $L^\infty(\Omega_1, \Sigma_1, \mu_1)^+$ and $\|T\| \leq 1 + \epsilon$, then T is also 3ϵ-isometric on the whole $L^\infty(\Omega_1, \Sigma_1, \mu_1)$.

    In fact, Theorem 4.2 holds for all AM-spaces with order units, hence the answer to Problem (A_1) is the same as that to Problem (A) for the spaces $B(C(K), C(S))$ and $B(L^\infty(\Omega_1), L^\infty(\Omega_2))$ to some degree.

    Recently, the author improved the result of Theorem 4.1 as follows:

Theorem 4.3.[19] Let $(\Omega_1, \Sigma_1, \mu_1)$, $(\Omega_2, \Sigma_2, \mu_2)$ be measure spaces with $(\Omega_1, \Sigma_1, \mu_1)$ σ–finite. Then the answers to Problems (A) and (B_2) for the space $B(L^\infty(\Omega_1), L^\infty(\Omega_2))$ are affirmative.

    Using the representation theorem[20] of integrals of linear ope- rators in $B(C[0, 1])$, Xiang Guangping got

Theorem 4.4.[21] The answer to Problem (B_2) for $B(C[0, 1])$ is affirmative.

Recently, Xiao Yuanhui extended Theorem 4.4 by using some knowledge of Banach lattice as follows:

Theorem 4.5.[22] Let K, S be compact T_2–spaces with K metrizable. Then the answer to Problem (B$_2$) for $\mathbb{B}(C(K), C(S))$ is affirmative.

From above we can ask

Question 4. What about Problem (A) or (B$_2$) for the space $B(L^\infty(\Omega_1), L^\infty(\Omega_2))$ (where $(\Omega_1, \Sigma_1, \mu_1)$, $(\Omega_2, \Sigma_2, \mu_2)$ are arbitrary measure spaces)?

5. Some counterexamples

The first counterexample for Problem (A) was constructed by Wang Yaoting, he got:

Theorem 5.1.[23] The answers to Problems (A) and (A$_1$) are negative for $\mathbb{B}(l^1, l^\infty)$.

By means of this result, Huang Senzhong got the following two results:

Theorem 5.2.[24] There exists a (non-separable) Banach lattice E such that for the space $B(E)$ the answers to both of Problems (A) and (A$_1$) are negative. (For example, $E = l^1 \times l^\infty$).

Theorem 5.3.[24] There exists a separable (B)-space E such that the answer to Problem (A) is negative for $B(E)$.

Wang Risheng obtained the following

Theorem 5.4.[25] Let Γ be an infinite index and Ω be a compact T_2–space. The answer to Problem (A) for the space $B(l^1(\Gamma), C(\Omega))$ is negative iff $l^1(\Gamma)$ can be isometrically embedded in $C(\Omega)$.

Corollary.[25] For $B(l^1, C[0,1])$ the answer to Problem (A) is negative.

The author got a general result which is stronger than Theorem 5.1 and Theorem 5.4, and the proof is quite different.

Theorem 5.5.[26] Let μ, ν be measures and Δ an interval of $\mathbb{R}$. If $L^1(\mu)$ is infinitely dimensional and can be isometrically embedded in $L^\infty(\nu)$ and $C_b(\Delta)$, then the answers to both of Problems (A) and (A$_1$) for $\mathbb{B}(L^1(\mu), L^\infty(\nu))$ and $\mathbb{B}(L^1(\mu), C_b(\Delta))$ are negative.

Key to Proof: We say only on the space $B(L^1(\mu), L^\infty(\nu))$. Let $\{A_{ki}\}$ and $\{B_{ki}\}$ be trees in the measure spaces (A, μ) and (B, ν) respectively, $V_{ki} : L^1(A_{ki}^c) \to L^\infty(B_{ki})$ be an isometry $(1 \le i \le 2^k, k \in \mathbb{N})$. For each $k \in \mathbb{N}$, define $T_k \in \mathbb{B}(L^1(\mu), L^\infty(\nu))$ by

$$T_k f = \sum_{i=1}^{2^k} V_{ki}(f|_{A_{ki}^c}), \quad \text{for all } f \in L(\mu).$$

Then T_k is "$\frac{1}{2^k}$-isometric", but $\|T_k - S\| \ge 1$ for all isometric operators $S \in B(L^1(\mu), L^\infty(\nu))$.

6. Other conclusions

By now we have given some results about specific spaces. In this section we will give two theorems about general "uniformly smooth" spaces , their methods are enlightening.

The following conclusions are got by Wang Risheng:

Theorem 6.1.[27] Let X be a uniformly smooth Banach space and Γ be an index set, suppose that the pair $(X, l^\infty(\Gamma))$ is proper, then the answer to Problem (A) is affirmative for $B(X, l^\infty(\Gamma))$ iff $\dim(X) = 1$ or ∞.

It is easy to see each $T \in \mathbb{B}(X, l^\infty(\Gamma))$ is represented by a family $(f_t)_{t \in \Gamma}$ of continuous linear forms on X in the fashion: $T(x)(t) = f_t(x)$ for $x \in X$. The key to the proof of the theorem is to show that $(f_t)_{t \in \gamma}$ is a $\delta(\epsilon)$-net of the unit sphere of X^* (where $\delta(\epsilon) \to 0$ as $\epsilon \to 0$) if T is ϵ-isometric.

By means of some knowledge of vector-valued measures Liu Fei proved

Theorem 6.2.[28] Let (Ω, Σ, μ) be a purely non-atomic finite measure space and X a uniformly smooth and separable Banach space. Then the answer to Problem (A) is affirmative for $\mathbb{B}(X, L^\infty(\mu))$.

Naturally, we ask

Question 5. What about Problem (A) for $\mathbb{B}(L^p[0, 1], C[0, 1])$, or generally for $\mathbb{B}(X, C[0, 1])$ where X is a uniformly smooth Banach space?

Finally we end this paper with an interesting open problem put forward by Huang Senzhong.

Question 6.[2] Does there exist a (B)-space $E = (E, \|\circ\|)$ such that for any renormed space $E_1 = (E, \|\circ\|_1)$ there is an isometry in $\mathbb{B}(E, E_1)$, or weakly, the pair (E, E_1) is proper?

References

[1] Michael, E. and Pelczynski, A., Separable spaces which admit approximations, Israel J. Math. 4(1966), 189–198.

[2] Huang Senzhong, Renorm and IAP, Chinese Ann. of Math. 9A(1988), 488–497.

[3] Luo Yaohu, Some notes About ϵ−isometric operators, J. Shanxi Univ., 1(1984), 24–30.

[4] Li Shengjia, There is no isometric operator in $\mathbb{B}(C[0, 1], L^p[0, 1])$,_ _ _ _, 31–36.

[5] Wang Sichun, The ϵ−isometric operators in $\mathbb{B}(l_n^\infty \to L(X, \mu))$ and $\mathbb{B}(l_n^1 \to L(X, \mu))$, (to appear).

[6] Schechtman, G., Fine imbedding of finite dimensional subspaces of L_p, $1 \le p < 2$, into l_1^∞, Proc. A.M.S. 94 (1985), 617–623.

[7] Wang Risheng, Almost isometric imbeddings of two dimensional real (B)-spaces into l_1, (to appear).

[8] Silvio Machado, Functional analysis, Holomorphy, and Approximation Theory, Lecture Notes in Math., 843, Springer-Verlag, 1981.

[9] Kothe,G. Topological Vector Spaces II, Springer-Verlag, Berlin, 1979.

[10] Ding Guanggui, Isometric and Almost Isometric Operators of $\mathbb{B}(L^1)$, Acta. Math. Sinica, 1(1985), 126–140.

[11] Schaefer, H.H., Banach Lattices and Positive Operators, Springer-Verlag, Berlin, 1974.

[12] Huang Senzhong, On Operators that are almost isometric on the positive cones of L^p−spaces, $1 < p < \infty$, Proc. A.M.S., 106 (1989), 1039–1047.

[13] Alspach, D.E., Small into isomorphism on L^p spaces, Illinois J. Math., 27(1983), 300–314.

[14] Huang Senzhong, Small into isomorphism on L_p spaces, $0 < p \le 1$, (to appear).

[15] Lacey, H.E., Isometric Theory of Classical Banach Spaces, Springer-Verlag Berlin, 1974.

[16] Benyamini, Y., Small into isomorphisms between spaces of Continue functions, Proc. A.M.S., 33(1981), 479–485.

[17] Benyamini, Y., Small into-isomorphisms between spaces of Continuous functions II, Trans. A.M.S., 277(1983), 825–833.

[18] Ding Guanggui, The Approximation problem of almost isometric operators by isometric operators, Acta. Math. Sinica, 8(1988), 361–372.

[19] Ding Guanggui, Small into isomorphisms on L_∞ space, (to appear in "Acta Math. Sinica").

[20] Dunford, N. & Schwatz, J.T. Linear Operators I, New York, 1958.

[21] Xiang Guangping, Approximation of some operators in $\mathbb{B}(C, C)$, Acta. Sci. Nat. Univ. Nankai, 2(1985), 27–33.

[22] Xiao Yuanhui, The isometric approximation of positive operators from $C(K)$ into $C(S)$, Acta Sci. Nat. Univ. Nankai, 3(1990), 16–22.

[23] Wang Yaoting, A counterexample of approximating an almost isometric operator by an isometric operator in $\mathbb{B}(l^1 \to l^\infty)$, J. of Shanxi Univ., 1(1984), 19–23.

[24] Huang Senzhong, Constructing operators which cannot be isometrically approximated, Acta. Math. Sci., 6:2.(1986), 195–200.

[25] Wang Risheng, The problem of isometric approximation on the space $\mathbb{B}(l^1(\Gamma), C(\Omega))$, Chinese Sci. Bull., 10(1990), 975–978.

[26] Ding Guanggui, On Almost isometries From $L^1(\mu)$ into $L^\infty(\nu)$ or $C_b(\Delta)$, Acta. Math. Sci., 12:3(1992), 308–311.

[27] Wang Risheng. Isometric Approximations From Uniformly Smooth Spaces into $l^\infty(\Gamma)$ Type spaces, Acta. Math. Sci., 9:1(1989), 27–32.

[28] Liu Fei, The isometric approximation problem from uniformly smooth separable spaces to $L^\infty(\mu)$, (to appear).

[29] Diestel,J. & Uhl,J.J. Vector measures, Mathematical Surveys 15, American Mathematical Society, Providence, Phode Island, 1977.

Several Results on Nonsmooth Analysis

Xianling Fan

Department of Mathematics, Lanzhou University, Lanzhou

In this paper we summarize the author's several results on nonsmooth analysis.

1. The C^1-admissible approximation for Lipschitz functions and the Hamiltonian inclusions

Throughout this paper R^n denotes the n-dimensional Euclidean space,C^1 and C^{1-0} denote the classes of continuously differentiable maps and of locally Lipschitz maps respectively. For $f \in C^{1-0}(R^n, R)$ and $x \in R^n$,$\partial f(x)$ denotes the Clarke's generalized gradient of f at x (see [10]).

In [11, 19] we gave the following definition and theorem of the so-called C^1-admissible approximation for Lipschitz functions.

Definition 1.1. Let D be an open subset of R^n and $f \in C^{1-0}(D, R)$. Suppose that $\varepsilon : D \to (0, \infty)$ and $g \in C^1(D, R)$. g is called an $\varepsilon(x)$-admissible approximation for f on D if g satisfies:

1. $|g(x) - f(x)| \leq \varepsilon(x)$ for $x \in D$.
2. For any $x \in D$ there is $y \in D$ such that $|x - y| \leq \varepsilon(x)$ and

$$dist(g'(x), \partial f(y)) \leq \varepsilon(x).$$

In particular, we call g an ε-admissible approximation for f on D if $\varepsilon(x) \equiv \varepsilon$.

Theorem 1.1 ([11, 15, 19]). Let D be an open subset of R^n and $f \in C^{1-0}(D, R)$. Then for any continuous function $\varepsilon : D \to (0, \infty)$ there is an $\varepsilon(x)$-admissible approximation g for f on D and $g \in C^\infty(D, R)$.

1991 Mathematics Subject Classification: 49J52, 47H04, 47H11, 58F05

Research supported by the National Science Foundation of China

Definition 1.1 and Theorem 1.1 are very useful. For example let us consider the Hamiltonian inclusion

$$\dot{x}(t) \in J\partial H(x(t)) \tag{1}$$

where $H \in C^{1-0}(D, R)$, D is an open subset of R^{2n}, J is the standard sympletic $2n \times 2n$ matrix.

It is well known that , if $H \in C^1$, then any solution $x(t)$ of (1) is necessarily conservative, i.e. $H(x(t)) \equiv$ constant. If $H \in C^{1-0}$, F. Clarke [10] proved that any solution of (1) is also conservative provided H is regular. However if $H \in C^{1-0}$ is not regular, then a solution of (1) need not be conservative. The following theorem which is obtained easily from [2, Theorems 0.3.4 and 1.4.1] provides a method to obtain the conservative periodic solutions of (1).

Theorem 1.2 ([11, 15, 19]). Let D be an open subset of R^{2n} and $H \in C^{1-0}(D, R)$. Suppose that for each $k = 1, 2, \cdots, H_k$ is an ε_k-admissible approximation for H on D and $\varepsilon_k \to 0$ as $k \to \infty$. Suppose that for $k = 1, 2, \cdots, x_k(\cdot)$ is a T_k-periodic solution of the Hamiltonian system

$$\dot{x}_k(t) = JH'_k(x_k(t)).$$

If $\{T_k : k = 1, 2, \cdots\}$ is bounded and $\{x_k(t) : t \in R, k = 1, 2, \cdots)$ is contained in a compact subset of D , then $\{x_k\}$ has a subsequence which converges uniformly to a conservative periodic solution of (1).

Using Theorems 1.1 and 1.2 we can extend some known existence results on the periodic solutions of Hamiltonian systems with $H \in C^1$ to the case $H \in C^{1-0}$. For example in [12, 15, 20] we extend the corresponding results of [23, 21, 4, 1, 3] to Hamiltonian inclusions.

2. The necessary and sufficient condition for Lipschitz local homeomorphism

Let U be an open subset of R^n and $f : U \to R^n$ a map . If $f : U \to f(U) = V$ is a homeomorphism, $f \in C^1(C^{1-0})$ and its inverse $f^{-1} \in C^1(V, U)$ (resp. $f^{-1} \in C^{1-0}(V, U)$), then we say that f is a C^1 (resp. C^{1-0}) homeomorphism on U. Let $x_0 \in U$. If there is a neighborhood $\Omega \subset U$ of x_0 such that f is a C^1 (resp. C^{1-0}) local homeomorphism on Ω, then we say that f is a C^1 (resp. C^{1-0}) local homeomorphism at x_0.

The following two inverse function theorems are well known.

Proposition 2.1 ([8, 9]). Let $f \in C^1(U, R^n)$ and $x_0 \in U$. Then f is a C^1 local homeomorphism at x_0 if and only if $f'(x_0)$ is invertible.

Proposition 2.2 ([9, 10]). Let $f \in C^{1-0}(U, R^n)$ and $x_0 \in U$. If the generalized Jacobian $\partial f(x_0)$ is invertible (i.e. each element of $\partial f(x_0)$ is invertible), then f is a C^{1-0} local homeomorphism at x_0.

In [14] we gave an example, in which $f \in C^{1-0}(R^2, R^2)$ and $f : R^2 \to R^2$ is a C^{1-0} homeomorphism, but $\partial f(0)$ is not invertible. The example shows that the condition that $\partial f(x_0)$ is invertible is not necessary for Lipschtz local homeomorphism at x_0.

In [14] we proved the following theorem which gives the necessary and sufficient conditions for Lipschitz local homeomorphism.

Theorem 2.1 ([14]). Let $f \in C^{1-0}(U, R^n)$ and $x_0 \in U$. Then f is a C^{1-0} local homeomorphism at x_0 if and only if there is a neighborhood $\Omega \subseteq U$ of x_0 such that

1. There is a positive constant L such that

$$\|\delta f(x, h)\| \geq L \quad for \quad x \in \Omega \quad and \quad h \in S,$$

where $S = \{x \in R^n : \|x\| = 1\}$ and

$$\delta f(x, h) = \{y \in R^n : \; There \; is \; a \; sequence \; t_n \to 0^+ \; such \; that$$

$$\lim t_n^{-1}(f(x + t_n h) - f(x)) = y\}.$$

2. $\det f'(x) > 0$ (or < 0) for $x \in \Omega \backslash \Omega_f$, where

$$\Omega_f = \{x \in \Omega : f \; is \; not \; differentiable \; at \quad x\}.$$

3. $f(x) \neq f(x_0)$ for $x \in \Omega \backslash \{x_0\}$ and $deg(f, \Omega, f(x_0)) = 1$, (resp. -1).

In [14] we gave also the examples which show that any one of the conditions 1-3 in Theorem 3.1 can not be dropped.

3. Essential critical points of Lipschitz functionals

In this section, X denotes a Banach space.

Let U be an open subset of X, $f \in C^{1-0}(X, R)$ and $x_0 \in U$. By the usual definition x_0 is called a critical point of f if $0 \in \partial f(x_0)$. Otherwise x_0 is called a regular point of f.

The Clarke's generalized gradient possesses good analytical properties. As an example, the following generalized Fermat theorem holds.

If x_0 is an extreme point of f, then $0 \in \partial f(x_0)$.

Nevertheless it is well known that the Clarke's generalized gradient has also some shortcomings which are caused by that the set $\partial f(x)$ is too big sometimes. In nonsmooth optimization the points x satisfying condition $0 \in \partial f(x)$ are too many sometimes (see [23]). For this reason it is necessary to improve the concept of critical points of C^{1-0} functionals.

This problem will become more clear if we consider it in C^{1-0} category.

Let M be a C^{1-0} Banach manifold and $f \in C^{1-0}(M, R)$. How ought we to define the concept of critical points of f?

Naturally a rational definition of this concept ought to be independent of the choice of C^{1-0} local coordinate systems. Thus we must answer the following question.

Question: Let U, V be open subsets of $X, f \in C^{1-0}(U, R)$, $T : U \to V$ a C^{1-0} homeomorphism, $x_0 \in U$ and $T(x_0) = y_0$. Set $g = fT^{-1} : V \to R$. We ask:

Is the conclusion, $0 \in \partial f(x_0) \Leftrightarrow 0 \in \partial g(y_0)$, certainly valid ?

In [16] we gave a negative answer to the above -mentioned question by an example and proved the following result.

Theorem 3.1. Let $X = R^n$ with $n > 1$ and let $f \in C^{1-0}(X, R)$. Then for any point $x_0 \in X$ there are a neighborhood U of x_0 and a C^{1-0} homeomorphism $T : U \to V$ such that $T(x_0)$ is a critical point of fT^{-1}.

In virtue of the above-mentioned fact the following definitions given in [16] are natural.

Definition 3.1 ([16]). Suppose that $f \in C^{1-0}(X, R)$ and x_0 is a critical point of f. We call x_0 an essential critical point of f if for any C^{1-0} local homeomorphism $T : U \to V$ at x_0, $T(x_0)$ is a critical point of fT^{-1}. Otherwise, we call x_0 a nonessential critical point of f.

Definition 3.2 ([16]). Suppose that M is a C^{1-0} Banach manifold with a C^{1-0} atlas $\Gamma = \{(U_\alpha, \phi_\alpha)\}, f \in C^{1-0}(M, R)$ and $x_0 \in M$. Then x_0 is called an essential critical point of f if there is a chart $(U, \phi) \in \Gamma$ at x_0 such that $\phi(x_0)$ is an essential critical point of $f\phi^{-1}$.

Obviously each extreme point of f is surely an essential critical point of f.

The local topological property of f at a nonessential critical point is the same as that at a regular point .

In [16] we discussed some elementary properties of the essential critical points of C^{1-0} functionals .

In [17] we showed that using the minimax principle and the method indicated by K. C. Chang [7] we can obtain the existence and multiplicity results of the essential critical points of C^{1-0} functionals.

4. Nonsmooth Fredholm maps and topological degree

It is well known that the usual concept of Fredholm maps is only concerned with C^1 maps. In [18, 13] we generalized this concept to the cases of continuous maps and of upper semi-continuous set-valued maps and established the corresponding degree theory for the case of index zero.

Here we give only the definition of continuous Fredholm maps of index zero.

Definition 4.1 ([18]). Let M and N be C^0-Banach manifolds with atlases Γ and Γ' modelled on a Banach space X. Let $f : M \to N$ be a C^0(i.e. continuous) map. f is called a C^0-Fredholm map of index zero if for each point $x \in M$ and for any chart $(V, \Psi) \in \Gamma'$ at $f(x)$ there is a chart $(U, \phi) \in \Gamma$ at x such that $f(U) \subseteq V$ and the map $\psi f \phi^{-1} : \phi(U) \to \psi(V)$ has the form

$$\psi f \phi^{-1} = I - K$$

where I is the identity map and the range of K is contained in a finite-dimensional subspace of X.

From [5] it follows that if in Definition 4.1 we replace C^0 by C^1, then we obtain an equivalent definition of C^1-Fredholm maps of index zero.

Using the finite-dimensional reduction method we can define the topological degrees for nonsmooth (single-valued or set-valued) Fredholm maps of index zero and for their compact perturbations (see [18, 13]).

References

[1] Ambrosetti A. , Coti-Zelati V. ,Closed orbits of fixed energy for singular Hamiltonian systems, Arch. Rat. Mech. Anal.,112(1990),339-362.

[2] Aubin J.P. , Cellina A., Differentiale Inclusions, Springer-Verlag.Berlin,1984.

[3] Benci V. , Giannoni F.,Periodic solution of prescribed energy for a class of Hamiltonian systems with singular potentials, J. Diff. Eq., 82(1989), 60-70.

[4] Berestycki H., Lasry J., Mancini G. , Ruf B., Existence of multiple periodic orbits in star-shaped Hamiltonian surfaces, Comm. Pure Appl. Math., 38(1985), 253-289.

[5] Borisovich Yu., Zviagin V. , Sapronov Yu., Nonlinear Fredholm maps and Leray-Schauder theory. Uspekhi Mat. Nauk,32:4(1977),3-54.

[6] Chang K.C., Critical Point Theory and Its Applications, Shanghai Science and Technology Press, Shanghai,1986.(Chinese)

[7] Chang K.C., Variational methods for nondifferentiable functionals and their application to PDE., J. Math. Anal. Appl., 80(1981), 102-129.

[8] Chen W.Y., Nonlinear Functional Analysis, Gansu Renmin Press, Lanzhou, 1982. (Chinese)

[9] Chen W.Y., Fan X.L., Implicit Function Theorems, Lanzhou Univ. Press, Lanzhou, 1986.(Chinese)

[10] Clarke F.H., Optimization and Nonsmooth Analysis, John Wiley , Sons, Inc., 1983.

[11] Fan X.L., The C^1-admissible approximation for Lipschitz functions and the Hamiltonian inclusions, (Research Announcements) Advances in Math.(China), 20:1 (1991), 127-128.

[12] Fan X.L., A Viterbo-Hofer-Zehnder type result for Hamiltonian inclusions, Ann. Faculte Sci. Toulouse Math., 12:3(1991), 365-372.

[13] Fan X.L., The set-valued Fredholm mappings and topological degree, Chinese Quarterly J. Math., 6:4 (1991),68-71. (Chinese)

[14] Fan X.L., The necessary and sufficient conditions for Lipschitz local homeomorphism, Chin. Ann. of Math., 13 B:1(1992),40-45.

[15] Fan X.L., Existence of multiple periodic orbits on star-shaped Lipschitz-Hamiltonian surfaces, J. Diff. Eq., 98:1(1992),91-110.

[16] Fan X.L.,The essential critical points of lipschitz functionals, J. Math. Research Exposition,14:4 (1994),557-560.(Chinese)

[17] Fan X.L. , Liu B.S., The essential critical points of minimax type of Lipschitz functionals, J. Lanzhou Univ. (Natur. Sci.), 31:1(1995),1-4 . (Chinese)

[18] Fan X.L. , Qin C.L, The C^0-Fredholm maps and topological degree,J. Lanzhou Univ. (Natur. Sci.),30:4(1994),1-4. (Chinese)

[19] Fan X.L. , Liu B.S., The C^1-admissible approximation for Lipschitz functions and the Hamiltonain inclusions, J. Lanzhou Univ. (Natur. Sci.),29:3(1993),38-42. (Chinese)

[20] Fan X.L. , Wang B.X., Remarks on periodic solutions of prescibed energy for singular Hamiltonian systems,Houston J. Math., 17:3 (1991),385-393.

[21] Hofer H. , Zehnder E., Periodic solutions on hypersurfaces and a result by C. Viterbo, Invent. Math., 90 (1987) ,1-9.

[22] Rabinowitz P.H., Periodic solutions of Hamiltonian systems, Comm. Pure Appl. Math.,31 (1978), 157-184.

[23] Shi S.Z., Nonsmooth analysis, Adv. in Math. (China), 15:1 (1986),9-21.(Chinese)

[24] Viterbo C., A proof of the Weinstein conjecture in R^{2n} , Ann. Inst. H. Poincare, Anal. non Lineaire, 4 (1987), 337-356.

New Progress on Operator Semigroups and Linear Systems in Abstract Spaces

Dexing Feng

Institute of Systems Science, Academia Sinica, Beijing

Since the fifties, operator semigroups have attracted much attention because they are closely related to Cauchy problems which are very important to applications, and a complete theory on them is established. For the general theory of C_0 semigroups, refer to [1, 2]. Since the seventies, especially since the eighties, Chinese mathematicians have obtained a series of important results in the field of operator semigroups and abstract Cauchy problems. This note briefly outlines the main results in this area accomplished in China during this period. Of course, it can not include all the results obtained. The contents are divided into three parts: stability of C_0 semigroups, integrated semigroups and C-semigroups, and higher order differential equations in abstract spaces.

Throughout the paper, $\mathbb{N}$, $\mathbb{R}$ and $\mathbb{C}$ denote the natural number set, real number field and complex number field, respectively. For a linear operator A in a Banach space X, $\mathcal{D}(A)$, $\mathcal{R}(A)$, $\rho(A)$, $\sigma(A)$, $\sigma_p(A)$, $\sigma_c(A)$, $\sigma_r(A)$, $R(\lambda; A)$ denote the domain, range, resolvent set, spectrum, point spectrum, continuous spectrum, redsiual spectrum and resolvent of A, respectively. We denote by X^* the dual space of a Banach space X, and denote by A^* the adjoint of a linear operator A. For two Banach spaces X and Y, denote by $\mathcal{L}(X, Y)$ the Banach space of all bounded linear operators from X to Y and denote $\mathcal{L}(X) = \mathcal{L}(X, X)$ for short.

1. Stability of C_0 semigroups

It is known that many problems in mathematical physics can be solved by operator semigroup theory. Let X be a Banach space, consider the following first order evolution

1991 Mathematics Subject Classification: 46S99

Project supported by the National Natural Science Foundation of China and in part by the Laboratory of Systems and Control

equation in X:

$$\begin{cases} \dfrac{dx(t)}{dt} = Ax(t), & t \geq 0, \\ x(0) = x, \end{cases} \tag{1.1}$$

where A is a densely defined, closed linear operator in X and generates a C_0 semigroup $T(t)$ in X, i.e., $T(t), t \geq 0$ is a strongly continuous family of bounded linear operators in X, such that

$$T(0) = I, \quad T(t+s) = T(t)T(s), \; \forall t, s \geq 0.$$

Set

$$\omega(A) \overset{\triangle}{=} \inf\{\omega \in \mathbb{R} \mid \exists M \geq 1, \|T(t)\| \leq Me^{\omega t}, \forall t \geq 0\},$$

which is called the exponential growth order or the stability index of $T(t)$ (or of A).

We say that $T(t)$ is exponentially stable if there exist constants $M \geq 1$ and $\omega \geq 0$ such that

$$\|T(t)\| \leq Me^{-\omega t}, \quad \forall t \geq 0,$$

that $T(t)$ is L^p stable for some $p \geq 1$ if

$$\int_0^\infty \|T(t)\|^p dt < \infty,$$

that $T(t)$ is strongly stable if for any $x \in X$,

$$\|T(t)x\| \to 0 \quad \text{as } t \to \infty,$$

that $T(t)$ is weakly stable if for any $x \in X$ and $x^* \in X^*$,

$$\langle T(t)x, x^* \rangle \to 0 \quad \text{as } t \to \infty,$$

and that $T(t)$ is weakly L^p stable for some $p \geq 1$ if

$$\int_0^\infty |\langle T(t)x, y \rangle|^p dt < \infty, \quad \forall x, y \in X.$$

Obviously, $T(t)$ is exponentially stable if and only if $\omega(A) < 0$.

In the following, $s(A) \overset{\triangle}{=} \sup\{\operatorname{Re} \lambda \mid \lambda \in \sigma(A)\}$ is called the spectral bound of A. In general, when $s(A) < 0$ and even if A has a compact resolvent, the C_0 semigroup generated by A is not necessarily exponentially stable. The following four important theorems concerned with exponential stability for C_0 semigroups are given for the first time by Huang [3] in 1985, where there is the frequency criterion for the exponential stability of C_0 semigroups in Hilbert spaces. For contraction C_0 semigroups, the frequency criterion of exponential stability in Hilbert spaces was obtained by Gearhart [4] in 1978.

Theorem 1.1. Let A generate a C_0 semigroup $T(t)$ in a Hilbert space H. Then $s(A) = \omega(A)$ iff for any $s > s(A)$, it holds true that

$$\sup\{\|(\lambda - A)^{-1}\| \mid \operatorname{Re}\lambda \geq s\} < \infty.$$

The condition $\omega(A) = s(A)$ is usually called the spectrum determined growth assumption.

Theorem 1.2. A C_0 semigroup $T(t)$ generated by A in a Hilbert space is exponentially stable iff $s(A) < 0$ and there exists $s \in (s(A), 0)$ such that

$$\sup\{\|(\lambda - A)^{-1}\| \mid \operatorname{Re}\lambda \geq s\} < \infty,$$

or iff $s(A) < 0$ and

$$\sup\{\|(\lambda - A)^{-1}\| \mid \operatorname{Re}\lambda \geq 0\} < \infty.$$

Theorem 1.3. Let $T(t)$ be a uniformly bounded C_0 semigroup generated by A in a Hilbert space. Then $T(t)$ is exponentially stable iff $i\mathbb{R} \subset \rho(A)$ and

$$\sup\{\|(i\omega - A)^{-1}\| \mid \omega \in \mathbb{R}\} < \infty.$$

Theorem 1.4. Let A generate a C_0 semigroup $T(t)$ in a Hilbert space H. Then

$$\omega(A) = \inf\left\{ s \in \mathbb{R} \mid s > s(A), \sup\{\|(\lambda - A)^{-1}\| \mid \operatorname{Re}\lambda \geq s\} < \infty \right\}.$$

In terms of the resolvent of a C_0 semigroup, Yao and Feng in [6] obtained the criterion of the exponential stability of C_0 semigroups in Hilbert spaces.

Theorem 1.5. Let A generate a C_0 semigroup $T(t)$ in a Hilbert space H, then the C_0 semigroup $T(t)$ is exponentially stable if and only if
(1) $\{\lambda \in \mathbb{C} \mid \operatorname{Re}\lambda > 0\} \subset \rho(A)$;
(2) $\lim_{\sigma \to 0+} \sigma \sup_{\tau \in \mathbb{R}} \|R(\sigma + i\tau; A)\| = 0$.

Different from the case of weak L^p-stability, in [6] they impose conditions on the generator A itself rather than on $T(t)$ to obtain the following theorem.

Theorem 1.6. Let $T(t)$ be a C_0 semigroup on a Hilbert space H with the generator A. Then a C_0 semigroup $T(t)$ is exponentially stable if and only if
(1) $\{\lambda \in \mathbb{C} \mid \operatorname{Re}\lambda > 0\} \subset \rho(A)$;

(2) there exists p with $1 < p < \infty$ such that $\forall\, x, y \in H$,

$$\int_{-\infty}^{\infty} |\langle R(\sigma + i\tau; A)x, y\rangle|^p d\tau \leq C(x, y), \quad \forall\, \sigma > 0$$

with a positive constant $C(x, y)$ only dependent on x, y.

Pritchard and Zabczyk in [7] showed that exponential stability is equivalent to L^p-stability for C_0 semigroups. Obviously, exponential stability implies weak L^p-stability for any $p \geq 1$. They have proposed an open problem: Is the converse statement true? or does the weak L^p-stability for some $p \geq 1$ imply the exponential stability of a C_0 semigroup? In Hilbert spaces, the answer is positive. Huang in [3] for $p = 1$ and Huang and Liu in [5] for $p > 1$ proved the following result.

Theorem 1.7. A C_0 semigroup on a Hilbert space is weakly L^p-stable for some $p \geq 1$ iff it is exponentially stable.

In [6], the same conclusion is obtained as a corollary of Theorem 1.5.

In general, the above criterion of exponential stability of C_0 semigroups in Banach spaces does not hold. (For a counterexample, see [2]) However, it is possible to obtain weaker results in this case. Huang in [8] obtained the following results concerning the stability of C_0 semigroups in Banach spaces.

Theorem 1.8. A C_0 semigroup $T(t)$ in a Banach space X with generator A is weakly L^1-stable if and only if $s(A) < 0$ and there exists $\sigma \in (s(A), 0)$ such that

$$\sup\{\|R(\lambda; A)\| \mid \operatorname{Re} \lambda \geq \sigma\} < \infty. \tag{1.2}$$

Specially, when the above condition (1.2) is satisfied, it holds true that

$$\int_0^{\infty} e^{-\sigma t} |\langle T(t)x, x^*\rangle| dt < \infty, \quad \forall\, x \in X,\, x^* \in X^*.$$

Theorem 1.9. Let $T(t)$ be the C_0 semigroup in a Banach space X with generator A. Then the following statements are equivalent:
(1) there is $\sigma > 0$ such that $\displaystyle\int_0^{\infty} e^{\sigma t} |\langle T(t)x, x^*\rangle| dt < \infty, \quad \forall\, x \in X,\, x^* \in X^*$;
(2) $\{\lambda \in \mathbb{C} \mid \operatorname{Re} \lambda \geq 0\} \subset \rho(A)$, and $\sup\{\|R(\lambda; A)\| \mid \operatorname{Re} \lambda \geq 0\} < \infty$;
(3) $\displaystyle\int_0^{\infty} |\langle T(t)x, x^*\rangle|^p dt < \infty, \quad \forall\, x \in X,\, x^* \in X^*$.

Let k be a nonnegative integer. A C_0 semigroup $T(t)$ is called A-exponentially stable on $\mathcal{D}(A^k)$ if there exist constants M and ω such that

$$\|T(t)x\| \leq Me^{-\omega t} \|A^k x\|, \quad \forall\, x \in \mathcal{D}(A^k),\, t \geq 0.$$

M. Slemrod in [11] proved the following conclusion.

Proposition 1.1. Let $T(t)$ be a C_0 semigroup in a Banach space with generator A. If $s(A) < 0$ and $\|R(\sigma + i\tau; A)\| = O(|\tau|^k), \forall \sigma > s(A)$, then $T(t)$ is A^{k+2}-exponentially stable, i.e., there exist constants $\delta >$ and $M > 0$ such that

$$\|T(t)x\| \leq Me^{-\delta t}\|A^{k+2}x\|, \quad \forall x \in \mathcal{D}(A^{k+2}).$$

It will be seen that this result can be strengthened in the case of Hilbert space.

In [8], Huang also proved the following result.

Theorem 1.10. Let $T(t)$ be a C_0 semigroup in a Banach space with generator A. If $s(A) < 0$ and there exists $\sigma \in (s(A), 0)$ such that (1.2) holds, then $T(t)$ is A-exponentially stable.

To consider deeply the stability of C_0 semigroups in Hilbert spaces, first we define the order function $\mu(\sigma)$ of $R(\lambda; A)$ on $(s(A), \infty)$. For $\sigma \in (s(A), 0)$, denote by $\mu(\sigma)$ the infimum of all $\xi \in \mathbb{R}$ such that

$$\|R(\sigma + i\tau; A)\| = O(|\tau|^\xi), \quad \text{as } |\tau| \to \infty;$$

if there is no $\xi \in \mathbb{R}$ such that the above estimate holds, then define $\mu(\sigma) = \infty$. $R(\lambda; A)$ is said to be of finite order at σ if $\mu(\sigma) < \infty$. Set

$$\mu_1 \stackrel{\triangle}{=} \inf\{\omega \mid \omega > s(A), \text{ such that } \mu(\sigma) < \infty, \forall \sigma > \omega\},$$

$$\mu_0 \stackrel{\triangle}{=} \inf\{\omega \mid \omega > s(A), \text{ such that } \mu(\sigma) < 0, \forall \sigma > \omega\}.$$

By using the generalized Paley-Wiener theorem, Feng and Yao in [9] proved the following theorem:

Theorem 1.11. Let $T(t)$ be a C_0 semigroup in a Hilbert space with generator A. Then there exist constants $\delta > 0$ and $M > 0$ such that

$$\|T(t)x\| \leq Me^{-\delta t}\|A^k x\|, \quad \forall x \in \mathcal{D}(A^k)$$

if and only if there exists σ_0 with $0 > \sigma_0 > \mu_1$ such that $\|R(\sigma + i\tau; A)\| = O(|\tau|^k) \forall \sigma \geq \sigma_0$.

Denote

$$\omega_1 = \inf\left\{\omega \in \mathbb{R} \,\middle|\, \int_0^\infty e^{-\omega t}\|T(t)x\|dt < \infty, \forall x \in \mathcal{D}(A)\right\};$$

ω_1 defined above is called the exponential growth order for solutions of Cauchy problem (see [12]). $T(t)$ is said to be exponentially stable for solutions of Cauchy problem if $\omega_1 < 0$. The following corollary is a direct result of Theorem 1.11.

Corollary 1.12. Let $T(t)$ and A be as above. Then $T(t)$ is exponentially stable for solutions of Cauchy problem if and only if there exists σ_0 with $0 > \sigma_0 \geq \mu_1$ such that $\|R(\sigma + i\tau; A)\| = O(|\tau|)$.

Recently, Luo and Feng in [10] extended the some results on stability of single C_0 semigroup to a family of C_0 semigroups in Hilbert space. The main results are the following Theorems 1.13 and 1.14

Theorem 1.13. Let $\{T_h(t) \mid h \in \Delta\}$ be an equi-exponentially bounded family of C_0 semigroups in a Hilbert space X, i.e. there exist constants $M \geq 1$, $\omega \in \mathbb{R}$ such that

$$\|T_h(t)\| \leq Me^{\omega t}, \quad \forall h \in \Delta, \forall t \geq 0. \tag{1.3}$$

Denote by A_h the generator of C_0 semigroup $T_h(t)$ for every $h \in \Delta$. Then the following statements are equivalent:

(1) $\{T_h(t) \mid h \in \Delta\}$ is equi-exponentially stable, i.e., ω in (1.3) is negative;

(2) $\sup_{h \in \Delta} \sup\{\|R(\lambda; A_h)\| \mid \operatorname{Re}\lambda \geq 0\} < \infty$;

(3) $\lim_{\sigma \to 0+} \sigma \sup_{h \in \Delta} \sup\{\|R(\lambda; A_h)\| \mid \operatorname{Re}\lambda \geq \sigma\} = 0$;

(4) there exists $p \geq 1$ such that $\sup_{h \in \Delta} \sup_{\sigma > 0} \int_{-\infty}^{\infty} |\langle R(\sigma + i\tau; A_h)x, y\rangle|^p d\tau < \infty, \ \forall x, y \in X$;

(5) there exists $p \geq 1$ such that $\sup_{h \in \Delta} \int_0^{\infty} |\langle T_h(t)x, y\rangle|^p dt < \infty, \quad x, y \in X$.

Theorem 1.14. Let $\{T_h(t) \mid h \in \Delta\}$ be an equi-exponentially bounded family of C_0 semigroups in a Banach space X, satisfying (1.3). Let $\sigma_0 < 0$ and $R(\lambda; A_h)$ be of finite exponential order on $\Sigma_{\sigma_0} = \{\lambda \in \mathbb{C} \mid \operatorname{Re}\lambda \geq \sigma_0\}$ for every $h \in \Delta$, i.e., for every $h \in \Delta$, there exists a constant $\beta_h > 0$ such that

$$R(\lambda; A_h) = O(e^{|\lambda|^{\beta_h}}), \quad |\lambda| \to \infty, \operatorname{Re}\lambda \geq \sigma_0.$$

Let k be a nonnegative integer. Then the following three statements are equivalent:

(1) there exist $\sigma \in [\sigma_0, 0)$ and $M > 0$ such that

$$\|T_h(t)x\| \leq Me^{\sigma t}\|A_h^k x\|, \forall x \in \mathcal{D}(A_h^k), \ t \geq 0, h \in \Delta;$$

(2) there exists $\sigma \in [\sigma_0, 0)$ such that

$$\sup_{h \in \Delta} \sup_{\tau \in \mathbb{R}} |\sigma + i\tau|^{-k}\|R(\sigma + i\tau; A_h)\| < \infty;$$

(3) there exist $p \geq 1$, $\sigma \in [\sigma_0, 0)$ and $\lambda_0 \in \mathbb{C}$ with $\mathrm{Re}\,\lambda_0 > \omega$ such that

$$\sup_{h \in \Delta} \int_0^\infty e^{\sigma t} |\langle T_h(t) R^k(\lambda_0; A_h) x, y \rangle|^p dt < \infty, \quad x, y \in X.$$

Moreover, when $k \geq 1$, the above three statements are equivalent to the following (4), separately:

(4) there exist $\lambda_0, \lambda_1 \in \mathbb{C}$ with $\mathrm{Re}\,\lambda_0 \in [\sigma_0, 0)$, $\mathrm{Re}\,\lambda_1 > \omega$, such that

$$\sup_{h \in \Delta} \sup_{a \geq 0} \left\| \int_0^a e^{\lambda_0 t} T_h(t) R^{k-1}(\lambda_1; A_h) x \, dt \right\| < \infty, \quad \forall x \in X.$$

Now we turn to the strong stability of C_0 semigroups. Huang in [13] proved the following several theorems.

Theorem 1.15. Let $T(t)$ be a uniformly bounded C_0 semigroup with generator A in a Banach space X. If $\mathrm{Re}\,\lambda < 0$ for all $\lambda \in \sigma(A)$, then $T(t)$ is strongly stable. Conversely, if $T(t)$ is strongly stable, then $T(t)$ is uniformly bounded, $\mathrm{Re}\,\lambda \leq 0$ for all $\lambda \in \sigma(A)$ and $i\mathbb{R} \cup \sigma_p(A) = \emptyset$, $i\mathbb{R} \cup \sigma_r(A) = \emptyset$ where, as usual, $\sigma_p(A)$ and $\sigma_r(A)$ represent the point spectrum and residual spectrum of A, respectively.

Theorem 1.16. Let $T(t)$ be a uniformly bounded C_0 semigroup with generator A in a Banach space X. Suppose A has a compact resolvent; then $T(t)$ is strongly stable if and only if $T(t)$ is uniformly bounded and $\mathrm{Re}\,\lambda < 0$ for all $\lambda \in \sigma(A)$.

Theorem 1.17. Let $T(t)$ be a uniformly bounded C_0 semigroup with generator A in a Banach space X. If $\mathrm{Re}\,\lambda < 0$ for all $\lambda \in \sigma(A)$, then $T(t)$ is strongly stable if and only if $T(t)$ is weakly stable.

To further describe the strong stability of C_0 semigroups, first the condition (H) is introduced by Huang in [14]:

(H) There exists a Banach space $Y \subset X$ with norm $\| \cdot \|_Y$ which is stronger than the norm $\| \cdot \|_X$ in X, and Y is dense in X. Moreover, $T(t)Y \subset Y$ for all $t \geq 0$ and $T(t)$ is also a C_0 semigroup in Y with generator B.

The following theorem can be found in [14].

Theorem 1.18. Let A be the generator of a uniformly bounded C_0 semigroup $T(t)$ in a Banach space X. Assume that condition (H) holds such that $\sigma(B) \cup i\mathbb{R}$ is contained in $\sigma_c(B)$ and is countable; then $T(t)$ is strongly stable.

Therefore Theorem 2.4 in [17] is obtained as a corollary of Theorem 1.18:

Corollary 1.19. Let $T(t)$ be a uniformly bounded C_0 semigroup in X with generator A and assume that $\sigma(A) \cup i\mathbb{R}$ is contained in $\sigma_c(A)$ and is countable. Then $T(t)$ is strongly stable.

2. Integrated semigroups and C-semigroups

Integrated semigroups, C-semigroups and their applications have been studied by many authors since 1987. The concepts of C-semigroups and exponentially bounded C-semigroups were introduced in [18, 19]. Let X be a Banach space and C be a bounded injective operator in X. We say that $\{S(t) \mid t \geq 0\}$ is a C-semigroup if $\{S(t) \mid t \geq 0\}$ is a strongly continuous family of bounded linear operators in X satisfying

(C1) $S(0) = C$,

(C2) $S(t)S(s) = S(t+s)C, \forall t, s \geq 0$.

Besides, if it also satisfies

(C3) $\|S(t)\| \leq Me^{\omega t}, \forall t \geq 0$,

where $M > 0$ and $\omega \in \mathbb{R}$ are constants, then $\{S(t) \mid t \geq 0\}$ is called an exponentially bounded C-semigroup.

Since the concept of C-semigroups is not very efficient for applications, the concept of integrated semigroups was proposed by W. Arendt in [20] in 1987. Let n be a natural number. $\{S(t) \mid t \geq 0\}$ is called an n-times integrated semigroup if it is a a strongly continuous family of bounded linear operators in X satisfying $S(0) = C$ and

$$S(t)S(s) = \frac{1}{(n-1)!}\left[\int_t^{s+t}(s+t-r)^{n-1}S(r)dr - \int_0^s(s+t-r)^{n-1}S(r)dr\right].$$

Similarly, an integrated semigroup $\{S(t) \mid t \geq 0\}$ is called exponentially bounded if there exist $M > 0$ and $\omega \in \mathbb{R}$ such that $\|S(t)\| \leq Me^{\omega t}, \forall t \geq 0$.

Throughout this section, X and Y denote Banach spaces. All operators involved are linear. Let A be a linear operator in X, $Z \subset X$ be a subspace of X. We write $A|_Z$ (resp. A_Z) for the restriction (resp. part) of operator A in Z, $\hat{A}$ for $A_{\overline{\mathcal{D}(A)}}$, C for an injective in $\mathcal{L}(X)$, $\rho_C(A)$ for all complex λ such that $\mathcal{R}(C) \subset \mathcal{R}(\lambda - A)$ and $\lambda - A$ is injective, $[\mathcal{D}(A^m)]$ for the Banach space $\left(\mathcal{D}(A^m), \|x\| + \|Ax\| + \cdots + \|A^m x\|\right)$, and M (resp. ω) for positive (resp. real) constant. $(A, S(t))$ (or $A) \in G(M, \omega, C, X)$ means that A generates the C-semigroup $S(t)$ with $\|S(t)\| \leq Me^{\omega t} (t \geq 0)$. Moreover, we denote $G(C, X) = \cup\{G(M, \omega, C, X) \mid M > 0, \omega \in \mathbb{R}\}$. Similar notations $G_n(X), \overline{G}(C, X)$ and $\overline{G}_n(X)$ denote n-times integrated semigroups, C-cosine functions and n-times integrated cosine functions, respectively.

The following results on perturbation, approximation, analyticity and adjoint semigroups of C-semigroups are obtained by Zheng in [21, 22].

Theorem 2.1. Let $A \in G(C, X)$ and let $B \in \mathcal{L}(X)$ commute with A, C (or B in $\mathcal{L}(X)$ map $\overline{\mathcal{D}(A)}$ to $\mathcal{R}(C)$). If $\hat{C}$ in $\mathcal{L}(X)$ is injective, $\mathcal{R}(\hat{C}) \subset \mathcal{L}(C)$ and $A + B \subset \hat{C}^{-1}(A + B)\hat{C}$, then $\hat{C}^{-1}(A + B) \in G(\hat{C}, X)$. In particular, $A + B \in G(\hat{C}, X)$ when $\rho(A + B) \neq \emptyset$.

Theorem 2.2. Let $(A, S(t)) \in G(M, \omega, C, X)$ and suppose that

(1) $\mathcal{D}(A) \subset \mathcal{D}(B)$, $B : \mathcal{D}(A) \to C\overline{\mathcal{D}(A)}$ and $B(\lambda_0 - A)^{-1}$ is closed for some $\lambda_0 > \omega$;

(2) $C^{-1}BS(\cdot)x \in C([0, \infty), X)$, $\forall x \in \mathcal{D}(A)$ and $\gamma_\infty \triangleq \lim_{\lambda \to \infty} \gamma_\lambda < \infty$, where

$$\gamma_\lambda = \sup \left\{ \int_0^\infty e^{-\lambda t} \|C^{-1}BS(t)x\| dt \ \middle| \ x \in \mathcal{D}(A), \|x\| \leq 1 \right\}.$$

If $|\varepsilon| < 1/\gamma_\infty$ and there exists an injective operator $C_\varepsilon \in \mathcal{L}(\overline{\mathcal{D}(A)})$ such that $\mathcal{R}(C_\varepsilon) \subset C\overline{\mathcal{D}(A)}$ and $C_\varepsilon^{-1}(\hat{A} + \varepsilon B)C_\varepsilon \subset \hat{A} + \varepsilon B$, then $C_\varepsilon^{-1}(\hat{A} + \varepsilon B)C_\varepsilon \in G(C_\varepsilon, X)$.

Theorem 2.3. Let $(A_n, S_n(t)) \in G(M, \omega, C_n, X)$ $(n \in \mathbb{N})$. If C_n and $(\lambda - A_n)^{-1}C_n$ (for every $\lambda > \omega$) converge to injective operators C and L_λ respectively, then there exists a closed operator A such that

(1) A is the generator of an integrated C-semigroup $T(t)$ and $L_\lambda = (\lambda - A)^{-1}C$ for $\lambda > \omega$.

(2) $(\hat{A}, S(t)) \in G(M, \omega, C|_{\overline{\mathcal{D}(A)}}, \overline{\mathcal{D}(A)})$, where $S(t)x = T'(t)x$ for $x \in \overline{\mathcal{D}(A)}$ and $t \geq 0$.

(3) For every $x \in \overline{C\mathcal{D}(A)}$, $y \in X$ and $T \geq 0$,

$$\lim_{n \to \infty} \sup_{0 \leq t \leq T} \|S_n(t)x - S(t)x\| = \lim_{n \to \infty} \sup_{0 \leq t \leq T} \left\| \int_0^t S_n(s)y - T(t)y \right\| = 0.$$

If, in addition, $\overline{\mathcal{R}(C)} = \overline{\{x \in X \mid Cx \in \mathcal{R}(L_{\lambda_0})\}}$ for some $\lambda_0 > \omega$, then $\overline{\mathcal{D}(A)} = \overline{C\mathcal{D}(A)} = X$. In particular, $(A, S(t)) \in G(M, \omega, C, X)$.

From the relations between integrated semigroups and C-semigroups, the corresponding results of integrated semigroups can be deduced. For example, we have (see [23])

Theorem 2.4. Let $A \in G_n(X)$ and let $B \in \mathcal{L}(X)$ commute with A (or $B : \overline{\mathcal{D}(A)} \to \mathcal{D}(A^n)$ be bounded, or $B \in \mathcal{L}([\mathcal{D}(A)])$ map $\mathcal{D}(A)$ to $\mathcal{D}(A^{n+1})$). Then $A + B \in G_n(X)$.

The following result is proved in [24].

Theorem 2.5. Let $r \in \rho_C(A)$. Then A generates an integrated C-semigroup $T(t) \Longleftrightarrow A$ generates a C_1-semigroup $S(t)$, where $C_1 = (r - A)^{-1}C$. Moreover, in this case,

$$T(t) = C_1 + r \int_0^t S(s)ds - S(t) \ (t \geq 0),$$

and

$$S(t) = e^{rt} C_1 - r \int_0^t e^{r(t-s)} T(s)ds - T(t) \ (t \geq 0).$$

The basic theory of integrated cosine functions and C-cosine functions is established by Zheng in [25, 26].

Theorem 2.6. Let $\omega \geq 0$. Then $A \in \overline{G}(M, \omega, C, X) \iff (\omega^2, \infty) \subset \rho_C(A)$, $A = C^{-1}AC$, and there exists a strongly continuous family $T(t)$ with $\|T(t)\| \leq Me^{\omega t}$ $(t \geq 0)$ such that

$$\lambda(\lambda^2 - A)^{-1}\dot{C} = \int_0^\infty e^{-\lambda t} T(t)dt \quad (\lambda > 0).$$

Theorem 2.7. Let $r \in \rho(A)$. Then $(A, C(t)) \in \overline{G}_{2n}(X) \iff (A, T(t)) \in \overline{G}(C, X)$, where $C = R(r; A)^n$. Moreover, $T(t) = C^{(2n)}(t)R(r, A)^n$ and

$$C(t) = (r - A)^n \int_0^t (t - t_1) \int_0^{t_1} (t_1 - t_2) \cdots \int_0^{t_{n-1}} (t_{n-1} - t_n)T(t_n)dt_n \cdots dt_2 dt_1.$$

The following two results concerning operator matrices are shown in [28].

Theorem 2.8. Let $\overline{\mathcal{D}(A)} = X$. For every eigenvalue α of complex matrix $M \overset{\triangle}{=} (a_{ij})_{n \times n}$, we assume

(1) $\alpha A \in G_m(X)$;

(2) αA generates an analytic m-times integrated semigroup when α is not simple;

(3) $A \in \mathcal{L}(X)$ when $\alpha = 0$ is not simple.

Thus $\mathcal{A} \overset{\triangle}{=} (a_{ij}A)_{n \times n} \in G_m(X^n)$. Conversely, if $\mathcal{A} \in G_m(X^n)$, then *(1)*, *(3)* and the following *(2')* hold:

(2') the m-times integrated semigroup $S(t)$ generated by αA satisfies $\|S'(t)\| \leq t^{-1}\mu e^{\omega t}$ $(t \geq 0)$ for some constants μ, ω, if α is not simple.

Theorem 2.9. Let $(A, S(t)) \in G_n(X)$ and $(D, T(t)) \in G_m(Y)$. Let $\mathcal{D}(D) \subset \mathcal{D}(B)$ and $BR(\lambda; D) \in \mathcal{L}(X, Y) \, (\lambda \in \rho(D))$. Then

(1) $\mathcal{A} \overset{\triangle}{=} \begin{bmatrix} A & B \\ 0 & D \end{bmatrix}$ generates the $(m + k + 1)$-times integrated semigroup:

$$\begin{bmatrix} \dfrac{1}{m!} \int_0^t (t-s)^m S(s)ds & \int_0^t S(t-s)B \int_0^s T(r)drds \\ 0 & \dfrac{1}{n!} \int_0^t (t-s)^n T(s)ds \end{bmatrix}, \quad \forall t \geq 0.$$

(2) In the case $\overline{\mathcal{D}(D)} = Y$, $\mathcal{A} \in G_{m+k}(X \times Y) \iff \overline{R_t}$ is strongly continuous and exponentially bounded, where $R_t = \int_0^t S(t-s)BT(s)ds$ with $\mathcal{D}(R_t) = \mathcal{D}(D)$. Moreover, the $(m+k)$-times integrated semigroup generated by $\mathcal{A}$ is

$$
\left[
\begin{array}{cc}
\dfrac{1}{(m-1)!} \displaystyle\int_0^t (t-s)^{m-1} S(s)ds & \overline{R_t} \\[2ex]
0 & \dfrac{1}{(n-1)!} \displaystyle\int_0^t (t-s)^{n-1} T(s)ds
\end{array}
\right], \quad \forall t \geq 0;
$$

(3) In the case $B \in \mathcal{L}(\mathcal{D}(D), \mathcal{D}(A))$, $\mathcal{A} \in G_{m+k}(X \times Y)$.

Let iA_k, $1 \leq k \leq n$, be commuting generators of bounded strongly continuous groups and $p(A) = \sum_{|\alpha| \leq m} a_\alpha A^\alpha$ with $A^\alpha = A_1^{\alpha_1} \cdots A_n^{\alpha_n}$. It is shown in [29] that $p(A)$ generates an analytic (or integrated or C-) semigroup under some conditions, that is,

Theorem 2.10. Let $p(A) = \sum_{|\alpha| \leq m} a_\alpha \xi^\alpha$, $\xi \in \mathbb{R}^n$, and $\omega = \sum\{\operatorname{Re} p(\xi) \mid \xi \in \mathbb{R}^n\} < \infty$.

(1) If $p(\xi)$ is strongly elliptic, i.e., m is even and $\sum_{|\alpha|=m} a_\alpha \xi^\alpha < 0$ for all $\xi \neq 0$, then $\overline{p(A)}$ generates an analytic strongly continuous semigroup of angle $\pi/2$.

(2) If $p(\xi)$ is elliptic, i.e., m is even and $\sum_{|\alpha|=m} a_\alpha \xi^\alpha = 0$ and $j > n/2 \, (j \in \mathbb{N})$, then $C^{-1}\overline{p(A)}C \in G(C,X)$, where $C = \overline{(\omega' - p(A))}^{-1} (\omega' > \omega)$. If, in addition, $\rho(\overline{p(A)}) \neq \emptyset$ (or $\rho(A) \neq \emptyset$, or $m > n/2$), then $\overline{p(A)} \in G(C,X)$ (or $p(A) \in G_j(X)$, or $\overline{p(A)} \in G_j(X)$).

(3) If $j > n/2$, $j \in \mathbb{N}$, then $C^{-1}\overline{p(A)}C \in G(C,X)$, where $C = \left(1 + \overline{\sum_{k=1}^n A_k^2}\right)^{-jm/2}$. If, in addition, $\rho(\overline{p(A)}) \neq \emptyset$, then $\overline{p(A)} \in G(C,X)$.

Now consider the problems:

$$x'(t) = Ax(t), \quad t \geq 0, \qquad x(0) = x, \qquad\qquad (ACP)$$

and

$$x(t) = \int_0^t x(s)ds + t^k x/k!, \quad t \geq 0. \qquad\qquad (ACP^k)$$

The following two theorems are obtained in [30], also by Tanake, Miyadera and deLaubenfels.

Theorem 2.11. Let A be closed, $\rho_C(A) \neq \emptyset$ and $A = C^{-1}AC$. Then the following statements are equivalent:

(1) A generates a C-semigroup (not necessarily exponentially bounded).

(2) (ACP) has a unique solution for every $x \in \mathcal{R}((\lambda - A)^{-1}C)$ $(\lambda \in \rho_C(A))$.

(3) (ACP^0) has a unique solution for every $x \in \mathcal{R}(C)$.

Theorem 2.12. Let $r \in \rho(A)$ and denote $(ACP^{-1}) \triangleq (ACP)$. Then the following statements are equivalent:

(1) A generates an n-times nondegenerate integrated semigroup (not necessarily exponentially bounded).

(2) (ACP^k) has a unique solution for every $x \in \mathcal{D}(A^{n-k})$, $-1 \le k \le n$.

(3) (ACP^k) has a unique solution for every $x \in \mathcal{D}(A^{n-k})$, some $-1 \le k \le n$.

3. Higher order linear differential equations in abstract spaces

Let X be a complex Banach space with norm $\| \cdot \|$, $A_0, A_1, \cdots, A_{n-1}$ be closed linear operators on X. Consider the following Cauchy problem:

$$\begin{cases} u^{(n)}(t) + \displaystyle\sum_{k=0}^{n-1} A_k u^{(k)}(t) = 0, \quad \forall t \ge 0, \\ u^{(k)}(0) = u_k, \quad 0 \le k \le n-1. \end{cases} \qquad (ACP_n)$$

The problem $(ACP_n)\,(n \ge 2)$ has been extensively studied in various aspects by many authors, and a lot of results have been obtained. In recent years, T. J. Xiao investigated some properties of (ACP_n), specially of (ACP_2), by the direct treatment method, different from the usual approach to transform (ACP_n) into a first order system and then to use the operator semigroup theory.

Throughout this section, we write $\mathbb{R}^+ = [0, \infty)$, denote by $C^{(k)}(\mathbb{R}^+, X)$ the set of all k-time continuously differentiable X-valued functions on $\mathbb{R}^+$, and write $C(\mathbb{R}^+, X) \triangleq C^{(0)}(\mathbb{R}^+, X)$ for short.

If $\left(\lambda^n + \Sigma_{k=0}^{n-1} \lambda^k A_k \right)$ is invertible, we denote

$$R_\lambda = \left(\lambda^n + \sum_{k=0}^{n-1} \lambda^k A_k \right)^{-1}, \quad \rho(A_0, A_1, \cdots, A_{n-1}) = \{ \lambda \in \mathbb{C} \mid R_\lambda \in \mathcal{L}(X) \},$$

$$D_k = \bigcap_{j=0}^{k} \mathcal{D}(A_j), \quad B_k = A_k |_{D_k}, \; 0 \le k \le n.$$

The solutions, well-posedness, propagators $S_k(\cdot)$ $(0 \le k \le n-1)$ of (ACP_n) are defined as in [31].

We say that (ACP_n) is strongly well posed if it is well posed, $S_k(\cdot)u \in C^{(k)}(\mathbb{R}^+, X)$, $S_{n-1}^{(k-1)}(t)u \in \mathcal{D}(A_k)$ $(t \ge 0)$ and $A_k S_{n-1}^{(k-1)}(\cdot)u \in C(\mathbb{R}^+, X)$ for each $u \in X$, $1 \le k \le n-1$. In particular, when $n = 2$ and $A_1 \in \mathcal{L}(X)$, strong well posedness is equivalent to well posedness (see [31]).

The following two theorems on well posedness for (ACP_2) are obtained in [32].

Theorem 3.1. Let (ACP_2) be strongly well posed and $S_1'(t) \subset \mathcal{D}(A_1)$ for $t > 0$. Then $-A_0$ is the generator of a strongly continuous cosine family on X.

Theorem 3.2. Let $A_1 \in \mathcal{L}(X)$. Then (ACP_2) is well posed iff $-A_1$ is the generator of a strongly continuous cosine family on X.

Theorem 3.2 is also obtained independently by F. Neubrander [33].

In [34], a theorem of Hille-Yosida-Phillips type is established for (ACP_2) to be well posed. Furthermore, a similar theorem in Frechet space setting is obtained in [35]:

Let F be a complex Frechet space, its topology being induced by a separating family of seminorms $\Gamma = \{p_n \mid n \in \mathbb{N}\}$, and A_0, A_1 be closed densely defined linear operators on F.

Theorem 3.3 (Hille-Yosida-Phillips type). For (ACP_2) in the above Fréchet space F, the following statements are equivalent:

(1) (ACP_2) is strongly well posed (defined similarly as in Banach spaces, see [35]). There exist positive numbers μ, ω such that for each $p_n \in \Gamma$, there corresponds a continuous seminorm q_n, satisfying

$$p_n(S'(t)u), \; p_n(BS(t)u), \; p_n(C(t)u) \leq \mu e^{\omega t} q_n(u), \; \forall t \geq 0, u \in F,$$

where $C(t), S(t)$ are the propagators of (ACP_2);

(2) $\mathcal{D}(A_1) \cap \mathcal{D}(A_0)$ is dense in F, there exist positive numbers μ, ω such that for $\lambda \in \mathbb{C}$ with $\operatorname{Re} \lambda > \omega$, $R_\lambda \in \mathcal{L}(X)$, $R_\lambda A_0$ is closable and for each $p_n \in \Gamma$, there is a continuous seminorm q_n such that for $k = 0, 1, 2 \cdots$,

$$p_n([\lambda R_\lambda u]^{(k)}), \; p_n([A_1 R_\lambda u]^{(k)}) \leq \mu k! (\operatorname{Re} \lambda - \omega)^{-k-1} q_k(u), \; \forall u \in F,$$

$$p_n([R_\lambda A_1 u]^{(k)}) \leq \mu k! (\operatorname{Re} \lambda - \omega)^{-k-1} q_n(u), \; \forall u \in \mathcal{D}(A_1) \cup \mathcal{D}(A_0).$$

In [36, 37], the almost periodicity of propagators or solutions are considered for incomplete or complete (ACP_2). In the following three theorems assume $A_1 = 0$.

Theorem 3.4. $\{S_1(t) \mid t \in \mathbb{R}\}$ is almost periodic iff $\{S_1(t) \mid t \in \mathbb{R}\}$ is uniformly bounded and the set of eigenvectors of A_0 is total in X.

Theorem 3.5. Let $\{S_1(t) \mid t \in \mathbb{R}\}$ be almost periodic. Then every solution of (ACP_2) is almost periodic.

Theorem 3.6. Let $\{S_1(t) \mid t \in \mathbb{R}\}$ be almost periodic and for some $\lambda \in \mathbb{R}$, $-\lambda^2$ be an isolated point of the spectrum of A_0. Then $-\lambda^2$ lies in the point spectrum of A_0 and it is a simple pole of the resolvent of A_0 with residue $2i\lambda P_\lambda$, where

$$P_\lambda u = \lim_{\lambda \to \infty} \frac{1}{2t} \int_{-t}^{t} e^{-i\lambda s} S_1(s) u \, ds, \quad \forall u \in X.$$

Let Y be a complex Banach space, $\omega > 0$, $f : (\omega, \infty) \to Y$ be infinitely differentiable. We say $f \in \pi(Y, \omega)$ if

$$\sup\{\|(\lambda - \omega)^{m+1} f^{(m)}(\lambda)/m!\| \mid \lambda > \omega, m = 0, 1, \cdots\} < \infty.$$

The following four theorems for (ACP_n) are obtained in [38].

Theorem 3.7 (Hille-Yosida-Phillips type theorem of higher order differential equations). (ACP_n) is strongly well posed iff D_{n-1} is dense in X, there exists a constant $\omega > 0$ such that for each $\lambda > \omega$, $R_\lambda \in \mathcal{L}(X)$, $R_\lambda B_k$ $(0 \leq k \leq n-1)$ are closable, and for $k = 1, 2, \cdots, n-1$,

$$\lambda^{n-1} R_\lambda, \; \lambda^{k-1} A_k R_\lambda, \; \lambda^{k-1} \overline{R_\lambda B_k} \in \pi(\mathcal{L}(X), \omega).$$

Moreover, if (ACP_n) is strongly well posed, then the propagators of (ACP_n) have the following expressions: for $t > 0$,

$$S_k(t)u = \lim_{n \to \infty} \frac{(-1)^m}{m!} \left(\frac{m}{t}\right)^{m+1} \left(\lambda^{-k-1} \sum_{j=k+1}^{n} \lambda^j \overline{R_\lambda B_j}\right)^{(m)}\bigg|_{\lambda = \frac{m}{t}},$$

$$\forall u \in X, \; 0 \leq k \leq n-1;$$

$$S_0(t) = w - \lim_{\lambda \to \infty} \sum_{m=1}^{\infty} (-1)^{m+1} \frac{1}{m!} e^{m\lambda t} R_{m\lambda} B_0 u, \quad \forall u \in D_0,$$

and for every $\sigma > \omega$,

$$S_0(t)u = \frac{1}{2\pi i} \int_{\sigma - i\infty}^{\sigma + i\infty} e^{\lambda t}(\lambda^{-1} - \lambda^{-1} R_\lambda B_0)u d\lambda, \quad \forall u \in D_0,$$

$$S_k(t)u = \frac{1}{2\pi i} \int_{\sigma - i\infty}^{\sigma + i\infty} e^{\lambda t}\left(\lambda^{-k-1} \sum_{j=k+1}^{n} \lambda^j \overline{R_\lambda B_j} u\right) d\lambda, \quad \forall u \in X, 1 \leq k \leq n-1.$$

Theorem 3.8. Let $\omega > 0$. If D_{n-1} is dense in X, for every $\lambda > om$, $R_\lambda \in \mathcal{L}(X)$ and $\lambda^{k-1} A_k R_\lambda \in \pi(\mathcal{L}(X), \omega)$, $(1 \leq k \leq n)$, then

(1) (ACP_n) has a solution for every $u_0, u_1, \cdots, u_{n-1} \in D_0 \times \cdots \times D_{n-1}$;

(2) for every initial value, (ACP_n) has at most one solution.

Theorem 3.9. For $\theta \in (0, \pi/2]$, $\omega \in \mathbb{R}$, the following are equivalent:

(1) (ACP_n) is strongly well posed. The propagators $S_k(\cdot)$ $(0 \leq k \leq n-1)$ of (ACP_n) can be extended analytically to $\Sigma_\theta = \{z \in \mathbb{C} \mid z \neq 0, |\arg z| < \theta\}$, $S_{n-1}^{(k-1)}(z)X \subset \mathcal{D}(A_k)$ and $A_k S_{n-1}^{(k-1)}(\cdot)$ are analytic in Σ_θ for $1 \leq k \leq n-1$. For every $\theta' \in (0, \theta)$ and $u \in X$, as $z \to 0$, $(z \in \Sigma_\theta)$,

$$S_k^{(k)}(z)u \to u \, (0 \leq k \leq n-1), \quad A_k S_{n-1}^{(k-1)}(z)u \to 0 \, (1 \leq k \leq n-1),$$

and there exists $\mu_{\theta'} > 0$ such that for every $z \in \Sigma_{\theta'}$,

$$\|S_k^{(k)}(z)\|, \quad \|A_k S_{n-1}^{(k-1)}(z)\| \leq \mu_{\theta'} e^w \operatorname{Re} z \quad (1 \leq k \leq n-1).$$

(2) D_{n-1} is dense in X. For every $\theta' \in (0, \theta)$, there exist $M_{\theta'} > 0$ such that for every $\lambda \in \mathbb{C}$ with $\lambda \neq w$ and $|\arg(\lambda - \omega)| < \pi/2 + \theta'$, $R_\lambda \in \mathcal{L}(X)$, $R_\lambda B_k$ $(0 \leq k \leq n-1)$ are closable and for every k with $1 \leq k \leq n-1$,

$$\|\lambda^{n-1} R_\lambda\|, \quad \|\lambda^{k-1} A_k R_\lambda\|, \quad \|\lambda^{k-1} \overline{R_\lambda B_k}\| \leq M_{\theta'} |\lambda - w|^{-1}.$$

Moreover, in this case, we have for $z \in \Sigma_{\theta'}$,

$$S_{n-1}^{(n)}(z) + \sum_{j=0}^{n-1} A_j S_{n-1}^{(j)}(z) = 0,$$

$$S_k^{(n)}(z)u + \sum_{j=0}^{n-1} A_j S_{n-1}^{(j)}(z)u = 0, \quad \forall u \in D_k, 0 \leq k \leq n-2,$$

and for every $\theta' \in (0, \theta)$, $0 \leq k \leq n-2$, as $z \to 0$, $z \in \Sigma_{\theta'}$,

$$S_k^{(j)}(z)u \to 0, \quad \forall u \in D, 0 \leq j \leq n-1, j \neq k,$$

$$S_{n-1}^{(j)}(z)u \to 0, \quad \forall z \in X, 0 \leq k \leq n-2.$$

Notation $[A_k; 0 \leq k \leq n-1] \in \Omega$ (resp. $[A_k; 0 \leq k \leq n-1] \in \Phi$) means that the equivalent conditions of Theorem 3.7 (resp. the equivalent conditions of Theorem 3.9) hold; notation $[A_k; 0 \leq k \leq n-1] \in \Omega_0$ means that the assumptions of Theorem 3.8 hold. Assume $A_n = I$. The following perturbation theorem is proved in [38].

Theorem 3.10 (Perturbation Theorem). Suppose for $0 \leq k \leq n-1$, A_{0k} is a closable linear operator, $A_k + A_{0k}$ is a closed operator and there exists an index i_k with $k+1 \leq i_k \leq n$ such that $\mathcal{D}(A_{i_k}) \subset \mathcal{D}(A_{0k})$. Let $A_{0n} = 0$.

(1) Assume for every $0 \leq k \leq n-2$, there exists a $\lambda_k \in \rho(A_{i_k})$ such that $(\lambda_k - A_{i_k})^{-1} A_{0k}$ can be extended to a bounded linear operator on X. Then the fact that $[A_k; 0 \leq k \leq n-1] \in \Omega$ and $R_\lambda A_{i_k}$ $(\lambda > \omega, 0 \leq k \leq n-2)$ are closable implies $[A_k + A_{0k}; 0 \leq k \leq n-1] \in \Omega$; and the fact that $[A_k; 0 \leq k \leq n-1] \in \Phi$ and $R_\lambda A_{i_k}$ $(\lambda \in \Sigma(\theta, w), 0 \leq k \leq n-2)$ are closable implies $[A_k + A_{0k}; 0 \leq k \leq n-1] \in \Phi$.

(2) If $[A_k; 0 \leq k \leq n-1] \in \Omega_0$, then $[A_k + A_{0k}; 0 \leq k \leq n-1] \in \Omega_0$.

The following two theorems are shown in [40].

Theorem 3.11. The following statements are equivalent:

(1) There exist positive constants μ and ω such that $(\omega, \infty) \subset \rho(A_0, A_1, \cdots, A_{n-1})$ and for $\lambda > \omega$, $m \in \mathbb{N}$, $0 \le j \le n-2$,

$$\|(\lambda - \omega)^{m+1}(\lambda^{n-1}R_\lambda)^{(m)}\|, \quad \|(\lambda - \omega)^{m+1}(\lambda^{n-2}A_j R_\lambda)^{(m)}\| \le \mu m!.$$

(2) B_{n-1} is the generator of a strongly continuous semigroup.

A set of operators $B_j, 0 \le j \le n-1$, is said to be n-closed if $x_k^j \to x^j$ $0 \le j \le n-1$, and $\Sigma_{j=0}^{n-1} B_j x_k^j \to y$ as $k \to \infty$ imply $x^j \in \mathcal{D}(B_j)$ and $\Sigma_{j=0}^{n-1} B_j x^j = y$. (see [41] for the concept of biclosedness) Set

$$N = \begin{bmatrix} 0 & I & \cdots & 0 \\ \cdots & \cdots & \cdots & \cdots \\ 0 & 0 & \cdots & I \\ B_0 & B_1 & \cdots & B_{n-1} \end{bmatrix}, \quad \mathcal{D}(N) = D_n = \mathcal{D}(B_0) \times \cdots \times \mathcal{D}(B_{n-1}).$$

Clearly, N is a closed operator in X^n if $\{B_j \mid 0 \le j \le n-1\}$ is n-closed.

Theorem 3.12. If $\{B_j \mid 0 \le j \le n-1\}$ is n-closed and there exist positive constants μ, ω such that $(\omega, \infty) \subset \rho(B_0, B_1, \cdots, B_{n-1})$ and for $\lambda > \omega$, $m \in \mathbb{N}$, $0 \le j \le n-2$,

$$\|(\lambda^{n-1}R_\lambda)^{(m)}\|, \quad \|(\lambda^{n-2}A_j R_\lambda B_k)^{(m)}\| \le \mu m!(\lambda - \omega)^{-m-1},$$

where $R_\lambda B_k$ has a bounded extension $\overline{R_\lambda B_k}$ for every $0 \le k \le n-2$, then $B_0, \cdots, B_{n-2} \in \mathcal{L}(X)$ and B_{n-1} is the generator of a C_0 semigroup.

References

[1] Hille, E. and R. S. Phillips, Functional Analysis and Semi-groups, Amer. Math. Soc. Colloq. Publ., Vol. 31Providence R.I., 1957.

[2] Pazy, A., Semigroups of Linear Operators and Applications to Partial Differential Equations, Springer-Verlag, New York-Berlin-Heidelberg-Tokyo, 1983.

[3] Huang, F. L., Characteristic conditions for exponential stability of linear dynamical systems in Hilbert spaces, Ann. of Diff. Eqs., 1(1985), 1, 43–55.

[4] Gearhart, L., Spectral theory for contraction semigroups on Hilbert space, Trans. Amer. Math. Soc., 236(1978), 385–394.

[5] Huang, F. L. and K. S. Liu, On the exponential stability of linear dynamical systems in Hilbert spaces, Kexue Tingbao, (1987), 3, 161–163.

[6] Yao, P. F. and D. X. Feng, Characteristical conditions for exponential stability of C_0 semigroups, Chinese Bulletin of Science, 37(1993),

[7] Pritchard, A. J. and J. Zabczyk, Stability and stabilizability of infinite dimensional systems, SIAM Review, 23(1981), 1, 25–52.

[8] Huang, F. L., Exponential stability of linear systems in Banach spaces, Chin. Ann. of Math., 10B(1989), 3, 332–340.

[9] Feng, D. X. and P. F. Yao, Paley-Wiener theorem and asymptotic behavior of C_0 semigroups, accepted for publication in Acta Math. Appl. Sinica, 1995.

[10] Luo, Y. H. and D. X. Feng, On the equi-exponential stability for a family of C_0 semigroups, to appear.

[11] Slemrod, M., Asymptotic behavior of C_0 semigroups as determined by the spectrum of the generator, Indiana Math. J., 25(1976), 8, 783–792.

[12] Clément, Ph., H. J. A. M. Heijmans et al., One-Parameter Semigroups, North-Holland, Amsterdam - New York - Oxford - Tokyo, 1987.

[13] Huang, F. L., Asymptotic stability for linear dynamical systems in Banach spaces, Kexue Tongbao, 10(1983), 584–586.

[14] Huang, F. L., Spectral properties and stability of one-parameter semigroups, J. Diff. Eqs., 104(1993), 182–195.

[15] Huang, F. L., Strong asymptotic stability of linear dynamical systems in Banach spaces, J. Diff. Eqs., 104(1993), 182–195.

[16] Huang, F. L., On the mathematical model for linear elastic systems with analytic damping, SIAM J. Control & Optimization, 26(1988), 3, 714–724.

[17] Arendt, W. and C. J. K. Batty, Tauberian theorems and stability of one-parameter semigroups, Trans. Amer. Math. Soc., 306(1988), 837–852.

[18] Da Prato, G., Semigruppi regolarizzabili, Ricerche di Matematica, 15(1966), 2, 223–248.

[19] Davies, E. B. and M. M. H. Pang, The Cauchy problem and a generalization of the Hille-Yosida theorem, Proc. London Math. Soc., 55(1987), 1, 181–208.

[20] Arendt, W., Proc. London Math. Soc., 54(1987), 321–349.

[21] Zheng, Q. and Y. Lei, Exponentially bounded C-semigroups and integrated semigroups with nondensely defined generators, I: Approximation; II: Perturbation; III: Analyticity, Acta Math. Sci., 13(1993),3; 13(1993),4; 14(1994),1.

[22] Zheng, Q. and Y. Lei, Adjoint semigroups of exponentially bounded C-semigroups (in Chinese), Math. Appl., 6(1993), Special Issue.

[23] Zheng, Q., Perturbation and approximation of integrated semigroups, Acta Math. Sinica (new series), 9(1993), 3.

[24] Zheng, Q. and Y. Lei, On the integrated C-semigroups (in Chinese), J. Huazhong Univ. Sci. Tech., 20(1992), 5, 181–187.

[25] Zheng, Q., Integrated cosine functions, to appear.

[26] Zheng, Q., Higher order abstract Cauchy problems with n-closeness, Math. Japan, 38(1993), 531–539.

[27] Zheng, Q., Some operator matrices as integrated semigroup generators (in Chinese), Acta Math. Sinica, 36(1993), 456–467.

[28] Zheng, Q., Some operator matrices as integrated semigroup operators (in Chinese), Acta Math. Sinica, 36(1993), 456–467.

[29] Zheng, Q., Y. Lei and W. Yi, Semigroups of operators and polynomials of generators of bounded strongly continuous groups, Proc. London Math. Soc., to appear.

[30] Zheng, Q. and Y. Lei, *C*-semigroups, integrated semigroups and integrated solutions of abstract Cauchy problem, in: Applied Functional Analysis Ser., Vol. 1, ed. M.Z.Yang, World Publishing Corporation, Beijing, 1993.

[31] Fattorini, H. O., Extension and behavior at infinity of solutions of certain linear operational differential equations, Pacific J. Math., 33(1970), 583–615.

[32] Xiao, T. J., The well posedness of a class of complete second order differential equations in Banach spaces, J. Sichuan Univ. Nat. Sci. Edi., 25(1988), 421–425.

[33] Neubrander, F., Integrated semigroups and their applications to complete second problems, Semigroup Forum, 38(1989), 233-251.

[34] Xiao, T. J. and J. Liang, On complete second order differential linear equations in Banach spaces, Pacific J. Math., 142(1990), 175–197.

[35] Xiao, T. J., The well posedness of Cauchy problem of complete second order differential equations in Fréchet spaces, Acta Mathematica Sinica, 35(1992), 354–363.

[36] Xiao, T. J., Second order linear differential equations with almost periodic solutions, Acta Mathematica Sinica, New Series, 7(1991), 354–359.

[37] Xiao, T. J., J. Yunnan Norm. Nat. Sci. Edi., 12(1992), 4, 10–13.

[38] Xiao, T. J., The Cauchy problems of higher order abstract differential equations, Chin. Ann. of Math., 14A(1993), 5, 23–35.

[39] Xiao, T. J., Complete second order linear differential equations with almost periodic solutions, J. Math. Anal. Appl., 163(1992), 136–146.

[40] Xiao, T. J., On some worthy problems in abstract differential equations, J. Yunnan Norm. Univ. Nat. Sci. Edi., 11(1991), 4, 9–15.

[41] Sova, M., Ann. Scuola Norm. Sup. Pisa, 22(1968), 67–100.

[42] Xiao, T. J., A note on the propagators of second order linear differential equations in Hilbert spaces, Proc. Amer. Math. Soc., 113(1991), 663–667.

[43] Xiao, T. J., Complete second order linear differential equations, Kexue Tongbao, 33(1988), 1274–1275.

Some Results on Nonlinear Functional Analysis

Dajun Guo

Department of Mathematics, Shandong University, Jinan

Abstract. This paper gives a survey over some results on nonlinear functional analysis including fixed points of mixed monotone operators in ordered Banach spaces, nonzero fixed points and eigenvectors for nonlinear operators by means of computation of topological degree, critical point theorems by means of global variational method and applications to nonlinear integral equations. These results improve some results due to J. Cronin, V. Lakshmikantham, P.H. Rabinowitz, C.A. Stuart and others.

1. Fixed points of mixed monotone operators

Let a real Banach space E be partially ordered by a cone P in E, i.e. $x \leq y$ iff $y - x \in P$. Let $D \subset E$. Operator $A : D \times D \to E$ is said to be mixed monotone if $A(x, y)$ is nondecreasing in x and nonincreasing in y, i.e. $x_1 \leq x_2$ and $y_1 \geq y_2$ ($x_1, x_2, y_1, y_2 \in D$) imply $A(x_1, y_1) \leq A(x_2, y_2)$. Element $x^* \in D$ is called a fixed point of A if $A(x^*, x^*) = x^*$. Recall that cone P is said to be solid if the interior $\mathrm{int}(P)$ of P is nonempty, and P is said to be normal if there exists a positive constant N such that $\theta \leq x \leq y$ implies $\|x\| \leq N\|y\|$, where θ denotes the zero element of E. N is called the normal constant of P. If $y - x \in \mathrm{int}(P)$, we write $x \ll y$. For details on cone theory, see [31].

Theorem 1.1 [10]. Let P be solid and normal, and $A : \mathrm{int}(P) \times \mathrm{int}(P) \to \mathrm{int}(P)$ be a mixed monotone operator. Suppose that there exists an $0 \leq \alpha < 1$ such that

$$A(tx, t^{-1}y) \geq t^{\alpha} A(x, y), \quad \forall x, y \in \mathrm{int}(P), \quad 0 < t < 1. \tag{1.1}$$

Then A has exactly one fixed point $x^* \in \mathrm{int}(P)$, and by constructing successively sequences

$$x_n = A(x_{n-1}, y_{n-1}), \quad y_n = A(y_{n-1}, x_{n-1}) \quad n = 1, 2, \cdots \tag{1.2}$$

1991 Mathematics Subject Classification: 47H10, 45G10

Research supported by NNSF-China and NECDF-China

for any initial $(x_0, y_0) \in \text{int}(P) \times \text{int}(P)$, we have

$$\|x_n - x^\star\| \to 0 \quad and \quad \|y_n - x^\star\| \to 0 \quad \text{as } n \to +\infty \tag{1.3}$$

with convergence rate

$$\|x_n - x^\star\| = O(1 - r^{\alpha^n}) \text{ and } \|y_n - x^\star\| = O(1 - r^{\alpha^n}),$$

where $0 < r < 1$ and r depends on (x_0, y_0).

When $A(x, y)$ does not depend on y and x, we get the following two corollaries.

Corollary 1.2 [10]. Let P be solid and normal, and $A : \text{int}(P) \to \text{int}(P)$ be a nondecreasing operator. Assume that there exists an $0 \le \alpha < 1$ such that

$$A(tx) \ge t^\alpha Ax, \quad \forall x \in \text{int}(P), \quad 0 < t < 1.$$

Then A has exactly one fixed point $x^\star \in \text{int}(P)$, and by constructing a sequence

$$x_n = Ax_{n-1}, \quad n = 1, 2, \cdots$$

for any initial $x_0 \in \text{int}(P)$, we have

$$\|x_n - x^\star\| = O(1 - r^{\alpha^n}), \tag{1.4}$$

where $0 < r < 1$ depends on x_0.

Corollary 1.3 [10]. Let P be solid and normal, and $A : \text{int}(P) \to \text{int}(P)$ be a nonincreasing operator. Assume that there exists an $0 \le \alpha < 1$ such that

$$A(tx) \le t^{-\alpha} Ax, \quad \forall x \in \text{int}(P), \quad 0 < t < 1.$$

Then A has exactly one fixed point $x^\star \in \text{int}(P)$, and (1.4) holds for any iterative sequence

$$x_n = Ax_{n-1}, \quad n = 1, 2, \cdots$$

with any initial $x_0 \in \text{int}(P)$, where $0 < r < 1$ depends on x_0.

Theorem 1.4 [10]. Let the assumption of Theorem 1.1 be satisfied, and let $x_\lambda^\star$ be the unique solution in $\text{int}(P)$ of the equation

$$A(x, x) = \lambda x \quad (\lambda > 0).$$

Then $x_\lambda^\star$ is continuous with respect to λ, i.e. $\|x_\lambda - x_{\lambda_o}^\star\| \to 0$ as $\lambda \to \lambda_0 (\lambda_0 > 0)$. If, in addition, $\alpha \in \left[0, \frac{1}{2}\right)$, then $x_\lambda^\star$ is strongly decreasing with respect to λ, i.e. $0 < \lambda_1 < \lambda_2$ implies $x_{\lambda_1}^\star \gg x_{\lambda_2}^\star$, and $\|x_\lambda^\star\| \to 0$ as $\lambda \to \infty$, $\|x_\lambda^\star\| \to \infty$ as $\lambda \to 0^+$.

Remark 1.5. It should be pointed out that in Theorem 1.4 we do not require operator A to be continuous or compact.

We now apply Theorem 1.1 and Theorem 1.4 to the following initial value problem

$$\begin{cases} x' = \sum_{i=1}^{n} a_i(t)x^{\alpha_i} + \left(\sum_{j=1}^{m} b_j(t)x^{\beta_j} \right)^{-1} , \quad \text{a.e. } t \in J, \\ x(0) = x_0, \end{cases} \tag{1.5}$$

where $J = [0,T]$ $(T > 0)$, $0 < \alpha_i < 1$, $0 < \beta_j < 1$ $(i = 1,2,\cdots,n;\ j = 1,2,\cdots,m)$, $x_0 > 0$, $a_i(t)$ are non-negative bounded measurable functions on J and $b_j(t)$ are non-negative measurable functions such that

$$\inf_{t \in J} \sum_{j=1}^{m} b_j(t) > 0.$$

The set of all absolutely continuous functions from J into R^1 is denoted by $AC(J, R^1)$. A function $x(t)$ on J is said to be a solution of IVP(1.5) if $x \in AC(J, R^1)$ and $x(t)$ satisfies (1.5). It is clear that $x \in AC(J, R^1)$ is a positive solution of IVP(1.5) iff $x \in C(J, R^1)$ is a positive solution of the following integral equation

$$x(t) = x_0 + \int_0^t \left(\sum_{i=1}^{n} a_i(s)[x(s)]^{\alpha_i} \right) ds + \int_0^t \left(\sum_{j=1}^{m} b_j(s)[x(s)]^{\beta_j} \right)^{-1} ds, \quad \forall t \in J. \tag{1.6}$$

Eq.(1.6) can be written in the form $x = A(x, x)$ in space $E = C(J, R^1)$ with $A(x, y) = A_1 x + A_2 y$, where

$$(A_1 x)(t) = x_0 + \int_0^t \left(\sum_{i=1}^{n} a_i(s)[x(s)]^{\alpha_i} \right) ds, \text{ and } (A_2 x)(t) = \int_0^t \left(\sum_{j=1}^{m} b_j(s)[x(s)]^{\beta_j} \right)^{-1} ds.$$

Applying Theorem 1.1 and Theorem 1.4, we get the following two theorems.

Theorem 1.6 [10]. Under conditions mentioned above, IVP(1.5) has exactly one positive solution $x^*(t)$. Moreover, by constructing successively a sequence of functions

$$x_k(t) = x_0 + \int_0^t \left(\sum_{i=1}^{n} a_i(s)[x_{k-1}(s)]^{\alpha_i} \right) ds + \int_0^t \left(\sum_{j=1}^{m} b_j(s)[x_{k-1}(s)]^{\beta_j} \right)^{-1} ds,$$

$$k = 1, 2, 3, \cdots,$$

for any initial positive function $x_0(t)$, sequence $\{x_k(t)\}$ converges to $x^*(t)$ uniformly on J.

Theorem 1.7 [10]. Let the hypotheses of Theorem 1.6 be satisfied. Denote by $x_\lambda^*(t)$ the unique positive solution of the IVP

$$
\begin{cases}
\lambda x' = \sum_{i=1}^{n} a_i(t)x^{\alpha_i} + \left(\sum_{j=1}^{m} b_j(t)x^{\beta_j} \right)^{-1}, & \text{a.e. } t \in J, \\
\lambda x(0) = x_0, \quad \lambda > 0.
\end{cases}
$$

Then $x_\lambda^*(t)$ converges to $x_{\lambda_0}^*(t)$ as $\lambda \to \lambda_0 (\lambda_0 > 0)$ uniformly in $t \in J$. If, in addition, $0 < \alpha_i < \frac{1}{2}$ and $0 < \beta_j < \frac{1}{2}$ $(i = 1, 2, \cdots, n; j = 1, 2, \cdots, m)$, then

$$
0 < \lambda_1 < \lambda_2 \implies x_{\lambda_1}^*(t) > x_{\lambda_2}^*(t), \; \forall t \in J
$$

and

$$
\max_{t \in J} x_\lambda^*(t) \to 0 \text{ as } \lambda \to \infty, \quad \max_{t \in J} x_\lambda^*(t) \to \infty \text{ as } \lambda \to 0^+.
$$

Now we are going to weaken condition (1.1) in some sense.

Theorem 1.8 [11]. Let P be solid and normal, and $A : P \times P \to P$ be a mixed monotone operator. Suppose that (a) there exist $v \in \text{int}(P)$ and $c > 0$ such that $\theta < A(v, \theta) \leq v$ and $A(\theta, v) \geq cA(v, \theta)$, and (b) for any a, b satisfying $0 < a < b < 1$ and any bounded set $B \subset P$, there is an $\eta = \eta(a, b, B) > 0$ such that

$$
A(tx, t^{-1}y) \geq t(1+\eta)A(x, y), \quad \forall a \leq t \leq b, \quad x, y \in B. \tag{1.7}
$$

Then A has exactly one fixed point $x^* \in P$ and $\theta < x^* \leq v$. Moreover, by constructing sequence (1.2) for any initial $(x_0, y_0) \in P \times P$, (1.3) holds.

As in Theorem 1.1 and Theorem 1.4, Theorem 1.8 does not require the operator to be continuous or compact. Theorem can be applied to the following two-point BVP

$$
\begin{cases}
- x'' = \sum_{i=0}^{n} a_i(t)x^{\alpha_i} + \left(\sum_{j=0}^{m} b_j(t)x^{\beta_j} \right)^{-1}, & t \in [0, 1]; \\
ax(0) - bx'(0) = x_0, \quad cx(1) + dx'(1) = x_1,
\end{cases} \tag{1.8}
$$

where $0 = \alpha_0 < \alpha_1 < \alpha_2 < \cdots < \alpha_{n-1} < \alpha_n = 1$, $0 = \beta_0 < \beta_1 < \beta_2 < \cdots < \beta_{m-1} < \beta_m = 1$, $x_0 \geq 0$, $x_1 \geq 0$, $a \geq 0$, $b \geq 0$, $c \geq 0$, $d \geq 0$, and $\delta = ac + ad + bc > 0$, and $a_i(t)$, $b_j(t)(i = 0, 1, \cdots, n; j = 0, 1, \cdots, m)$ are non-negative continuous functions on $J = [0, 1]$. It is well known that $x \in C^2(J, R^1)$ is a non-negative solution of BVP(1.8) iff $x \in C^1(J, R^1)$ is a non-negative solution of the following integral equation:

$$
x(t) = z_0(t) + \int_0^1 G(t, s) \left\{ \left(\sum_{i=0}^{n} a_i(s)[x(s)]^{\alpha_i} \right) + \left(\sum_{j=0}^{m} b_j(s)[x(s)]^{\beta_j} \right)^{-1} \right\} ds, \tag{1.9}
$$

where

$$G(t,s) = \begin{cases} \delta^{-1}(at+b)(c(1-s)+d), & t \le s; \\ \delta^{-1}(as+b)(c(1-t)+d), & t > s, \end{cases}$$

and

$$z_0(t) = \delta^{-1}\{(c(1-t)+d)x_0 + (at+b)x_1\}.$$

(1.9) can be regarded as an equation of the form $x = A(x,x)$ in space $C(J, R^1)$ with $A(x,y) = A_1x + A_2y$, where

$$(A_1x)(t) = z_0(t) + \int_0^1 G(t,s)\left\{\left(\sum_{i=0}^n a_i(s)[x(s)]^{\alpha_i}\right)\right\} ds,$$

$$(A_2x)(t) = \int_0^1 G(t,s)\left\{\left(\sum_{j=0}^m b_j(s)[x(s)]^{\beta_j}\right)^{-1}\right\} ds.$$

Applying Theorem 1.8 to (1.9), we obtain the following

Theorem 1.9 [11]. Let the conditions mentioned above be satisfied. Assume that $a_0(t) > 0$, $b_0(t) > 0$ for $t \in J$ and

$$\int_0^1 a_n(t)\, dt < \delta[(a+b)(c+d)]^{-1}.$$

Then BVP(1.8) has exactly one non-negative nontrivial C^2 solution $x^\star(t)$ on J. Moreover, by constructing successively the sequence of functions ($k = 1, 2, \cdots$)

$$x_k(t) = z_0(t) + \int_0^1 G(t,s)\left\{\left(\sum_{i=0}^n a_i(s)[x_{k-1}(s)]^{\alpha_i}\right) + \left(\sum_{j=0}^m b_j(s)[x_{k-1}(s)]^{\beta_j}\right)^{-1}\right\} ds$$

for any initial non-negative continuous function $x_0(t)$, sequence $\{x_k(t)\}$ converges to $x^\star(t)$ uniformly on J.

One may try to replace condition (1.1) or (1.7) by a more general condition

$$A(tx, t^{-1}y) \ge \varphi(t)A(x,y),$$

where $\varphi(t)$ is a suitable function. For related results on fixed points of mixed monotone operators, see [4], [5], [12] [13] [32] and [49].

2. Nonzero fixed points and eigenvectors

Let E be a real Banach space and $\Omega \subset E$. The closure and boundary of Ω are denoted by $\bar{\Omega}$ and $\partial\Omega$ respectively. There is a fundamental lemma which was proved by the author in [14].

Lemma 2.1 [14]. Let E be infinite-dimensional, $\Omega \subset E$ be bounded and open, and let operator $A : \bar{\Omega} \to E$ be completely continuous (i.e. continuous and compact). Assume that (a) $\inf_{x \in \partial\Omega} \|Ax\| > 0$ and (b) $Ax \neq \mu x$, $\forall x \in \partial\Omega$, $\mu \in (0, 1]$. Then the Leray-Schauder degree $\deg(I - A, \Omega, \theta) = 0$, where I denotes the identical operator and θ denotes the zero element of E.

Using this lemma, we get the following theorem and corollary, which are concerned with the existence of nonzero fixed points and the global structure of eigenvectors.

Theorem 2.2 [14]. Let E be infinite-dimensional and Ω_1, Ω_2 be two bounded open sets in E such that $\theta \in \Omega_1$ and $\bar{\Omega}_1 \subset \Omega_2$. Let operator $A : \bar{\Omega}_2 \backslash \Omega_1 \to E$ be completely continuous. Suppose that one of the two conditions

$$\|Ax\| \leq \|x\|, \quad \forall x \in \partial\Omega_1; \quad \|Ax\| \geq \|x\|, \quad \forall x \in \partial\Omega_2 \qquad (2.1)$$

and

$$\|Ax\| \geq \|x\|, \quad \forall x \in \partial\Omega_1; \quad \|Ax\| \leq \|x\|, \quad \forall x \in \partial\Omega_2 \qquad (2.2)$$

is satisfied. Then A has at least one fixed point $x^\star \in \bar{\Omega}_2 \backslash \Omega_1$ (and so $x^\star \neq \theta$).

Remark 2.3. Theorem 2.2 is not true for a finite dimensional E.

Corollary 2.4 [14]. Let E be infinite-dimensional and $A : E \to E$ be completely continuous and $A\theta = \theta$. Suppose that one of the two conditions

$$\lim_{\|x\| \to 0} \frac{\|Ax\|}{\|x\|} = 0, \quad \lim_{\|x\| \to \infty} \frac{\|Ax\|}{\|x\|} = \infty \qquad (2.3)$$

and

$$\lim_{\|x\| \to 0} \frac{\|Ax\|}{\|x\|} = \infty, \quad \lim_{\|x\| \to \infty} \frac{\|Ax\|}{\|x\|} = 0 \qquad (2.4)$$

is satisfied. Then the following two conclusions hold:
(a) every $\mu \neq 0$ is an eigenvalue of A, i.e. there exists $0 \neq x_\mu \in E$, such that $Ax_\mu = \mu x_\mu$;
(b) $\lim_{\mu \to \infty} \|x_\mu\| = \infty$ under (2.3) and $\lim_{\mu \to \infty} \|x_\mu\| = 0$ under (2.4).

Remark 2.5. Corollary 2.4 is an improvement of a theorem due to Cronin(Theorem 4 in [6]). Cronin's theorem asserts only that one of the two numbers μ and $-\mu$ must be an eigenvalue of A for every $\mu \in (0, 1]$ when (2.3) is satisfied, and Cronin's theorem does not contain the conclusion (b).

Example 2.6. Consider the nonlinear integral equation

$$\mu\varphi(x) = \int_G k(x,y)|\varphi(y)|^p \, dy, \tag{2.5}$$

where $p > 1$, G is a bounded closed domain in R^N and k(x,y) is non-negative and continuous on $G \times G$ and

$$\int_G k(x,y) \, dx > 0, \quad \forall y \in G.$$

Clearly, $\varphi(x) \equiv 0$ is the trivial solution of (2.5) for any μ.

Conclusion: For every $\mu \neq 0$, (2.5) has a nontrivial continuous solution $\varphi_\mu(x)$, which satisfies

$$\lim_{\mu \to \infty} \|\varphi_\mu\|_C = \infty \text{ and } \lim_{\mu \to 0} \|\varphi_\mu\|_C = 0. \tag{2.6}$$

Remark 2.7. This conclusion improves a theorem obtained by Cronin[6]. Cronin's theorem asserts only that for any $\mu \in [-1,0) \cup (0,1]$, (2.5) has a nontrivial continuous solution. Cronin's theorem does not contain the conclusion (2.6), and the hypothesis about $k(x,y)$ is stronger than that mentioned above: $k(x,y)$ is continuous and $k(x,y) > 0$ for all $(x,y) \in G \times G$.

For cone maps, we have the following theorem which is similar to Theorem 2.2.

Theorem 2.8 [16]. Let P be a cone of E and Ω_1, Ω_2 be two bounded open sets in E such that $\theta \in \Omega_1$ and $\bar\Omega_1 \subset \Omega_2$. Let operator $A : P \cap (\bar\Omega_2 \backslash \Omega_1) \to P$ be completely continuous. Suppose that one of the two conditions

$$\|Ax\| \leq \|x\|, \quad \forall x \in P \cap \partial\Omega_1; \quad \|Ax\| \geq \|x\|, \quad \forall x \in P \cap \partial\Omega_2$$

and

$$\|Ax\| \geq \|x\|, \quad \forall x \in P \cap \partial\Omega_1; \quad \|Ax\| \leq \|x\|, \quad \forall x \in P \cap \partial\Omega_2$$

is satisfied. Then A has at least one fixed point $x^* \in P \cap (\bar\Omega_2 \backslash \Omega_1)$(and so $x^* > \theta$).

Remark 2.9. It is different from Theorem 2.2 in that Theorem 2.8 is true for any real Banach space E including the finite-dimensional case.

Example 2.10. Consider the two-point boundary value problem

$$\begin{cases} -x'' = x^\alpha + x^\beta, & t \in [0,1] \\ x(0) = x'(1) = 0, \end{cases} \tag{2.7}$$

where $\beta > 1 > \alpha > 0$. Clearly, $x(t) \equiv 0$ is the trivial solution.

Conclusion: BVP(2.7) has two nontrivial C^2 solutions x_1, x_2 satisfying $x_1(t) > 0$, $x_2(t) > 0$ for $t \in (0, 1]$ and $r < \max_{t \in [0,1]} x_2(t) < 1 < \max_{t \in [0,1]} x_2(t) < R$, where

$$r = \left[\frac{\alpha^\alpha}{(\alpha + 2)^{\alpha+2}} \right]^{\frac{1}{2(1-\alpha)}}, \quad \text{and } R = \left[\frac{(\beta + 2)^{\beta+2}}{\beta^\beta} \right]^{\frac{1}{2(\beta-1)}}$$

Lemma 2.1, Theorem 2.2 and Theorem 2.8 were extended in some sense to more general cases. They were generalized to multivalued maps and were applied to differential inclusions. For example, we have the following

Theorem 2.11 [9]. Let E be infinite-dimensional and Ω_1, Ω_2 be two bounded open sets in E such that $\theta \in \Omega_1$ and $\bar{\Omega}_1 \subset \Omega_2$. Let $A : \bar{\Omega}_2 \backslash \Omega_1 \to kE$ be completely continuous(i.e. upper semi-continuous and compact), where kE denotes the collection of all non-empty compact convex subsets of E. Suppose that one of the two conditions

$$\sup_{y \in Ax} \|y\| \leq \|x\|, \quad \forall x \in \partial\Omega_1; \qquad \inf_{y \in Ax} \|y\| \geq \|x\|, \quad \forall x \in \partial\Omega_2$$

and

$$\inf_{y \in Ax} \|y\| \geq \|x\|, \quad \forall x \in \partial\Omega_1; \qquad \sup_{y \in Ax} \|y\| \leq \|x\|, \quad \forall x \in \partial\Omega_2$$

is satisfied. Then A has at least one fixed point $x^* \in \bar{\Omega}_2 \backslash \Omega_1$(and so $x^* \neq \theta$).

They are generalized in some sense to strict-set-contractions and condensing maps by J. Sun [46] and Y. Sun [50]. For example, we have the following

Lemma 2.12 [46]. Let E be infinite-dimensional, $A : \bar{B}_r \to E$ be a k-set contraction with $k \in [0, 1)$, where $B_r = \{x \in E : \|x\| < r\}$. Assume that (a) there exists a $\delta > 0$such that $\|Ax\| \geq (k + \delta)\|x\|$ for all $x \in \partial B_r$ and (b) $Ax \neq \mu x$, $\forall x \in \partial B_r$, $\mu \in (0, 1]$. Then the topological degree $\deg(I - A, B_r, \theta) = 0$, where I denotes the identical operator and θ denotes the zero element of E.

Lemma 2.13 [50]. Let E be infinite-dimensional, $\Omega \subset E$ be bounded and open, and let $A : \bar{\Omega} \to E$ be a k-set contraction with $k \in [0, 1)$. Assume that (a') $\inf_{x \in \partial\Omega} \|Ax\| > 3k \left(\sup_{x \in \partial\Omega} \|x\| \right)$ and (b') $Ax \neq \mu x$, $\forall x \in \partial\Omega$, $\mu \in (0, 1]$. Then $\deg(I - A, \Omega, \theta) = 0$.

Remark 2.14. Condition (a) is weaker than condition (a'), but Lemma 2.12 is proved only for ball B_r. I do not know whether Lemma 2.13 is true or not if the condition (a') is replaced by a weaker one: $\inf_{x \in \partial\Omega} \|Ax\| > k \left(\sup_{x \in \partial\Omega} \|x\| \right)$.

They were generalized in some sense to P_γ-compact maps by Lafferiere and Petryshyn [40]. For related results, see [2, 17, 18, 34, 38, 39, 47, 51, 53].

3. Some critical point theorems

Let E be a real Banach space, $f : E \to R^1$ be a C^1 functional satisfying the Palais-Smale condition: Any sequence $\{x_n\} \subset E$ for which $\{f(x_n)\}$ is bounded and $f'(x_n) \to \theta (n \to \infty)$ possesses a convergent subsequence. For a real number c, let $K_c = \{x \in E : \quad f(x) = c$ and $f'(x) = \theta\}$.

Theorem 3.1 [36]. Let $D \subset E$ be open and $x_0 \in D$, $x_1 \in E \backslash \bar{D}$. Assume that

$$\inf_{x \in \partial D} f(x) \geq \max\{f(x_0), \quad f(x_1)\}. \tag{3.1}$$

Define

$$c = \inf_{h \in \Phi} \max_{0 \leq t \leq 1} f(h(t)), \tag{3.2}$$

where $\Phi = \{h : [0,1] \to E, h$ is continuous and $h(0) = x_0, h(1) = x_1\}$. Then we have (a) $K_c \backslash \{x_0, x_1\} \neq \emptyset$, (b) $K_c \cap \partial D \neq \emptyset$ if $c = \inf\limits_{x \in \partial D} f(x)$.

Corollary 3.2 [36]. Under the hypothesis of Theorem 3.1, if f(x) is bounded below, then $f(x)$ has at least three critical points.

Remark 3.3. When (3.1) is replaced by a stronger one:

$$\inf_{x \in \partial D} f(x) > \max\{f(x_0), \quad f(x_1)\}, \tag{3.3}$$

Theorem 3.1 becomes the well known mountain pass lemma obtained by Ambrosetti and Rabinowitz[1] in 1973. Theorem 3.1 also improves a result due to Pucci and Serrin[43], in which the following stronger condition is assumed:

$$\inf_{x \in N_r} f(x) \geq \max\{f(x_0), \quad f(x_1)\},$$

where $N_r = \{x \in E : \text{dist}(x, \partial D) < r\}$ $(r > 0)$.

Let $Q \subset E$ be a Banach manifold with boundary ∂Q and let S be a closed set in E. We say that ∂Q is linking with S if $\partial Q \cap S = \emptyset$ and $\varphi(Q) \cap S \neq \emptyset$ for any $\varphi \in C(Q, E)$ with $\varphi|_{\partial Q} = id|_{\partial Q}$.

Theorem 3.4 [7]. Assume that ∂Q is linking with S, $\sup_{x \in Q} f(x) < \infty$, and that there exists an $\alpha \in R^1$ such that

$$f(x) \leq \alpha \text{ for } x \in \partial Q; \quad f(x) \geq \alpha \text{ for } x \in S. \tag{3.4}$$

Define

$$c = \inf_{\varphi \in \Gamma} \sup_{x \in Q} f(\varphi(x)), \tag{3.5}$$

where $\Gamma = \{\varphi \in C(Q, E) : \varphi|_{\partial Q} = id|_{\partial Q}\}$. Then we have (a) $c \geq \alpha$, (b) $K_c \backslash \partial Q \neq \emptyset$ and (c) $K_c \cap S \neq \emptyset$ if $c = \alpha$.

Remark 3.5. When condition (3.4) is replaced by

$$f(x) \leq \alpha \text{ for } x \in \partial Q; \quad f(x) \geq \beta \text{ for } x \in S; \quad \alpha < \beta,$$

Theorem 3.4 becomes a well-known result due to Ni[42](see also [3]). On the other hand, Theorem 3.1 is a special case of Theorem 3.4.

Theorem 3.6 [7]. Let the conditions of Theorem 3.4 be satisfied. Define

$$d = \inf_{\varphi \in \Gamma^*} \sup_{x \in S} f(h(x)), \tag{3.6}$$

where $\Gamma^* = \{h \in C(E, E) : h \text{ is a homeomorphism of } E \text{ onto } E \text{ with } h|_{\partial Q} = id|_{\partial Q}\}$. Then we have (a) $c \geq d \geq \alpha$, (b) $K_d \backslash \partial Q \neq \emptyset$ and (c) $K_d \cap S \neq \emptyset$ if $d = \alpha$, where c is defined by (3.5).

Corollary 3.7 [7]. Under the assumption of Theorem 3.4, if (3.4) is replaced by

$$f(x) \leq \alpha \text{ for } x \in \partial Q; \quad f(x) > \alpha \text{ for } x \in S,$$

then we have (a) $c \geq d > \alpha$, (b) $K_d \neq \emptyset$ and $K_c \neq \emptyset$ where c and d are defined by (3.5) and (3.6) respectively.

Remark 3.8. Corollary 3.7 improves a result due to Chang[3], where the conclusion was proved under the condition that $f \in C^{2-0}$ and that every critical value of f corresponds to finitely many critical points.

In the following, we consider the mountain pass lemma on a closed convex set M in a Hilbert space H. Let $f : H \to R^1$ be a C^1 functional satisfying the Palais-Smale condition on M and

$$f(x) = \frac{1}{2}\|x\|^2 - g(x), \quad g'(x) = Ax.$$

We say that $f(x)$ satisfies the Schauder condition on M if $A(M) \subset M$.

Theorem 3.9 [36]. Let $D \subset E$ be an open subset in M and $x_0 \in D$, $x_1 \in M \backslash \bar{D}$. Assume that $f(x)$ satisfies the Schauder condition on M and

$$\inf_{x \in \partial D} f(x) \geq \max\{f(x_0), \quad f(x_1)\}.$$

Define

$$c = \inf_{h \in \Phi} \max_{0 \leq t \leq 1} f(h(t)),$$

where $\Phi = \{h : [0,1] \to M, h$ is continuous and $h(0) = x_0, h(1) = x_1\}$. Then $K_c \backslash \{x_0, x_1\} \neq \emptyset$.

For some related results see [44, 48].

4. Applications to nonlinear integral equations

(a) Integral equations of polynomial type

Consider the nonlinear integral equation

$$u(x) = \int_G k(x, y) f(y, u(y)) \, dy, \tag{4.1}$$

where G is a bounded closed set in R^N, and

$$f(x, u) = \sum_{i=1}^{n} a_i(x) u^{\alpha_i}, \quad \alpha_i > 0, \quad i = 1, 2, \cdots, n. \tag{4.2}$$

Obviously, $u(x) \equiv 0$ is the trivial solution of (4.1). For the sake of convenience, we formulate the following condition:

(H) $k \in C(G \times G, R_+)$ and there exist a closed set $G_0 \subset G$ with $mesG_0 > 0$ and $0 < \epsilon_0 < 1$ such that

$$k(x, y) > 0, \quad \forall (x.y) \in G_0 \times G_0,$$

$$k(x, y) \geq \epsilon_0 k(z, y), \quad \forall x \in G_0, \quad y, z \in G.$$

Theorem 4.1 [19, 20]. Suppose that (a) the kernel k(x,y) satisfies condition (H), (b) $a_i \in L(G, R_+)(i = 1, 2, \cdots, n)$ and among $\alpha_i(i = 1, 2, \cdots, n)$ there exist $\alpha_{i_0}(x) < 1$ and $\alpha_{i_1} > 1$ such that $\inf_{x \in G_0} a_{i_0} > 0$ and $\inf_{x \in G_0} a_{i_1}(x) > 0$ and (c) $\sum_{i=1}^{n} \|a_i\|_L < \|k\|_C^{-1}$. Then, (4.1) has at least two nontrivial non-negative continuous solutions.

Example 4.2. Consider the BVP

$$\begin{cases} - \quad u'' = f(x,u), \ x \in [0,1]; \\ au(0) - bu'(0) = 0, \quad cu(1) + du'(1) = 0, \end{cases} \tag{4.3}$$

where $f(x,u)$ is given by (4.2), $a \geq 0$, $b \geq 0$, $c \geq 0$, $d \geq 0$, $\delta = ac + ad + bc > 0$. Evidently, $u(x) \equiv 0$ is the trivial solution. It is well known that u(x) is a C^2 solution of BVP(4.3) iff $u(x)$ is a C^1 solution of the following integral equation

$$u(x) = \int_0^1 G(x,y)f(y,u(y)) \, dy, \tag{4.4}$$

where

$$G(x,y) = \begin{cases} \delta^{-1}(ax+b)(c(1-y)+d), & x \leq y; \\ \delta^{-1}(ay+b)(c(1-x)+d), & x > y. \end{cases}$$

It is easy to see that

$$\max_{0 \leq x,y \leq 1} G(x,y) \leq \gamma, \quad \gamma = \begin{cases} \delta^{-1}(4ac)^{-1}, & \text{if } ac \neq 0; \\ \max\{\delta^{-1}(bc+bd), \delta^{-1}(ad+bd)\}, & \text{if } ac = 0 \end{cases} \tag{4.5}$$

and that $G(x,y)$ satisfies the condition (H) for $G_0 = [\alpha,\beta]$ with any α,β satisfying $0 < \alpha < \beta < 1$ and

$$\epsilon_0 = \min\left\{ \frac{a\alpha + b}{a+b}, \frac{c(1-\beta)+d}{c+d} \right\}.$$

Applying Theorem 4.1 we get the following

Conclusion: Let $a_i(x)$ $(i = 1, 2, \cdots, n)$ be non-negative and continuous on $[0,1]$ and

$$\sum_{i=1}^n \int_0^1 a_i(x) \, dx < \gamma^{-1},$$

where γ is defined by (4.5). Suppose that among α_i $(i = 1, 2, \cdots, n)$ there exist $\alpha_{i_0} < 1$ and $\alpha_{i_1} > 1$ such that $a_{i_0}(x)a_{i_1}(x)$ is not identically zero for $t \in [0,1]$. Then, BVP(4.3) has at least two nontrivial non-negative C^2 solutions.

(b) Some nonlinear integral equations arising in science.

We first consider the equation

$$1 = \psi(x) + \psi(x) \int_0^1 \frac{R(x,y)}{x^2 - y^2} \psi(y) \, dy, \quad 0 \leq x \leq 1, \tag{4.6}$$

which is of interest in nuclear physics (see [45]). The solution $\psi(x)$ satisfying $0 < \psi(x) \leq 1$ can be interpreted as associating a probability $\psi(x)$ with the point $x \in [0,1]$.

Theorem 4.3 [21, 37]. Suppose that (a) $R(x, y)$ is continuous on $0 \leq x, y \leq 1$ and $R(x, y) \geq 0$ for $x > y$ and $R(x, y) \leq 0$ for $x < y$, (b) there exists a $\nu > 0$ such that

$$|R(x, y)| \leq c|x - y|^{\nu} S(x, y), \quad \forall 0 \leq x, y \leq 1, \quad x \neq y,$$

where c is a constant and $S(x, y)$ is a bounded and non-negative function on $0 \leq x, y \leq 1$ which satisfies

$$\overline{\lim}_{x, y \to 0+} \sup \frac{S(x, y)}{x + y} < \infty.$$

Then (4.6) has exactly one positive continuous solution $\psi^*(x)$. Moreover, $0 < \psi^*(x) \leq 1$ for $x \in [0, 1]$, and by constructing successively the sequence of functions

$$\psi_n(x) = \left[1 + \int_0^1 \frac{R(x, y)}{x^2 - y^2} \psi_{n-1}(y) \, dy\right]^{-1} \quad n = 1, 2, \cdots$$

for any initial continuous function $\psi_0(x)$ satisfying $0 < \psi_0(x) \leq 1$ for $x \in [0, 1]$, we have

$$\|\psi_n - \psi^*\|_C = \max_{x \in [0,1]} |\psi_n(x) - \psi^*(x)| \to 0 \text{ as } n \to \infty.$$

Remark 4.4. Theorem 4.3 is an improvement of a result due to Stuart[45]. It is easy to point out some elementary functions $R(x, y)$, which satisfy all conditions of Theorem 4.3. For example,

$$R(x, y) = (x - y)^{\frac{1}{3}} (x + 3y - \sin x + 2x^2 y)$$

and

$$R(x, y) = 2(x - y)^{\frac{1}{6}} \log(1 + x + y).$$

Next, we consider the equation

$$x(t) = \int_{t-\tau}^t f(s, x(s)) \, ds, \quad (4.7)$$

which can be interpreted as a model for the spread of certain infectious diseases with periodic contact rate that varies seasonally(see [41]). Here, $x(t)$ represents the proportion of infectives in the population at time t, $f(t, x(t))$ is the proportion of new infectious per unit time $(f(t, 0) = 0)$, and τ is the length of time an individual remains infectious. Obviously, $x(t) \equiv 0$ is the trivial solution. We formulate some conditions for $f(t, x)$:

(H_1) $f(t, x)$ is non-negative and continuous for $t \in (-\infty, \infty)$ and $x \geq 0$, $f(t, 0) = 0$ for $t \in (-\infty, \infty)$ and there exists an $\omega > 0$ such that $f(t + \omega, x) = f(t, x)$ for all $t \in (-\infty, \infty)$ and $x \geq 0$.

(H_2) $\overline{\lim}_{x \to 0+} \frac{f(t, x)}{x} = a_0(t)$ uniformly with respect to $t \in [0, \omega]$ and $\sup_{t \in [0, \omega]} a_0(t) < \tau^{-1}$.

(H_3) $\overline{\lim}_{x \to \infty} \frac{f(t,x)}{x} = a_1(t)$ uniformly with respect to $t \in [0, \omega]$ and $\sup_{t \in [0, \omega]} a_1(t) < \tau^{-1}$.

Theorem 4.5 [33]. Let the conditions (H_1)–(H_3) be satisfied. Assume that there exist an $a > 0$ and a non-negative continuous function $b(t)$ with period ω such that

$$f(t, x) \geq b(t), \quad \forall t \in [0, \omega], \quad x \geq a$$

and

$$\int_{t-\tau}^{t} b(s) \, ds > a, \quad \forall t \in [0, \omega].$$

Then (4.7) has at least two nontrivial non-negative and continuous solutions $x_1(t)$ and $x_2(t)$ with period ω and

$$\inf_{t \in (-\infty, \infty)} x_1(t) > a.$$

Remark 4.6. Theorem 4.5 shows that under certain circumstances the spread of the infectious disease may have two possible states. It is easy to give some elementary functions which satisfy all conditions of Theorem 4.5, see [33].

(c) Nonlinear integral equations in banach spaces.

Consider the integral equation

$$x(t) = \int_J H(t, s, x(s)) \, ds, \tag{4.8}$$

where $J = [a, b]$ is a compact interval, $H \in C(J \times J \times P, P)$ and P is a cone in the real Banach space E.

Theorem 4.7 [22]. Suppose that (a) $H(t, s, x)$ is uniformly continuous on $J \times J \times P_r$ for any $r > 0$, where $P_r = \{x \in P : \|x\| \leq r\}$, and there exists an $M \geq 0$ with $2M(b - a) < 1$ such that $\alpha(H(t, s, B)) \leq M\alpha(B)$ for $t, s \in J$ and a bounded $B \subset P$, where α denotes the Kuratowski's measure of noncompactness, (b) $\|H(t, s, x)\|/\|x\| \to 0$ as $x \in P$, $\|x\| \to \infty$ uniformly in $t, s \in J$ and (c) there are $x_0 > \theta$ and $k \in C(J \times J, R_+)$ such that $H(t, s, x) \geq k(t.s)x_0$ for $x \geq x_0$ and $\int_J k(t, s) ds \geq 1$ for $t \in J$. Then Eq.(4.8) has a solution $x^* \in C(J, P)$ satisfying $x^*(t) \geq x_0$ for all $t \in J$.

Example 4.8. Consider the infinite system of nonlinear integral equations

$$x_n(t) = \frac{1}{n} \int_0^1 (2 - ts) \left[1 + x_{n+1}(s) + x_{2n}(s)\right]^{\frac{1}{3}} ds - \frac{1}{n+1} \int_0^1 t^2 s \sin(t + s - x_n(s)) ds \tag{4.9}$$

$$(n = 1, 2, 3, \cdots)$$

This system can be regarded as an equation of the form (4.8), where $J = [0, 1]$, $E = \{x = (x_1, \cdots, x_n, \cdots) : x_n \to 0\}$ with norm $\|x\| = \sup_n |x_n|$, $P = \{x = (x_1, \cdots, x_n, \cdots) \in E : x_n \geq 0, n = 1, 2, \cdots\}$ and $H(t, s, x) = (H_1(t, s, x), \cdots, H_n(t, s, x), \cdots)$ with $x = (x_1, x_2, \cdots, x_n, \cdots)$ and

$$H_n(t, s, x) = \frac{1}{n}(2 - ts)\left[1 + x_{n+1} + x_{2n}\right]^{\frac{1}{3}} - \frac{1}{n+1}t^2 s \sin(t + s - x_n).$$

We can show that all conditions of Theorem 4.7 are satisfied for $M = 0$, $x_0 = (1, 0, \cdots, 0, \cdots)$ and $k(t, s) = 2 - ts - t^2 s$, and so the following conclusion holds:

Conclusion: System (4.9) has a continuous solution $\{x_1^*(t), \cdots, x_n^*(t), \cdots\}$ satisfying $x_n^*(t) \to 0$ as $n \to \infty$ and $x_1^*(t) \geq 1$, $x_n^*(t) \geq 0$ ($n = 2, 3, \cdots$) for all $t \in [0, 1]$.

For some related results see [8,23–30, 35].

References

[1] Ambrosetti, A., and Rabinowitz, P.H., Dual variational methods in critical point theory and applications, J. Funct. Anal. 14(1973), 349–381.

[2] Bai, J., Eigenvectors of nonlinear integral operators of Hammerstein type, Kexue Tong-bao, 29(1984), 704(In Chinese).

[3] Chang, K.C., Critical point theory and its applications, Shanghai Sci. and Tech. Press, 1986, (In Chinese).

[4] Chang, S.S., and Ma, Y.H., Coupled fixed points for mixed monotone condensing operators and existence theorem of the solutions for a class of functional equations arising in dynamic programming, J. Math. Anal. Appl., 160(1991), 468–479.

[5] Chen, Y.Z., Existence theorems of coupled fixed points, J. Math. Anal. Appl., 154(1991), 142–150.

[6] Cronin, J., Eigenvalues of some nonlinear operators, J. Math. Anal. Appl., 38(1972), 659–667.

[7] Du, Y., Deformation lemma and its applications, Chin. Sci. Bull., 36(1990), 103–105.

[8] Erbe, L.H., Guo, D. and Liu, X., Positive solutions of a class of nonlinear integral equations and applications, J. Integral Eq. Appl., 4(1992), 179–195.

[9] Erbe, L.H., Krawcewicz, W., and Guo, D., Some fixed point theorems for multivalued maps and applications, Results in Math., 21(1992), 42–64.

[10] Guo, D., Fixed points of mixed monotone operators with applications, Appl. Anal., 31(1988), 215–224.

[11] Guo, D., Existence and uniqueness of positive fixed points for mixed monotone operators and applications, Appl. Anal., 46(1992), 91–100.

[12] Guo, D., Fixed points and eigenvectors of some classes of concave and convex operators, Kexue Tongbal, 30(1985), 1132–1135(In Chinese).

[13] Guo, D., Positive eigenvectors of decreasing operators and applications, Northeastern Math. J., 1(1985), 101–109.

[14] Guo, D., Eigenvalues and eigenvectors of nonlinear operators, Chin. Ann. of Math. (Eng. Issue), 2(1981), 65–80.

[15] Guo, D., A new fixed point theorem, Acta Math. Sinica, 24(1981), 444–450(In Chinese).

[16] Guo, D., Some fixed point theorems on cone maps, Kexue Tongbao, 28(1983), 1217–1219(In Chinese).

[17] Guo, D., Some fixed point theorems and applications, Nonlinear Anal. TMA, 10(1986), 1293–1302.

[18] Guo, D., Some fixed point theorems of expansion and compression type with applications, Nonlinear Analysis and Applications(edited by V. Lakshmikatham) Marcel Dekker, Inc., New York, 1987, 213–221.

[19] Guo, D., The number of positive solutions of Hammerstein nonlinear integral equations, Acta Math. Sinica, 22(1979), 584–595(In Chinese).

[20] Guo, D., Nonlinear functional analysis (In Chinese), Shandong Sci. and Tech. Press, Jinan, China, 1985.

[21] Guo, D., The solution of a nonlinear integral equations in nuclear physics, Kexue Tongbao, 23(1978), 27–31(In Chinese).

[22] Guo, D., Multiple solutions of nonlinear Fredholm integral equations in Banach spaces, Chin. Ann. of Math., 10B(1989), 513–519.

[23] Guo, D., Proper values of Hammerstein nonlinear integral equations, Acta Math. Sinica, 20(1977), 99–108 (In Chinese).

[24] Guo, D., The solution for a nonlinear integral equation in the neutron transport theory, Acta Math. Sinica, 22(1979), 231–236 (In Chinese).

[25] Guo, D., Eigenvalues and eigenfunctions of Hammerstein integral equations, Acta Math. Sinica, 25(1982), 419–426 (In Chinese).

[26] Guo, D., Positive solutions of Hammerstein integral equations of polynomial type with applications, Chin. Ann. of Math. 4(1983), 645–656 (In Chinese).

[27] Guo, D., Positive solutions of nonlinear operator equations with applications to nonlinear integral equations, Adv. in Math. 13(1984), 294–310 (In Chinese).

[28] Guo, D., The number of nontrivial solutions to Hammerstein nonlinear equations, Chin. Ann. of Math., 7B(1986), 191–204.

[29] Guo, D., Non-negative solutions of two-point boundary value problems for nonlinear second order integro-differential equations in Banach spaces, J. Appl. Math. Stoch. Anal., 4(1991), 47–69.

[30] Guo, D., Multiple positive solutions of impulsive nonlinear Fredholm integral equations and applications, J. Math. Anal. Appl., 173(1993), 318–324.

[31] Guo, D. and Lakthmikantham, V., Nonlinear problems in abstract cones, Academic Press Inc., Boston & New York, 1988.

[32] Guo, D. and Lakshmikantham, V., Coupled fixed points of nonlinear operators with applications, Nonlinear Anal. TMA, 11(1987), 623–632.

[33] Guo, D. and Lakshmikantham, V., Positive solutions of nonlinear integral equations arising in infectious diseases, J. Math. Anal. Appl., 134(1988), 1–8.

[34] Guo, D. and Sun, J., Some global generalizations of the Birkhoff-Kellogg theorem and applications, J. Math. Anal. Appl., 129(1988), 231–242.

[35] Guo, D. and Sun, J., Nonlinear integral equations, Shandong Sci. and Tech. Press, Jinan, China, 1987 (In Chinese).

[36] Guo, D., Sun, J., and Qi, G., Some extensions of the mountain pass lemma, Diff. and Integ. Equations, 1(1988), 351–358.

[37] Guo, D., and Zhang, Q., On the uniqueness of solution of a nonlinear integral equation in nuclear physcis, Kexue Tongbao, 24(1979), 678–681 (In Chinese).

[38] Han, Z., An extension of Guo's theorem and its applications, Northeastern Math. J., 7(1991), 480–485.

[39] Huang, C., A generalization of the Guo Dajun theorem, Kexue Tongbao, 29(1984), 1341(In Chinese).

[40] Lafferiere, B., and Petryshyn, W.V., New positive fixed point and eigenvalue results for P_γ-compact maps and some applications, Nonlinear Anal. TMA, 13(1989), 1427–1440.

[41] Leggett, R.W., and Williams, L.R., A fixed point theorem with application to an infectious disease model, J. Math. Anal. Appl., 76(1980), 91–97.

[42] Ni, W.M., Some minimax principles and their applications in nonlinear elliptic equations, J. d'Analyses Math., 37(1980), 248–275.

[43] Pucci, P., and Serrin, J., A mountain pass lemma, J. Diff. Equations, 60(1985), 142–149.

[44] Qi, G., Extension of the mountain pass lemma, Kexue Tongbao, 31(1986), 724–727(In Chinese).

[45] Stuart, C.A., Positive solutions of a nonlinear integral equation, Math. Ann., 192(1972), 119–124.

[46] Sun, J., A generalization of Guo's theorem and applications, J. Math. Anal. Appl., 126(1987), 566–573.

[47] Sun, J., Some fixed point theorems for set-contractive operators, Kexue Tongbao, 31(1986), 728–729(In Chinese).

[48] Sun, J., Schauder condition in critical point theory, Kexue Tongbao, 31(1986), 328–331(In Chinese).

[49] Sun, Y., A fixed point theorem for mixed monotone operators with applications, J. Math. Anal. Appl., 156(1991), 240–252.

[50] Sun, Y., An extension of Guo's theorem on domain compression and expansion, Numer. Funct. and Optimiz., 10(1989), 607–617.

[51] Wang, N., The computation of topological degree and applications, Chin. Ann. of Math., 8A(1987), 311–318(In Chinese).

[52] Williams, L.E., and Leggett, R.W., Nonzero solutions of nonlinear integral equations modeling infectious disease, SIAM J. Math. Anal., 13(1982), 112–121.

[53] Zhao, Z., Further generalizations of Guo's theorem, J. of Math., 12(1992), 272–280(In Chinese).

Theory of Ba Spaces and Applications

Yuzan He and Peizhu Luo

Academia Sinica, Beijing

1. Introduction

The concept of Ba spaces was first coined by Ding Xiaxi and Luo Peizhu in 1980. The Ba space is a class of new function spaces extracted from a series of important research works by Ding and Luo themselves. It is a very natural generalization of the L^p spaces and includes some important Orlicz spaces, Sobolev spaces and Orlicz-Sobolev spaces, etc. As the investigation is based on L^p, the methods are more natural, the results are more concrete. It possesses not only many properties which are similar to those in Orlicz spaces but also the refined results which are different from those of Orlicz spaces. In the 14 years period since Ding-Luo paper appeared, several authors have found the Ba space to be a useful and interesting concept. This class of spaces has been effectively applied to partial differential equations, to harmonic analysis and function theory. By using the theory of Ba spaces Ding et al. proved the existence of the generalized solution for the strong non-linear parabolic equations, and they also found out that the solution for the strong non-linear variational problem just belongs to such a new function space. The boundedness of some operators on Ba spaces has yielded deep and complete results. In the function theory, some interesting relations between Ba space and BMOA (or Bloch space) were pointed out, some results on the estimate of approximation order in Ba space were obtained. In this survey we'll introduce the theory of Ba spaces with its applications which has been achieved by the Chinese scholars.

2. Ba Spaces

2.1. Ba Space and aB space [7, 18, 19]

1991 Mathematics Subject Classification: 46A45, 46B45

Supported in part by the State Natural Science Fund of China

Definition 2.1. Let $\{B_m\}_{m=1}^{\infty}$ be a sequence of linear normed spaces, and B be a linear normed space with $B_m \subset B$ (i.e. B_m is contained and embedded in B). Suppose that $\varphi(z) = \sum_{m=1}^{\infty} a_m z^m$ is a non-zero entire function with $a_m \geq 0$. For $u \in \prod_{m=1}^{\infty} B_m$, set

$$I(u, z) := \sum_{m=1}^{\infty} a_m \|u\|_{B_m}^m z^m, \tag{2.1}$$

where $\| \cdot \|_{B_m}$ stands for the norm in B_m. Then

$$Ba := \left\{ u, u \in \prod_{m=1}^{\infty} B_m, d_u > 0 \right\},$$

where d_u denotes the radius of convergence of (2.1).

The norm in Ba is defined by

$$\|u\|_{Ba} := \inf \left\{ \frac{1}{|\alpha|}, I(u, |\alpha|) \leq 1 \right\}. \tag{2.2}$$

Theorem 2.2. If $B_m (m = 1, 2, \cdots)$ and B are Banach spaces, then Ba is a Banach space.

One of the important subspaces of Ba is called the aB space which is defined as follows:

Definition 2.3. for a given Ba space,

$$aB := \{u, u \in Ba, d_u = \infty\}.$$

We have [5, 19, 24]

Theorem 2.4. For a given Ba space, then
  (i) aB is a closed subspace of Ba;
  (ii) aB is separable;
  (iii) Ba is separable if and only if $Ba = aB$.

2.2. Examples

Example 1. For arbitrary Banach space $\hat{B}$. In Definition 2.1, we take $B = \hat{B}, B_m = \hat{B}$ for all m and $\varphi(z) \equiv z$, then the corresponding Ba space is nothing but $\hat{B}$ itself.

Example 2. Put $B_m = L^{P_m}(\mathbb{R}^n, dx) := \left\{ u, \|u\|_{L^{P_m}}^{P_m} = \int_{\mathbb{R}^n} |u|^{P_m} dx < \infty \right\}, p_m \geq 1, m = 1, 2, \cdots$ and $B = L^1(\mathbb{R}^n, dx) \oplus L^{\infty}(\mathbb{R}^n, dx) := \{u = u_1 + u_2, u_1 \in L^1(\mathbb{R}^n, dx), u_2 \in$

$L^\infty(\mathbb{R}^n, dx)\}, \|u\|_B := \inf\{\|u_1\|_{L^1} + \|u_2\|_{L^\infty}, u = u_1 + u_2, u_1 \in L^1(\mathbb{R}^n, dx), u_2 \in L^\infty(\mathbb{R}^n, dx)\}$,

where R^n denotes the n-dimensional real Euclidean space, dx is the element of n-dimensional

Lebesgue measure on $\mathbb{R}^n$, $\varphi(z) = \sum_{m=1}^{\infty} a_m z^m$ is an entire function with $a_m \geq 0$. Then we get

a Ba space denoted by $L^{Ba}(\mathbb{R}^n, dx)$.

Example 3. Take $B_m = H^{p_m}(D) := \{f, \text{ analytic in the unit disk D},$

$\sup_{0 \leq r < 1} \int_0^{2\pi} |f(re^{i\theta})|^{p_m} d\theta < \infty\}$ and $B = H^1(D)$. Suppose that $\varphi(z) = \sum_{m=1}^{\infty} a_m z^m$ is an

entire function with finite order ρ and mean type σ, the sequence $\{p_m\}_{m=1}^{\infty}$ satisfies $1 \leq p_1 < p_2 < \cdots \nearrow \infty$ and

$$\varlimsup_{m \to \infty} \frac{p_m}{m^{\frac{1}{\rho}}} < \infty \qquad (2.3)$$

Then we have a Ba space denoted by $H^{Ba}(D)$.

We'll show that there is an interesting relation between $H^{Ba}(D)$ and $BMOA(D)$ [11].

2.3. Some further results

Zhuang Yadong and Yu Xintai[24] discussed the strict and uniform convexity of Ba space.

Let Ba be an abstract Ba space, $I(u) = \sum_{m=1}^{\infty} a_m \|u\|_{B_m}^m$ for $u \in Ba$, and set $\widehat{Ba} := \{u, u \in Ba, I(u) < \infty\}$. They proved

Theorem 2.5. If $\varlimsup \sqrt[m]{a_m}\|u\|_{B_m} < 1$ (as $m \to \infty$) for each $u \in Ba$, then Ba is strictly convex if and only if $I(u)$ is strictly convex, i.e. for $u, v \in \widehat{Ba}$ and $u \neq v, I\left(\frac{u+v}{2}\right) < \frac{1}{2}(I(u) + I(v))$.

In [2], Chen Guangrong and Meng Boqin studied the interpolation theorems of linear operator in Ba space. They gave the following theorem

Theorem 2.6. For Ba space $L^{Ba}(\mathbb{R}^n, dx)$, if $\{a_m^{\frac{1}{m}}\}$ and $\{a_m^{-\frac{1}{m}}\} \in l^\infty$, then $L^{Ba}(\mathbb{R}^n, dx)$ is an interpolation space with repect to $L^{p^*}(\mathbb{R}^n, dx)$ and $L^{p_0}(\mathbb{R}^n, dx)$ of exponent θ, in which $p^* = \sup_{m \geq 1}\{p_m\}$ and $p_0 = \inf_{m \geq 1}\{p_m\}$.

In [19] Ma Jigang constructed a non-separable Ba space satisfying the necessary and sufficient condition of boundedness for the Hilbert transform, but in Orlicz space the corresponding condition implies that the Orlicz space should be separable. We'll show it in 3.2. Ma also introduced and studied the Ba sequence space in great detail in [18].

3. Applications

3.1 Applications to nonlinear partial differential equations

In [8] Ding Xiaxi, Luo Peizhu, Gu Yonggeng and Fang Huizhong studied the calculus of variations with strong nonlinearity:

$$I(u) = \inf_{v \in K} I(v), I(v) := \int_{\Omega} F(x, v, v_x) dx,$$

$$K = \{v, v \in W^1 L_p(\Phi), v|_S = \varphi(S)\},$$

$$|F(x, v, v_x)| \leq \mu(|v|)(L(x) + \Phi(\beta|v|^p)).$$

where $\mu(t)$ is increasing, $L(x) \in L_1, \Phi(v)$ is any entire function. Using the trace theorem deduced in [9], and by a procedure of Ritz's type, they solved the above problem proposed by O.A. Ladyzenskaja in [15].

Before giving the following results let us introduce some notations of the function spaces.

1. Spaces $\overset{o}{W}{}^{1,0}_{m,2}(Q_T)$ are obtained by a completion in the norm

$$\|u\|_{m,2,Q_T} := \|u\|_{2,Q_T} + \|u_x\|_{m,Q_T}$$

of all smooth functions that are equal to zero on S_T where

$$\|u\|_{2,Q_T} := \left(\int_{Q_T} |u|^2 dx dt\right)^{1/2}, \|u_x\|_{m,Q_T} := \sum_{i=1}^{n} \left(\int_{Q_T} |u_{x_i}|^m dx dt\right)^{1/m}$$

where $Q_T = \Omega \times (0, T)$ is a cylindrical domain of the upper half space $\mathbb{R}^{n+1}_+$; $S_T = \partial\Omega \times [0, T]$ is the lateral surface of Q_T, Ω being a bounded domain of $\mathbb{R}^n$. The boundary $\partial\Omega$ is sufficiently smooth and the measure of Ω is denoted by mesΩ.

2. Spaces $\overset{o}{W}{}^{1,0} L_\gamma(\Phi, 2; Q_T)$ are composed of functions $u(x, t) \in L_2(Q_T)$ with generalized derivatives $u_{x_i} \in L_\gamma(\Phi; Q_T)$ and $u(x, t)|_{S_T} = 0$; their norms are defined by

$$\|u\|_{\Phi, 2; Q_T} := \|u\|_{2, Q_T} + \|u_x\|_{L_\gamma(\Phi; Q_T)} \tag{3.0}$$

where

$$\|u_x\|_{L_\gamma(\Phi; Q_T)} := \inf_{\lambda \geq 0} \left\{ \lambda^{-\frac{1}{\gamma}} | \int_{Q_T} \Phi(\lambda|u_x|^\gamma) dx dt \leq 1 \right\}.$$

In addition, we need the spaces $W^{1,0} E_\gamma(\Phi, 2; Q_T)$. They are composed of functions $u(x, t) \in L_2(Q_T)$ with generalized derivatives $u_{x_i} \in E_\gamma(\Phi; Q_T)$ and $u(x, t)|_{S_T} = 0$; their norms are defined as in (3.0). According to the above definitions, spaces $L_\gamma(\Phi; Q_T)$ and $E_\gamma(\Phi; Q_T)$ are two Orlicz spaces with N-function $\Phi(z^\gamma), \gamma > \frac{1}{2}$. The complementary N-function of $\Phi(z^\gamma)$ is denoted by $\psi_\gamma(z)$. It is easy to see that $\Psi_r(z)$ satisfies the so-called Δ_2-condition.

Suppose that

$$I(u) = \inf_{v \in W^1 E(\Phi)} I(v) \tag{3.1}$$

$$I(u) \geq f(\|u\|_{W^1 E(\Phi)})^{-\nu}, \forall u \in \overset{o}{W}{}^1 E(\Phi) \tag{3.2}$$

For any $x \in \overline{\Omega}, u \in L^\infty(\Omega), q, \overline{q} \in E^n$, there are

$$\frac{\partial F_{qi}}{\partial q_j} \xi_i \xi_j \geq \nu |\xi|^2, \tag{3.3}$$

$$|F(x, u, q)| \leq \mu(|u|)[g(x) + \Phi(\beta|q|)], \tag{3.4}$$

$$|F_{q_i}(x, u, q)| \leq \mu(|u|)[h(x) + \Psi^{-1}\Phi(\beta|q|)], \tag{3.5}$$

$$|F_u(x, u, q)| \leq \mu(|u|)[h(x) + \Psi^{-1}\tilde{\Phi}(\beta|q|)]. \tag{3.6}$$

where $\nu, \beta > 0$ are constants, $f(\tau)$ is a nonnegative continuous function over $[0, \infty)$ with $\lim_{\tau \to \infty} f(\tau) = +\infty, \mu(\tau)$ is a nonnegative nondecreasing function over $[0, \infty), g(x) \in L_1(\Omega)$, $h(x) \in E(\Psi)$ and the N-function $\tilde{\Phi}(t) \prec\prec \Phi(t)$, i.e. for any positive k_ϵ there exists t_0 such that for $t > t_0$ we have $\tilde{\Phi}(t) \leq \Phi(k_\epsilon t)$. They have got

Theorem 3.1. Under the conditions (3.1)–(3.6), there exists at least one extreme function $u(x) \in \overset{o}{W}{}^1 L(\phi)$ of the problem

$$L(u) = \inf_{v \in \overset{o}{W}{}^1 E(\Phi)} L(v),$$

and this function $u(x)$ can be obtained by Ritz method.

Here the space $\overset{o}{W}{}^1 L(\Phi)$ is a Ba space.

In [10] Gu Yonggeng, Luo Peizhu and Ding Xiaxi considered the first boundary value problem of nonlinear parabolic equations

$$\begin{cases} \frac{\partial u}{\partial t} - \frac{d}{dx_i} a_i(x, t, u, u_x) + a(x, t, u, u_x) = 0, & \text{in } Q_t \tag{3.7} \\ u|_{S_T} = 0, u|_{t=0} = \psi_0(x). \tag{3.8} \end{cases}$$

In [15] Ladyzenskaja has discussed the existence of generalized solution of nonlinear parabolic equations (3.7) and (3.8). (see Theorem 6.7 of Chapter V in [15], 466–475). But the proof is incorrect, because the authors used an incorrect Lemma 2.2 of Chapter II in [15]. In [10], Gu, Luo and Ding investigated the existence of the generalized solution of problem (3.7) and (3.8) by a different method. They proved

Theorem 3.2. Suppose that the functions $a_i(x, t, u, x), i = 1, \cdots, n$ and $a(x, t, u, s)$ satisfy the conditions

$$|a_i(x, t, v, s)| \leq h(x, t) + c\Psi^{-1}[\Phi(\beta'|v|^{p_1})] + c\Psi^{-1}[\Phi(\beta|s|)], i = 1, \cdots, n. \tag{3.9}$$

$$|a(x,t,v,s)| \le h_1(x,t) + c\Psi_{p_1}^{-1}[\Phi(\beta'|v|^p)] + c\Psi_{p_1}^{-1}[\Phi(\beta|s|)], \qquad (3.10)$$

$$\int_\Omega [a_i(x,t,v,v_x)v_{x_i} + a(x,t,v,v_x)v]dx \ge \quad \nu \int_\Omega \Phi(\beta|v_x|)dx - \zeta(t)\int_\Omega (1+v^2)dx$$

$$\qquad (3.11)$$

$$\forall v \in \overset{\circ}{W}{}^1 E(\Phi,\Omega)$$

$$\frac{\partial a_i(x,t,v,s)}{\partial s_j}\xi_i\xi_j \ge \mu(t)|\xi|^2, \forall \xi \in R^n. \qquad (3.12)$$

Then for any $\Psi_0(x) \in L_2(\Omega)$, the first boundary value problem (3.7) and (3.8) has at least one generalized solution $u(x,t) \in \overset{\circ}{W}{}^{1,0} L(\Phi,2;Q_T)$, i.e. for any smooth function $\eta(x,t)$ which is zero on S_T, the integral relation

$$\int_{Q_T}[-u\eta_t + a_i(x,t,u,u_x)\eta_{x_i} + a(x,t,u,u_x)\eta]dxdt + \int_\Omega u(x,t)\eta(x,t)|_{t=0}^{t=T}dx = 0 \qquad (3.13)$$

holds, where the constants $\nu,\beta,\beta' > 0, p = 1 + \frac{2}{n}, 0 < p_1 < p, h(x,t) \in L(\Psi;Q_T), h_1(x,t) \in L(\Psi_{p_1};Q_T); \zeta(t) \ge 0$ and $\int_0^T \zeta(t)dt \le \zeta_0 < \infty, \mu(t) \ge \mu_0 > 0$.

Here the space $\overset{\circ}{W}{}^{1,0} L(\Phi,2;Q_T)$ is a Ba space.

3.2. On the boundedness of some operators in Ba spaces

It is well known that the Hilbert transform

$$H(u)(x) = \lim_{\varepsilon\to 0}\int_{|x-y|>\varepsilon}\frac{u(y)dy}{y-x}$$

is bounded in $L^p(\mathbb{R}, dx), 1 < p < \infty$. A natural question to ask is: Is the operator H bounded in $L^{Ba}(\mathbb{R}, dx)$? It turns out that this is not true in general. Deng Yaohua and Gu Yonggeng gave a necessary and sufficient condition of the boundedness for the Hilbert transform, they proved

Theorem 3.3. The Hilbert transform H is bounded in $L^{Ba}(\mathbb{R}, dx)$ if and only if there exist constants α,β with $1 < \alpha < \beta < \infty$ such that

$$\alpha \le p_m \le \beta, m = 1, 2, \cdots \qquad (3.14)$$

Let x, y denote the points $(x_1, \cdots, x_n), (y_1, \cdots, y_n) \in R^n$, for $f \in L^p(\mathbb{R}^n, dx), 1 \le p < \infty$. The n-dimensional singular integral

$$(R_j f)(x) = \lim_{\varepsilon\to 0} C_n \int_{|y|\ge\varepsilon}\frac{y_i}{|y|^{n+1}}f(x-y)dy, j = 1, 2, \cdots, n$$

is known as the n-dimensional Riesz transform of f, where $c_n = \Gamma\left(\frac{n+1}{2}\right)/\frac{(n+1)\pi}{2}$. Deng Yaohua et al. proved the n-dimensional Riesz transform $R_j(j = 1, 2, \cdots, n)$ is bounded in $L^{Ba}(\mathbb{R}^n, dx)$ if and only if the sequence $\{p_m\}$ satisfies the conditions (3.14).

Ma Jigang [17] considered the Littlewood-Paley theorem in Ba space. Let $f(x)$ be a measurable function on $\mathbb{R}, u_f(x,y) = \frac{1}{\pi} \int_{\mathbb{R}} \frac{yf(t)dt}{(x-t)^2 + y^2}$ denote the Poisson integral of $f(x)$. Then the Littlewood-Paley g-function is defined by

$$g(f)(x) = \left\{ \int_{\mathbb{R}} |\nabla u_f(x,y)|^2 y dy \right\}^{1/2}$$

where $|\nabla u_f|^2 = \left| \frac{\partial u_f}{\partial x} \right|^2 + \left| \frac{\partial u_f}{\partial y} \right|^2$. Ma proved that the g-function is bounded in $L^{Ba}(\mathbb{R}, dx)$ if and only if the sequence $\{p_m\}$ satisfies the condition (3.14).

It is well known that the Hilbert transform H is bounded in an Orlicz space if and only if the N-function and the complementary N-function satisfy the $\triangle_2$-condition which implies that the Orlicz space should be separable. Ma Jigang [19] constructed a Ba space which satisfies the condition (3.14) but is non-separable. Let $p_1 = \frac{3}{2}, p, = 2 - \frac{1}{m!}, m \geq 2$ and $\varphi(z) = \sum_{m=1}^{\infty} (m!2)^{-\frac{m}{P_m}} z^m$. Then we denote the corresponding Ba space and aB space by $L_0^{Ba}(\mathbb{R}, dx)$ and $L_0^{aB}(\mathbb{R}, dx)$ respectively. It is easy to show that $u(x) = \begin{cases} x^{-\frac{1}{2}}, & x \in [0,1] \\ 0, & x \overline{\in} [0,1] \end{cases}$ belongs to $L_0^{Ba}(\mathbb{R}, dx) \backslash L_0^{aB}(\mathbb{R}, dx)$. By Theorem 2.3, $L_0^{Ba}(\mathbb{R}, dx)$ is non-separable. On the other hand, we have $1 < \frac{3}{2} \leq \inf\{p_m\} \leq \sup\{p_m\} \leq 2 < \infty$, by Theorem 3.3, the Hilbert transform is bounded in $L_0^{Ba}(\mathbb{R}, dx)$.

3.3. H^{Ba} and BMOA

The space BMO of functions of bounded mean oscillation was introduced by John and Nirenberg [14] in 1961. The space BMOA (analytic BMO) has a wide connection with other subjects in complex analysis, such as univalent functions, quasiconformal mapping, value distribution, etc. We know that there are a large number of charaterizations of BMOA [1] [20], one of which is the following: Suppose that $f(z)$ is an analytic function in D. We define the set $\mathcal{M}(f)$ by

$$\mathcal{M}(f) := \{g(z), g(z) = f \circ S(z) - f \circ S(0), S \in \mathcal{M}\},$$

where $\mathcal{M} := \{S(z), S(z) = e^{i\alpha} \frac{z+a}{1+\bar{a}z}, a \in D, e^{i\alpha} \in \partial D\}$. Then

$$BMOA := \left\{ f(z), \sup_{g \in \mathcal{M}(f)} \{\|g\|_{H^1}\} < \infty \right\}.$$

The norm $\| \cdot \|_*$ in BMOA is defined by

$$\|f\|_* := \sup_{g \in \mathcal{M}(f)} \{\|g\|_{H^1}\}.$$

In [11] He Yuzan found out an interesting relation between BMOA(D) and $H^{Ba}(D)$. He proved

Theorem 3.4. Let $f(z)$ be an analytic function in D. Then $f(z) \in BMOA(D)$ if and only if the set $\mathcal{M}(f)$ is bounded in $H^{Ba}(D)$, i.e. there exist two constants α and β such that

$$\alpha \|f\|_* \leq \sup_{g \in \mathcal{M}(f)} \{\|g\|_{Ba}\} \leq \beta \|f\|_*$$

In Definition 2.1, we take $B_m = L_a^{P_m}(D, d\sigma_z) = \{f(z),$ analytic in D and $\int\int_D |f(z)|^{P_m} d\sigma_z < \infty\}$, where $d\sigma_z$ is an element of 2-dimensional Lebesgue measure on D, $B = L_a^1(D, d\sigma_z), \varphi(z)$ and $\{p_m\}$ are the same as in $H^{Ba}(D)$. Then we get a Ba space denoted by $L_a^{Ba}(D, d\sigma_z)$. In [12] He Yuzan and Ouyang Caihen gave a similar relation between $L_a^{Ba}(D, d\sigma_z)$ and the Bloch space $\mathcal{B}(D)$, they proved that $f(z) \in \mathcal{B}(D)$ if and only if there exist two constants α and β such that

$$\alpha \|f\|_{\mathcal{B}} \leq \sup_{g \in \mathcal{M}(f)} \{\|g\|_{L_a^1}\} \leq \beta \|f\|_{\mathcal{B}}$$

where

$$\|f\|_{L_a^1} = \int\int_D |g(z)| d\sigma_z, \|f\|_{\mathcal{B}} = \sup_{z \in D} \{|f'(z)|(1 - |z|^2)\}.$$

The above results were generalized to the Bloch space $\mathcal{B}$ and L_a^{Ba} on the unit ball in $\mathbb{C}^n$ [12], to BMOA and H^{Ba} on Riemann surfaces [13] and other more general cases [22, 23].

3.4. Applications to the approximation of analytic functions

In [19] Ma Jigang discussed the approximation of analytic functions in weighted Ba-norm and obtained some interesting results on the estimate of approximation order in Ba space.

Let G be a bounded set in complex plane $\mathbb{C}$. The complementary $\Omega = \hat{\mathbb{C}} \backslash G$ is a simply connected domain. We denote by $z = \psi(w)$ and R the conformal mapping from $\{w, |w| > 1\}$ onto Ω with $\psi(\infty) = \infty$ and $\psi'(\infty) > 0$ and the transfinite radius of G respectively. Let $L_R := \{z, z = \psi\left(\frac{w}{R}\right), |w| = r > R\}$ and G_R be the bounded domain with boundary L_R. We define the set $L_a^p(G, \nu)$ by

$$L_a^p(G, \nu) := \left\{f(z), \text{ analytic in } G \text{ and } \int\int_G |f(z)|^p \nu(z) d\sigma_z < \infty\right\}$$

where $\nu(z)$ is a non-negative function and $\nu(z) \in L^1(G, d\sigma_z), \nu^{-1}(z) \in L^\infty(G, d\sigma_z)$. Now we take $B_m = L_a^{P_m}(G, \nu), B = L_a^1(G, \nu) \oplus L_a^\infty(G, \nu)$, then we get a Ba space denoted by $L_a^{Ba}(G, \nu)$.

For $f \in L_a^{Ba}(G, \nu)$, the n-degree minimal error of f in $L_a^{Ba}(G, \nu)$ is defined by

$$\triangle_n(f) := \inf_{P \in \mathcal{P}_{n-1}} \|f - P\|_{Ba}$$

where $\mathcal{P}_{n-1} := \{P(z), P(z) \text{ is a polynomial with degree} \leq n-1\}$

Then Ma proved [18]

Theorem 3.5. Let G be a bounded simply connected domain with transfinite radius $R > 0$. Then for $f \in H(R_0)$,

$$\overline{\lim}\{\triangle_n(f)\}^{\frac{1}{n}} = \frac{R}{R_0}, \quad (R_0 > R)$$

where $H(D_{R_0}) := \{f(z), f \text{ is analytic in } G_{R_0} \text{ and has singularity at } L_R\}$.

References

[1] A.Baernstein II, Analytic functions of bounded mean oscillation. Aspects of Contemp Complex Analysis (Durham, 1979), 3–36, Academic Press, London, 1980.

[2] Chen Guangrong and Meng Boqin, Interpolation of *Ba* spaces, Theory of *Ba* Spaces and its Applications, Science Press, Beijing 1992, 147–152.

[3] Deng Yaohua and Gu Yonggeng, A new class of function spaces and Hilbert transform, J. Math. Anal. Appl., 108(1985), 1, 99–108.

[4] Deng Yaohua, Chang Wendong and Li Yanyan, On boundedness of Riesz transform and the Riesz potential in *Ba* spaces, Theory of *Ba* Spaces and its Applications, Science Press, Beijing, 1992, 63–72.

[5] Ding Xiaxi, On a new class of function spaces, Kexue Tongbao, 26(1981), 973–976.

[6] Ding Xiaxi, He Yuzan and Luo Peizhu, The Theory of *Ba* Spaces and its Applications, Science Press Beijing, China, 1992.

[7] Ding Xiaxi and Luo Peizhu, *Ba* spaces and some estimates of the Laplace operator, J. sys. Sci. &. Math. Sci., 1(1980), 1, 9–33.

[8] Ding Xiaxi, Luo Peizhu, Gu Yonggeng and Fang Huizhong, Calculus of variations with strong nonlinearity, Sci. Sinica, 23(1980) 946–955.

[9] Ding Xiaxi, Luo Peizhu, Fang Huizhong and Gu Yonggeng, The trace theorem of Orlicz-Sobolev Spaces and calculus of variations with strong nonlinearity, Kexue Tongbao 25(1980), No.15.

[10] Gu Yonggeng, Luo Peizhu and Ding Xiaxi, Generalized solutions of strongly nonlinearity parabolic equations, Sci. Sinica, 31(1983), No.11.

[11] He Yuzan, A class of special Hardy-Orlicz spaces and the space of BMOA functions, Ann. Polonici Math. 48(1988), 3:217–226.

[12] He Yuzan and Ouyang Caiheng, A class of function spaces, Proceed of Sym. on Complex Analysis 1987, World Scientific Singapore (1988), 229–250.

[13] He Yuzan and Zhao Ruhan, BMOA and *Ba* spaces on compact bordered Riemann surfaces, Kexue Tongbao 35(1990), 1131–1134.

[14] F.John and L.Nirenberg, On functions of bounded mean oscillation. Comm. Pure Appl. Math. 14(1961), 415–426.

[15] O.A. Ladyzenskaja, V. A. Solonnikov and N.N. Uralceva, Linear and quasilinear Equations of parabolic Type, Copyright by the American Math. Society 1968.

[16] O.A. Idayzenskaja and N.N. Uralceva, Linear and quasilinear Equations of elliptic Type, (Russian) Science Press, 1973.

[17] Ma Jigang, Little-wood-Paley theorem in the Ba Spaces, Kexue Tongbao 33(1988), 20, 1529–1533.

[18] Ma Jigang, Ba sequance spaces and some basic properties, Theory of Ba Spaces and its Applications, Science Press Beijing 1992, 113–1134.

[19] Ma Jigang, Ba Spaces and related Problems, 1989, Thesis.

[20] T.A. Metzger, Bounded mean Oscillation and Riemann Surfaces, Bounded mean Oscillation in Complex Analysis (Joensuu, 1989), 79–100.

[21] L.A. Ruivivar, On the boundedness of the Polysingular integral operator in the $L^{Ba}(\mathbb{R}^m)$ Space, SEA. Bull. of Math. Special Issue 1993, 137–146.

[22] Xiao Jie, A class of weighted Orlicz-Bergman Spaces, Journ. of Inner Mongolia Teach Univ. 1991, No.4, 1–12.

[23] Xiao Zhijing, Weighted Bergman spaces, Bloch space and Ding space, Acta Math. Sci. 9(1989), 3, 265–276.

[24] Zhuang Yadong and Xu Xintai, Some properties of Ba spaces, Theory of Ba Spaces and its Applications, Science Press, Beijing 1992, 153–163.

On Completely Irreducible Operators

Chunlan Jiang and Shanli Sun

Institute of Mathematics, Jilin University, Changchun

1. Introduction

Looking back to Linear Algebra, to transform a matrix into a Jordan standard form is really situated at the centre in the theory of linear transformation. It is also a prototype in the spectral theory of bounded linear operators on infinite dimensional space. One naturally asks: What is a suitable analogue of Jordan blocks in the set of all bounded linear operators on an infinite dimensional Hilbert space? We consider that a suitable analogue is the completely irreducible operator, or BIR operator in brief. In 1970's Z.J. Jiang [JZJ, 1, 2] indicated the importance of investigating BIR operators and proposed some questions concerning BIR operators and the operator structure at the functional analysis seminar of Jilin University. In the past fifteen years some results were obtained by the seminar's collective efforts. In this paper we briefly sum up the works on BIR operators that have been done by our seminar.

In this paper we let H denote a complex Hilbert space, $\mathcal{L}(H)$ the set of all bounded linear operators on H, Lat T the lattice of all invariant subspaces of $T \in \mathcal{L}(H)$, $\mathcal{C}$ the complex plane.

Definition 1.1. Let $T \in \mathcal{L}(H)$, $\mathcal{M}$ be a closed subspace of H. If $\mathcal{M}$, $\mathcal{M}^{\perp} \in L$ at T, then $\mathcal{M}$ is called a reduced subspace of T. If T has a nontrivial reduced subspace, then T is called a reducible operator, or a HR operator in brief. Otherwise T is called an irreducible operator, or a HIR operator in brief.

At the end of the 1960's Halmos [Ha, 1] proposed a seemingly simple question: Is any bounded linear operator the limit of reducible operators in operator norm? Unexpectedly, to answer this question Voiculescu [V, 1] obtained the famous non-commutative Weyl-von Neumann theorem. From a lot of its applications we see that questions concerning reducibility are important to Operator Theory.

1991 Mathematics Subject Classification: 41A10, 47A55

From Linear Algebra we know that the Jordan standard form of a matrix is invariant under similar transformation, but with irreducible operators this is not the case. For example, Gilfeather [Gi, 1] proved

Theorem Gi. Let H be a separable Hilbert space, $N \in \mathcal{L}(H)$ a normal operator. If $\sigma_p(N)$, the point spectrum of N, is empty, then N must be similar to an irreducible operator.

Z.J. Jiang [JZJ, 3] gave the following example.

Example 1.1. There exists an irreducible operator $A \in \mathcal{L}(H)$ that is similar to a compact selfadjoint operator on H.

This example not only improves the example due to Hoover [Ho, 1] but also shows that the condition $\sigma_p(N) = \emptyset$ in the above Theorem Gi is not necessary. Thus Z.J. Jiang proposed that we should characterize the spectrum of a normal operator similarly to an irreducible operator. Now this is finished by C.K. Fong and C.L. Jiang [FJ, 2].

It is seen from Theorem Gi and Example 1.1 that we need the following concept.

Definition 1.2. Let $T \in \mathcal{L}(H)$, $\mathcal{M} \in L$ at T. If there is an $\mathcal{N} \in \text{Lat } T$ such that $H = \mathcal{M} + \mathcal{N}$ and $\mathcal{M} \cap \mathcal{N} = \{0\}$, writing $H = \mathcal{M}\dot{+}\mathcal{N}$, then $\mathcal{M}$ is called a completely reduced subspace of T. If T has a nontrivial completely reduced subspace then T is called a completely reducible operator, or a BR operator in brief. Otherwise T is called a completely irreducible operator, or a BIR operator in brief.

It follows from the definition that completely irreducible operators are invariant under similar transformation. This is a basic difference between BIR and HIR operators. In addition, it follows from the knowledge of Linear Algebra that BIR operators on finite dimensional space are equivalent to Jordan blocks. Therefore the concept of BIR operators is a natural generalization of Jordan blocks to infinite dimensional space. If Jordan blocks are regarded as building blocks in $\mathcal{L}(X)$, the set of all linear transformations on finite dimensional space X, then each $T \in \mathcal{L}(X)$ can be built by using the blocks. Each $T \in \mathcal{L}(X)$ may be represented as a direct sum of several Jordan blocks to be exact. A natural question is what is an analogue of Jordan blocks in $\mathcal{L}(H)$ for infinite dimensional space H, or what are the building blocks in infinite dimensional $\mathcal{L}(H)$. From the works finished by the Russian scholars it seems that they regard unicellular operators as an analogue of Jordan blocks in $\mathcal{L}(H)$. Though all unicellular operators are completely irreducible we will see, in the latter part of this paper, a suitable analogue of Jordan blocks in $\mathcal{L}(H)$ is the completely irreducible operators but not unicellular operators.

2. Completely irreducibility of several classes of operators

In this section we clarify which of the several well-known classes of operators are completely irreducible operators.

Weighted shifts on Hilbert space are the best known and important operators. Kelley [K, 1] investigated unilateral weighted shifts in his Ph.D. Dissertation in 1966. He proved that, in fact, all injective unilateral weighted shifts are completely irreducible. In 1969 Gellar [Ge, 1] investigated the commutant of bilateral weighted shifts. In fact, he proved also

Theorem Ge. Let $T \in \mathcal{L}(H)$ be an injective bilateral weighted shift. If $\sigma(T)$ does not degenerate into a circle then T is completely irreducible.

Following the work of Gellar, G.H. Gong [Go, 1] investigated bilateral weighted shift T with $\sigma(T)$ degenerating into a circle, and gave some conditions for T to be completely irreducible. The main results are the following

Theorem 2.1. Let $T \in \mathcal{L}(H)$ be an injective bilateral weighted shift with a positive weight sequence $\{w_n\}$ and $\sigma(T) = \{\lambda; |\lambda| = 1\}$. If anyone of the following conditions holds then T is completely irreducible:

(1) $\displaystyle\sum_{n=0}^{\infty} \frac{1}{\beta^2(n)} < \infty$,

(2) $\displaystyle\sum_{n=0}^{\infty} \frac{1}{\beta^2(-n)} < \infty$,

(3) $\displaystyle\sum_{n=0}^{\infty} \beta^2(n) < \infty$,

(4) $\displaystyle\sum_{n=0}^{\infty} \beta^2(-n) < \infty$,

(5) $\displaystyle\sum_{n=0}^{\infty} \inf_k \left\{ \min \left\{ \frac{\beta(n+k)}{\beta(k)}, \frac{\beta(k)}{\beta(n+k)} \right\} \right\} < \infty$,

where

$$\beta(0) = 1, \ \beta(n) = \begin{cases} w_0 w_1 \cdots w_{n-1}, & n > 0, \\ (w_{-1} w_{-2} \cdots w_n)^{-1}, & n < 0. \end{cases}$$

L.F. Liu and N. Meng [LM, 1] investigated also the irreducibility of weighted shifts.

Contractions of the class C_0 are the most successful one in the theory of contraction operators due to Sz-Nagy and Foias. It is significant to clarify which of the contractions of class C_0 are completely irreducible. C.Z. Zou [Z, 1] proved

Theorem 2.2. Let $T \in \mathcal{L}(H)$ be a completely irreducible contraction of the class C_0. Then the spectrum of T is a single set.

Toeplitz operator is an important class of operators. Which of Toeplitz operators are completely irreducible is an unsolved and significant question. Z.P. Wang [W, 1] gave the following

Theorem 2.3. Whole subnormal analytic Toeplitz operator is completely irreducible, where an operator is said to be whole subnormal if each operator in its commutant is subnormal.

Furthermore, Z.S. Yu [Y, 1], F. Xu and C.Z. Zou [XZ, 1], S.L. Sun [S, 2] also discussed some questions concerning completely irreducible operators.

3. Completely irreducible operators and approximation

To build each operator in $\mathcal{L}(H)$ by using completely irreducible operators, we attach importance to investigating BIR operators. Hence we should investigate the approximation of BIR operators in operator norm, their direct sum or the density of finite direct sums of BIR operators in $\mathcal{L}(H)$, since all completely irreducible operators are not dense in $\mathcal{L}(H)$.

G.H. Gong (in a personal letter to Z.J. Jiang, 1987) proved

Theorem 3.1. The set $\mathcal{G}$ of all BIR operators is a nowhere dense set in $\mathcal{L}(H)$.

This theorem answers a question raised by Z.J. Jiang in [JZJ, 1]. Then G.H. Gong conjectured that the closure of $\mathcal{G}$ equals all operators with connected spectrum. D.A. Herrero and C.L. Jiang [HJ, 1] proved that this conjecture is true.

Theorem 3.2. The closure of $\mathcal{G}$ equals all operators with connected spectrum.

In fact, this theorem tells us that an operator with connected spectrum is nearly a BIR operator. Hence it is an approximate converse of the famous Riesz decomposition theorem.

Suppose that BIR operators are regarded as building blocks in $\mathcal{L}(H)$. Then for each $T \in \mathcal{L}(H)$ there exists a sequence $\{T_n\}_{n=1}^{\infty}$ of BIR operators such that the most natural forms of building T seem to be

$$\lim_{n} \|T_n - T\| = 0,$$

and

$$T = \dotplus_{n=1}^{\infty} T_n,$$

where $\dotplus$ denotes the direct sum. Apparently, the first form is impossible. By the following theorem due to Z.J. Jiang, the second form is impossible, either.

Theorem 3.3. Let $A \in \mathcal{L}(H)$ be a selfadjoint operator. If $\sigma_p(A) = \emptyset$, then A can not be represented as a direct sum of BIR operators.

Because the above building forms are not successful, Z.J. Jiang conjectured that the finite direct sums of BIR operators should approximate to each bounded linear operator in 1987. The conjecture has been proved by C.L. Jiang [JCL, 1], D.A. Herrero and C.L. Jiang [HJ, 1].

Theorem 3.4. Let $T \in \mathcal{L}(H)$. Then there exist BIR operators $T_{k1}, ..., T_{kn_k}$, $k = 1, 2, ...,$ such that

$$\lim_k \left\| (\dotplus_{j=1}^{n_k} T_{kj}) - T \right\| = 0.$$

In fact, this is a theorem concerning the operator structure. It should be mentioned that

(1) Russian scholars regard unicellular operators as an analogue of Jordan blocks. But we can prove that any Cowen-Douglas operator cannot be approximated by a finite direct sum of unicellular operators, (see [JZJ, 4], [JCL, 2]).

(2) D.A. Herrero (in a personal letter to C.L. Jiang, 1989) regarded the operators with form $\lambda + S$, where λ is a complex number, S or S^* is an injective unilateral weighted shift, as building blocks in $\mathcal{L}(H)$ and conjectured that

$$\mathrm{LID}(H) = \{T; \quad T \text{ is similar to } \oplus_{i=1}^{\infty} (\lambda_i + S_i), \quad \text{where } \lambda_i \in \mathcal{C},$$

$$S_i \text{ or } S_i^* \text{ is an injective weighted shift}\}$$

is dense in $\mathcal{L}(H)$. But C.L. Jiang [JCL, 1], C.K. Fong and C. L. Jiang [FJ, 1] proved that LID(H) is not dense in $\mathcal{L}(H)$. Hence Herrero's conjecture is not true.

To sum up, BIR operators are a suitable analogue of Jordan blocks in infinite dimensional $\mathcal{L}(H)$.

4. Several questions concerning completely irreducible operators

4.1 Is any invariant subspace of BIR operators hyperinvariant?

It is well known that all invariant subspaces of each unicellular operator are hyperinvariant. One naturally asks: Is any invariant subspace of a BIR operator hyperinvariant? Z.P. Wang [W, 1] proved

Theorem 4.1. Let $\phi \in H^\infty$. Then the following assertions are equivalent:
 (1) ϕ is a weak* generator of H^∞,
 (2) $W_{T_\phi} = \{T_\phi\}'$,
 (3) each invariant subspace of T_ϕ is hyperinvariant,
where W_{T_ϕ} is the weak closed algebra generated by T_ϕ and I, $\{T_\phi\}'$ is the commutant of T_ϕ.

Then he showed that $T_{e^{\mathbf{a}z}}$, a BIR analytic Toeplitz operator, has an invariant subspace that is not hyperinvariant.

L.F. Liu and N. Meng [LM, 1] proved

Theorem 4.2. For each invertible bilateral weighted shift there exists a nontrivial invariant subspace that is not hyperinvariant.

If T is an invertible bilateral weighted shift and $\sigma(T)$ does not degenerate into a circle, then T is a BIR operator. It follows from Theorem 4.2 that T has an invariant subspace that is not hyperinvariant.

4.2 Cyclic vector for BIR operators

All unicellular operators have cyclic vectors. Each unilateral weighted shift has cyclic vector, too. Z.J. Jiang [JZJ, 3] raised a question whether all BIR operators have cyclic vectors. Then he gave the following example to show that a general BIR operator does not have any cyclic vector.

Example 4.1. Let $\{e_n\}$ be an orthonormal basis of H and define a bilateral weighted shift as follows:
$$Te_n = \begin{cases} (-n)^{-1/2}e_{n+1}, & \text{if } n < 0, \\ e_{n+1}, & \text{if } n \geq 0. \end{cases}$$

It follows from Theorem Ge that T is a BIR operator. It is easy to prove that T does not have any cyclic vector.

L.F. Liu and N. Meng [LM, 1] proved

Theorem 4.3. Let $T \in \mathcal{L}(H)$ be an injective unilateral weighted shift, $\sigma(T^*) = \{0\}$. If T is represented as a multiplication operator M_z on $H^2(\beta)$, $f \in H^2(\beta)$ is a cyclic vector of T, then pf is also a cyclic vector of T for any polynomial p with a nonzero constant term.

Corollary 4.1. If T is the same as in Theorem 4.3, then every polynomial with a nonzero constant term is a cyclic vector of T.

4.3 Inheritance of BIR operators

It is clear that the restriction of a unicellular operator onto any invariant subspace is still unicellular. Hence unicellularity is inherited. Is complete irreducibility inherited? This is another question raised by Z.J. Jiang [JZJ, 1]. In 1987, G.H. Gong [Go, 1] proved

Theorem 4.4. If $T \in \mathcal{L}(H)$ is a decomposable operator whose spectrum is not a single set, then there exists an invariant subspace $\mathcal{M}$ of T such that $T|_{\mathcal{M}}$ is a BR operator.

It follows from this theorem and a result due to S.L. Sun [S, 1] that there exist a bilateral weighted shift T and an invariant subspace $\mathcal{M}$ of T such that T is BIR but $T|_{\mathcal{M}}$ is BR.

4.4 BIR operators and quasisimilar transformation

Definition 4.1. Let T, $S \in \mathcal{L}(H)$. If there exist two injective operators A, $B \in \mathcal{L}(H)$ with dense range such that $AT = SA$ and $TB = BS$, then T is said to be quasisimilar to S.

The most important difference between quasisimilarity and similarity is that the quasisimilar transformation does not preserve the spectrum. Furthermore, maybe quasisimilar transformation loses more properties. Complete irreducibility is invariant under similar transformation. But Z.J. Jiang [JZJ, 1] conjectured quasisimilar transformation does not preserve complete irreducibility. C.L. Jiang [JCL, 2] proved that the conjecture is true.

Theorem 4.5. Let $S \in \mathcal{L}(H)$ be the unilateral shift. Then

$$\begin{pmatrix} S^* + I & 0 \\ 0 & -(S^* + I) \end{pmatrix} \in \mathcal{L}(H \oplus H)$$

is quasisimilar to a BIR operator on $H \oplus H$.

References

Fong, C.K. and Jiang, C.L.

[1] Approximation by Jordan type operators, Houston J. Math., **19**(1993), 51–62.

[2] Normal operators similar to irreducible operators, Acta Math. Sinica, **10**(1994), 132–135.

Gellar, R.

[1] Operators commuting with a weighted shift, Proc. Amer. Math. Soc, **23**(1969), 538–545.

Glfeather, F.

[1] Strong reducibility of operators, Ind. Univ. Math. J., **22**(1972), 393–397.

Gong, G.H.

[1] Banach reducibility of bilateral weighted shifts, Northeastern Math. J., **3**(1987), 17–36.

Halmos, P. R.

[1] Irreducible operators, Mich. Math. J., **15**(1968), 215–223.

Herrero, D.A. and Jiang, C.L.

[1] Limits of strongly irreducible operators and the Riesz decomposition theorem, Mich. Math. J., **37**(1990), 283–291.

Hoover, T.B.

[1] Quasisimilarity of operators, Illinois J. Math., **16**(1972), 678–686.

Jiang, C.L.

[1] Approximation of direct sum of strongly irreducible operators, Northeastern Math. J., **5**(1989), 253–254.

[2] Strongly irreducible operators and Cowen-Douglas operators, Northeastern Math. J., **7**(1991), 1–3.

Jiang, Z.J.

[1] Some questions concerning BIR operators, I, II, The reports at the functional analysis seminar of Jilin Univ., Changchun, 1979.

[2] Some aspects on the spectral theory of operator, The report at the second Chinese functional analysis conference, Jinan, 1979.

[3] On linear operator structure, The report at Chinese operator theory conference, Jiujiang, 1981.

[4] On completely irreducible operators, The report at the fifth Chinese functional analysis conference, Nanjing, 1990.

Kelley, P.R.

[1] Weighted shifts on Hilbert space, Diss., Univ. Mich. Ann. Arbor. Mich., 1966.

Liu, L.F. and Meng, N.

[1] Weighted shifts and BIR operators, J. Math. Research and Exposition, **5**(1985), No. 4, 35–38.

Sun, S.L.

[1] Spectral decomposition of weighted shift operators, Acta Math. Sinica, New Series, **2**(1986), 367–376.

[2] The composition operators, and reducing subspaces of certain analytic Toeplitz operators, Advances in Math., **22**(1993), No., 5, 422–434.

Voiculescu, D.

[1] A non-commutative Weyl-von Neumann theorem, Rev. Roum. Math. Pures Appl., **21**(1976), 97–113.

Wang, Z.P.

[1] Banach reduction of Hilbert space operators, Acta Sci. Nat. Univ. Jilinensis, (1980), No.3, 49–55.

Xu, F. and Zou, C.Z.

[1] Banach reducibility of decomposable operators, Dongbeishida Xuebao, (1983), No.4, 61–69.

Yu, Z.S.

[1] On Banach reduced operators, Acta Sci. Nat. Univ. Jilinensis, (1980), No.4, 1–12.

Zou, C.Z.

[1] Banach irreducibility of contraction operators of the class C_0, Acta Sci. Nat. Univ. Jilinensis, (1981), No.2, 43–49.

Multiplication of Distributions via Harmonic Representations

Banghe Li and Yaqing Li

Institute of Systerms Science, Academia Sinica, Beijing

There were many methods to give products of 1-dimensional distributions. We can briefly divide them into two kinds. The first kind satisfies (1) commutativity, (2) coincidence with the ordinary multiplication of continuous functions. The second kind does not satisfy at least one of (1) and (2). Most of the methods belong to the first kind, as examples, cf. Y. Hirata and A. Ogata [HO], J. Mikusinski [M], M. Itano [It1], H. J. Bremermann and L. Durand [BD] and B. Fisher [F]. Examples of the second kind are those of H. Konig [K] and W. Guttinger [G] which do not satisfy (1), and that of H. J. Bremermann [B] which does not satisfy (2). Although the methods were of a great variety , their common point was to attempt to obtain some finite values in some senses. But any of these methods can not give products for any two distributions. This intrinsic difficulty shows that we must go beyond the scope of finite numbers and generalize the concept of distributions.

The first author in [L] gave a method to define product for any two 1- dimensional distributions, which generalized the method of [T] and [BD] given by analytic representations of distributions. In [LL1], we proved that the multiplication defined in [L] generalized also the method of Shiraishi [S] given by restricted symmetric δ-sequences, hence included the most well-known multiplications of the first kind.

By harmonic representations of distributions, we [LL2] generalized the multiplication defined in [L] to any dimension. The multiplication defined in [L] and [LL2] possesses the character of being able to calculate concretely. And we have gotten systematic results about the products of pairs of well-known singular 1-dimensional distributions, such as $H(x)$, $\delta^{(n)}$, $x_+^{\pm m}$, x_+^{λ}, $x_+^{\lambda} \ln^q x_+$, x_-^{μ}, $x_-^{\mu} \ln^p x_-$, p_+^{λ} and x^{-n}, $(x \pm i0)^{\pm n} \ln^k (x \pm i0)$. For higher dimensional distributions, the products $\delta \circ \delta$ (in any dimension) and $\delta \circ r^{-1}$ (in dimension 2 and 3) and $\delta(x_1, \ldots, x_n) \circ \delta(x_1) \otimes I(x_2, \ldots, x_n)$ are obtained.

1991 Mathematics Subject Classification: 46F05, 46F10, 46F20, 03H10

Project supported by the National Natural Science Foundation of China

1. Representations of distributions

Suppose T is a distribution on R. Then there is an analytic function $\hat{T}(z)$ on $C\backslash R$ such that

$$T = \lim_{y \to 0_+} (\hat{T}(x + iy) - \hat{T}(x - iy)) \qquad (\mathcal{D}'(R)).$$

$\hat{T}(z)$ is called an analytic representation of T.

The difference of any two analytic representations of T is an entire function ([LL3]).

Let

$$\mathcal{H} = \left\{ h \ \middle| \ \begin{array}{l} h \text{ is analytic on } C\backslash R, \text{ and for any finite closed interval } I, \\ \text{there exist } M \in R \text{ and a finite natural number } n \text{ such that} \\ |h(x + iy)| < M/|y|^n \quad \text{for any } x \in I, 0 < |y| < 1 \end{array} \right\}.$$

Then $\mathcal{H}$ is just the space consisting of all analytic representations of distributions.

For any distribution T on R^n, there is a harmonic function $\hat{T}(x_1, \ldots, x_n, y)$ on $R^n \times R_+$, such that

$$\lim_{y \to 0_+} \hat{T}(x_1, \ldots, x_n, y) = T(x_1, \ldots, x_n) \qquad (\mathcal{D}'(R^n)),$$

and the difference of any two harmonic representations of T extends to a harmonic function on R^{n+1} which is skew-symmetric for y.

A harmonic function f on $R^n \times R_+$ is a harmonic representation of some distribution on R^n if and only if $f \in \mathcal{H}_+$, where

$$\mathcal{H}_+ = \left\{ u \ \middle| \ \begin{array}{l} u \text{ is a harmonic function on } R^n \times R_+, \text{ and for any compact set} \\ K \subset R^n, \text{any } \eta > 0, \text{there are constants } M \text{ and } m \text{ such that} \\ |u(x, y)| \leq My^{-m}, \text{ for any } (x, y) \in K \times (0, \eta] \end{array} \right\}.$$

For distributions on R, harmonic representations and analytic representations coincide. More precisely, given a harmonic representation $u(x, y)$ of T, there must be an analytic representation of T satisfying

$$u(x, y) = \hat{T}(x + iy) - \hat{T}(x - iy),$$

and conversely. Therefore, for $n > 1$, the harmonic representations of distributions on R^n may be also regarded as a generalization of analytic representations of distributions on R ([LL3]).

2. Definition of the multiplication

When $n = 1$, Bremermann and Durand [BD] have given a definition of multiplication (suggested also by Tillman [T]) as follows: If $\hat{T}(x + iy)$ and $\hat{S}(x + iy)$ are analytic representations of distributions T and S respectively, and if

$$\lim_{y \to 0_+} (\hat{T}(x + iy) - \hat{T}(x - iy))(\hat{S}(x + iy) - \hat{S}(x - iy)) \qquad (\mathcal{D}'(R))$$

exists, then the limit is called the product of T and S.

By using analytic representations, some concrete calculations have been done mainly by Itano and Ivanov, e.g. in [It2], [Iv1] and [Iv2].

Itano [It1] showed by an example that the multiplication defined above can not be generalized to higher dimensions by using analytic representations.

In [L], using nonstandard analysis and analytic representations of distributions, the first author generalized the multiplication above to any two 1-dimensional distributions. And in [LL2], we generalized further the multiplication to any dimension, by using harmonic representations of distributions. As mentioned above, analytic representations coincide with harmonic representations for 1-dimensional distributions. We describe our multiplication only for the one given by harmonic representations.

Denote by C the complex field, $^*C$ a nonstandard model of C, R the real field and $\rho \in {}^*R$ a positive infinitesimal. Let

$$^\rho C \quad = \{x \in {}^*C| \text{ for some finite integer } n, |x| < \rho^{-n}\}$$

$$\theta \quad = \text{the set of all infinitesimals in } {}^*C$$

$$^\rho C' \quad = {}^\rho C/\theta.$$

Then $^\rho C'$ is a complex vector space, and we call a complex linear functional of $\mathcal{D}(R^n) \longrightarrow {}^\rho C'$ a hyperdistribution on R^n.

Suppose $S, T \in \mathcal{D}'(R^n), \hat{S}, \hat{T}$ are harmonic representations of S and T, and $^*\hat{S}, {}^*\hat{T}$ are the nonstandard extensions of $\hat{S}$ and $\hat{T}$ respectively. We have

Lemma 1. For any $\varphi \in \mathcal{D}(R^n)$, $< {}^*\hat{S}(x, \rho){}^*\hat{T}(x, \rho), {}^*\varphi(x) > \in {}^\rho C$.

Lemma 2. Let $\hat{S}_1, \hat{S}_2$ and $\hat{T}_1, \hat{T}_2$ be two harmonic representations of S and T respectively. Then for any $\varphi \in \mathcal{D}(R^n)$, we have

$$< {}^*\hat{S}_1(x, \rho){}^*\hat{T}_1(x, \rho), {}^*\varphi(x) > - < {}^*\hat{S}_2(x, \rho){}^*\hat{T}_2(x, \rho), {}^*\varphi(x) > = \text{ infinitesimal}.$$

From Lemma 1 and Lemma 2, $< \hat{S}(x, \rho)\hat{T}(x, \rho), \varphi(x) >$ modulo infinitesimals is independent of the choice of harmonic representations, and hence determines a unique element in $^\rho C'$.

Let $\psi : {}^{\rho}C \longrightarrow {}^{\rho}C'$ be the homomorphism modulo θ. Then

$$\varphi \longrightarrow \psi(< \hat{S}(x,\rho)\hat{T}(x,\rho), \varphi(x) >)$$

defines a complex linear functional of $\mathcal{D}(R^n) \longrightarrow {}^{\rho}C'$.

Now, we introduce the following definition.

For any $S, T \in \mathcal{D}'(R^n)$ and any of their harmonic representations $\hat{S}, \hat{T}$, we call the hyperdistribution on R^n given by

$$< S \circ T, \varphi > = \psi(< \hat{S}(x,\rho)\hat{T}(x,\rho), \varphi(x) >), \qquad \varphi \in \mathcal{D}(R^n)$$

the product of S and T, denoted by $S \circ T$.

3. Local representations and the multiplication

Suppose U is a set in R, $\tilde{U}$ is an open set in C, $\tilde{U} \supset U$ and $f(z)$ is an analytic function on $\tilde{U} \backslash R$. For any $T \in \mathcal{D}'(R)$, if

$$\lim_{y \longrightarrow 0_+} < f(x+iy) - f(x-iy), \varphi(x) > = < T, \varphi >$$

holds for any $\varphi \in \mathcal{D}(U)$, we call f an analytic representation of $T|_U$.

If T and S are merely distributions on U, i.e. $T, S \in \mathcal{D}'(U)$, we may define their analytic representations in the above way and use analytic representations to define their product $T \circ S$ as in the case of $\mathcal{D}'(R)$.

Suppose $T, S \in \mathcal{D}'(R)$, $\hat{T}$ and $\hat{S}$ are analytic representations of T and S, $\tilde{T}$ and $\tilde{S}$ are analytic representations of $T|_U$ and $S|_U$ respectively. Then

$$< \hat{T}_{\rho}\hat{S}_{\rho}, {}^*\varphi > - < \tilde{T}_{\rho}\tilde{S}_{\rho}, {}^*\varphi > = \text{ infinitesimal.}$$

Hence, we may use local analytic representations to calculate $T \circ S$. And in general, to find local analytic representations is much easier than to find analytic representations.

For distributions on R^n, we may use local harmonic representations to define the products.

Example. Let $P(x)$ be a polynomial with real coefficients such that $\{x \mid P(x) > 0\}$ being nonempty, define a distribution P_+^{λ} by

$$< P_+^{\lambda}, \varphi > = \int_{P(x)>0} (P(x))^{\lambda}\varphi(x)\, dx, \qquad \varphi \in \mathcal{D}(R)$$

where the integral is defined for those $\lambda \in C$ such that the integral exists, and for other λ, it is determined by analytic continuation. It seems difficult to find the analytic representations of P_+^{λ}. But we may find local ones [LL1] and so the products of $P_+^{\lambda} \circ P_-^{\mu}$.

4. Fundamental properties of the multiplication

1. Commutativity: $S \circ T = T \circ S$.
2. Bilinearity: For any complex numbers $\alpha_1, \alpha_2, \beta_1, \beta_2$,

$$(\alpha_1 S_1 + \alpha_2 S_2) \circ (\beta_1 T_1 + \beta_2 T_2) = \sum_{k,m=1}^{2} \alpha_k \beta_m (S_k \circ T_m),$$

3. Localizability: Let U be any open set in R^n, if $S_1|_U = S_2|_U, T_1|_U = T_2|_U$, then

$$(S_1 \circ T_1)|_U = (S_2 \circ T_2)|_U,$$

3'. $\mathrm{Supp}(S \circ T) \subset \mathrm{Supp}\, S \cap \mathrm{Supp}\, T$.
4. Coincidence with the multiplication of continuous functions :
    Let $f(x)$ and $g(x)$ be continuous functions on R^n. Then $f \circ g = fg$.
5. Coincidence with the multiplication by multipliers in the space $\mathcal{D}'(R^n)$:
    Let $f \in C^\infty(R^n)$, $T \in \mathcal{D}'(R^n)$, fT denote the product of f (as a multiplier) and T.
Then

$$f \circ T = fT.$$

6. Coincidence with the multiplication by a multiplier in the space $(\mathcal{D}^m(R^n))'$:
    Let $f \in C^m(R^n)$, $T \in (\mathcal{D}^m(R^n))'$, fT be similar to 5. Then $f \circ T = fT$.
7. Let U be an open set in R^n, $S|_U = f$, $T|_U = g$, and f, g be L^p, L^q functions on U
    respectively with $1 \le p, q \le \infty$ and $\frac{1}{p} + \frac{1}{q} = 1$, $(fg)(x) = f(x)g(x)$. Then $(S \circ T)|_U = fg$.
8. The Leibnitz formula :   For any $S, T \in \mathcal{D}'(R^n)$,

$$\frac{\partial}{\partial x_i}(S \circ T) = \frac{\partial S}{\partial x_i} \circ T + S \circ \frac{\partial T}{\partial x_i}.$$

([LL1] [LL])

5. Two theorems

   L. Schwartz had an idea that two distributions could be multiplied if the degree of niceness of one is bigger than the degree of badness of the other. The properties 5 and 6 in §4 may be regarded as the realization of Schwartz's thought. We refine this idea by introducing a localization of niceness and badness.

   Let T be a distribution on R^n, and U a bounded open set in R^n. If $T|_U \in (\mathcal{D}_U^m)'$, but $T|_U \notin (\mathcal{D}_U^{m-1})'$, we ([L] [LL4]) define that the order of distribution T on U is m denoted by

$$O_T(U) = m.$$

For a point $x \in R^n$, let $\mathcal{U}_x$ be a set of bounded open sets containing x. Then

$$O_T(x) = \min_{U \in \mathcal{U}_x} O_T(U)$$

is defined to be the order of distribution T at a point x.

We say that a 1-dimensional distribution T is m-order-continuously -differentiable at a point x, if there exists $U \in \mathcal{U}_x$ such that $T^{(m)}|_U$ is a locally integrable function and is continuous at x.

Let

$$d_T(x) = \begin{cases} m, & \text{if } m \text{ is the minimal nonnegative integer such that} \\ & T \text{ is m-order-continuously-differentiable at x}, \\ -\infty, & \text{otherwise.} \end{cases}$$

We have in ([L])

Theorem 5.1. Let T and S be two 1-dimensional distributions. If for any $x \in R$, either

$$d_T(x) \geq O_S(x) \qquad \text{or} \qquad d_S(x) \geq O_T(x)$$

holds, then $S \circ T$ is a distribution.

Next, we consider a multiplication via the restricted δ-sequence given first by Shiraishi [Sh]. A sequence $\{\delta_n(x)\}$, $x \in R$, is called a restricted δ-sequence, if it satisfies
1) $\delta_n(x) = 0$, for $|x| \geq a_n \longrightarrow 0$,   2) $\delta_n(x) \geq 0$,
3) $\delta_n(x) = \delta_n(-x)$,   4) $\int_{-a_n}^{a_n} \delta_n(x)dx = 1$,
5) $\int_{-a_n}^{a_n} |x|^p |\delta_n^{(q)}(x)|dx \leq M_p$ (M_p being independent of n).
For two 1-dimensional distributions S and T, if the distributional limit

$$\lim_{n \to \infty} (S * \delta_n)(T * \delta_n)$$

exists for any restricted δ-sequence and is independent of the sequence, then the limit is denoted by $S \bullet T$.

We proved in 1979 the following result in [LL1]:

Theorem 5.2. For any two 1-dimensional distributions S and T, if $S \bullet T$ exists, then

$$S \circ T = S \bullet T.$$

As mentioned in [CO], this result was also obtained by Obergugenberger in his paper [O] published in 1988.

6. Some nonassociativity formulas of the products

L.Schwartz pointed earlier that multiplication of distributions is nonasssociative in genaral, so is ours. We got some nonassociativity formulas in [LL1] for some special cases:

Theorem 6.1. Let λ, μ be any complex numbers, σ, τ denote $+$ or $-$, and $\alpha(x), \beta(x) \in C^{\infty}(R)$. Then
$$(\alpha(x) \, x_\alpha^\lambda) \circ (\beta(x) \, x_\tau^\mu) =$$

$$
= \begin{cases}
0, & \text{Re } (\lambda + \mu) > -1 \\
\sum_{j=0}^{[-\text{Re}\,(\lambda+\mu)]-1} \sum_{r+s=j} \frac{\alpha^{(j)}(0)}{r!} \frac{\beta^{(s)}(0)}{s!} [x_\sigma^{\lambda+r} \circ x_\tau^{\mu+s} - x^j \, (x_\sigma^\lambda \circ x_\tau^\mu)], & \text{Re } (\lambda + \mu) \leq -1.
\end{cases}
$$

Thorem 6.2. Let $\alpha(x)$ be any C^{∞} function, T a distribution, the order of T at point 0 be m. Then
$$(\alpha(x)\delta^{(n)}) \circ T - \alpha(x)(\delta^{(n)} \circ T) =$$

$$
= \sum_{j=1}^{n} \binom{n}{j} \alpha^{(j)}(0)(-1)^j \, (\delta^{(n-j)} \circ T) - \sum_{j=1}^{m+n} \frac{\alpha^{(j)}(0)}{j!} \, x^j (\delta^{(n)} \circ T)
$$

and for any $T \in \mathcal{D}'$,
$$(\alpha(x)\delta^{(n)}) \circ T = \alpha(x)(\delta^{(n)} \circ T)$$

if and only if for any $s \geq 1$, $\alpha^{(s)}(0) = 0$.

7. Some formulas of products

In this section, we only choose to list some formulas of products of distributions on two principles : the products have not been obtained by other authors in any definition of multiplication to our knowledge; and they are not too complicated.

To fit the second principle, we write mainly the finite part of $S \circ T$, denoted by $\mathbf{Pf} S \circ T$ which is obtained from $S \circ T$ by delecting

$$\sum_{k=1}^{N} c_k \rho^{\mu_k} \ln^{\gamma_k} \rho,$$

where $c_k \in C$, $\mu_k \leq 0$, $\gamma_k \geq 0$, and $\gamma_k - \mu_k > 0$.

1. Products of $\delta^{(m)}$ and $\delta^{(n)}$ ([L] [LL2]) :

$$(1.1) \qquad \delta^{(m)}(x) \circ \delta^{(n)}(x) = \sum_{k=0}^{[\frac{m+n}{2}]} (-1)^{m+n} \rho^{-1-2k} b_{m,n,k} \, \delta^{(m+n-2k)}(x), \qquad ([L])$$

$$(1.2) \qquad \delta'(x) \circ \delta'(x) = \frac{1}{8\pi} \rho^{-1} \delta^{(2)}(x) + \frac{1}{4\pi} \rho^{-3} \delta(x),$$

$$(1.3) \qquad \delta(x) \circ \delta^{(5)}(x) = \frac{1}{64\pi} \rho^{-1} \delta^{(5)}(x) - \frac{5}{16\pi} \rho^{-3} \delta^{(3)}(x) + \frac{15}{8\pi} \rho^{-5} \delta'(x),$$

$$(1.4) \qquad \delta(x_1, \ldots, x_n) \circ \delta(x_1, \ldots, x_n) =$$

$$= \sum_{j=0}^{[\frac{n}{2}]} \rho^{2j-n} \sum_{s_1 + \cdots + s_n = j} C_{s_1, \ldots, s_n} \frac{\partial^{2j}}{\partial^{2s_1} x_1 \cdots \partial^{2s_n} x_n} \delta(x_1, \ldots, x_n),$$

where

$$C_{s_1, \ldots, s_n} = \frac{2 \prod_{i=1}^{n} \Gamma(s_i + \frac{1}{2})}{c_n^2 \Gamma(\frac{n}{2} + j) \prod_{i=1}^{n} (2s_i)!} \int_0^{\infty} \frac{t^{2j+n-1}}{(1+t^2)^{n+1}} dt,$$

$$c_n = \pi^{\frac{1}{2}(n+1)} / \Gamma(\frac{n+1}{2}).$$

$$(1.5) \qquad \mathbf{Pf}\delta(x_1, \ldots, x_n) \circ \delta(x_1, \ldots, x_n) =$$

$$= \begin{cases} 0, & \text{if } n \text{ is odd,} \\[2mm] \sum_{s_1 + \cdots + s_n = \frac{n}{2}} C_{s_1, \ldots, s_n} \dfrac{\partial^n \delta(x_1, \ldots, x_n)}{\partial^{2s_1} x_1 \cdots \partial^{2s_n} x_n} \neq 0, & \text{if } n \text{ is even.} \end{cases}$$

([LL2])

In particular, let $\triangle$ be the Laplacian operator. We have

$$(1.6) \qquad \delta(x) \circ \delta(x) = \frac{1}{2\pi\rho} \delta(x), \qquad ([L])$$

$$(1.7) \qquad \delta(x_1, x_2) \circ \delta(x_1, x_2) = \frac{1}{32\pi} \triangle \delta(x_1, x_2) + \frac{1}{8\pi\rho^2} \delta(x_1, x_2), \qquad ([LL2])$$

$$(1.8) \qquad \mathbf{Pf}\delta(x_1, \ldots, x_4) \circ \delta(x_1, \ldots, x_4) =$$

$$\frac{3}{2^{12}\pi^2} \sum_{i=1}^{4} \frac{\partial^4 \delta(x_1, \ldots, x_4)}{\partial^4 x_i} + \frac{3}{2^{11}\pi^2} \sum_{i<j} \frac{\partial^4 \delta(x_1, \ldots, x_4)}{\partial^2 x_i \partial^2 x_j}, \qquad ([LL2])$$

$$(1.9) \qquad \mathrm{Pf}(\delta(x_1, \ldots, x_n) \circ \delta(x_1) \otimes I(x_2, \ldots, x_n)) = \frac{1}{2\pi\rho} \delta(x_1, \ldots, x_n), \qquad ([LK] [Ly7])$$

where $I(x_2, \ldots, x_n)$ is a function in variables $x_2, \ldots, x_n$ constantly equal to 1.

2. Products of $\delta^{(m)}$ and x^{-n} ([L]) :

(2.1)
$$\delta^{(m)}(x) \circ x^{-n} = \sum_{k=0}^{[\frac{m+n}{2}]} (-1)^{m+n} \rho^{-2k} \tilde{b}_{m,n,k} \, \delta^{(m+n-2k)}(x),$$

(2.2)
$$Pf(\delta^{(m)}(x) \circ x^{-n}) = \frac{(-1)^n}{(n-1)!} B_{\frac{1}{2}}(n, m+1) \delta^{(m+n)}(x),$$

where $B_x(n, m) = \int_0^x t^{n-1}(1-t)^{m-1} \, dt$ is the incomplete Beta function. Particularly,

(2.3)
$$Pf(\delta(x) \circ x^{-n}) = \frac{(-1)^n}{n! 2^n} \delta^{(n)}(x),$$

(2.4)
$$Pf(\delta^{(m)}(x) \circ x^{-1}) = \frac{(\frac{1}{2})^{m+1} - 1}{m+1} \delta^{(m+1)}(x),$$

(2.5)
$$\delta^{(n-1)} \circ x^{-n} = Pf(\delta^{(n-1)} \circ x^{-n}) = \frac{(-1)^n (n-1)!}{(2n-1)! 2} \delta^{(2n-1)}(x),$$

(2.6) $\quad \delta(x_1, x_2) \circ (x_1^2 + x_2^2)^{\frac{-1}{2}} = \dfrac{1}{2\rho} \delta(x_1, x_2), \cdot$ ([LL2])

(2.7) $\quad \delta(x_1, x_2, x_3) \circ (x_1^2 + x_2^2 + x_3^2)^{\frac{-1}{2}} = \dfrac{1}{\pi\rho} \delta(x_1, x_2, x_3).$ ([LL2])

3. Products of $\delta^{(m)}(x)$ and $(x \pm i0)^{\pm n} \ln^k(x \pm i0)$ ([Ly2]) :

(3.1)
$$\delta(x) \circ \ln^2(x + i0) = [-\frac{\pi^2}{3} + \ln^2 2 + i\pi \ln 2 + (2\ln 2 + \pi i) \ln \rho + \ln^2 \rho] \delta(x),$$

(3.2)
$$\delta(x) \circ \ln^2(x - i0) = [-\frac{\pi^2}{2} + \ln^2 2 + 3i\pi \ln 2 + (2\ln 2 + 3\pi i) \ln \rho + \ln^2 \rho] \delta(x),$$

(3.3)
$$\delta(x) \circ \ln(x + i0) = [\frac{\pi i}{2} + \ln 2 + \ln \rho) \delta(x),$$

(3.4)
$$\delta(x) \circ \ln(x - i0) = [\frac{3\pi i}{2} + \ln 2 + \ln \rho) \delta(x),$$

4. Products of $\delta^{(n)}(x)$ and $x_+^m \ln^k x_+$:

(4.1) For $m, n = 1, 2, \ldots,$ ([Ly5]),

$$\mathbf{Pf}\,(\delta^{(m+n)}(x) \circ x_+^m \ln x_+)$$

$$= (-1)^n (m+n)! \left[\left(\frac{1}{n!2} \left(\ln 2 + \sum_{k=1}^m \frac{1}{k} \right) + \sum_{j=1}^n \frac{m!(-1)^{j+1}}{(m+j)!(n-j)!2^j j} \right) \right] \delta^{(n)}(x).$$

(4.2) $$\delta(x) \circ \ln x_+ = \frac{1}{2}(\ln 2 + \ln \rho)\delta(x), \quad ([Ly2])$$

(4.3) $$\delta(x) \circ \ln^2 x_+ = (\frac{1}{2} \ln^2 \rho + 2 \ln \rho \ln 2 + \frac{1}{2} \ln^2 2 + \frac{\pi^2}{8})\delta(x), \quad ([Ly5])$$

(4.4) $$\mathbf{Pf}\,(\delta'(x) \circ x_+ \ln x_+) = \frac{1}{2}(\ln 2 + 1)\delta(x),$$

(4.5) $$\mathbf{Pf}\,(\delta''(x) \circ x_+ \ln x_+) = -(\ln 2 + 1\frac{1}{4})\delta'(x),$$

(4.6) $$\mathbf{Pf}\,(\delta^{(3)}(x) \circ x_+ \ln x_+) = (\frac{3}{2} \ln 2 + \frac{35}{16})\delta''(x),$$

(4.7)

$$\mathbf{Pf}\,\delta^{(m)}(x) \circ x_+ \ln x_+ = (-1)^{m+1} [\frac{m}{2} \ln 2 + \frac{m^2}{8} + \frac{3}{8}m + \sum_{j=3}^m \binom{m}{j} \frac{(-1)^j}{2^j (j-1)}]\delta^{(m-1)}(x),$$

(4.8) $$\mathbf{Pf}\,\delta^{(m)}(x) \circ x_+^n = \frac{(-1)^n m!}{2(m-n)!}\delta^{(m-n)}(x), \quad m > n. \quad ([Ly2])$$

5. Products of $\delta^{(n)}(x)$ and $x_+^{-m} \ln^k x_+$ ([Ly5]):

(5.1) $$\mathbf{Pf}(\delta^{(m)}(x) \circ x_+^{-n} \ln^k x_+) = \sum_{q=0}^{m+n} \sum_{j=0}^k \ln^j \rho a(k, m, -n, j, q)\rho^{q-m-n}\delta^{(q)}(x),$$

(5.2) $$\mathbf{Pf}\,\delta'(x) \circ x_+^{-1} \ln x_+ = -(\frac{1}{4} \ln 2 + \frac{1}{3})\delta'(x).$$

6. Products of $\delta^{(m)}(x)$ and x_+^{-n} ([Ly2]):

(6.1)
$$\mathbf{Pf}(\delta^{(m)}(x) \circ x_+^{-n}) = \frac{(-1)^n}{2^{n+1}(n-1)!} \sum_{p=0}^{m} \binom{m}{p} \frac{(-1)^p}{(n+p)2^p} \, \delta^{(m+n)}(x),$$

(6.2)
$$\mathbf{Pf}\delta'(x) \circ x_+^{-1} = -\frac{3}{16}\delta^{(2)}(x),$$

(6.3)
$$\mathbf{Pf}\delta^{(2)}(x) \circ x_+^{-1} = -\frac{7}{48}\delta^{(3)}(x),$$

(6.4)
$$\mathbf{Pf}\delta^{(2)}(x) \circ x_+^{-2} = \frac{11}{3 \times 2^7}\delta^{(4)}(x).$$

7. Products of Heaviside function $H(x) = x_+^0$ and x^{-n} $n = 1, 2, 3, \ldots$, ([L]):

$$\mathbf{Pf}H(x) \circ x^{-n} = x_+^{-n} + (A_n + B_n - \frac{1}{(n-1)!}\ln 2)(-1)^{n-1}\delta^{n-1}(x).$$

8. Products of x_+^m and x_-^{-m-k} ([Ly1]) :

(8.1). For $m = 0, 1, 2, \ldots, k = 1, 2, \ldots,$

$$x_+^m \circ x_-^{-(m+k)} = \sum_{j=0}^{k-1} \sum_{N=0}^{1} \rho^{j+1-k} \ln^N \rho \, a_{j,N}(m, -m-k)\delta^{(j)}(x),$$

$$\mathbf{Pf}x_+^m \circ x_-^{-(m+k)} = \frac{(-1)^{m+k}(k+m)!}{2^{k+m+1}(k-1)!(k+m-1)!} \, a_{k-1,0}(m, -m-k) \, \delta^{(k-1)}(x),$$

where

$$a_{k-1,0}(m, -m-k)$$
$$= \sum_{n=0}^{[\frac{k-1}{2}+m]} \frac{(-1)^n(2k+2m-2n-3)!!}{2^n n!(2n+1)(k+2m-2n-1)} (\sum_{q=1}^{n} \frac{2}{2q-1} - \sum_{q=1}^{k+m-1} \frac{1}{q} - 2\ln 2).$$

(8.2)
$$H(x) \circ x_-^{-1} = (\frac{\ln\rho}{2} + \frac{\ln 2}{2})\delta(x),$$

(8.3)
$$H(x) \circ x_-^{-2} = -\frac{\delta(x)}{2\pi\rho}(\ln\rho + 1 + \ln 2) - \delta'(x)[\frac{\ln\rho}{2} + \frac{1}{4}(2\ln 2 + 1)].$$

9. Products of x_+^{-m} and x_-^{-n} $m, n = 1, 2, \ldots,$ ([Ly1]):

(9.1)
$$\mathbf{Pf} x_+^{-n} \circ x_-^{-n} = 0,$$

(9.2)
$$\mathbf{Pf} x_+^{-m} \circ x_-^{-n} = \frac{(-1)^n}{2\pi(m+n-1)!} sgn(n-m)\, b(m,n)\, \delta^{(m+n-1)}(x),$$

where

$$b(m,n) = \sum_{p=0}^{[\frac{|n-m|-1}{2}]} (-1)^p \binom{|m-n|}{2p+1} D(m+n+|m-n|-2p-2, 2p),$$

and

$$D(2j, p) = \frac{1}{2} B(\frac{p+1}{2}, j+\frac{1}{2})(\psi(\frac{p+1}{2}) - \psi(\frac{p+2j+2}{2})),$$

and B is Beta function, $\psi(x) = \frac{d}{dx}\ln\Gamma(x)$.

10. Products of x_+^λ and x_-^μ, where $\lambda, \mu \in C$ and are not integers :

(10.1) $\mathrm{Re}(\lambda + \mu) < -1$

$$x_+^\lambda \circ x_-^\mu = \frac{1}{\sin\lambda\pi\sin\mu\pi} \sum_{j=0}^{[-\mathrm{Re}(\lambda+\mu)]-1} \frac{(-1)^j}{j!} \rho^{\lambda+\mu+j+1} a(\lambda, \mu, j)\, \delta^{(j)}(x),$$

(10.2) $\mu = -\lambda - n, n > 0$

$$x_+^\lambda \circ x_-^{-\lambda-n}$$

$$= \frac{-\delta^{(n-1)}(x)\pi}{2\sin\lambda\pi\,(n-1)!} + \begin{cases} \dfrac{(-1)^{\frac{n}{2}}}{2\sin^2\lambda\pi} \sum_{j=0}^{\frac{n-2}{2}} \dfrac{b(\lambda, n, 2j)}{(2j)!} \rho^{2j+2-n}\delta^{(2j)}(x) & \text{if } n \text{ is even,} \\[3mm] \dfrac{(-1)^{\frac{n+1}{2}}}{2\sin^2\lambda\pi} \sum_{j=0}^{[\frac{n-2}{2}]} \dfrac{b(\lambda, n, 2j+1)}{(2j+1)!} \rho^{2j+1-n}\delta^{(2j+1)}(x) & \text{if } n \text{ is odd,} \end{cases}$$

$$\mathbf{Pf}(x_+^\lambda \circ x_-^{-\lambda-n}) = \frac{-\delta^{(n-1)}(x)\pi}{2\sin\lambda\pi\,(n-1)!}.$$

11. Products of x_+^m and x_+^{-m-k} ([Ly4]):

(11.1)
$$x_+^m \circ x_+^{-m-k} = x_+^{-k} + \sum_{j=0}^{k-1}\sum_{N=0}^{1} \rho^{j+1-k}\ln^N\rho\; c_{j,n}(m, -m-k)\delta^{(j)}(x),$$

(11.2)
$$H(x) \circ x_+^{-2} = x_+^{-2} + (2\pi\rho)^{-1}(1 + \ln 2 + \ln\rho)\,\delta(x) + \frac{1}{4}(1 + 2\ln 2 + 2\ln\rho)\,\delta'(x),$$

(11.3) $$H(x) \circ x_+^{-1} = x_+^{-1} - \frac{1}{2}(ln2 + ln\rho)\delta(x).$$

12. Products of x_+^{-m} and x_+^{-n} ([Ly4]) :

(12.1) $$x_+^{-m} \circ x_+^{-n} = x_+^{-m-n} + \sum_{j=0}^{m+n-1}\sum_{N=0}^{2} \rho^{j+1-m-n} ln^N \rho d_{j,n}(-m,-n)\delta^{(j)}(x),$$

(12.2) $x_+^{-1} \circ x_+^{-1} = x_+^{-2} + (2\pi\rho)^{-1}(\ln^2 \rho + \ln^2 2 + 2\ln 2 \ln\rho + \frac{5\pi^2}{12})\delta(x) + \frac{1}{2}(\ln 2 + \ln \rho)\delta'(x).$

13. Products of x_+^λ and x_+^μ, where $\lambda, \mu \in C$ and are not integers, ([Ly4]):

(13.1) For $\text{Re}\lambda + \mu = -m$, m being an integer,

$$x_+^\lambda \circ x_+^{-\lambda-m} = x_+^{-m} + \sum_{j=0}^{m-1} \rho^{j+1-m}[\ln\rho\, b_{1,j}(-m,\lambda) + b_{0,j}(-m,\lambda)]\delta^{(j)}(x),$$

(13.2) $$x_+^{-\frac{1}{2}} \circ x_+^{-\frac{1}{2}} = x_+^{-1} + (ln2 - ln\rho)\delta(x),$$

(13.3) $$\mathbf{Pf}(x_+^{\frac{1}{2}} \circ x_+^{-\frac{3}{2}} - x_+^{-\frac{1}{2}} \circ x_+^{-\frac{1}{2}}) = -2\delta(x),$$

(13.4) $$\mathbf{Pf}(x_+^{-\frac{1}{2}} \circ x_+^{-\frac{5}{2}} - x_+^{-\frac{3}{2}} \circ x_+^{-\frac{3}{2}}) = -\frac{1}{3}\delta''(x),$$

(13.5) $$\mathbf{Pf}(x_+^{-\frac{1}{2}} \circ x_+^{-\frac{3}{2}}) = x_+^{-2} - (ln2 - 1)\delta'(x).$$

14. Product of $x_+^\lambda \ln^p x_+$ and $x_-^\mu \ln^q x_-$:

(14.1) For $\lambda, \mu \in C$ being not integers, $m = 1, 2, \ldots$, and $p, q = 0, 1, 2, \ldots$,   ([Ly6])

$$x_+^\lambda \ln^p x_+ \circ x_-^\mu \ln^q x_- = \sum_{n=0}^{m-1} \rho^{\lambda+\mu+n+1} \sum_{j=0}^{p+q} \ln^j \rho a(\lambda, \mu, j, p, q, n)\delta^{(n)},$$

where $Re(\lambda + \mu) = -m - c, 0 \leq c < 1$.

(14.2) For $\lambda \in R$ being not an integer, $\lambda + \mu = -m$,

$$\mathbf{Pf}(x_+^{-\lambda} \ln^p x_+ \circ x_-^{\lambda-m} \ln^q x_-) = \frac{1}{(m-1)!} a(\lambda, -m, p, q)\delta^{(m-1)}(x).$$

In particular,

(14.3)

$$a(\lambda, -1, 1, 0) = \frac{\pi}{4 \sin \lambda \pi} (-2 \ln 2 + \pi \operatorname{ctg} \lambda \pi + 2\psi(1) - \psi(1 + \lambda) - \psi(1 - \lambda) - \frac{1}{\lambda}),$$

$$a(\lambda, -2, 1, 0) = a(\lambda, -1, 1, 0) - \frac{\pi}{4 \sin \lambda \pi} [\frac{1 + 2\lambda}{\lambda(1 + \lambda)} - \frac{1}{\lambda}],$$

$$a(\lambda, -3, 1, 0) = a(\lambda, -1, 1, 0) - \frac{\pi}{4 \sin \lambda \pi} [\frac{6\lambda^2 + 13\lambda + 4}{2\lambda(1 + \lambda)(2 + \lambda)} - \frac{1}{\lambda}],$$

where $\psi(x) = \frac{d}{dx} \ln \Gamma(x)$.

8. u−Generalized Function

For any subset S of R^n we define many kinds of new generalized functions in [LL5], every kind forms an algebra on a nonachimedian algebraically closed field. One of them is very close to a simplified version [Bi] of Colombeau's generalized functions $G_s(\Omega)$.

We know from [LZ] that there are many sets **u** consisting of nonnegative elements in $^*R$, with the following properties:

I) $a, b \in \mathbf{u}, a < b$ implies $x \in \mathbf{u}$, if $a < x < b$,

(**)   II) $a, b \in \mathbf{u}$ implies $ab \in \mathbf{u}$,

III) $2 \in \mathbf{u}$.

Let

$$\mathbf{u}^{-1} = \{x \in {}^*R \mid x \geq 0, \text{ and } x^{-1} > a, \text{ for any } a \in \mathbf{u}\}.$$

The algebra of u−generalized functions $G\mathbf{u}(S)$ is defined by

$$G\mathbf{u}(S) = F\mathbf{u}(S)/F_{\mathbf{u}^{-1}}(S),$$

where $F\mathbf{u}(S)$ and $F_{\mathbf{u}^{-1}}(S)$ are sets of all $^*C-$ valued nonstandard C^∞ functions defined on an open internal subset of $^*R^n$ containing

$$O(S) = \{x \in {}^*S \mid \text{ there exsists } y \in S \text{ such that } ||x - y|| \text{ is infinitesimal}\}$$

with the following properties respectevely

$$f \in F\mathbf{u}(S) \qquad\qquad\qquad ||D^\alpha f(x)|| \in \mathbf{u},$$
$$\text{iff for any } x \in O(S), \alpha \in N^n,$$
$$f \in F_{\mathbf{u}^{-1}}(S) \qquad\qquad\qquad ||D^\alpha f(x)|| \in \mathbf{u}^{-1}.$$

We discussed in [LL5] the property of u-generalized functions, and the relations between our new generalized functions $G_u(S)$ and C^∞−functions, Colombeau's generalized functions, Schwartz distributions and hyperfunctions. Partial results are quoted in the following:

Theorem 8.1. $G_{\mathbf{u}}(S)$ is an algebra over $C_{\mathbf{u}}$, where $C_{\mathbf{u}}$ is an algebraically closed extension of C.

Theorem 8.2.

a. $C^{\infty}(S)$ is a subset of any $G_{\mathbf{u}}(S)$, and if Ω is an open set in R^n, then $C^{\infty}(\Omega) = G_O(\Omega)$, where $O = \{x| \in {}^*R, x \geq 0 \text{ is finite }\}$.

b. Let Ω be open set of R^n. For any $\mathbf{u} \supset M_\rho$, the association-equivalence classes of $G_{\mathbf{u}}(\Omega)$ include Schwartz distribution space $\mathcal{D}'(\Omega)$, where f is associated with g in $G_{\mathbf{u}}(\Omega)$, if for any $\phi \in \mathcal{D}(\Omega)$,

$$\int_{{}^{\bullet}\Omega} (f(x) - g(x))^* \phi(x) dx = \text{ infinitesimal.}$$

c. When $\mathbf{u} = M_\rho = \{x \in {}^*R | 0 \leq x < \rho^{-n}$, for some standard natural number $n\}$, then $G_{M_\rho}(\Omega)$ is closely related to Colombeau's new generalized functions $G'_s(\Omega)$.

For a distribution, there are many $u-$generalized functions associated with it. We would ask: Up to association, what can be the products of two new generalized functions associated with distribution zero ? Some answer is in [Ly8-9], e.g.,

Theorem 8.3. Every Lebesgue locally integrable function is associated with the product of two elements in $G_{\bullet R_+}(R)$ associated with distribution zero.

Theorem 8.4. Let T be a distribution with compact support, and $\mathbf{u}$ be as above (**). Then there are two elements f_1, f_2 in $G_{\mathbf{u}}(R)$ associated with distribution zero such that

$$< T, \phi > - \int_U f_1(x) f_2(x) \phi(x)\, dx = \text{ an infinitesimal}$$

for any $\phi \in \mathcal{D}(R)$, that is, the product $f_1 f_2$ in $G_{\mathbf{u}}(R)$ is associated with T.

Reference

[B] Bremermann. H. J., Some remark on analytic representations and products of distributions, SIAM. J. Appl. Math., 1967, 15, 929–943.

[Bi] Biagioni, H. A., Introduction to a nonlirear theory of generalized functions, Notes de Matematica, State Univ. of Campinas, SP Brazil

[BD] Bremermann. H. J. & Durand. L., On analytic continuation, multiplication and Fourier transformations of Schwarts distributions, J. Math. Phys., 1961, 2, 240–258.

[C] Colombeau, J. F., New Generalized Functions and Multiplication of distributions, North Holland, Amsterdam, Oxford, New York 1984

[CO] Colombeau, J. F. & Oberguggenberger, M., Generalized functions and products of distributions,

[F] Fisher, B., The product of distributions, Quart. J. Math., 1971, 22, 291–298.

[G] Guttinger, W., Product of improper operators and the renormalization problem of quantum field theory, Prog. Theoret. Phys., 1955, 13, 612–626.

[HO] Hirata, Y. & Ogata, H., On the exchange formula for distributions, J. Sci. Hiroshima Univ.,Ser., A-1, 1958, 22, 147–162.

[It1] Itano, M., On the theory of the multiplicative product of distributions, J. Sci. Hiroshima Uni., Ser., A-1, 1966, 30, 151–181.

[It2] Itano, M., On the multiplicative products of x_+^α and x_+^β, J. Sci. Hiroshima Uni., Ser., A-1, 1965, 225–241.

[Iv1] Ivanov, B. K., Multiplication of distributions and regularization of divergent integrals, Izv. Vyss. Ucebn. Zaved. Mate., 1971, 106, 41–49.

[Iv2] Ivanov, B. K., Relations among products of generalized functions, Izv. Vyss. Ucebn. Zaved. Mate., 1979, 209, 38–46.

[K] Konig, H., Multiplikation von Distributionen, Math. Ann., 1955, 128, 420–452.

[L] Li Bang-He, Nonstandard analysis and multiplication of distributions, Scientia Sinica, 1978, 5, 561–585.

[Lb1] Li Bang-He, On distributions with parameted and their analytic representations.Chin. Ann. of Math.,2 (4) 1981, 399–405.

[Lb2] Li Bang-He, On the moire problem from distributional point of view, J. Sys. Sci. & Math. Scis., 6, 4, 1986, 263–268.

[LL1] Li Bang-He & Li Ya-Qing, On multiplications of distributions, Acta scientiarum naturalium universitatis Jilinenesis, 1981, 13–30.

[LL2] Li Bang-He & Li Ya-Qing, Nonstandard analysis and multiplication of distributions in any dimension, Scientia Sinica, 1985, 28, 716–726.

[LL3] Li Bang-He & Li Ya-Qing, On the harmonic and analytic representations of distributions, Scientia Sinica, 1985, 28, 923–937.

[LL4] Li Bang-He & Li Ya-Qing, Distributions and their harmonic and analytic representations, 1992, National fedence industry press.

[LL5] Li Bang-He & Li Ya-Qing, New generalized functions in nonstandard framwork, Acta Math. Sci., 1992, 3, 260–269.

[LK] Li Ya-Qing & Kuribayashi,Y., Products $\delta(x_1) \otimes I(x_2, \cdots, x_n)$ and $\delta(x_1, \cdots, x_n)$, J. Fac. Educ. Tottori Univ. (Nat. Sci.), 42, (1994) 109–117.

[Ly1] Li Ya-Qing, The products of generalizes functions x_+^λ and x_-^μ, Scientia Sinica, 1979, Special issue (I) on Math., 103–123.

[Ly2] Li Ya-Qing, The products of singular distributions, Kexue Tongbao, 1980, 4, 149–155.

[Ly3] Li Ya-Qing, Some results on products of distributions, Kexue Tongbao, 1980, Special issue of Math. Phys. & Chem., 16–19.

[Ly4] Li Ya-Qing, Products of x_+^λ and x_+^μ, Kexue Tongbao, 1980, 6, 454–461.

[Ly5] Li Ya-Qing, The products $\delta^{(m)} \circ x_+^n \ln^k x_+$ and $\delta^{(m)} \circ x_+^{-n} \ln^k x_+$, J. SyS. Sci. & Math., 1984, 4, 81–86.

[Ly6] Li Ya-Qing, The products $x_+^\lambda \ln^p x_+$ and $x_-^\mu \ln^q x_-$, J. Sys. Sci. & Math., 1985, 5, 241–250.

[Ly7] Li Ya-Qing, A short formula for the finite part of products $\delta(x_1) \otimes I(x_2, \cdots, x_n)$ and $\delta(x_1, \cdots, x_n)$, Preprint.

[Ly8] Li Ya-Qing, What can a product of distribution zero and itself be? Sys. Sci.& Math. Sci. 1990, 381–383.

[Ly9] Li Ya-Qing, A property of products of new generalized functions, Preprint.

[LZ] Li Bang-He & Zhang Ji-Jiang, On the Dekind completion of $^\#R$, J. Sys.Sci. & Math. Sci. 1988, 29–39.

[M] Minkusinski, J., Criteria of the existence and of the associativity of the product of distributions, Studia Math., 1962, 21, 253–259.

[O] Obergugenberger, M., Products of distributions : nonstandard methods, Zeitschr. Anal. Anw., 1988, 7, 347–365.

[Sh] Shiraishi, R., On the value of distributions at a point and the multiplicative products, J. Sci. Hiroshima Univ., 1967, 31, 89–104.

[T] Tillmann, H. G., Darstellung der Schwartzschen Distributioned durch analytische Funktionen, Math. Zeit. 1961, 77, 2, 106–124.

C^*-Crossed Products $C(X) \times_\alpha \mathbb{Z}$ and $C(X) \times_\alpha \mathbb{Z}_n$ with $\alpha^n = id$

Bingren Li and Qing Lin

Institute of Mathematics, Academic Sinica, Beijing

The C^*-crossed product $C(X) \times_\alpha \mathbb{Z}_n$ has been studied for a long time. For instance, see Effros and Hahn [1].

This paper is a summary of our works on C^*-crossed products $C(X) \times_\alpha \mathbb{Z}$ and $C(X) \times_\alpha \mathbb{Z}_n$, where X is a compact Hausdorff space, α is a homeomorphism of X with $\alpha^n = id$, and n is a fixed positive integer.

1. C^*-crossed products $A \times_\alpha \mathbb{Z}$, $A \times_\alpha \mathbb{Z}_n$ and mapping torus

Consider a C^*-algebra with a periodic action.

Let A be a C^*-algebra, $\alpha \in \mathrm{Aut}(A)$ (the $*$ automorphism group of A), and $\alpha^n = id$, where n is a fixed positive integer. Then we have two C^*-dynamical systems:

$$(A, \mathbb{Z}, \alpha) \quad \text{and} \quad (A, \mathbb{Z}_n, \alpha),$$

and two C^*-crossed products $A \times_\alpha \mathbb{Z}$ and $C(X) \times_\alpha \mathbb{Z}_n$.

As is well known, $A \times_\alpha \mathbb{Z}$ is the C^*-enveloping algebra of Banach $*$ algebra $l^1(A, \mathbb{Z}, \alpha)$. Let $\mathcal{A}$ be the closure of

$$l_n^1(A, \mathbb{Z}, \alpha) = \{f \in l^1(A, \mathbb{Z}, \alpha) | f(k) = 0, \forall k \not\equiv 0 (\mathrm{mod}\ n)\}$$

in $A \times_\alpha \mathbb{Z}, \lambda \in l^1(A \dot{+} \mathbb{C}, \mathbb{Z}, \alpha)$ such that

$$\lambda(k) = 0, \quad \forall k \neq 1, \quad \text{and} \quad \lambda(1) = 1.$$

1991 Mathematics Subject Classification: 46L30

Partially supported by NNSF of China

Then $\mathcal{A}$ is a C^*-subalgebra of $A \times_\alpha \mathbb{Z}$, and λ is invertible in $(A\dot{+}\mathbb{C}) \times_\alpha \mathbb{Z}$, and we have ([2])

$$\mathcal{A} \cong A \times_{id} \mathbb{Z} \cong C(T, A),$$

$$A \times_\alpha \mathbb{Z} = \mathcal{A}\dot{+}\mathcal{A}\lambda\dot{+}\cdots\dot{+}\mathcal{A}\lambda^{n-1}$$

$$\cong C(T, A) \times \cdots \times C(T, A) \quad (n \text{ times}),$$

where $T = \{z \in \mathbb{C} | |z| = 1\}$.

On the other hand, $l^1(A, \mathbb{Z}_n, \alpha)$ is C^*-equivalent ([3]) and

$$A \times_\alpha \mathbb{Z}_n = l^1(A, \mathbb{Z}_n, \alpha) \cong A \times \cdots \times A \quad (n \text{ times}).$$

Consider the dual C^*-system $(A \times_\alpha \mathbb{Z}_n, \hat{\mathbb{Z}}_n = \mathbb{Z}_n, \hat{\alpha})$, where

$$\hat{\alpha}((a_j)_{0 \le j \le n-1}) = (e^{-2\pi i j/n} a_j)_{0 \le j \le n-1},$$

$a_j \in A, 0 \le j \le n-1$, and $(a_j)_{0 \le j \le n-1} \in A \times_\alpha \mathbb{Z}_n$. Then we have the following relation between $A \times_\alpha \mathbb{Z}$ and $A \times_\alpha \mathbb{Z}_n$ immediately.

Theorem 1.1 (Mapping torus [2,4]). Let A be a C^*-algebra, $\alpha \in \text{Aut}(A)$ and $\alpha^n = id$. Then

$$A \times_\alpha \mathbb{Z} \cong M_{\hat{\alpha}}(A \times_\alpha \mathbb{Z}_n),$$

where $M_{\hat{\alpha}}(A \times_\alpha \mathbb{Z}_n)$ is the mapping torus of $\hat{\alpha}$ defined as follows:

$$M_{\hat{\alpha}}(A \times_\alpha \mathbb{Z}_n) = \left\{ (F_j(t))_{0 \le j \le n-1} \left| \begin{matrix} t \to F_j(t) : [0, 1] \to A \text{ continuous,} \\ \forall j, \text{ and } (F_j(1))_j = \hat{\alpha}((F_j(0))_j) \end{matrix} \right. \right\}.$$

Furthermore, from $A \times_\alpha \mathbb{Z} \cong C(T, A) \times \cdots \times C(T, A)$ and $A \times_\alpha \mathbb{Z}_n \cong A \times \cdots \times A$ (n times), we have the following matrix represntations.

Theorem 1.2 ([5]). Let A be a C^*-algebra, $\alpha \in \text{Aut}(A)$ and $\alpha^n = id$. Then $A \times_\alpha \mathbb{Z}$ and $A \times_\alpha \mathbb{Z}_n$ can be naturally embedded into $C(T, A) \otimes M_n$ and $A \otimes M_n$ respectively, i.e.,

$$(f_j)_{0 \le j \le n-1} \in [C(T, A)]^n = A \times_\alpha \mathbb{Z}$$

$$\longrightarrow (\alpha^{-i}(f_{i-j}(z)) \cdot z^{\kappa(i-j)})_{0 \le i, j \le n-1} \in C(T, A) \otimes M_n, \quad \forall z \in T,$$

where the foot index of any integer is understood in the sense of (mod n), and the function $\kappa(\cdot)$ on $\mathbb{Z}$ is defined as follows:

$$\kappa(k) = \begin{cases} 1, & \text{if } k < 0, \\ 0, & \text{if } k \ge 0, \end{cases}$$

and

$$(a_j)_{0 \le j \le n-1} \in A^n = A \times_\alpha \mathbb{Z}_n$$

$$\longrightarrow (\alpha^{-i}(a_{i-j}))_{0 \le i, j \le n-1} \in A \otimes M_n.$$

Moreover, let $B(z) = (b_{ij}(z))_{0 \leq i,j \leq n-1} \in C(T, A) \otimes M_n$, and $B = (b_{ij})_{0 \leq i,j \leq n-1} \in A \otimes M_n$. Then $B(\cdot) \in A \times_\alpha \mathbb{Z}$ and $B \in A \times_\alpha \mathbb{Z}_n$ (under the above embeddings), if and only if

$$\alpha B(z) = A_n(z) B(z) A_n(z)^*, \quad \forall z \in T$$

and

$$\alpha B = A_n B A_n^*,$$

where

$$\alpha B(z) = (\alpha(b_{ij}(z)))_{0 \leq i,j \leq n-1}, \quad \alpha B = (\alpha(b_{ij}))_{0 \leq i,j \leq n-1},$$

and

$$A_n(z) = \begin{pmatrix} 0 & \cdots & \cdots & z \\ 1 & 0 & & \vdots \\ & \ddots & \ddots & \vdots \\ 0 & & 1 & 0 \end{pmatrix}, \quad z \in T, \quad n \times n \text{ matrix},$$

and $A_n = A_n(1)$.

2. Pure state spaces of $C(X) \times_\alpha \mathbb{Z}$ and $C(X) \times_\alpha \mathbb{Z}_n$

Let X be a compact Hausdorff space, α be a homeomorphism of X and $\alpha^n = id$, where n is a fixed positive integer. If we define $\alpha(f) = f \circ \alpha, \forall f \in C(X)$, then we have the abelian $C*$-dynamical systems:

$$(C(X), \mathbb{Z}, \alpha) \quad \text{and} \quad (C(X), \mathbb{Z}_n, \alpha),$$

and by Section 1, we have the following embeddings:

$$C(X) \times_\alpha \mathbb{Z} \hookrightarrow C(Y) \otimes M_n,$$

and

$$C(X) \times_\alpha \mathbb{Z}_n \hookrightarrow C(X) \otimes M_n,$$

where $Y = X \times T$ is also a compact Hausdorff space.

Now let φ be a pure state on $C(X) \times_\alpha \mathbb{Z}$. By $C(X) \times_\alpha \mathbb{Z} \hookrightarrow C(Y) \otimes M_n, \varphi$ can be extended to a pure state on $C(Y) \otimes M_n$, and has the following form:

$$\varphi = \chi \otimes \psi,$$

where χ is a pure state on $C(Y)$, and ψ is a pure state on M_n. Furthermore, it is easy to see that

$$\varphi = \varphi_{x,z,\lambda},$$

i.e.,

$$\varphi_{x,z,\lambda}(f) \; = \langle f(x,z)\lambda, \lambda \rangle$$

$$= \sum_{i,j=0}^{n-1} \overline{\lambda}_i \lambda_j f_{i-j}(\alpha^{-i}x, z) z^{\kappa(i-j)}, \tag{1}$$

$\forall f = (f_j)_{0 \le j \le n-1} \in [C(Y)]^n = C(X) \times_\alpha \mathbb{Z}$, where $x \in X, z \in T, \lambda \in \mathbb{C}^n$ and $|\lambda|^2 = \sum_{i=0}^{n-1} |\lambda_i|^2 = 1$. Of course, not each $\varphi_{x,z,\lambda}$ is pure on $C(X) \times_\alpha \mathbb{Z}$.

Definition 2.1. $x \in X$ is called a *p-degree point* (of α), if p is the minimal positive integer such that $\alpha^p(x) = x$.

Since $\alpha^n = id$, it follows that $p|n$.

Theorem 2.2 ([5]). Let X be a compact Hausdorff space, α be a homeomorphism of X and $\alpha^n = id$.

1) Each pure state on $C(X) \times_\alpha \mathbb{Z}$ has the form like (1).

2) Let $x \in X$ be a point of n-degree, and $z \in T$. Then each $\varphi_{x,z,\lambda}$ is a pure state on $C(X) \times_\alpha \mathbb{Z}, \forall \lambda \in \mathbb{C}^n$ and $|\lambda| = 1$. Moreover, these pure states $\{\varphi_{x,z,\lambda} | \lambda \in \mathbb{C}^n, |\lambda| = 1\}$ are unitarily equivalent to each other, and the common primitive ideal is as follows:

$$J_0^{(n)}(x,z) = \; \{f = (f_j)_{0 \le j \le n-1} \in [C(Y)]^n = C(X) \times_\alpha \mathbb{Z}|$$

$$f_i(\alpha^j x, z) = 0, 0 \le i, j \le n - 1\},$$

and

$$C(X) \times_\alpha \mathbb{Z} / J_0^{(n)}(x,z) \cong M_n.$$

3) Let $x \in X$ be a point of 1-degree, and $z \in T$. Then in $\{\varphi_{x,z,\lambda} | \lambda \in \mathbb{C}^n, |\lambda| = 1\}$ there are only n pure states $\{\varphi_j | 0 \le j \le n - 1\}$ on $C(X) \times_\alpha \mathbb{Z}$ (indeed, each φ_j is multiplicative), where by (1) and $\alpha(x) = x$,

$$\varphi_j(f) \; = \varphi_{x,z,\lambda_j^{(n)}(z)}(f)$$

$$= \sum_{k=0}^{n-1} f_k(x,z) \langle A_n(z) \lambda_k^{(n)}(z), \lambda_k^{(n)}(z) \rangle$$

$$= \sum_{k=0}^{n-1} r_j^{(n)}(z)^k f_k(x,z),$$

$\forall f = (f_j)_{0 \le j \le n-1} \in [C(Y)]^n = C(X) \times_\alpha \mathbb{Z}, 0 \le j \le n - 1, \{\lambda_j^{(n)}(z) | 0 \le j \le n - 1\}$ and $\{r_j^{(n)}(z) | 0 \le j \le n - 1\}$ are the eigenvectors and eigenvalues of $A_n(z)$ respecively, i.e.,

$$A_n(z) \lambda_j^{(n)}(z) = r_j^{(n)}(z) \lambda_j^{(n)}(z), \quad 0 \le j \le n - 1,$$

and $\{\lambda_j^{(n)}(z)|0 \le j \le n-1\}$ is also a normalized orthogonal basis of $\mathbb{C}^n$. Moreover, the primitive ideal corresponding to φ_j is as follows:

$$J_j^{(1)}(x, z) = \left\{ f \left| \sum_{k=0}^{n-1} r_j^{(n)}(z)^k f_k(x, z) = 0 \right. \right\}$$

and

$$C(X) \times_\alpha \mathbb{Z}/J_j^{(1)}(x, z) \cong \mathbb{C}, \quad 0 \le j \le n-1.$$

4) Let $x \in X$ be a point of p-degree, $n = pl$, and $z \in T$. Then in $\{\varphi_{x,z,\lambda}|\lambda \in \mathbb{C}^n, |\lambda| = 1\}$ there are only l unitarily equivalent classes of pure states on $C(X) \times_\alpha \mathbb{Z}$ as follows:

$$\tilde{\varphi}_j = \{\varphi_{x,z,\mu\otimes\lambda_j^{(l)}(z)}|\mu \in \mathbb{C}^p, |\mu| = 1\}, \quad 0 \le j \le l-1,$$

where $(\mu \otimes \delta)_{tp+r} = \mu_r \delta_t, 0 \le t \le l-1, 0 \le r \le p-1, \mu = (\mu_0, \cdots, \mu_{p-1}) \in \mathbb{C}^p, \delta = (\delta_0, \cdots, \delta_{l-1}) \in \mathbb{C}^l$, and $\mu \otimes \delta \in \mathbb{C}^n$; $\{\lambda_j^{(l)}(z)|0 \le j \le l-1\}$ and $\{r_j^{(l)}(z)|0 \le j \le l-1\}$ are the eigenvectors and eigenvalues of

$$A_l(z) = \begin{pmatrix} 0 & \cdots & \cdots & z \\ 1 & \ddots & & \vdots \\ & \ddots & \ddots & \vdots \\ 0 & & 1 & 0 \end{pmatrix} \quad (l \times l \text{ matrix}),$$

i.e. $A_l(z)\lambda_j^{(l)}(z) = r_j^{(l)}(z)\lambda_j^{(l)}(z), 0 \le j \le l-1$, and $\{\lambda_j^{(l)}(z)|0 \le j \le l-1\}$ is also a normalized orthogonal basis of $\mathbb{C}^l$. If $\mu_0 = (1, 0, \cdots, 0) \in \mathbb{C}^p$, then $\varphi_j = \varphi_{x,z,\mu_0\otimes\lambda_j^{(l)}(z)} \in \tilde{\varphi}_j$, and

$$\varphi_j(f) = \sum_{t=0}^{l-1} r_j^{(l)}(z)^t f_{tp}(x, z),$$

$\forall f = (f_i)_{0 \le i \le n-1} \in [C(Y)]^n = A \times_\alpha \mathbb{Z}, 0 \le j \le l-1$. Moreover, denoting by $J_j^{(p)}(x, z)$ as the common primitive ideal correspoinding to $\tilde{\varphi}_j$, then

$$C(X) \times_\alpha \mathbb{Z}/J_j^{(p)}(x, z) \cong M_p, \quad 0 \le j \le l-1.$$

However, other $\varphi_{x,z,\lambda}'s(\notin \tilde{\varphi}_j, 0 \le j \le l-1)$ are not pure on $C(X) \times_\alpha \mathbb{Z}$.

Similarly, we also have the following

Theorem 2.3 ([6]). Let X be a compact Hausdorff space, α be a homeomorphism of X and $\alpha^n = id$.

1) Each pure state on $C(X) \times_\alpha \mathbb{Z}_n$ has the following form:

$$\varphi_{x,\lambda}(f) = \langle f(x), \lambda, \lambda \rangle = \sum_{i,j=0}^{n-1} \overline{\lambda}_i \lambda_j f_{i-j}(\alpha^{-i}x) \tag{2}$$

$\forall f = (f_j)_{0 \le j \le n-1} \in [C(X)]^n = C(X) \times_\alpha \mathbb{Z}_n, x \in X, \lambda \in \mathbb{C}^n$ and $|\lambda| = 1$.

2) Let $x \in X$ be a point of n-degree. Then each $\varphi_{x,\lambda}$ is a pure state on $C(X) \times_\alpha \mathbb{Z}_n, \forall \lambda \in \mathbb{C}^n, |\lambda| = 1$. Moreover, these pure states $\{\varphi_{x,\lambda} | \lambda \in \mathbb{C}^n, |\lambda| = 1\}$ are unitarily equivalent to each other, and the common primitive ideal is as follows:

$$J_0^{(n)}(x) = \{ \quad f = (f_j)_{0 \le j \le n-1} \in C(X)^n = C(X) \times_\alpha \mathbb{Z}_n |$$

$$f_i(\alpha^j x) = 0, \quad 0 \le i, j \le n-1\},$$

and

$$C(X) \times_\alpha \mathbb{Z}_n / J_0^{(n)}(x) \cong M_n.$$

3) Let $x \in X$ be a point of 1-degree. Then in $\{\varphi_{x,\lambda} | \lambda \in \mathbb{C}^n, |\lambda| = 1\}$ there are only n pure states $\{\varphi_j | 0 \le j \le n-1\}$ on $C(X) \times_\alpha \mathbb{Z}_n$ (indeed, each φ_j is multiplicative), where

$$\varphi_j(f) \quad = \varphi_{x,\lambda_j^{(n)}}(f)$$

$$= \sum_{k=0}^{n-1} f_k(x) \langle A_n \lambda_k^{(n)}, \lambda_k^{(n)} \rangle = \sum_{k=0}^{n-1} r_j^{(n)k} f_k(x),$$

$\forall f = (f_j)_{0 \le j \le n-1} \in [C(X)]^n = C(X) \times_\alpha \mathbb{Z}_n, \{\lambda_j^{(n)} | 0 \le j \le n-1\}$ and $\{r_j^{(n)} | 0 \le j \le n-1\}$ are the eigenvectors and eigenvalues of A_n respectively, i.e., $A_n \lambda_j^{(n)} = r_j^{(n)} \lambda_j^{(n)}, 0 \le j \le n-1$, and $\{\lambda_j^{(n)} | 0 \le j \le n-1\}$ is also a normalized orthogonal basis of $\mathbb{C}^n$. Moreover, the primitive ideal corresponding to φ_j is as follows:

$$J_j^{(1)}(x) = \left\{ f \left| \sum_{k=0}^{n-1} r_j^{(n)k} f_k(x) = 0 \right. \right\}$$

and

$$C(X) \times_\alpha \mathbb{Z}_n / J_j^{(1)}(x) \cong \mathbb{C}, \quad 0 \le j \le n-1.$$

4) Let $x \in X$ be a point of p-degree, and $n = pl$. Then in $\{\varphi_{x,\lambda} | \lambda \in \mathbb{C}^n, |\lambda| = 1\}$ there are only l unitarily equivalent classes of pure states on $C(X) \times_\alpha \mathbb{Z}_n$ as follows:

$$\tilde{\varphi}_j = \{\varphi_{x,\mu \otimes \lambda_j^{(l)}} | \mu \in \mathbb{C}^p, |\mu| = 1\}, \quad 0 \le j \le l-1,$$

where $\{\lambda_j^{(l)} | 0 \le j \le l-1\}$ and $\{r_j^{(l)} | 0 \le j \le l-1\}$ are the eigenvectors and eigenvalues of $A_l = A_l(1)(l \times l$ matrix), i.e., $A_l \lambda_j^{(l)} = r_j^{(l)} \lambda_j^{(l)}, 0 \le j \le l-1$, and $\{\lambda_j^{(l)} | 0 \le j \le l-1\}$ is also a normalized orthogonal basis of $\mathbb{C}^l$. If $\mu_0 = (1, 0, \cdots, 0) \in \mathbb{C}^p$, then $\varphi_j = \varphi_{x,\mu_0 \otimes \lambda_j^{(l)}} \in \tilde{\varphi}_j$ and

$$\varphi_j(f) = \sum_{t=0}^{l-1} r_j^{(l)t} f_{tp}(x),$$

$\forall f = (f_i)_{0 \le i \le n-1} \in [C(X)]^n = C(X) \times_\alpha \mathbb{Z}_n, 0 \le j \le l-1$. Moreover, denoting by $J_j^{(p)}(x)$ the common primitive ideal corresponding to $\tilde{\varphi}_j$, then

$$C(X) \times_\alpha \mathbb{Z}_n / J_j^{(p)}(x) \cong M_p, \quad 0 \le j \le l-1.$$

However, other $\varphi_{x,\lambda}'s(\notin \tilde{\varphi}_j, 0 \le j \le l-1)$ are not pure on $C(X) \times_\alpha \mathbb{Z}_n$.

Denote by $\overline{P(A)}$ the w^*-closure of the pure state space $P(A)$ of a C^*-algebra A in A^*.

Theorem 2.4 ([7]). Let X be a compact Hausdorff space, α be a homeomorphism and $\alpha^n = id$. If the n-degree points are dense in X, then we can introduce an equivalent relation $\sim$ in $X \times P(\mathbb{C}^n)$ such that

$$X \times P(\mathbb{C}^n)/\sim \; \cong \; \overline{P(C(X) \times_\alpha \mathbb{Z}_n)}.$$

Proposition 2.5 ([7]). If α, β act on X, Y freely (i.e., each point is n-degree) and $\alpha^n = id, \beta^n = id$, and $H^2(X/\alpha, \mathbb{Z})$ has no element annihilated by n, then $C(X) \times_\alpha \mathbb{Z}_n \cong C(Y) \times_\beta \mathbb{Z}_n$ if and only if $X/\alpha \cong Y/\beta$.

3. The case of $n = 2$

Let X be a compact Hausdorff space, α be a homeomorphism of X and $\alpha^2 = id$.
If μ is a regular Borel measure on X, then we define the following operators on $L^2(X, \mu)$:

$$\begin{cases} M_f\xi = f\xi, \\ U\xi = \xi \circ \alpha, \end{cases} \quad \forall f \in C(X), \xi \in L^2(X, \mu).$$

$(M., U, L^2(X, \mu))$ will be a covariant representation of the C^*-system $(C(X), \mathbb{Z}_2, \alpha)$, i.e.,

$$UM_fU^* = M_{f\circ\alpha}, \quad \forall f \in C(X).$$

Theorem 3.1 ([3]). Let μ be an α-invariant probability measure on X, and supp $\mu = X$. If $X^\alpha = \{x \in X | \alpha(x) = x\}$ is nowhere dense in X (in particular, α is free, i.e., $\alpha(x) \ne x, \forall \in X$), then $C(X) \times_\alpha \mathbb{Z}_2$ has a faithful representation on $L^2(X, \mu)$:

$$\|(f_0, f_1)\| = \|M_{f_0} + M_{f_1}U\|,$$

$\forall (f_0, f_1) \in [C(X)]^2 = C(X) \times_\alpha \mathbb{Z}_2$.

About projections in $C(X) \times_\alpha \mathbb{Z}_2$, we have the following

Theorem 3.2 ([3]). If X is connected, then each projection in $C(X) \times_\alpha \mathbb{Z}_2(\hookrightarrow C(X) \otimes M_2)$ has the form:

$$\begin{pmatrix} f_0 & f_1 \\ \bar{f}_1 & 1 - f_0 \end{pmatrix}$$

where $f_0, f_1 \in C(X), 0 \leq f_0 \leq 1$ and $|f|^2 = f_0(1 - f_0)$.

As a Banach space, $C(X) \times_\alpha \mathbb{Z}_2 = [C(X)]^2$. Hence, each element of $(C(X) \times_\alpha \mathbb{Z}_2)^*$ has the form (μ, ν), where μ, ν are bounded measures on X. Now we have the following

Theorem 3.3 ([3]). Let $F = (\mu, \nu) \in (C(X) \times_\alpha \mathbb{Z}_2)^*$. Then F is a tracial state on $C(X) \times_\alpha \mathbb{Z}_2$, if and only if μ is an α-invariant probability measure on X, and $\nu = h \cdot \mu$, where h is an α-invariant real μ-measurable function on X, $|h| \leq 1$ and $h(t) = 0, \forall t \in X \backslash X^\alpha$, a.e. μ.

In the rest of this section, let $A = C(X) \times_\alpha \mathbb{Z}_2$. By Theorem 2.3,

$$\text{Prim}(A) = \{J_x | x \notin X^\alpha\} \sqcup \{J_x^+, J_x^- | x \in X^\alpha\},$$

where $X^\alpha = \{x | \alpha(x) = x\}$,

$$J_x = \{(f_0, f_1) | f_0, f_1 \in C(X), f_i(\alpha^j x) = 0, 0 \leq i, j \leq 1\},$$

$\forall x \notin X^\alpha$, and

$$J_x^\pm = \{(f_0, f_1) | f_0, f_1 \in C(X), f_0(x) \pm f_1(x) = 0\},$$

$\forall x \in X^\alpha$.

It is easy to see that any closed two-sided ideal I of A has the following form;

$$I = J_F \cap J_{F_+} \cap J_{F_-},$$

where $F, F_\pm$ are closed subsets of X such that $F = \alpha(F), F_\pm \sqsubset X^\alpha$, and

$$J_F = \bigcap_{x \in F} J_x = \{(f_0, f_1) | f_0 = f_1 = 0 \text{ on } F\},$$

$$J_{F_+} = \bigcap_{x \in F_+} J_x^+, \quad J_{F_-} = \bigcap_{x \in F_-} J_x^-.$$

Recall that the Jacobson topology in $\text{Prim}(A)$ is as follows: a subset U of Prim (A) is open if and only if there exists a closed two-sided ideal I of A such that

$$U = \{J \in \text{Prim}(A) | J \not\supseteq I\}$$

$$= \{J \in \text{Prim}(A) | J \not\supseteq J_F \cap J_{F_+} \cap J_{F_-}\}.$$

Theorem 3.4 ([8]). i) For any $x, y \in X$ and $q(x) \neq q(y)(q : X \to X/\alpha$ quotient map), J_x (if $x = \alpha(x)$, then $J_x = J_x^+$ or J_x^-) and J_y (if $y = \alpha(y)$, then $J_y = J_y^+$ or J_y^-) are separated by neighborhoods in Prim (A).

ii) If $x = \alpha(x)$, then J_x^+ and J_x^- can be separated by neighborhoods in $\text{Prim}(A)$ if and only if $x \in \overset{\circ}{X}{}^\alpha$.

Cosequently, Prim (A) is Hausdorff if and only if $\partial X^\alpha = (X^\alpha \backslash \overset{\circ}{X}{}^\alpha)$ is empty.

4. Differential structure of certain $C(X) \times_\alpha \mathbb{Z}$

In this section, we assume X to be a compact smooth manifold without boundary, $\alpha \in$ Diffeo (X) with period n and $^\#\{\alpha^i(x)|0 \le i \le n-1\} = n, \forall x \in X$.

By Theorem 2.2,

$$P(C(X) \times_\alpha \mathbb{Z}) = \{\varphi_{x,z,\lambda}|(x,z,[\lambda]) \in X \times T \times P(\mathbb{C}^n)\}.$$

It is not hard to show that

$$\varphi_{x,z,\lambda} = \varphi_{x',z',\lambda'} \Longleftrightarrow z = z', x = \alpha^k(x') \quad \text{and } [\lambda] = [A_n^k(z)\lambda'].$$

Thus the smooth mapping diagram

$$
\begin{array}{ccc}
(x, z, [\lambda]) & \overset{\varphi}{\longrightarrow} & \varphi_{x,z,\lambda} \\
\downarrow p & \swarrow \pi & \\
(q(x), z) & &
\end{array}
$$

factors a continuous projection π from $P(C(X) \times_\alpha \mathbb{Z})$ onto $(X/\alpha) \times T$. The map φ above induces a natural differential strucutre on $P(C(X) \times_\alpha \mathbb{Z})$ which makes π smooth, and $P(C(X) \times_\alpha \mathbb{Z}) \overset{\pi}{\longrightarrow} (X/\alpha) \times T$ carries a natural PU_n-bundle structure.

Theorem 4.1 ([9]). Under our assumption and dim $X \le 2$, $C(X) \times_\alpha \mathbb{Z} \cong C(Y) \times_\beta \mathbb{Z} \Longleftrightarrow \exists$ a diffeomorphism $\psi : (X/\alpha) \times T \to (Y/\beta) \times T$ such that

$$\psi^* P_\beta \cong P_\alpha \quad \text{as} \quad PU_n\text{-bundles.}$$

Corollary 4.2 ([9]). $C(X) \times_\alpha \mathbb{Z} \cong C(Y) \times_\beta \mathbb{Z}$, then there is a diffeomorphism $\psi : (X/\alpha) \times T \to (Y/\beta) \times T$ such that

$$\psi^* c_1(E_\beta) = c_1(E_\alpha) + nc_1(L)$$

for some complex line bundle L over $(X/\alpha) \times T$.

Theorem 4.3 ([9]). Suppose $X = T$ and α is the rational rotation defined by

$$\alpha(z) = e^{2\pi i/n} \cdot z, \quad \forall z \in T.$$

Let $P_\alpha = P(C(X) \times_\alpha \mathbb{Z})$ and $\mathbb{P}(E_\alpha) = P_\alpha$. Then

$$\chi(\wedge^n E_\alpha) \equiv 1 \pmod{n}.$$

Now we can give a simple proof to a well-known result.

Theorem 4.4 ([9] Classification of rational rotation algebras). If $\theta_1, \theta_2 \in [0, 1] \cap \mathbb{Q}$, then $A_{\theta_1} \cong A_{\theta_2}$ if and only if $\theta_1 = \theta_2$ or $(1 - \theta_2)$, where $A_\theta = C(T) \times_\theta \mathbb{Z}$.

References

[1] Effros, E. and Hahn, F., Locally compact transformation groups and C^*-algebras, Mem. Amer. Math. Soc., 75(1967).

[2] Li, B. and Lin, Q., Notes on the mapping torus of C^*-algebra, Chin. Sci. Bull., 38(1993), 97–99.

[3] Li, B. and Lin, Q., Elembentary properties of $C(X) \times_\alpha \mathbb{Z}_2$, Acta Math. Sinica, 35(1992), 563–569.

[4] Blackadar, B., K-theory of operator algebras, Springer-Verlag, 1986.

[5] Li, B. and Lin, Q., Pure states on $C(X) \times_\alpha \mathbb{Z}$ with $\alpha^n = id$, Lecture Notes, Inst. of Math., Acad. Sinica, 2(1991), 147–162.

[6] Li, B. and Lin, Q., Pure state space of $C(X) \times_\alpha \mathbb{Z}_n$, to appear.

[7] Li, B. and Lin, Q., Pure state approach to $C(X) \times_\alpha \mathbb{Z}_n$, Chin. Ann. of Math., 168(1995), 75–86.

[8] Li, B. and Lin, Q., The ideal structure of $C(X) \times_\alpha \mathbb{Z}_2$, Sci. in China, 36(1993), 1054–1062.

[9] Lin, Q., Differential structure of certain $C(X) \times_\alpha \mathbb{Z}$, Acta Math. Sinica, 11(1995), 113–120.

Advance of Operator Theory
in Fudan University

Shaokuan Li

Chinese Textile University, Shanghai

Xiaoman Chen

Institute of Mathematics, Fudan University, Shanghai

1. Introduction

In the past ten years the theory of operators on Hilbert space made great progress. The fundamental problems are: (1) To determine the spectrum of an operator and discuss the properties of its resolvent. (2) To determine the relations of the spectra between operators. (3) To determine the structure of an operator or to give a concrete model of a class of operators. These problems are difficult and complex. We discuss these problems in some special classes of operators.

In these areas, our research mainly focuses on the theory of non-normal operators. At first, Prof. Xia Daoxing established the singular integral models of hyponormal and semi-hyponormal operators on Hilbert space. We have made researches systematically in the theory of hyponormal and semi-hyponormal operators and gotten a series of results. We introduced the symbol operators and proved the representation theorem of hyponormal and semi-hyponormal operators. In the Putnam-Fuglede Theorem of nonnormal operators, we gave the equivalent condition for the P-F theorem to hold true. We proved the P-F theorem valid for hyponormal, semi-hyponormal operators, and so on. Meanwhile, we also studied the tuple of operators on Hlilbert space and got some beautiful results. In this note we mainly give an account of researches.

1991 Mathematics Subject Classification: 47B20

2. Theory of hyponormal and semi-hyponormal operators

In this section we present the work on the theory of hyponormal and semi hyponormal operators.

Definition 2.1. An operator A on Hilbert space H is called a hyponormal operator if it satisfies

$$A^*A - AA^* = [A^*, A] = D_A \geq 0.$$

An operator A on Hilbert space H is called a semi-hyponormal operator if it satisfies

$$(A^*A)^{1/2} - (AA^*)^{1/2} = D_{1/2} \geq 0.$$

Note: If $A = UP$ is its polar decomposition, it is obvious that $(A^*A^{1/2} - (AA^*)^{1/2} = P - UPU^*$. Thus, for semi-hyponormality the polar decomposition is useful.

Example 2.1. Example of hyponormal operator:

Let $\mathbb{R}^1$ be real line, $\mathcal{B}$ be the family of all Borel subsets of $\mathbb{R}^1$ and m be Lebesgue measure on $\mathcal{B}$. Denote $Q = (\mathbb{R}^1, \mathcal{B}, m)$. Let $\mathfrak{D}$ be a Hilbert space, $L^2(Q, \mathfrak{D})$ be Hilbert space of all measurable functions $f(\cdot)$ with values in $\mathfrak{D}$ defined on $\mathbb{R}^1$ such that $\int \|f(t)\|^2 dm < \infty$. Define operator P by

$$(Pf)(t) = s - \lim_{\varepsilon \to 0+} \frac{1}{2\pi i} \int_{-\infty}^{+\infty} \frac{f(s)ds}{t - (s + i\varepsilon)}.$$

Let $\triangle$ be a bounded closed subset of $\mathbb{R}^1$ with $m(\triangle) > 0$, and let ν be a singular measure on $\mathcal{B}_\triangle$ and $\mu = m + \nu$, i.e. there is a measurable set $F \subset \triangle$ such that $m(F) = 0, \nu(\triangle \backslash F) = 0$. Let $R(\cdot)$ be a measurable function of projection value with $R(\cdot) \in B(\mathfrak{D})$. Denote $\hat{H} = L^2(Q_\triangle, \mathfrak{D}, R(\cdot)) = \{f(\cdot) \mid f(\cdot) \text{ is a measurable function defined on } \triangle \text{ with values in } \mathfrak{D} \text{ such that } f(t) \in R(t)\mathfrak{D} \text{ and } \int_\triangle \|f(t)\|^2 d\mu < \infty\}$, where $Q_\triangle = (\triangle, \mathcal{B}_\triangle, \mu)$.

Define

$$(\hat{u}f)(t) = tf(t), \ t \in \triangle.$$

Let $\alpha(\cdot)$ and $\beta(\cdot)$ be uniformly bounded measurable operator-valued functions defined on $\triangle$ such that

$$\alpha(\cdot)R(\cdot) = R(\cdot)\alpha(\cdot) = \alpha(\cdot)$$

$$\beta(\cdot)R(\cdot) = R(\cdot)\beta(\cdot) = \beta(\cdot)$$

$$\beta(\cdot) = \beta(\cdot)^*$$

and $\alpha(t) = 0$ for $t \in F$.

Thus we can define

$$(\hat{T}f)(t) = (t + i((t))f(t) + i\alpha(t)^* P(\alpha(\cdot)f(\cdot)).$$

It is a hyponormal operator on Hilbert space $\hat{H}$.

Example 2.2. Example of semi-hyponormal operators.

If we let $C_1 = \{e^{i\theta} | \theta \in [0, 2\pi)\}$ replace $\mathbb{R}^1$ in Example 2.1 and assume $\beta(\cdot) \geq 0$,

$$(P_r f)(e^{i\theta}) = s - \lim_{r \to 1-0} \frac{1}{2\pi i} \int_{C_1} \frac{f(\xi)d\xi}{\xi - re^{i\theta}}$$

and define

$$(\hat{T}f)(e^{i\theta}) = e^{i\theta}[\alpha(e^{i\theta})^* P_r(\alpha(\cdot)f(\cdot))$$
$$+\beta(e^{i\theta})f(e^{i\theta})].$$

then $\hat{T}$ is a semi-hyponormal operator on Hilbert space $\hat{H}$.

The above example is the models of hyponormal and semi-hyponormal operators, respectively. In fact, we have

Theorem 2.1[26] [27] [28]. If operator T on a separable Hilbert space H is a hyponormal operator, then T is unitarily equivalent to $\hat{T}$ in Example 2.1.

If T is a semi-hyponormal operator, then T is unitarily equivalent to $\hat{T}$ in Example 2.2.

For hyponormal and semi-hyponormal operators the corresponding symbol operators are introduced as

Definition 2.2. Let $T = X + iY$ be a hyponormal operator on Hilbert space H, where X, Y are self-adjoint operators. Then the limits

$$s - \lim_{t \to \pm\infty} e^{itX} T e^{-itX} = T_\pm$$

exist. We say that $T_\pm$ are symbol operators of T, and $T_{(k)} = kT_+ + (1 - k)T_-$ $(0 \leq k \leq 1)$ are general symbol operators of T.

Let $T = UP$ be a semi-hyponormal operator, where U is an isometric operator and $P \geq 0$. Then the limits

$$s - \lim_{n \to \pm\infty} U^{[n]} T U^{[-n]} = T_{[\pm]}$$

exist, where $U^{[n]} = \begin{cases} U^n & n \geq 0 \\ U^{*(n)} & n < 0. \end{cases}$ We say that $T_{[\pm]}$ are (polar) symbol operators of T, and $T_{[k]} = kT_{[+]} + (1 - k)T_{[-]}$ $(o \leq k \leq 1)$ are general (polar) symbol operators of T.

In Example 2.1., hyponormal operator $\hat{T}$ has symbol operator

$$(\hat{T}_{(k)}f)(x) = [x + i(\beta(x) + k\alpha(x)^*\alpha(x))]f(x).$$

In Example 2.2, semi-hyponormal operator $\hat{T}$ has (polar) symbol operator

$$(\hat{T}_{[k]}f)(e^{i\theta}) = (e^{i\theta}[\beta(e^{i\theta}) + k\alpha(e^{i\theta})]f(e^{i\theta}).$$

For the symbol operator we prove the representation theorem.

Theorem 2.2[29]. If T is a hyponormal operator on Hilbert space H, then

$$\sigma(T) = \cup_{0 \le k \le 1}\sigma(T_{(k)}).$$

If T is semi-hyponormal operator, then

$$\sigma(T) = \cup_{0 \le K \le 1}\sigma(T_{[k]}).$$

Theorem2.3[10]. Let T be hyponormal on Hilbert space H. If $x_0, x_1, \cdots, x_n \in H$ such that for all $k \in [0,1]$

$$\sum_{i=0}^{n} T_{(k)}^i x_i = 0$$

then it holds true that $\sum_{i=0}^{n} T^i x_i = 0$.

For symbol operator there is an invariant theorem.

Theorem 2.4[5]. Let $T = X + iY$ be a complete nonnormal hyponormal operator on Hilbert space H. If $\varphi \in \beta(\sigma(X))$, where $\beta(E) = \{\varphi(t)|\varphi'(t) > 0 \text{ on } E \text{ and } \varphi''(t) \in L'(E)\}$, then

$$s - \lim_{t \to \pm\infty} e^{i+\varphi(x))}Te^{-it\varphi(x)} = T_{\pm}.$$

Likewise, for semi-hyponormal operator we have

Theorem 2.5[5]. Let $T = UP$ be a complete nonnormal semi-hyponormal operator on Hilbert space, where U is a unitary operator and $P \ge 0$. If

$$\varphi(e^{i\theta}) = e^{ig(\theta)}, \ g(\theta) \in \beta([0, 2\pi)).$$

then

$$\lim_{b] \to \pm\infty} \varphi(U)^{[n]}T\varphi(U)^{[-n]} = T_{[\pm]}.$$

For hyponormal and semi-hyponormal operators we study some spectral functional transformation.

Definition 2.3. Let $E \subset \mathbb{R}^1$ be a bounded closed subset. Denote $M(E)$ the set of all strict monotone increasing continuous functions on E.

For $\varphi \in M(E)$, we define singular integral operator K_φ on $L^2(E)$ by

$$(K_\varphi f)(x) = s - \lim_{\varepsilon \to 0+} \frac{1}{2\pi} \int \frac{\varphi(x) - \varphi(y)}{x - (y + i\varepsilon)} f(y) dy$$

Thus we denote $S(E) = \{\varphi \in M)E)|K_\varphi \geq 0\}$. This is an important class of functions for hyponormal operator.

For $E \subset C_1 = \{e^{i\theta}|\theta \in (0, 2\pi)\}$, denote $M(E) = \{\varphi(\cdot)|\varphi(e^{i\theta}) = e^{ig(\theta)} : g(\cdot) \in M_0(\overset{\circ}{E})\}$, where $\overset{\circ}{E} = \{\theta \in [0.2\pi)|e^{i\theta} \in E\}$.

For $\varphi \in M(E)$, we define K_φ by

$$(K_\varphi f)(e^{i\theta}) = s - \lim_{r \to 1^-} \frac{1}{2\pi i} \int \frac{1 - \varphi(e^{i\theta})\varphi(e^{i\eta})}{1 - re^{i\theta})e^{-i\eta}} d\eta$$

and $S(E) = \{\varphi \in M(E) : K_\varphi \geq 0\}$. Finally, we set

$$L(E) = \{\varphi \in M(E)| \sup_{\substack{x_1 \neq x_2 \\ x_1, x_2 \in E}} \frac{\varphi(x_1) - \varphi(x_2)}{x_1 - x_2} = M_\varphi < \infty,$$

$$\inf_{\substack{x_1 \neq x_2 \\ x_1, x_2 \in E}} \frac{\varphi(x_1) - \varphi(x_2)}{x_1 - x_2} = m_\varphi > 0\}$$

Definition 2.4. Let $T = X + iY$ be an operator on Hilbert space H, where X, Y are self-adjoint operators, and let $\varphi.\psi \in M(\mathbb{R}^1)$. We define

$$\tau_{\varphi.\psi}(T) = \varphi(X) + i\psi(Y).$$

and

$$\tau_{\varphi.\psi}(z) = \varphi(x) + i\psi(y)$$

for complex number $z = x + iy$.

Let $T = UP$ be an operator on Hilbert space H, where U is a unitary operator and $P \geq 0$, and $\varphi \in M(C_1), \psi \in M(\mathbb{R}^+)$. We define

$$U_{\varphi,\psi}(T) = \varphi(U)\psi(P).$$

and

$$\nu_{\varphi\psi}(z) = \varphi(e^{i\theta})\psi(r)$$

for complex number $z = re^{i\theta}$.

Theorem 2.6 [29][2][7][31][32]. Let $T = X + iY$ be a hyponormal operator on Hilbert space H, and φ, ψ be scale functions on $\mathbb{R}$ such that

1° $\varphi_1 \in M(\sigma(X))$, $\varphi_2 \in S(\sigma(Y))$ or
2° $\varphi_1 \in S(\sigma(X))$, $\varphi_2 \in M(\sigma(Y))$ or
3° $\varphi_1 \in L(\sigma(X))$, $\varphi_2 \in L(\sigma(Y))$ and

$$(M_{\varphi_1} - m_{\varphi_1})(M_{\varphi_2} - m_{\varphi_2}) < 16 m_{\varphi_1} m_{\varphi_2}.$$

Then $\sigma_*(\tau_{\varphi_1,\varphi_2}(T)) = \tau_{\varphi_1,\varphi_2}(\sigma_*(T))$, where $\sigma_*(\cdot)$ denotes $\sigma_a(\cdots), \sigma_r(\cdot)$, or $\sigma(\cdot)$, and

$$\|(T^*T - TT^*)\| \leq \frac{1}{\pi} \int \int_{\sigma(T)} d\varphi_1^{-1}(x) d\varphi_2^{-1}(y).$$

Theorem 2.7[29][2][7][31][32]. Let $T = UP$ be a semi-hyponormal operator, where U is a unitary operator and $P \geq 0$. If $\varphi_1 \in S(\sigma(U))$, $\psi \in M(\mathbb{R}^+)$ such that $\psi(0) = 0$, $\frac{\psi(s)}{s}$ is a monotone decreasing function, then $\nu_{\varphi,\psi}(\sigma_*(T)) = \sigma_*(\nu_{\varphi,\psi}(T))$, where $\sigma_*(\cdot)$ denotes $\sigma_a(\cdot), \sigma_r(\cdot)$, or $\sigma(\cdot)$.

3. Other classes of nonnormal operators

In the theory of nonnormal operators many classes are introduced.

Definition 3.1. Let T be an operator on Hilbert space H.
    If T satisfies

$$\varphi(T^*T) - \varphi(TT^*) = D_\varphi \geq 0$$

we say T is φ-hyponormal operator, where φ is a scale function on $\mathbb{R}^+$.

    For example, if

$$(T^*T)^\alpha - (TT^*)^\alpha \geq 0$$

we call T an α-hyponormal operator.
    If for any scale function $\varphi(T^*T) - \varphi(TT^*) \geq 0$, then we call T a complete hyponormal operator.
    If there is $M > 0$, such that

$$\|(T - z)^* f\| \leq M \|(T - z)f\|$$

for any $z \in \mathbb{C}$, $f \in H$. we call T a M-hyponormal operator.
    If for any $z \in \mathbb{C}$, there is $M(z)$ such that

$$\|(T - z)^* f\| \leq M(z) \|(t - z)f\|,$$

then we call T a dominant operator.

If for any z, there is $M(z)$ such that

$$\|(T-z)^n f\|^2 \leq M(z)\|(T-z)^{2n} f\|\,\|f\|, \ f \in H$$

we call T a power (N)-class operator.

If $T^*(T^*T - TT^*) = (T^*T - TT^*)T$ we call T a θ-class operator.

In this case, $\overline{R(T^*T - TT^*)}$ is invariant for T^* if $T^*|\overline{R(T^*T - TT^*)}$ is M-hyponormal; we say that T is a strong θ-class operator.

For M-hyponormal operator we prove

Theorem 3.1 [16]. An operator T on Hilbert space H is M-hyponormal if and only if for any $f \in H$ it holds true that

$$|(Tf, f)\|^2 \leq \frac{1}{M^2 - 1}([T^*, T]f, f)\|f\|^2 + \|Tf\|^2\|f\|^2.$$

For θ-class operators, we not only improve the important theorem ,but we get many new properties. In particular, we have

Theorem 3.2 [20]. T is a θ-class operator. If there is a polynomial $p(\cdot)$ such that $P(T)$ is a normal operator, then T is normal, too.

To study θ-class operators, we consider the solution of operator equation and get

Theorem 3.3[25]. An operator A on a Hilbert space satisfies $AA^* = \alpha A^*A + \beta A^{*2}$ if and only if one of the following holds:

1^0 $\beta = 0$ and A is normal operator.

2^0 $\beta \neq 0$, $\alpha \neq -1$, such that $|1 - \alpha| = |\beta|$. and $A_\theta = e^{i\theta/2}A$ is a selfadjoint operator.

3^0 $\alpha = -1$, and $|\beta| \geq 2$. and $H = H_1 \oplus H_2 \oplus H_3$ such that A has form $(\,0 \quad \lambda B^*B \quad 0\,) \oplus 0$ where $\lambda = \dfrac{|B|^2 - |\beta|\sqrt{|\beta|^2 - 4}}{2\beta}$.

For a strong θ-class operator we have proved

Theorem 3.4[33]. Let T be a complete nonnormal operator on Hilbert space H. Then T is a strong θ-class operator if and only if there is a Hilbert $\mathfrak{D}$ and a self-adjoint operator Q, and a positive operator P with dense range on $\mathfrak{D}$ such that $\hat{T} = \hat{Q} + V_D\hat{P}$ is an operator on $\hat{H} = \mathfrak{D} \oplus \mathfrak{D} \oplus \cdots$, where $\hat{Q} = Q \oplus Q \oplus \cdots, \hat{P} = P \oplus P \oplus \cdots$, and V_P is shift on $\hat{H}$, such that T is unitary equivalent to $\hat{T}$.

For a complete hyponormal operator we have proved

Theorem 3.5[7]. $T = UP$ is an operator on Hilbert space H, where U is a unitary operator and P is a positive operator with resolvent decomposition $\int_{\Delta}^{m} \lambda dE_{\lambda}$. Then T is complete hyponormal if and only if for any $t > 0$, $m_t = (I - E_t)H$ is an invariant subspace of U.

It is obvious that if T is a hyponormal operator such that T^*T and TT^* commute, then T is complete hyponormal. But the converse is false. We have the following example:

Example 3.1. Let $H = \oplus_{-\infty}^{+\infty} C^2$, and U be shift. P be diag.(P_n).

$$P_n = \begin{cases} C & n \leq 0 \\ D & n > 0. \end{cases}$$

where $C = \begin{pmatrix} 1 & 0 \\ 0 & 0 \end{pmatrix}$, $D = \begin{pmatrix} 2 & 1 \\ 1 & 2 \end{pmatrix}$. Then $T = UP$ is a complete hyponormal operator but TT^* and T^*T do not commute.

For a semi-hyponormal operator we prove two useful inequalities:

Theorem 3.6[8]. Let $T = UP$ be a semi-hyponormal operator on Hilbert space H. Then
1^0 $|(Tf \cdot f)| \leq (Pf \cdot f)$, $f \in H$,
2^0 $\|(T - z)^* f\| \leq (\frac{1}{\sqrt{r}}\|P^{1/2}\| + 1)\|(T - z)f\|$, $f \in H, z = re^{i\theta} \neq 0$.

For 1^0, we discuss the inverse inequality:

Theorem 3.7 [8]. Let $T = UP$ be an operator on Hilbert space. If it satisfies

$$(Pf, f) \leq |(Tf, f)|, \quad f \in H,$$

then there is a θ_0 such that $T = e^{i\theta_0} P$.

For 2^0, we introduce the concept of $C\backslash\delta$-dominant and finite order $C\backslash\delta$-dominant (see §4).

On the spectrum relation. We prove the following results:

Theorem 3.8[35].
1^0. If A and B are quasi-similar subnormal operators, then $\sigma_e(A) = \sigma_e(B)$.
2^0. If A and B are quasi-similar powr-(N) class operators, then $\sigma(A) = \sigma(B)$.

4. The Putnam-Fuglede theorem

It is well known for normal operators N_1, N_2, if $N_1 X = X N_2$, then $N_1^* X = X N_2^*$, and ker $X^\perp$ and $\overline{R(X)}$ reduce N_2 and N_1 to unitarily equivalent operators, respectively. This is Putnam-Fuglede Theorem. We plan to extend this theorem to the case of nonnormal operators. Professor Yan Shaozhong investigated the polar-product operator $A^{*-1}A$ of A and considered the Putanam-Fuglede Theorem which is of form $A \times B = X$.

Theorem 4.1[21][22][23][24]. Let N_1, N_2 be normal operators. If $N_1 X N_2 = X$, then $N_1^* X N_2^* = X$, and $\overline{R(X)}$, $\mathrm{Ker}(X)^\perp$ reduce N_1, N_2 to invertible operators.

Furthermore, we proved the equivalent theorem of the Putnam-Fuglede Theorem:

Theorem 4.2[21][22][23][24]. Let A and B be operators on Hilbert space, and let X be an operator with polar decomposition $X = WR$. Then $AXB = X$ and $A^* X B^* = X$ (or $AX = XB, A^* X = XB^*$) hold if and only if
(1) $\overline{R(X)}$, $\mathrm{Ker}(X)^\perp$ reduce A, B to invertible operators, respectively (or reduce A, B, resp.)
(2) W as a unitary operator from $\mathrm{Ker}(X)^\perp$ onto $\overline{R(X)}$ such that $(B|\mathrm{Ker}(X)^\perp)^{-1}$ (or $B|\mathrm{Ker}(X)^\perp$) and $A|_{\overline{R(X)}}$ are unitarily equivalent, and R commutes with B.

For a nonnormal operator, we get

Theorem 4.3[21][22][23][24]. If A and B^* are hyponormal or semi-hyponormal operators, and X is an operator such that $AXB = X$ (or $AX = XB$), then $A^* X B^* = X$ (or $A^* X = XB^*$), and $\overline{R(X)}$, $\mathrm{Ker}(X)^\perp$ reduce A, B to normal operators, respectively.

And we get the asymptotic form of the Putnam-Fuglede theorem:

Theorem 4.4[11]. Let A and B^* be hyponormal operators and $\{X_n\}$ be a uniformly bounded sequence of operators such that $\|AX_n - X_n B\| \to 0$ (or $\|AX_n - X_n\| \to 0$). Then we have $\|A^* X_n - X_n B^*\| \to 0$ (or $\|A^* X_n B^* - X_n\| \to 0$).

For the extension of Putnam-Fuglede Theorem we introduce the concept of a $C\backslash\delta$-dominant operator:

Definition 4.1.
(1) Let T be an operator on Hilbert space H, and δ be a closed subset of C. If for any $z \in C\backslash\delta$, there is $M(z)$ such that

$$\|(T-z)^* f\| \le M(z)\|(T-z)f\|, \ f \in H$$

we say that T is a $C\backslash\delta$-dominant operator.

(2) If there is k such that

$$M^2(z) \leq C(\mathrm{dis}(z,\delta)^{-k} + 1)$$

where C is a positive constant, we call T a $C\backslash\delta$-k-order dominant operator.

(3) If $\delta = (z_1, \cdots, z_\ell)$ and $k_1, \cdots, k_\ell$ such that

$$M^2(z) \leq C(\Pi_{i=1}^{\ell}|z - z_i|^{-k_i} + 1),$$

where C is a positive constant, we say that T is a finite polar dominant operator, and $z_1, \cdots, z_\ell$ are polar points of T with order $k_1, \cdots, k_\ell$, respectively.

Theorem 4.5[11][21][22][23][24]. Let S be a $C\backslash\{z_0', \cdots, z_m'\}$-finite polar dominant operator and $\mathrm{Ker}(S - z_i')(i = 0, 1, \cdots, m)$ reduce S. And let T be $C\backslash\{z_0, z_1 \cdots, z_m\}$-dominant operator and $\mathrm{Ker}(T - z_i)(i = 0, 1, \cdots, n)$ reduce T. If an operator X satisfies $TXS = S$ (or $TX = XS$), then $T^*XS^* = X$ (or $T^*X = XS^*$), and $\overline{R(X)}$, $\mathrm{Ker}(X)^\perp$ reduce T, S to be a normal operator, respectively.

In the Putnam-Fuglede theorem of restriction of X we have

Theorem 4.6[21][22][23][24]. If A and B^* are contraction operator and X is compact operator such that $AXB = X$, then $\mathrm{Ker}(X)^\perp$ and $\overline{R(X)}$ reduce B and A to a unitary operator, respectively, and we have $A^*XB^* = X$.

For the extension form of Putnam-Fuglede Theorem we have

Theorem 4.7[11][21][22][23][24]. Let A, B^*, D^* be hyponormal operators and $BD = DB$, $BD^* = D^*B$. If X is an injective operator such that $AXB = XD$, then B and D are normal operators, and $A^*XB^* = XD^*$.

Theorem 4.8[11]. Let $(N_1, \cdots, N_n)$ and $(M_1, \cdots, M_n)$ be commuting normal operator groups. If $N_1, \cdots, N_n$ can diagonablize simultaneously or $N_k = P_k(N)(K = 1, \cdots, n)$ where P_k are polynomials, then

$$\left\|\sum_{i=1}^{n} A_i X B_i\right\|_2 = \left\|\sum_{i=1}^{n} A_i^* X B_i^*\right\|_2,$$

where $\|\cdot\|_2$ is Hilbert-Schmidt norm.

5. The tuple of operators

The research of the tuple of operators is a new direction in the operator theory. First, we discuss the tuple of hyponormal operators:

Definition 5.1. Let $T = (T_1, \cdots, T_n)$ be a tuple of operators. If $T_i T_j = T_j T_i (i \neq j)$, we say that T is a commuting tuple. If $T_i T_j = T_j T_i (i \neq j), T_i T_j^* = T_j^* T_i (i \neq j)$, we say that T is a double commuting tuple.

For a double commuting tuple of hyponormal operators we find that many properties of a singular hyponormal operator can be extended to the case of a double commuting tuple of hyponormal operators. For example. we prove

Theorem 5.1[1][6]**.** Let $T = (T_1, \cdots, T_n)$ be double commuting tuple of hyponormal operators. Then

$$S_P(T) = \bigcup_{\substack{0 \leq k_i \leq 1 \\ i=1,\cdots,n}} S_P(T_{(k_1)}, \cdots, T_{(k_n)})$$

where $S_P(T_1, \cdots T_n)$ denotes the Taylor's spectrum of $(T_1, \cdots, T_n)$.

Theorem 5.2[1][6]**.** Let $T = (T_1, \cdots, T_n)$ be a double commuting tuple of hyponormal operators. If $\varphi = (\varphi_1, \cdots, \varphi_n)$ and $\psi = (\psi_1, \cdots, \psi_n)$ be tuples of scale functions, and $\varphi_i \in S(\mathbb{R})$ then

$$S_P(\tau_{\varphi_1, \psi_1}(T_1) \cdots \tau_{\varphi_n, \psi_n}(T_n)) = \{(\tau_{\varphi_1 \psi_1}(z_1) \cdots (\tau_{\varphi_n \psi_n}(z_1))|(z_1, \cdots, z_n) \in S_P(T)\}.$$

Theorem 5.3[12][14]**.** Let $T = (T_1, \cdots, T_n)$ be a double commuting tuple of hyponormal operators. $P(\cdot, \cdots, \cdot)$ be a polynomial of n-variants. If $\sigma(P(T_1, \cdots, T_n)) = \{0\}$, then $P(T_1, \cdots, T_n) = 0$.

In the theory of a tuple of operators the corresponding right and left multiplication operators are the important contents.

Definition 5.2. Let $A = (A_1, \cdots, A_n)$ be a tuple of operators on Hilbert space H. We denote $R_A = (R_{A_1}, \cdots, R_{A_n})$ where R_{A_i} is a right multiplication operator on $R(H)$ defined by $R_{A_i}(X) = A_i X$. Likewise, we denote $L_A = (L_{A_1}, \cdots, L_{A_n})$, where L_A is a left multiplication operator on $B(H)$ defined by $L_A : (X) = X A_i$.

And we denote $\triangle_{AB}(X) = \sum_{i=1}^{n} A_i X B_i$, we call $\triangle_{AB}$ the elementary operator corresponding to tuples A and B.

For the spectrum of (L_A, R_B) we prove

Theorem 5.4[13][15][18]. Let $A = (A_1, \cdots, A_n)$ and $B = (B_1, \cdots, B_n)$ be commuting tuples of operators on Hilbert space H. Then we have

$$S_P(L_A, R_B) = SP(\times S_P(B)).$$
$$S_{P_e}(L_A, R_B) = SP(A) \times S_{P_e}(B) \cup S_{P_e}(A) \times S_P(B).$$

For the elementary operator Δ_{AB} we have

Theorem 5.5[13][15][18]. Let $A = (A_1, \cdots, A_n)$ and $B = (B_1, \cdots, B_n)$ be two commuting tuples of operators on Hilbert space H. Then

$$\sigma(\Delta) = S_P(A) \circ S_P(B)$$
$$= \left\{ \sum_{i=1}^{n} \alpha_i \beta_i | \alpha = (\alpha_1, \cdots, \alpha_n) \in S_P(A), \beta = (\beta_1, \cdots, \beta_n) \in S_P(B) \right\}.$$
$$\sigma_e(\Delta) = S_P(A) \circ S_{P_e}(B) \cup S_{P_e}(A) \circ S_P(B).$$

For the tuple of operators we introduce the joint maximal range as

Definition 5.2. Let $A = (A_1, \cdots, A_n)$ be a tuple of operators on Hilbert space H. Denote

$$W_N(A) = \{(\lambda_{ij})_{ij=1}^n | \exists x_m \in H, \|x_m\| = 1,$$

$$\|A_i x_m\| \to \|A_i\| \ (m \to \infty), \text{ and}$$

$$\left(\frac{A_i x_m}{\|A - i\|}, \frac{A_j x_m}{\|A_j\|} \right) \to (\lambda_{ij}) : ij = 1, \cdots, n\}$$

$$W_N(A)^{-1} = \{(\lambda_{ij})_{ij=1}^n | \exists x_m^i \in H, \|\|x_m^i\|\| = 1,$$

$$\|A_i x_m^i\| \to \|A_i\|, \left\| \frac{A_i x_m^i}{\|A_i\|} - \frac{A_j x_m^j}{\|A_j\|} \right\|$$

$$\to 0 \ (m \to \infty), \text{ and} (x_m^i, x_m^j) \to \lambda_{ij}, ij = 1, \cdots, n\}.$$

We call $W_N(A)$. $W_N^{-1}(A)$ are the joint maximal range and the joint inverse maximal range of A.

We proved

Theorem 5.6[13][15][18]. $W_N^{-1}(A) = W_N(A^*))$.

Theorem 5.7[13][15][18]. Let $A = (A_1, \cdots, A_n)$ and $B = (B_1, \cdots, B_n)$ be two tuples of operators on Hilbert space H. Then $\|\Delta_{AB}\| = \sum_{i=1}^{n} \|A_1\| \, \|B_1\|$ if and only if $W_n^{-1}(A) \cap W_N(B) \neq \emptyset$.

For the closedness of the range of the elementary operator Δ_{AB}, we have

Theorem 5.8[13][15][18]. Let $A = (A_1, \cdots, A_n)$ and $B = (B_1 \cdots, B_n)$ be two commuting tuples of normal operators on Hilbert space such that $(S_P(A), S_P(B))$ has property (D), i.e. for any $\alpha \in S_P(A)$, the set $\{\beta | \beta \in S_P(B), \alpha \circ \beta = 0\}$ is a finite set only and for any $\beta \in S_P(B)$, the set $\{\alpha | \alpha \in S_P(A), \alpha \circ \beta = 0\}$ is finite, too. Then the $R(\Delta_{AB})$ is closed if and only if $\{\alpha | \alpha \in S_P(A), \exists \beta \in S_P(B)$ such that $\alpha \circ \beta = 0\}$ does not contain the closure point of $S_P(A)$ and $\{\beta | \beta \in S_P(B), \exists \alpha \in S_P(A)$ such that $\alpha \circ \beta = 0\}$ docs not contain the closure point of $S_P(B)$.

For the spectral relation of a tuple we prove

Theorem 5.9[13][15][18]. Let $A = (A_1, \cdots, A_n)$ and $B = (B_1, \cdots, B_n)$ be two tuples of operators on Banach space E such that $A_i B_j A_j = A_j B_j A_i$ and $B_i A_j B_j = B_j A_j B_i$ for all $i, j = 1, \cdots, n$. Then

$$S_P(A_1 B_1, \cdots, A_n B_n) \cup \{(0, \cdots, 0)\} =$$
$$S_P(A_1 B_1, \cdots, A_n B_n) \cup \{(0, \cdots, 0)\}.$$

References

[1] Chen Xiaoman, On the joint spectrum of noncommutiy tuples of self-adjoint operators, Act Math. Sinica, 29(1986), 661–665.

[2] Chen Xiaoman, Spectral Inequility for φ-Quasihyponormal Operator, Kexue Tongbao (Chinese), 28(1983).

[3] Chen Xiaoman, A Basic Problem of Semi-hyponormal Operator, Fudan Xuebao(Chinese), 26(1987).

[4] Chen Xiaoman, Mosaic and Pincus Functions for Hyponormal and Semi-hyponormal Operatorsm, Chin. Ann. of Math., 8A(2)(1987).

[5] Chen Xiaoman, Perturbation Invariance for Symbol Operators, Chin. Ann. of Math., 10A(3)(1989).

[6] Chen Xiaoman, Huang Chaocheng, On the representation of joint spectrum for a crmmuting tuples of nonnormal operators and several properties, Acta. Math. Sinica New Series, 3(1987).

[7] Li Shaokuan, A class of hyponormal operator-complete hyponormal operators, Fudan Jour, 20(1981), 229–232.

[8] Li Shaokuan, An equality of semi-hyponormal operator, Kexue Tongbao, 28(1983), 961–963.

[9] Li Shaokuan, Quasi-similiry of power (N)-class of operators, Chin. Math. Annal, 5(A)(1984), 165–168.

[10] Li Shaokuan, On the spectrum of hyponormal operators, Intergral Equ. Operator Theory, 11(1988), 536–556.

[11] Li Shaokuan, On some problems of elementary operators (II), Math. Acta Sinica, 34(1991), 365–371.

[12] Li Shaokuan, On the guess of the tuple of hyponormal operators, Kexue Tongbao, 1991, 844–886.

[13] Li Shaokuan, On the commuting properties of the Taylor spectrum, Kexue Tongbao, 1992, 681–684.

[14] Li Shaokuan, On the hyponormal tuples of operators, Chin. Math. Annal, 14A(1993), 11–15.

[15] Li Shaokuan, Jieyao, On the Taylor's spectrum of $2n$-tuples (L_A, R_B), Sci. Sinica, 1988, 919–928.

[16] Li Shaokuan, Chen xiaoman, On the M-hyponormal operators, Fudan Jour, 28(1989), 141–147.

[17] Li Shaokuan, Chen xiaoman, On the mapping theorem of spectrum of nonnormal operators, Fudan Jour, 23(1984), 149–156.

[18] Li Shaokuan, Guzexin, On the some problems of elementary operators, Chin. Math Anna, 9(A)(1988), 188–202.

[19] Li Shaokuan, Guzexin, On the spectral representation of the elementary operators, Acta. Math. Sinica, (A)(1984), 775–783.

[20] Yan Shaozhong, On θ-class of operators, Sci. Slnica, 8(1984), 677–684.

[21] Yan Shaozhong, Li Shaokuan, On the operator equation $A \times B = X$, Math. Acta. Sinica., 26(1983), 597–603.

[22] Yan Shaozhong, Li Shaokuan, On the $P - F$ theorem of nonnormal operators, Chin. Math. Ann., 4(B)(1983), 51–56.

[23] Yan Shaozhon,g Li Shaokuan, On the Putanm-Fuqlede theorem, Sci. Sinica, (A)(1984), 775–783.

[24] Yan Shaozhong, Li Shaokuan, On P_F theorem, Kexue Tongbao, 30(1985), 810–815.

[25] Yan Shaozhong, Zu jiezhong, On the operator equation $\lambda A^2 + \mu A^{*2} = \alpha A^* A + \beta A A^*$, Sci. Sinica, 11(1987), 1139–1146.

[26] Xia Daoxing, On nonnormal operators (I), Math. Acta. Sinica, 12(1962), 216–228.

[27] Xia Daoxing, On nonnormal operators (II), Math. Acta Sinica, 21(1978), 187–189.

[28] Xia Daoxing, On the non-normal opertors–semi-hyponormal operators, Sci. Sinica, 23(1980), 700–713.

[29] Xia Daoxing, On the spectrum of hyponormal and semi-hyponormal operators, Jour. Operator Theory, 5(1981), 257–266.

[30] Xia Daoxing, Spectral mapping of hyponormal and semi-hyponormal operators, Jour. Math. Analysis App., 79(1981), 409–427.

[31] Xia Daoxing, Li Shaokuan, On the functional transformation of hyponormal operators, Kexue Dongbao, 1980, 625–627.

[32] Xia Daoxing, Li Shaokuan, On the functional fransformation of semi-hyponormal operators, Chin. Math. Ann., 1(1980), 501–504.

[33] Yang Lemeng, The structure and properties of strong θ-class operators, Chin. Math. Ann., 10A(1989), 210–220.

[34] Yang Lemeng, Equality of essential spectra of quasisimilar subnormal operators, Integral Equation and Operator Theory, 13(1990), 433–441.

Some Aspects of Nonlinear Operators and Critical Point Theory

Shujie Li

Institute of Mathematics, Academia Sinica, Beijing

Abstract. In this paper we sum up the following results: An open problem stated by L. Nirenberg, Palacs-Smale condition and coercivity, some new existence theorems of critical point, applications to nonlinear differential equations.

1. An open problem of L. Nirenberg

In 1974 L. Nirenberg [N] stated the following open problem: Suppose T is a continuous expanding nonlinear operator from Hilbert space H into H, T maps a neighborhood of the origin onto a neighborhood of the origin. Does T maps H onto H? In 1982, K.C. Chang and Shujie Li [CL] answered positively the problem when T is differentiable.

Theorem 1.1[CL]. Suppose X, Y are Banach spaces, also suppose T is an expanding map from X into Y, T is Frechet-differentiable in X and $\forall x_0 \in X$ we have $\overline{\lim_{x \to x_0}} \|T'(x) - T'(x_0)\| < 1$, $T(0) = 0$ and T maps a neighborhood of the origin of X onto a neighborhood of the origin of Y. Then $TX = Y$.

Theorem 1.2[CL]. Suppose T is an expanding map from H into H and $T \in C^1$ is a potential operator. Then T maps H onto H.

If we consider αT instead of T for $\alpha > 0$, without loss of generality we may assume that $\|Tx - Ty\| \geq \alpha \|x - y\|$.

In Theorem 1.1 if T satisfies $\|Tx - Ty\| \geq \alpha \|x - y\|$ $\forall x, y \in X$ where $\alpha > 0$, and if $\overline{\lim_{x \to x_0}} \|T'(x) - T'(x_0)\| < \alpha, \forall x_0 \in X$. Then $TX = Y$.

In 1984, J. Morel and H. Steinlein [MS] gave an example in l^1. They constructed a continuous mapping T from l^1 into l^1 with

(1) $T(l^1) \neq l^1$.

1991 Mathematics Subject Classification: 58E05, 49J35, 47H15

(2) $\|Tx - Ty\| \geq \frac{1}{5}\|x - y\|$ for all $x, y \in l^1$.

(3) $Tx = x$ for $\|x\| \leq 1$.

(4) $Tx = \sigma x$ for $\|x\| \geq 2$

where σ is the shift operator, $\sigma : (x_1, x_2, \cdots) \to (0, x_1, x_2, \cdots)$. This example shows that if we consider the problem in l^1 and remove the differentiability on T, the answer is negative.

It remains open in the case where X is a Hilbert space or even a reflexive Banach space. I suspect the answer is negative if T is only continuous.

2. Palais-Smale condition and coercivity

Let X be a real Banach space. A Gateaux differentiable functions $f : X \to R$ satisfies the Palais-Smale condition, (PS) in short, if every sequence (x_n), such that $f(x_n)$ is bounded and $f'(x_n) \to 0$, contains a convergent subsequence. If we replace $f(x_n)$ being bounded by $f(x_n) \to c$, then we say f satisfies $(PS)_c$. A function $f : X \to R$ is coercive if $f(x) \to +\infty$ as $\|x\| \to \infty$.

By using a gradient flow Shujie Li [L1] and J.Q. Liu [Liu1] proved the following

Theorem 2.1. For a C^1 functional bounded from below on a Banach space, the Palais-Smale condition implies coercivity.

Shi Shu-zhong and K.C. Chang [SC] extended this result to nonsmooth functions. [SC] also included some interesting results such as the extension of Minimax Theorem due to Von Neuman, Sion and Ky Fan.

By using Ekeland's variational principle, L. Caklovic, Shujie Li and M. Willem [CLW] considered the coercivity of $|f|$ when f is not bounded from below.

Theorem 2.2[CLM]. Let X be a Banach space and let $f : X \to R$ be a Gateaux differentiable continuous function satisfying the (PS). If there exists $c \in R$ such that $f^{-1}(c)$ is bounded, then $|f|$ is coercive.

In [BN], H. Brezis and L. Nirenberg defined a number

$$\alpha = \lim_{\|x\| \to \infty} \inf f(x).$$

They proved that if α is finite then there exists a sequence (x_n) in X such that $\|x_n\| \to \infty$, $f(x_n) \to \alpha$ and $f'(x_n) \to 0$. This result implies Theorem 2.1.

Costa and Silva [CS] also gave a proof of Theorem 2.1 by using Ekeland's variational principle.

Recently Shujie Li and M. Willem [LW] generalized Theorem 2.1 to a more general case.

Let X be a real Banach space with a direct sum decomposition $X = X^1 \oplus X^2$. Consider the two sequences of subspaces

$$X^1_0 \subset X^1_1 \subset \cdots \subset X^1, \qquad X^2_0 \subset X^2_1 \subset \cdots \subset X^2$$

such that

$$X^j = \overline{\cup_{n \in N} X^j_n} \qquad j = 1, 2.$$

For every multi-index $\alpha = (\alpha_1, \alpha_2) \in N^2$, we denote by X_α the space $X^1_{\alpha_1} \oplus X^2_{\alpha_2}$. Let us recall that

$$\alpha \leq \beta \Longleftrightarrow \alpha_1 \leq \beta_1, \alpha_2 \leq \beta_2.$$

A sequence $(\alpha_n) \subset N^2$ is admissible if, for every $\alpha \in N^2$, there is $m \in N$ such that

$$n \geq m \Longrightarrow \alpha_n \geq \alpha.$$

We denote by f_α the functional f restricted to X_α.

Definition 2.3[LW]. Let $c \in R$ and $f \in C^1(X, R)$. The functional f satisfies the $(PS)^*_c$ condition if every sequence (x_{α_n}) such that α_n is admissible and

$$x_{\alpha_n} \in X_{\alpha_n}, \quad f(x_{\alpha_n}) \to c, \quad f'_{\alpha_n}(x_{\alpha_n}) \to 0$$

contains a subsequence which converges to a critical point of f. The functional f satisfies $(PS)^*$ if every sequence (x_{α_n}) such that (α_n) is admissible and

$$x_{\alpha_n} \in X_{\alpha_n}, \quad \sup f(x_{\alpha_n}) < \infty, \quad f'_{\alpha_n}(x_{\alpha_n}) \to 0$$

contains a subsequence which converges to a critical point of f. When $\dim X^2 < \infty$, the $(PS)^*$ condition was introduced in [LL1].

Theorem 2.4[LW]. Let $f \in C^1(X, R)$ be bounded below. If $(PS)^*_c$ holds for all $c \in R$, then f is coercive.

3. Some existence theorems of critical point

In critical point theory the typical questions are whether there exists at least one critical point of f, or whether there exists a nontrivial critical point if the trivial critical point is known. In particular, if the trivial critical point is degenerate then the existence of nontrivial critical point usual by needs more additional conditions. This is because from the point of view of Morse theory we can say almost nothing about the critical group $C_k(f, 0)$ in the range $k \in [\mu_0, \mu_0 + \nu_0]$ where μ_0 is the Morse index of 0 and ν_0 is the nullity of $f''(0)$.

First let us recall the definition of critical group. Let $x_0 \in K = \{x \in X, f'(x) = 0\}$ be an isolated critical point with $c = f(x_0)$. Denote $f^c = \{x \in X, f(x) \leq c\}$. Then the group $C_*(f, x_0) = H_*(f^c, f^c \backslash \{x_0\}, R)$ is called the critical group of f at x_0 with respect to $H_*(-, R)$, here $H_*(-, R)$ denoting singular homology with coefficients in a commutative ring R.

In order to find a nontrivial critical point we use the following local linking condition, see [LW]. Let X be a Banach space with a direct sum decomposition $X = X^1 \oplus X^2$. The functional f has a local linking at 0 with respect to (X^1, X^2) if, for some $\rho > 0$,

$$f(x) \geq 0 \quad x \in X^1 \quad \|x\| \leq \rho \tag{3.1}$$
$$f(x) \leq 0 \quad x \in X^2 \quad \|x\| \leq \rho. \tag{3.2}$$

It is easy to see that when 0 is a nondegenerate critical point of a functional of class C^2 defined on a Hilbert space and $f(0) = 0$, f has a local linking at 0.

The condition of local linking was used in [LL$_1$] under the stronger assumptions

$$f(x) \geq \beta > 0 \quad x \in X^1 \quad \|x\| = \rho \tag{3.3}$$
$$\dim X^2 < \infty. \tag{3.4}$$

In [BN], Brezis and Nirenberg first used this condition without (3.3).

The following theorem is related to a result of K.C. Chang [C1] contained in [LL1].

Theorem 3.5. Suppose that $f \in C^1(X, R)$ is bounded below, satisfies (3.1)-(3.4) and (PS), $f(0) = 0$ is not the minimal value of f. Then f has at least three distinct critical points.

In [BN], Brezis and Nirenberg proved that Theorem 3.5 still holds without condition (3.3).

Recently, Shujie Li and M. Willem pointed out that Theorem 3.5 still holds without conditions (3.3) and (3.4).

Theorem 3.6[LL1][LW]. Suppose that $f \in C^1(X, R)$ satisfies (3.1), (3.2) and (PS)*. For every $m \in N$ $f(x) \to -\infty$ $\|x\| \to \infty$ $x \in X_m^1 \oplus X^2$, then f has at least one nontrivial critical point.

In [LL2], Shujie Li and J.Q. Liu gave an existence theorem of nontrivial critical point for asymptotically quadratic function.

Theorem 3.7. Suppose that $f \in C^1(X, R)$ satisfies (3.1)–(3.4) and

(A$_1$) The gradient of f has the form

$$f'(x) = Ax + B(x)$$

where A is a bounded self-adjoint operator, 0 is not in the essential spectrum of A, and B is a nonlinear compact mapping.

(A_2) There exists a compact self-adjoint operator B_∞ such that

$$B(x) = B_\infty x + o(\|x\|) \qquad \|x\| \to \infty.$$

(A_3) $A + B_\infty$ is invertible.

Let X^- be the negative spectral space of $A + B_\infty$, $m = \dim X^-$, $n = \dim X^2$. Then f has at least one nontrivial critical point, provided $n \neq m$.

Let P_α denote the orthogonal projector from X onto X_α, and denote by $M^-(L)$ the Morse index of self-adjoint operator L. The following result is an extension of Theorem 3.7.

Theorem 3.8[LW]. Suppose that $f \in C^1(X, R)$ satisfies (3.1), (3.2) (A_1), (A_2), (A_3) and
(A_4) For infinitely many multi-indices $\alpha = (n, n)$,

$$M^-((A + P_\alpha B_\infty)|_{X_\alpha}) \neq \dim X_n^2.$$

Then f has at least one nontrivial critical point.

One can say something about the critical group $C_k(f, 0)$ if the local linking condition holds. The following theorem is due to J.Q. Liu [Liu2].

Theorem 3.9. Suppose that X splits as $X^1 \oplus X^2$ with

$$f(x) > 0 \quad \forall x \in X^1 \quad 0 < \|x\| \leq \rho$$

$$f(x) \leq 0 \quad \forall x \in X^2 \quad \|x\| \leq \rho.$$

If 0 is the unique critical point of f in $B(0, \rho)$, then $C_k(f, 0) \neq 0$ for $k = \dim X^2$.

An extension of Theorem 3.9 may be found in [LS1].

Recently, T. Bartsch and Shujie Li [BL] introduced an angle condition which allows further computations of critical groups and which can be verified in applications.

We shall assume the following:

(A_0):
$$\begin{cases} f \text{ has an isolated critical point 0 and is of class } C^2 \\ \text{near } 0, 0 \text{ is isolated in the spectrum of} A_0 = f''(0) \text{and} \\ \dim \text{Ker} A_0 < \infty. \end{cases}$$

We write V_0 is the kernel of A_0, and $W_0 = V_0^\perp$. Let ν_0 denote the nullity of A_0 and μ_0 denote the Morse index of 0.

Theorem 3.10[BL]. Suppose that f satisfies (A_0).

(a) If f satisfies the following angle condition (AC^+): There exist $\varepsilon > 0$ and $\theta \in (0, \frac{\pi}{2})$ such that

$$\langle f'(x), v \rangle \geq 0 \text{ for any } x = v + w \in X = V_0 \oplus W_0$$

with $\|x\| \leq \varepsilon$ and $\|w\| \leq \|x\| \cdot \sin\theta$, then $C_k(f, 0) \cong \delta_{k\mu_0} R$.

(b) If f satisfies the following angle condition (AC^-): There exists $\varepsilon > 0$ and $\theta \in (0, \frac{\pi}{2})$ such that

$$\langle f'(x), v \rangle \leq 0 \text{ for any } x = v + w \text{ with } \|x\| \leq \varepsilon$$

and $\|w\| \leq \|x\| \cdot \sin\theta$, then $C_k(f, 0) \cong \delta_{k\mu_0 + \nu_0} R$.

In [BL], Bartsch and Shujie Li introduced the critical group at infinity: $C_k(f, \infty) = H_k(X, f^a)$ if the set of critical values of f is bounded from below by $a \in R$. They show the following

Theorem 3.11. Suppose f satisfies $(PS)_c$ for every $c \in R$, 0 is an isolated critical point of f. If there exists some $k \geq 0$ such that $C_k(f, \infty) \neq C_k(f, 0)$, then f has a nontrivial critical point.

$(PS)_c$ can be replaced by the deformation condition $(D)_c$.

Thus one is interested in computing $C_*(f, \infty)$. The following theorem is an analog of Theorem 3.9 at infinity.

Theorem 3.12[BL]. Suppose X splits as $X = X^+ \oplus X^-$ such that f is bounded from below on X^+ and $f(x) \to -\infty$ for $x \in X^-$ as $\|x\| \to \infty$. Then $C_k(f, \infty) \neq 0$ for $k = \text{Im} X^-$.

One can get more information about $C_*(f, \infty)$ if f is asymptotically quadratic. More precisely, we need the following conditions.

$(A\infty)$

$$\begin{cases} f(x) = \frac{1}{2}\langle Ax, x \rangle + g(x) \text{ where } A : X \to X \text{ is a self-adjoint} \\ \text{linear operator such that } 0 \text{ is isolated in the spectrum of } A. \\ \text{The map } g \in C^1(X, R) \text{ is of class } C^2 \text{ in a neighborhood of} \\ \text{infinity and satisfies } g''(x) \to 0 \text{ as } \|x\| \to \infty. \text{ Moreover, } g \\ \text{and } g' \text{ map bounded sets to bounded sets. Finally, the} \\ \text{critical values of } f \text{ are bounded below and } f \text{ satisfies } (PS)_c \\ \text{for } c << 0 \text{ (or } (D)_c \text{ for } c << 0). \end{cases}$$

As a consequence we have $g(x) = o(\|x\|^2)$ and $g'(x) = o(\|x\|)$ as $\|x\| \to \infty$. Thus we may say that $A = f''(\infty)$ and f'' is continuous at infinity. We set $V = \text{Ker } A$ and $W = V^\perp$. Then W splits as $W = W^+ \oplus W^-$ with $W^\pm$ invariant under A, and $A|_{W^+}$ is positive definite,

$A|_{W^-}$ is negative definite. Let $\mu = \dim W^-$ be the Morse index of f at infinity and $\nu = \dim V$ the nullity of f at infinity. The next theorem is an analogue of Theorem 3.10 at infinity.

Theorem 3.13[BL]. If (A_∞) holds, then

(a) $C_k(f, \infty) \cong \delta_{k\mu} R$ provided f satisfies the following angle condition at infinity:

   (AC_∞^+) There exist $R \geq 0$ and $\theta \in (0, \frac{\pi}{2})$ such that $\langle f'(x), v \rangle \geq 0$ for any $x = v + w \in X = V \oplus W$ with $\|x\| \geq R$ and $\|w\| \leq \|x\| \cdot \sin \theta$.

(b) $C_k(f, \infty) \cong \delta_{k\mu+\nu} R$ provided f satisfies the following angle condition at infinity:

   (AC_∞^-) There exist $R \geq 0$ and $\theta \in (0, \frac{\pi}{2})$ such that $\langle f'(x), v \rangle \leq 0$ for any $x = v + w$ with $\|x\| \geq R$ and $\|w\| \leq \|x\| \cdot \sin \theta$.

Next we state an analog of the Gromall-Meyer Theorem [GM].

Theorem 3.14[BL]. If (A_∞) holds then

$$C_k(f, \infty) = 0 \quad \text{for } k \notin [\mu, \mu + \nu].$$

This is also true if $\mu = \infty$ or $\nu = \infty$. If $\mu < \infty$ and $\nu = 0$ then $C_\mu(f, \infty) \simeq R$, the coefficient ring.

In [BL], some kind of Morse lemma at infinity may be found.

4. Applications to nonlinear differential equations

In this section we give some applications of the abstract theorems stated in Section 3.

First, we give an application of Theorem 3.5 (without conditions (3.3) and (3.4)) to the problem

$$\begin{cases} -\Delta u - a(x)u = \lambda g(u) & \text{in } \Omega \\ u = 0 & \text{on } \partial \Omega \end{cases} \tag{4.1}$$

where $\Omega \subset R^N$ is a bounded domain whose boundary is smooth. We assume that $a \in L^\infty(\Omega)$, g is smooth,

$$g(u) = o(|u|) \quad |u| \to 0$$

$$\overline{\lim} \frac{g(u)}{u} < 0$$

$$G(u) > 0 \quad \text{for some } u \in R$$

where $G(u) = \int_0^u g(s)ds$. If 0 is an eigenvalue of $-\Delta - a$, we assume also that, for some $\delta > 0$, either $G(u) \geq 0$ for $|u| \leq \delta$ or $G(u) \leq 0$ for $|u| \leq \delta$.

Theorem 4.2. Under the above assumptions for every λ sufficiently large there are at least two nontrivial solutions of (4.1).

This theorem is due to Brezis and Nirenberg [BN] for the case $G(u) \geq 0$, to Shujie Li and M. Willem [LW] for the case $G(u) \leq 0$.

Second, we consider the problem

$$\begin{cases} -\Delta u - a(x)u = g(x, u) & \text{in } \Omega \\ u = 0 & \text{on } \partial\Omega. \end{cases} \tag{4.3}$$

We assume

(g_1) $a \in L^\infty(\Omega), g \in C(\overline{\Omega} \times R, R)$.

(g_2) There are constants $a_1, a_2 \geq 0$ such that

$$|g(x, u)| \leq a_1 + a_2|u|^s$$

where $0 \leq s < (N + 2)/(N - 2)$ if $N \geq 3$. If $N = 1$, (g_2) can be dropped. If $N = 2$ we assume that

$$|g(x, u)| \leq a_1 \exp g(u)$$

where $g(u)/u^2 \to 0$ as $|u| \to \infty$.

(g_3) $g(x, u) = o(|u|)$ $|u| \to 0$ uniformly on Ω.

(g_4) There are constants $\mu > 2$ and $R \geq 0$ such that for $|u| \geq R$,

$$0 < \mu G(x, u) \leq u \cdot g(x, u)$$

where $G(x, u) = \int_0^u g(x, s)ds$.

(g_5) For some $\delta > 0$ either

$$G(x, u) \geq 0 \text{ for } |u| \leq \delta \quad x \in \Omega,$$

or

$$G(x, u) \leq 0 \text{ for } |u| \leq \delta \quad x \in \Omega.$$

By using Theorem 3.7 (without conditions (3.3), (3.4)) Shujie Li and M. Willem [LW] proved the following

Theorem 4.4. Suppose that g satisfies (g_1)–(g_4). If 0 is an eigenvalue of $-\Delta - a$ (with Dirichlet boundary condition), assume also (g_5). Then problem (4.3) has at least one nontrivial solution.

Similar researches have also been made for Hamiltonian systems. We offer [L$_2$] [LS$_2$] [LW] for details. These works extend a result in [R$_1$].

Now we turn to the wave equation. Consider the problem

$$\begin{cases} \Box u = u_{tt} - u_{xx} = bu + g(t,x,u) & t \in R \quad 0 < x < \pi \\ u(t,0) = u(t,\pi) = 0 & t \in R \\ u(t + 2\pi, x) = u(t,x) & t \in R \quad 0 < x < \pi. \end{cases} \tag{4.5}$$

Suppose f satisfies the following conditions:

(f_1) There exists an $\varepsilon > 0$ such that

$$(f(t,x,\xi) - f(t,x,\eta))(\xi - \eta) \geq \varepsilon|\xi - \eta|^s \text{ for all } t,x,\xi,\eta_0.$$

(f_2) $g(t,x,\xi) = o(|\xi|)$ uniformly in t,x as $\xi \to 0$.

(f_3) There exists an $\bar{r} > 0$ such that

$$0 < sF(t,x,\xi) \leq \xi f(t,x,\xi) \text{ for all } t,x,\xi \text{ with } |\xi| \geq \bar{r}$$

where $F(t,x,\xi) = \int_0^{\xi} f(t,x,s)ds.$

(f_1)' The mapping $\xi \to f(t,x,\xi)$ is nondecreasing for all t,x,ξ. Denote the spectrum of the wave operator $\Box$ subject to the conditions in (4.5) by $\sigma(\Box)$. The following result may be found in [LS₃].

Theorem 4.6. Suppose that g satisfies (f_1)–(f_3). Then (4.5) has at least one nontrivial (weak) solution in each of the following cases:

(1) $b \notin \sigma(\Box)$ or $b = 0$.

(2) $b \in \sigma(\Box)$ and there exists a $\delta > 0$ such that $G(t,x,\xi) \geq 0$ for all $|\xi| \leq \delta, t \in R$ and $x \in (0,\pi)$.

(3) $b \in \sigma(\Box)$ and there exists a $\delta > 0$ such that $G(t,x,\xi) \leq 0$ for all $|\xi| \leq \delta, t \in R$ and $x \in (0,\pi)$.

(4) If $b = 0$ then condition (f_1) may be replaced by (f_1)'.

Remark. Case (4) is a result of P.Rabinowitz [R2].

Concerning the applications of Theorem 3.8 we still consider the problem (4.3). Let $\mu_1 \leq \mu_2 \leq \cdots \mu_j \leq \cdots$ be the eigenvalues of $-\Delta - a$ and $\mu_0 = -\infty$. We assume that

(g_6) $g(x,u) = g_\infty u + o(|u|) \ |u| \to \infty$, uniformly in Ω and $\mu_k < g_\infty < \mu_{k+1}$.

Theorem 4.7[LW]. Suppose that g satisfies (g_1), (g_3), (g_6) and one of the following conditions:

(a) $\mu_j < 0 < \mu_{j+1}, j \neq k$.

(b) $\mu_j = 0 < \mu_{j+1}, j \neq k$, and for some $\delta > 0$,

$$G(x,u) \geq 0 \text{ for } |u| \leq \delta, x \in \Omega.$$

(c) $\mu_j < 0 = \mu_{j+1}, j \neq k$, and for some $\delta > 0$,

$$G(x, u) \leq 0 \text{ for } |u| \leq \delta, x \in \Omega.$$

Then problem (4.3) has at least one nontrivial solution.

Similar researches for Hamiltonian systems and wave equations may be found in [LL3], [LS3]. All these works extend the results in [AZ1], [AZ2], [C2], [CWL].

Now we shall concentrate on a more difficult case-asymptotically linear Dirichlet problem with resonance both at 0 and at infinity.

Consider the problem

$$\begin{cases} -\Delta u = p(x, u) & \text{in } \Omega \\ u = 0 & \text{on } \partial\Omega \end{cases} \tag{4.8}$$

where Ω is the same as before. We assume that $p \in C^1(\overline{\Omega} \times R, R)$ satisfies $p(x, 0) = 0$ for all $x \in \Omega$, and the following limits exist:

$$a = \lim_{t \to 0} \frac{p(x, t)}{t} \text{ uniformly in } x \in \Omega,$$

$$a_\infty = \lim_{|t| \to \infty} \frac{p(x, t)}{t} \text{ uniformly in } x \in \Omega.$$

Let $q(x, t) = p(x, t) - at, Q(x, t) = \int_0^t q(x, s)ds$. Similarly, we set $q_\infty(x, t) = p(x, t) - a_\infty t, Q_\infty(x, t) = \int_0^t q_\infty(x, s)ds$. We consider the following hypotheses on q:

(q_1) There exist constants $c_1 > 0$ and $r \in (0, 1)$ such that

$$|q_\infty(x, t)| \leq C_1(|t|^r + 1) \text{ for } t \in R \text{ } x \in \Omega.$$

(q_2) There exist constants $c_2 > 0$ and $\alpha > 1$ such that either

$$Q_\infty(x, t) - \frac{1}{2}q_\infty(x, t) \geq C_2(|t|^\alpha - 1) \text{ for } t \in R, x \in \Omega.$$

or

$$\frac{1}{2}q_\infty(x, t) - Q_\infty(x, t) \geq C_2(|t|^\alpha - 1) \text{ for } t \in R, x \in \Omega.$$

(q_3) $\frac{\partial}{\partial t}q_\infty(x, t) \to 0$ as $|t| \to \infty$ uniformly in $x \in \Omega$.
($q_4^\pm$) $\pm Q(x, t) > 0$ for $|t| > 0$ being small, $x \in \Omega$.
Let $0 < \lambda_1 < \lambda_2 \leq \lambda_3 \leq \cdots$ denote the eigenvalues of the Laplacion $-\Delta$ on Ω with 0 boundary condition, counted with multiplicities.

Theorem 4.9[BL]. Suppose that the assumptions (q_1)-(q_3) are satisfied
   (a) If a is not an eigenvalue of $-\Delta$ then (4.8) has at least one nontrivial solution provided $a < \lambda_n < a_\infty$ or $a_\infty < \lambda_n < a$ for some $n \in N$.

(b) If $a = \lambda_n$ is an eigenvalue but (q_4^+) holds in addition then (4.8) has at least one nontrivial solution provided $a_\infty < a$ or $a < \lambda_m < a_\infty$ for some $m > n$.

(c) If $a = \lambda_n$ is an eigenvalue but (q_4^-) holds in addition then (4.8) has at least one nontrivial solution provided $a < a_\infty$ or $a_\infty < \lambda_m < a$ for some $m < n$.

Next we replace some of the assumptions of Theorem 4.9 by the following ones

(q_5) There exists $c_3 > 0$ such that

$$\left|\frac{\partial}{\partial t} q_\infty(x,t)\right| \leq C_3(|t|^{4/(N-2)} + 1) \text{ for } t \in R, x \in \Omega.$$

(q_6) $Q_\infty(x,t) \to \infty$ as $|t| \to \infty$ uniformly in $x \in \Omega$.

(q_7) Q_∞ is bounded above.

(q_8) g_∞ is bounded.

Assumption (q_6) implies the Landesman-Lazer condition but is much simpler to check.

Theorem 4.10 [BL]. Suppose that (q_1), (q_2) and (q_5) are satisfied

(a) If in addition (q_6) holds then (4.8) has at least one nontrivial solution provided $a < \lambda_n \leq a_\infty$ or $a_\infty < \lambda_n < a$ for some $n \in N$.

(b) If in addition (q_7) holds then (4.8) has at least one nontrivial solution provided $a < \lambda_n < a_\infty$ or $a_\infty \leq \lambda_n < a$ for some $n \in N$.

Theorem 4.11 [BL]. If (q_5), (q_6) and (q_8) hold then (4.8) has at least one nontrivial solution provided $a < \lambda_n \leq a_\infty$ or $a_\infty \leq \lambda_n < a$ for some $n \in N$.

For the application of the angle condition we consider the boundary value problem

$$\begin{cases} -u''(x) = -\dfrac{d^2}{dx^2}u = p(x,u) & \text{for } 0 < x < \pi \\ u(0) = u(\pi) = 0. \end{cases} \tag{4.12}$$

We keep the notation and assumptions taken from problem (4.8) but replace Ω by $(0, \pi)$. Moreover, we assume

(q_9) There exists a positive integer l such that $q(x,t) = \dfrac{q^{(2l+1)}(x,0)}{(2l+1)!}t^{2l+1} + o(|t|^{2l+1})$, as $t \to 0$ uniformly in $x \in [0, \pi]$.

Comparing the following result with Theorems 4.9–4.11 we see that here the conditions on a_∞ and a are sharper.

Theorem 4.13 [BL]. Suppose $a_\infty = n^2$ and $a = n_0^2$ for $n, n_0 \in N$. Then there exists a nontrivial solution of (4.12) in each of the following cases:

(i) The hypotheses (q_1), (q_2), (q_7) and (q_9) hold. In addition, either $q^{(2l+1)}(x,0) > 0$ for all $x \in [0, \pi]$ and $n \neq n_0$, or $q^{(2l+1)}(x,0) < 0$ for all $x \in [0, \pi]$ and $n \neq n_0 + 1$.

(ii) The same conditions as in (i) hold with assumptions (q_6) and (q_8) instead of (q_1), (q_2) and (q_7).

(iii) The hypotheses (q_1), (q_2), (q_6) and (q_9) hold. In addition, either $q^{(2l+1)}(x, 0) > 0$ for all $x \in [0, \pi]$ and $n \neq n_0 - 1$ or $g^{(2l+1)}(x, 0) < 0$ for all $x \in [0, \pi]$ and $n \neq n_0$.

References

[AZ1] H. Amann, E. Zehnder, Nontrivial solutions for a class of nonresonance problems and applications to nonlinear diff. eq. Annali scuola norm. Pisa (1980), 539–603.

[AZ2] H. Amann, E. Zehnder, Periodic solutions of asymptotic linear Hamiltonian System, Manuscripta, Math. 32(1980), 149–189.

[BL] T. Bartsch, S.J. Li, Critical point theory for asymptotically quadratic functional and applications to problems with resonance, to appear.

[BN] H. Brezis, L. Nirenberg, Remarks on finding critical points, Comm. Pure Appl. Math. 64(1991), 939–963.

[CLW] L. Caklovic, S.J. Li, M. Willem, A note on Palais-Smale condition and coercivity, Differential Integral Equations 3(1990), 799–800.

[C1] K.C. Chang, Infinite dimensional Morse theory and its applications, Presses de l'Université de Montréal, Montréal (1985).

[C2] K.C. Chang, Solutions of asymptotically linear equations via Morse theory, Comm, Pure Appl. Math 34(1981), 693–712.

[CL] K. C. Chang, S.J. Li, A remark on expanding maps, Proc. Amer. Math. Soc, Vol 85, N4(1982), 583–585.

[CWL] K.C. Chang, S.P. Wu, S.J. Li, Multiple periodic solutions for an asymptotically linear wave equation, Indiana Univ. Math. J. 31(1982), 721–731.

[CS] D.G. Costa, E. Silva, The Palais Smale condition versus coercivity, Nonlinear Anal., Theory, Methods & Appl. 16(1991), 371–381.

[GM] D. Gromoll, W. Meyer, On differentiable functions with isolated critical points. Topology 8(1969), 361–369.

[L1] S.J. Li, An existence theorem on multiple critical point and application in nonlinear P.D.E, Proceeding of the 1982 Changchun Symposium on differential geometry and differential equation, S.S. Chern, Wang Rou-huai, Chi Min-you (1986).

[L2] S.J. Li, Periodic solutions of non-autonomous second order systems with superlinear terms, Differential and Integral equations, Vol 5, N6(1992), 1419–1424.

[LL1] J.Q. Liu, S.J. Li, Some existence theorems on multiple critical points and their applications, Kexue Tongbao Vol.17 (1984).

[LL2] S.J. Li, J.Q. Liu, Nontrivial critical points for asymptotically quadratic function, J. Math. Anal. and Appl. Vol 165, N2(1992), 333–345.

[LL3] S.J. Li, J.Q. Liu, Morse theory and asymptotic linear Hamiltonian systems, J.Diff. Equ. 78(1989) 53–73.

[LS1] S.J. Li, A. Szulkin, Periodic solutions of an asymptotically linear wave equation, Topological methods in Nonlinear Analysis, Vol 1(1993), 211–230.

[LS2] S.J. Li, A. Szulkin, Periodic solutions for a class of nonautonomous Hamiltonian systems, J. Diff. Equ. Vol 112(1994), 226–238.

[LS3] S.J. Li, A. Szulkin, Periodic solutions for a class of nonautonomous wave equations, to appear.

[LW] S.J. Li, M. Willem, Applications of local linking to critical point theorem, J. Math. Anal and Appl, (to appear).

[Liu1] J.Q. Liu, Doctoral thesis.

[Liu2] J.Q. Liu, The Morse index of a saddle point. Syst. Sc. & Math. Sc. 2(1989), 32–39.

[MS] J. Morel, H. Steinlein, On a problem of Nirenberg concerning expanding maps, Journal of Functional Analysis 59(1984), 145–150.

[N] L. Nirenberg, Topics in nonlinear functional analysis, Lecture Notes, Courant Inst., New York Univ. New York, 1974.

[R1] P. Rabinowitz, Periodic solutions of Hamiltonian systems, Comm Pure Appl. Math. 31(1978), 157–184.

[R2] P. Rabinowitz, Free vibrations of a semilinear wave equation, Comm. Pure Appl. Math. 31(1978), 31–68.

[SC] S.Z. Shi, K.C. Chang, A local minimax theorem without compactness, Nonlinear and convex analysis Vol 107, Processding in honor of Ky Fan, 211–233.

Survey of Recent Progress
in Non-Commutative Tori

Qing Lin

Institute of Mathematics, Academia Sinica, Beijing

One of the fundamental relations describing the basic interaction in quantum physics is the Weyl relation:

$$[A_i, A_j] = \theta_{ij} I, \quad 1 \leq i < j \leq n,$$

where A'_js are self-adjoint operators.

The multiplicative version of such relations would be

$$e^{2\pi i A_i} \cdot e^{2\pi i A_j} \cdot e^{-2\pi i A_i} \cdot e^{-2\pi i A_j} = e^{2\pi i \theta_{ij}} \cdot I, \quad 1 \leq i < j \leq n.$$

Notice that $e^{2\pi i A_j}$ is a unitary operator.

Therefore, it is natural to consider the universal C^*-algebra generated by such unitaries.

Definition 1. A non-commutative n-torus is a universal C^*-algebra generated by n unitaries, $U_1, U_2, \cdots, U_n$, with multiplicative Weyl relations:

$$U_i U_j = e^{-2\pi i \theta_{ij}} U_j U_i, \quad i < j,$$

where θ'_{ij}s are real numbers. We denote this C^*-algebra by A_Θ, where $\Theta = (\theta_{ij})_{i,j}$ is an $n \times n$ anti-symmetric matrix and is called a symbol of this torus.

When $n = 2$, they are exactly the well-known rotation algebras.

As is expected, such C^*-algebras have a close connection with quantum physics (see [Be1], [Be2], [AP], [HR1] and [HR2]). On the other hand, they are one of the important classes among non-commutative differential manifolds (see [C], [CR] and [R2]). They play a certain universal role in several representation theories of groups (cf. [P1] and [P2]). For instance, D Poguntke proves that, for any connected Lie group and each irreducible unitary representation of it, the unique simple ideal (non-zero) in the C^*-algebra generated by the

1991 Mathematics Subject Classification: 46L50

Partially supported by NNSF of China

range of the representation is either a compact operator algebra (including matrix algebra) or a simple non-commutative torus tensored by compact operator algebra.

To understand the recent progress in the theory of non-commutative tori, let us review the past achievement briefly.

1. The past achievement

An excellent survey of the non-commutative tori achieved up to 1989 has been given by M. A. Rieffel ([R2]). We shall quote some results relevant to the recent progress from that article.

Let us use the notations as in Definition 1 above.

1. A_Θ is simple $\iff$ the bicharacter "$(X, Y) \to \exp 2\pi i (Y^T \Theta X)$", where X and Y are in $\mathbb{Z}^n$, is non-degenerate.

2. A_Θ has a canonical normalized trace; and if A_Θ is simple, this is the unique trace.

3. If all θ_{ij} are rational numbers (called the rational torus), A_Θ is strongly Morita equivalent to $C(T^n)$.

4. As groups, $K_0(A_\Theta) \cong K_1(A_\Theta) \cong \mathbb{Z}^{2^{n-1}}$

5. Denote by τ the canonical trace on a non-rational torus A_Θ. Then

$$K_0^+(A_\Theta) = \tau^{-1}(\tau(K_0(A_\Theta)) \cap \mathbb{R}^+) \cup \{0\}.$$

6. The ordinary n-torus T^n acts on A_Θ ergodically by

$$(z_1, \cdots, z_n) \cdot U_j = z_j U_j, \quad j = 1, 2, \cdots, n.$$

7. Two simple 2-tori A_{θ_1} and A_{θ_2} are isomorphic (where $\Theta_j = \begin{pmatrix} 0 & \theta_j \\ -\theta_j & 0 \end{pmatrix}$), if and only if $\theta_1 \equiv \pm\theta_2 (\mathrm{mod}\ \mathbb{Z})$.

There are several important progresses after 1989 in this field. In a paper in 1990, I. Putnam settled a long-standing conjecture that any simple 2-torus has stable rank one (i.e. the invertible elements are dense). Later on, B. Blackadar, A. Kumjian, and M. Rørdam proved that any simple non-commutative torus has stable rank one and real rank zero (i.e. any self-adjoint element can be approximated by some invertible self-adjoint one).

These results indicate that simple non-commutative tori should have nice structure and nice clssification theory.

At the Kingston Conference in 1980, E.Effros raised a question: Is any simple non-commutative 2-torus an inductive limit of a sequence of direct sums of matrix algebras over commutative C^*-algebras?

In 1985, S. Disney, G. A. Elliott, A. Kumjian, and I. Raeburn ([DEKR]) classified all rational tori. Their result is: if Θ_1 and Θ_2 are two $n \times n$ anti-symmetric rational matrices, then $A_{\Theta_1} \cong A_{\Theta_2} \iff \Theta_1$ and Θ_2 are integrally congruent (i.e. there is a $P \in GL(n, \mathbb{Z})$ such that $P^T \Theta_1 P = \Theta_2$).

In two subsequent papers ([CEGJ] and [BCEN]), a classification for the canonical smooth subalgebra of simple 3 and 4 dimensional tori was established. Recall that once we fix a standard unitary generator $\{U_1, \cdots, U_n\}$ of an n-torus A, we have a T^n-action on A through this generator and an associate canonical smooth subalgebra

$$A^\infty := \{x \in A : (z_1, \cdots, z_n) \to (z_1, \cdots, z_n) \cdot x \text{ is differentiable}\}.$$

The main result of the last paper is: if A_{Θ_1} and A_{Θ_2} are two simple 3-tori (or both are generic 4-tori) with specified canonical smooth subalgebras $A_{\Theta_1}^\infty$ and $A_{\Theta_2}^\infty$ respectively, then $A_{\Theta_1}^\infty \cong A_{\Theta_2}^\infty \iff \Theta_1$ and Θ_2 are integrally congruent.

Until recently, the classification of higher dimensional non-commutative tori has been open even in a single higher dimension in spite of considerable effort (cf. [R2] and [J]).

In 1989, G. A. Elliott initiated a classification program in another direction. He succeeded in classifying all real rank zero inductive limits of direct sums of circle algebras (referred as $M_n(C(T))$). Later on, this result is extended to all simple and real rank zero inductive limits of homogenuous C^*-algebras.

A simple version of such result says that if A and B are both unital, simple and real rank zero and inductive limits of direct sums of circle algebras, then

$$A \cong B \iff (K_0(A), K_0^+(A), [1_A]) \cong (K_0(B), K_0^+(B), [1_B]) \text{ and } K_1(A) \cong K_1(B).$$

In 1991, an important progress was made by Elliott and Evans ([EE]). They used some subtle machinary to prove that any simple 2-torus (i.e. irrational rotation algebra) is an inductive limit of direct sums of two circle algebras. This answers Effros' question mentioned above. One significance of this result is that Elliott's K-theoretical classification result on certain inductive limit C^*-algebras might be applied to more general C^*-algebras by a suitable inductive limit decomposition technique. One of such candidates is a higher dimensional non-commutative torus.

2. Recent progress

There is an equivalent expression for a non-commutative n-torus. Using the notation of Definition 1, let $e_1, \cdots, e_n$ be an integral basis of $\mathbb{Z}^n$, β be a bicharacter on $\mathbb{Z}^n$ (i.e. a bi-group map of $\mathbb{Z}^n \times \mathbb{Z}^n$ into T) such that

$$\beta(e_i, e_j)\overline{\beta(e_j, e_i)} = e^{-2\pi i \theta_{ij}}.$$

Then the twisted group C^*-algebra of $\mathbb{Z}^n$ with respect to β is isomorphic to A_Θ in a canonical way:

$$U_j \to \chi_{e_j}, \quad j = 1, 2, \cdots, n,$$

where χ_{e_j} is the characteristic function on $\mathbb{Z}^n$ supported at e_j.

Thus the group generated by all singly supported characteristic functions on $\mathbb{Z}^n$ and $T \cdot \chi_0$ (0 is the zero point in $\mathbb{Z}^n$) is naturally isomorphic to the group generated by all U_j's and $T \cdot 1$, where T is the unit circle and 1 is the identity of A_Θ. The last group will be called a generalized unitary generator of A_Θ.

In a paper of 1988 ([R1]), M. A. Rieffel described a representative in each equivalence class of finitely generated projective modules over A_Θ, where the symbol Θ is not rational. His construction is based on the above expression of a non-commutative torus, Heisenberg representation and the notion of strong Morita equivalence.

One significance of his construction is that fixing a generalized unitary generator of A_Θ (i.e. an expression of A_Θ as $C^*(\mathbb{Z}^n, \beta)$), there is a canonical way to get the dual generalized unitary generator of $pA_\Theta p$ for each non-zero projection p in A_Θ (more precisely, a representation of $pA_\Theta P$ as $C^*(\mathbb{Z}^n \times F, \beta')$, where F is a finite abelian group). With a carefully chosen p (one requirement is that p approximately commutes with each $U_j, j = 1, 2, \cdots, n$), $pU_j p$ is approximated by a unitary from that dual generalized unitary generator of $pA_\Theta p$.

This observation combined with the following idea provides the key to prove several theorems stated below. We outline the idea as follows. To approximate a finite subset S of a C^*-algebra A by a C^*-algebra of specified type, we may start from a projection p in A which approximately commutes with each element of S. Thus if

$$\|px - xp\| < \varepsilon/2, \quad x \in S,$$

we have that

$$\|x - (pxp \oplus (1-p)x(1-p))\| < \varepsilon.$$

Therefore the problem reduces to two similar problems: to approximate pSp in pAp and to appoximate $(1-p)S(1-p)$ in $(1-p)A(1-p)$. If, in addition, the two parts from cutting down S by p are both nearly "half" of S, that is,

$$pSp \text{``} \sim'' (1-p)S(1-p) \text{``} \sim'' \text{``} 1/2 \cdot S''$$

in an appropriate sense, the two reduced problems may be easier to solve.

In our case, the finite set S is just a standard unitary generator $\{U_1, \cdots, U_n\}$ of A_Θ. Since $\|pU_j - U_j p\| < \varepsilon$, $pU_j p$ is nearly a unitary in $pA_\Theta p$. Thus it is reasonable to consider $\{[pU_j p]_1 : 1 \leq j \leq n\}$ in $K_1(pA_\Theta p)$ and $\{[(1-p)U_j(1-p)]_1 : 1 \leq j \leq n\}$ in $K_1((1-p)A_\Theta(1-p))$. The ranks of these two sets provide a measure of the "half" of S. Such restriction on p combined with the approximate centrality of p can be reformulated as a non-standard Diophantine approximation to $\theta'_{ij}s$, which leads to an inductive limit decompostion. Moreover, a suitable modification of certain higher dimensional continued fraction approximation can be used to solve this non-standard Diophantine approximation.

The technique described above leads to the following progresses.

Theorem 1 (Q. Lin, [L]). Any simple 3-torus is an inductive limit of direct sums of four circle algebras. Therefore, they are classified by the classical K-theory.

Corollary. Any non-commutative 3-torus is either an inductive limit of four circle algebras, or the inductive limit of direct sums of two C^*-algebras, each of which is a matrix algebra over a rational 2-torus (including the commutative 2-torus), or a rational 3-torus. Moreover, up to a stable isomorphism, they are classified by the ordered K_0-group.

Theorem 2 (G. A. Elliott and Q. Lin, [EL2]). Any simple non-commutative n-torus which is a crossed product of a rational $n-1$ torus by $\mathbb{Z}$ is an inductive limit of 2^{n-1} circle algebras.

It might be worthy to point out that the method of proving these two theorems can be used in an arbitary simple non-commutative torus, except for that we don't know whether the associate non-standard Diophantine approximation problem is solvable.

In an earlier joint paper ([EL1]), the authors characterized another non-standard Diophantine approximation condition for θ'_{ij}s, which guarantees the existence of an expression of inductive limit of direct sums of circle algebras in dimension 3 tori. They indicated the possibility to generalize it to higher dimensional tori. Later on, F. P. Boca justified this indication. He showed that the symbols of all simple non-commutative n-torus which satisfy the last Diophantine approximation condition form a probability one set. This leads to the following result.

Theorem 3 (F. P. Boca, G. A. Elliott and Q. Lin, [Bo] and [EL1]). Almost all simple non-commutative tori are inductive limits of direct sums of circle algebras.

Applying a strong form of Elliott's classification theorem (cf. [E1]) on inductive limit of direct sums of circle algebras to the above results, we get that in many simple non-commutative tori (including all generic 3 dimensional tori), there are more than one isomorphic classes of the canonical smooth subalgebras. This situation is similar to the one on a topological 4-manifold.

In the other direction, there appear several results on fixed point algebras of certain finite order automorphisms.

Recall that if A is a simple unital C^*-algebra and α is a nontrivial torsion element in $\mathrm{Aut}(A)$ (e.g. $\alpha^n = id$), then $A^\alpha = \{a \in A : \alpha(a) = a\}$ and $A \times_\alpha \mathbb{Z}_n$ are strongly Morita equivalent. Thus we have a basic picture that

$$A^\alpha \subsetneq A \subsetneq A \times_\alpha \mathbb{Z}_n.$$

Let $\{U_1, U_2\}$ be a standard unitary generator of a simple 2-torus A. By the universal property of A, the map $U_i \to U_i^{-1}, i = 1, 2$ extends to an automorphism of A. We call such an automorphism a flip, it is obviously of order two. Notice that even when there are many distinct flips on A, their fixed point algebras are mutually isomorphic. Thus, we can speak of the flip fixed point algebra of A.

Theorem 4 (O. Brattili and A. Kishimoto, [BK]). The flip fixed point algebra of any simple 2-torus is an AF-algebra.

Later on, several people gave different proofs of this result ([S] and [B_0]). F. Boca's approach is based on the concrete inductive limit expressions in Theorems 1, 2 and 3. This enables him to show the next theorem.

Theorem 5 (F. P. Boca). The flip fixed point algebra of any simple 3-torus and any simple non-commutative torus considered in Theorems 2 and 3 is an AF-algebra.

The analogous problem for other natural automorphisms of finite order on simple non-commutative tori (e.g. the automorphisms induced by $U_1 \to U_2, U_2 \to U_1^{-1}$ on simple 2-tori) has been considered, but remains open. The related outstanding problem whether, in any simple 2-torus, the spectrum of $U_1 + U_2 + U_1^{-1} + U_2^{-1}$ is a Cantor set remains open.

The classification program (see [E2] and [K]) and the above progresses shake several branches in C^*-algebra theory. On the one hand, it shakes the difficult theory of nuclear and simple C^*-algebras. It suggests that there is a strong possibility that any finite, separable, unital, nuclear, quasidiagonal, simple and real rank zero C^*-algebra is an inductive limit of homogenuous C^*-algebras. Therefore, they are classified by the classical K-theory. On the other , since non-commutative tori are fundamental in the theory of twisted C^*-crossed product (see [J]), the above progresses bring a fresh air into this old theory.

References

[AP] D. Applebaum, Quantum diffusions on involutive algebras, Quantum Probability and Applications, 70–85, Lecture Notes in Math., 1442, Springer, 1990.

[Be1] J. Béllissard, K-theory of C^*-algebras in solid state physics, Statistical Mechanics and Field Theory, Mathematical Aspects, Lecture Notes in Physics 257(1986), 99–156.

[Be2] J. Béllissard, C^*-algebras in solid state physics, 2D electrons in a uniform magnetic fields, Operator Algebras and Applications, Vol.2, 49–76, London Math. Soc. Lecture Notes Ser., 136, Cambridge Univ. Press, 1988.

[B_0] F. P. Boca, Metric Diophantine approximation and the structure of non-commutative tori, preprint.

[BK] O. Bratteli, and A. Kishimoto, Non-commutative spheres, III, Comm. Math. Phys. 146(1992), 605–624.

[BCEN] B. Brenken, J. Cuntz, G. A. Elliott, and R. Nest, On the classification of non-commutative tori, III, Proc. of Conf. on Operator Algebras and Mathematical Physics, Univ. of Iowa (1985), Contemporary Math., 62(1987), 503–526.

[C] A. Connes, C^*-algebras et géométrie différentielle, C. R. Acad. SC. Paris, 290(1980), 599–604.

[CR] A. Connes and M. A. Rieffel, Yang-Mills for non-commutative two-tori, Proc. of Conf. on Operator Algebras and Mathematical Physics, Univ. of Iowa (1985), Contemporary Math., 62(1987), 237–266.

[CEGJ] J. Cuntz, G. A. Elliott, F. M. Goodman, and P. E. T. Jorgensen, On the classification of non-commutative tori, II, C. R. Math. Rep. Acad. Sci. Canada 7(1985), 189–194.

[DEKR] S. Disney, G. A. Elliott, A. Kumjian, and I. Raeburn, On the classification of non-commutative tori, C. R. Math. Rep. Acad. Sci. Canada 7(1985), 137–141.

[E1] G. A. Elliott, On the classification of C^*-algebras of real rank zero, J. Reine Angew. Math., 443(1993), 179–219.

[E2] G. A. Elliott, The classification problems for amenable C^*-algebras, preprint.

[EE] G. A. Elliott, and D. E. Evans, The structure of the irrational rotation C^*-algebra, Ann. of Math., 138(1993), 477–501.

[EL1] G. A. Elliott, and Q. Lin, Cut-down method in the inductive limit decomposition of non-commutative tori, J. London Math. Soc., to appear.

[EL2] G. A. Elliott, and Q. Lin, Cut-down method in the inductive limit decomposition of non-commutative tori, II: The degenerated case, preprint.

[HR1] R. L. Hudson, and P. Robinson, Quantum diffusions and the non-commutative torus, Letters Math. Phys. 15(1988), 47–53.

[HR2] R. L. Hudson, and P. Robinson, Quantum diffusions on the non-commutative torus and solid state physics, Proceedings XVII International Conference on Differential Geometric Methods in Theoretical Physics, ed. A. Solomon, World Scientific, Singapore, 1989.

[K] E. Kirchberg, The classification of purely infinite C^*-algebras using Kasparov's theory, preprint.

[L] Q. Lin, Cut-down method in the inductive limit decomposition of non-commutative tori, III: a complete answer in 3-dimension, preprint.

[P] J. A. Packer, Transformation group C^*-algebras: a selective survey, pp. 183–217, C^*-algebras: 1943–1993 A Fifty Year Celebration, Contemporary Math. 157, Amer. Math. Soc., Providence, 1994.

[P1] D. Poguntke, Simple quotiens of group C^*-algebras for two step nilpotent groups and connected Lie groups, Ann. Scient. Ec. Norm. Sup. 16(1983), 151–172.

[P2] D. Poguntke, The structure of twisted convolution C^*-algebras on abelian groups, preprint.

[R1] M. A. Rieffel, Projective modules over higher dimensional non-commutative tori, Canadian J. Math., 40(1988), 257–338.

[R2] M. A. Rieffel, Non-commutative tori–a case study of non-commutative differentiable manifolds, pp. 191–211, Contemporary Math., 105, Amer. Math. Soc., Providence, 1990.

[S] S. Walters, Inductive limit automorphisms of the irrational rotation algebra, Comm. Math. Phys., to appear.

Some Results on the Structure of Toeplitz Operators

Shunhua Sun

Department of Mathematics, Sichuan University, Chengdu

This is a survey of some Chinese operator theorists' works on the Toeplitz operators and related topics. Particularly, the characterizations for Toeplitz operator on Hardy space to be hyponormal weighted shift or in the θ-class are given. As direct applications of these charactcrizations, that the Bergman shift is not unitarily equivalent to a Toeplitz operator and that the answer to the Halmos' fifth question is "No" are obtained. Another application is to show that there exist Toeplitz operators that are unitarily equivalent but not inner-implemented. As to Toeplitz operators on Bergman space, some unitary equivalence characterizations for analytic Toeplitz operators are given too, and, in particular, the super-isometric dilation theory with applications is developed. Furthermore a duality approach is exhibited to unify the well-known results on the boundedness and compactness of Hankel operators both on Hardy and Bergman spaccs.

1. Halmos' 5th question and some related problems

In [17], Halmos asked (5th Question): If T_φ is subnormal, must T_φ be normal or analytic? M. B. Abrahamse proved in [1] that, roughly speaking, for "almost all Toeplitz Operators" the answer to Halmos' 5th Question is "Yes". Meanwhile, Abrahamse raised the following questions: (Problem 2) If A is a hyponormal weighted shift that is unitarily equivalent to a Toeplitz operator, must A be a scalar multiple of an isometry? (Problem 3) Is the Bergman shift unitarily equivalent to a Toeplitz operator?

The basic and classical deep results on Toeplitz operators (on Hardy space $H^2(\partial D)$) over the disc) can be found in [5], [14] and [16].

1991 Mathematics Subject Classification: 47B35, 47B37, 47B20, 47A20

Supported by NNSFC and a Grant from Education Mission of China

Theorem 1.1[21]. Let T be a (unilateral) weighted shift with weights $\{a_l\}_0^\infty$ satisfying $a_l < a_{l+1}, \forall l \geq 0$, and $\lim_{l \to \infty} a_l = 1$. If $(1 - a_1^2)^{1/2} \neq 1 - a_0^2$, then T is not unitarily equivalent to a Toeplitz operator.

This leads to a negative answer to Abrahamse's Problem 3.

Corollary 1.2[21]. Bergman shift is not unitarily equivalent to a Toeplitz operator.

The proof itself given in [21] together with some modified construction also leads to the following general characterization.

Theroem 1.3[22]. If T is a hyponormal weighted shift of norm one with weights $\{a_l\}_0^\infty$, then a necessary condition that T be unitarily equivalent to a Toeplitz operator is that $1 - |a_l|^2 = (1 - |a_0|^2)^{l+1}, \forall l \geq 0$.

Does the converse of Theorem 1.3 hold? Yes, it is. Actually, the proof itself given in [21] or [22] shows that the only possible symbol $\varphi (\in L^\infty(\partial D))$ must satisfy the condition below:

$$\begin{cases} (\varphi - \gamma(1 - |a_0|^2)^{1/2}\overline{\varphi}) \in H^\infty(\partial D), \\ |\varphi(t)| = 1, \forall \text{ a.e. } t \in \partial D, \end{cases} \qquad (*)$$

where γ is a constant of modulus one. Based upon analyzing the proofs given in [21] and [22], the following is derived.

Theorem 1.4[10],[23]. The converse of Theorem 1.3 is also true.

Therefore, the Abrahamse's Problem 2 is answered completely.

In a short letter in Aug. 1983, C. C. Cowen informed the author that if $1 - |a_0|^2 < 1/2$, then the converse of Theorem 1.3 is true and that the sequence given in Theorem 1.3 is subnormal. One month later, in a letter of Sept. 16, 1983, C. C. Cowen passed his original proof together with J. J. Long's proof of Theorem 1.4 to the author. In both their proofs it is mentioned that the subnormality of the sequence given in Theorem 1.3 is discovered by S. C. Power. By the way, the author would like to mention another note that the subnormality of the the sequence above is also essentially proved in [18] (cf. [19] for details) of which the original draft was reported by Ma Jipu at the Third National Conference on Functional Analysis held at Kunming in May, 1983.

Lemma 1.5[10],[18]. The sequence given in Theorem 1.3 is subnormal.

The answer to Halmos' 5th Question is thus derived from Theorem 1.4 and Lemma 1.5, as well as from Theorem 1.7 to be stated later, immediately.

Corollary 1.6[10],[23],[24]. There is a subnormal Toeplitz operator neither normal nor analytic.

We now come to characterizing the θ-class Toeplitz operators. The characterization itself of such Toeplitz operators will lead to the same conclusion about the answer to Halmos' 5th Question. An operator $T \in \mathcal{L}(\mathcal{H})$ is in the θ-class if T^*T commutes with $T + T^*$ (cf. [6] for the details).

The following is given in [24].

Theorem 1.7[24]. A Toeplitz operator T_φ on the Hardy space $H^2(\partial D)$ with its symbol $\varphi \in L^\infty(\partial D)$ is in the θ-class if and only if one of the following condition holds:

(I) T_φ is normal (cf. [5] for the symbol of this case);

(II) T_φ is analytic and $\varphi = \gamma\chi + \alpha$, where χ is an inner function and γ and α are constants with α real;

(III) There exist real constants α, β and γ, constant ρ with $\left|\frac{1+\rho}{\rho}\right| > 2$ and an inner function χ such that $\varphi(t) = \gamma\{\Phi(t) + i\alpha + \bar{\rho}\overline{\Phi(t)}\} + \beta, \forall$ a.e. $t \in \partial D$, where $\Phi = \sqrt{\frac{1+\bar{\rho}}{\rho}}\{\Omega(b_j\chi) - \Omega(b_j\chi(0))\}, b^2 = \left(\left|\frac{1+\rho}{\rho}\right| - 2\right) / \left(\left|\frac{1+\rho}{\rho}\right| + 2\right)$ and $\Omega(b_j \cdot)$ is the Riemann mapping of the unit disc D onto the interior domain bounded by an ellipse defined by $y^2 + \frac{x^2}{b^2} = 1$.

Remark 1.8. Since a Toeplitz operator T_φ belonging to the case (III) of Theorem 1.7 is obviously hyponormal, and hence subnormal (cf. [6]), thus Corallary 1.6 follows from Theorem 1.7 directly. This approach to Corollary 1.6 was discovered by Dr. Zheng Dechao first in late Septumber 1983.

Remark 1.9. So far, the structures of two classes of Toeplitz operators, one for hyponormal weighted shift and another for θ-class, have been completely characterized. How about the Toeplitz operators being a hyponormal weighted shift of multiplicity greater than one? The complete characteization is also done by Zheng Dechao in [33]. Furthermore, a charaterization of weighted shift Toeplitz operator that is not necessarily hyponormal is obtained in [30] by Yu Dahai. Moreover, much more important progress has been made by C. C. Cowen in [11] and [12] for subnormal and hyponormal Toeplitz operators.

We now turn to the unitary equivalence problem for Toeplitz operators. In [9], the following theorem is proved by C. C. Cowen:

Theorem C[9]. If $f = h \circ u$ and $g = h \circ v$, then the Toeplitz operators T_f and T_g are unitarily equivalent to each other, denoted by $T_f \cong T_g$, where $h \in L^\infty(\partial D)$, u and v are inner functions of the same order.

Also, Cowen raised the following questions: (I) When does the converse of Theorem C hold? (II) Does the converse of Theorem C hold for non-normal Toeplitz operators?

Basing on Theorem 1.4, the author provided a counterexample of Question (II) above in [25].

Theorem 1.10[25]. There exist some non-normal Toeplitz operators T_f and T_g such that $T_f \cong T_g$ but the condition of Theorem C is never satisfied.

Remark 1.11. As to Question (I), a necessary and sufficient condition was obtained also in [25] under the assumptions placed on the weak closed self-adjoint subalgebra of $L^\infty(\partial D)$ generated by the symbol f associated with the Toeplitz operator T_f. Though the assumptions given there seem serve, it is fairly satisfactory in the case of analytic Toeplitz operators.

Remark 1.12. In [20], the notion of near subnormality was introduced to show how large the gap there existing between hyponormality and subnormality for Hilbert space operator. Also, a new necessary and sufficient condition for subnormal operators is given there. As an application of the results given in [20], a new example of a hyponormal weighted shift that is not the trace class perturbation of a subnormal weighted shift is exhibited in [34].

2. Super-isometrical dilation and multiplication operators on Bergman space

As usual, the Bergman space on the disc is denoted by $L^2_a(D)$. For a given $\varphi \in H^\infty(D)$, define the multiplication operator M_φ on $L^2_a(D)$ (i.e. the analytic Toeplitz operator, still denoted by T_φ as usual, on $L^2_a(D)$) by $M_\varphi h = \varphi \cdot h, \forall \in L^2_a(D)$. Obviously, M_φ is subnormal and its minimal normal extension is simply a multiplication by φ on $L^2(D)$ (cf. [2], or [7] p.436). A natural question is: For given φ and ψ in $H^\infty(D)$, when is $M_\varphi \cong M_\phi$? In the case of Hardy space, the quesstion is completely solved by C. C. Cowen under a "finiteness" condition (cf.[9]). Of course, the unitary equivanlence problem for general analytic Toeplitz operators is still open (cf [7] p.274).

With significant differences from what was proved in [9], the following are given in [26] and [27] separately.

Theorem 2.1[26]. For f and φ both in $H^\infty(D)$, if the inner part of $(f - f(\alpha))$ is a finite Blaschke product for some $\alpha \in D$, then $M_f \cong M_\varphi$ iff there is an inner function η of order one such that $\varphi = \eta^N$.

Remark 2.3. For $N = 2$, the Theorem 2.2 was proved by Yu Guoliang in term of complex geometry (cf. [32]).

A natural question is now raised: Can we give any intrinsic characterization to show why and how there exist so much essential difference between analytic Toeplitz Operators on Hardy space and those on Bergman space. Motivated by the phenomena above and the theory of Hilbert module over function algebra (cf. [15]), the super-isometric dilation of certain operators is developed in [28].

Definition 2.4. An operator $T \in \mathcal{L}(\mathcal{H})$ is super-isometrically dilatable if there exist a Hilbert space $\mathcal{K}(\supset \mathcal{H})$ and a pair of doubly commuting pure isometries U and V in $\mathcal{L}(\mathcal{K})$ such that

$$\begin{cases} P_{\mathcal{H}} U^i V^j |\mathcal{H} = T^{i+j}, \forall i, j \geq 0, \\ U^{\cdot} |\mathcal{H} = V^{\cdot} |\mathcal{H} = T^{\cdot}, \end{cases} \tag{2.1}$$

where $P_{\mathcal{H}}$ is the orthogonal projection of $\mathcal{K}$ onto $\mathcal{H}$. In this case, $\{U, V\}$ is said to be the super-isometric dilation of T. Moreover, the triple $\{U, V, \mathcal{K}\}$ is said to be minimal if $\mathcal{K}$ is the minimal space containing $\mathcal{H}$.

Note that not all the contractions are super-isometrically dilatable. This illustrates the difference between the super-isometric dilation and the classical Nagy-Foias isometric dilation (cf. [35]). However, owing to the important role in the operator theory played by the Bergman shift (cf. [4]) which is super-isometrically dilatable as shown later, it seems that the theory of super-isometric dilation would provide an interesting approach to the study of general operator theory.

Proposition 2.5[28]. If T is super-isometrically dilatable, then the minimal super-isometric dilation $\{U, V, \mathcal{K}\}$ of T is unique up to a unitaty equivalence.

Proposition 2.6[28]. Let $T, S \in \mathcal{L}(\mathcal{H})$. Assume that T is super-isometrically dilatable and $T \cong S$. Then S is super-isometrically dilatable too.

Theorem 2.7[28]. Suppose that T is super-isometrically dilatable with its minimal super-isometric dilation $\{U, V, \mathcal{K}\}$. Then for an inner function $\Phi(\in A(D))$ of order $N(< +\infty)$, $\Phi(T)$ is super-isometrically dilatable too. Moreover, $\{\Phi(U), \Phi(V); \tilde{\mathcal{K}}\}$ is the minimal super-isometric dilation of $\Phi(T)$, where $\tilde{\mathcal{K}}=$span $\{\Phi^l(U)\Phi^k(V)\mathcal{H}, \forall l, k \geq 0\}$.

The following theorem is essentially due to R. G. Douglas and V. I. Paulsen (cf. Chap. 5 of [15]).

Theorem 2.8[15],[28]. Let $\mathcal{H} = \text{span} \{P_n(z,w) \equiv z^n + z^{n-1}\omega + \cdots + z\omega^{n-1} + \omega^n, n \geq 0\}(\subset H^2(\partial D \times \partial D)$, the Hardy space on the bidisc). Then

$$H^2(\partial D \times \partial D) = \mathcal{H} \oplus cl\{(z-\omega)H^2(\partial D \times \partial D)\}.$$

Moreover, $\{T_k, T_\omega, H^2(\partial D \times \partial D)\}$ is the minimal super-isometric dilation of the Bergman shift.

As a consequence of Theorems 2.7–2.8 and a result given in [8], we obtain, via computing some invariants associated with minimal super-isometric dilation, the following

Theorem 2.9[28]. Assume that φ, ψ and f are in $H^\infty(D)$ and that there exists some $\alpha \in D$ such that the inner part of $(f - f(\alpha))$ is a finite Blaschke product. Then, as multiplication operators on Bergman space, $M_\varphi \oplus M_\psi \cong M_f$ iff $\varphi = \psi = f =$const.

Remark 2.10. Theorem 2.9 answers a question asked in [25]. From the complex geometry point of view, all of Theorems 2.1, 2.2 and 2.9 tell us some strong "rigidity" of $H^\infty(D)$ viewed as the Bergman module in contrast to the Hardy module (cf. [15] for definitions). It seems very interesting to give new proofs of these theorems in term of either complex geometry (cf. [13]) or Hilbert modules over function algebras (cf. [15]).

Remark 2.11. In [31], the Fredholm theory of Toeplitz operators on Bergman space $L^2_a(D)$ with symbols being the harmonic extension of functions in $H^\infty + C(\partial D)$ is completely established.

3. Duality and Hankel operators

All the known results about Hankel operators (cf. [3], [36] and [37]) suggest that there should be a unified duality approach that would provide an intrinsic description of which Hankel operators are bounded, compact, or nuclear. Let us begin with some basic observations.

Observation (I)[29]. Let S_∞ be the compact operator ideal of $\mathcal{L}(\mathcal{H})$, where $\mathcal{H}$ is a separable Hilbert space, and let $\mathcal{A}$ be a normed function space and $L : \mathcal{A} \to S_\infty$ a bounded linear map, i.e. $\|L(g)\|_{S_\infty} \leq M\|g\|_{\mathcal{A}}, \forall g \in \mathcal{A}$ with M constant, independent of $g \in \mathcal{A}$. By dual pairing for $(\mathcal{A}, \mathcal{A}^*)$ we now define the dual the operator L^* and the second dual operator

L^{**} as follows:

(*)
$$\begin{cases} L^* : S_\infty^* = S_1 \to \mathcal{A}^* \text{ by } \langle L(g), A \rangle_{(S_\infty, S_1)} \\ \quad = \langle g, L^*(A) \rangle_{(\mathcal{A}, \mathcal{A}^*)}, \forall g \in \mathcal{A}, A \in S_1, \\ L^{**} : \mathcal{A}^\# \to \mathcal{L}(\mathcal{H}) = (S_1)^* \text{ by } \langle g, L^*(A) \rangle \\ \quad = \langle L^{**}(g), A \rangle, \forall g \in \mathcal{A}^{\#*}, A \in S_1, \end{cases}$$

where $\mathcal{A}^\# \equiv \| \cdot \|_{\mathcal{A}^*} - cl\{L^*(S_1)\}$, the closure of $L^*(S_1)$ w. r. t. the norm of $\mathcal{A}^*$, and S_1 the trace class opertors. With canonical embedding $\mathcal{A} \to \mathcal{A}^{\#*}$, it is evident that $L^{**}|\mathcal{A} = L$. And hence L^{**} can be viewed as an extension of L to $\mathcal{A}^\#$.

Observation (I)[29]. Let L be a linear bounded map of $S_\infty(\subset \mathcal{L}(\mathcal{H}))$ into $\mathcal{A}$, a normed space. By a similar construction, we can define

$$L^* : \mathcal{A}^* \to S_1 \quad \text{by} \quad \langle L(A), g \rangle = \langle A, L^*(g) \rangle, \forall A \in S_\infty, g \in \mathcal{A}^*. \tag{**}$$

With these observations, we have the following lemmas.

Lemma A[29]. Assume $\mathcal{A}^\# \equiv \| \cdot \|_{\mathcal{A}^{\#*}} - cl\{\mathcal{A}\}$, the closure of $\mathcal{A}$ w. r. t. the norm of the dual of $\mathcal{A}^\#$. Then $L(_\#\mathcal{A}^\#) \subset S_\infty$ and $L(\mathcal{A}^{\#*}) \subset \mathcal{L}(\mathcal{H})$.

Lemma B[29]. If L is a linear bounded mapping of S_∞ into $\mathcal{A}$, as given in Observation (I), then $L^*(\mathcal{A}^*) \subset S_1$.

Based upon these lemmas together with choosing the normed function space $\mathcal{A}$ and the associated mapping L properly, a unified approach is given in [29] to recapture the well known results on the boundedness and compactness of Hankel operators, with coanalytic symbols, both on Hardy and Bergman spaces. Meanwhile, a simplified proof of Peller's criterion for the nuclearity of Hankel operators on Hardy space is exhibited in [29] by means of the same approach. As another application of our approach, the following is also proved.

Theorem 3.1[29]. If $\bar{g} \in L_a^p(D)$ and $\bar{f} \in L_a^q(D)$ where $p, q > 1$ and $\frac{1}{p} + \frac{1}{q} = 1$, then $H_g^* H_f \in S_1$, i. e. $H_g^* H_f$ is nuclear.

Remark 3.2. The condition given in Theorem 3.1 is sharp but not necessary as pointed out by V. V. Peller.

References

[1] Abrahamse, M. B., Subnormal Toeplitz operators and functions of bounded type, Duke Math. J., 43(1976), 507–604.

[2] Axler, Sheldon, The Bergman space, the Bloch space, and commutators of multiplication operators, Duke Math. J., 53(1986), 315–332.

[3] Axler, Sheldon, Bergman space and their operators, Pitman Research Notes in Mathematics Serics 171(1988), 1–50.

[4] Bercovici, H., Foias, C. and C. Pearcy, Dual algebras with applications to invariant subspaces and dilation theory, CBMS, 56(1985).

[5] Brown, A. and P. R. Halmos, Algbraic properties of Toeplitz operators, J. Reine angew. Math., 213(1963-64), 89–102.

[6] Campbell, S. L., Linear operators for which T^*T and $T^* + T$ commute, III, Pacific J. Math., 79(1978), 17–19.

[7] Conway, J. B., Subnormal operators, Pitman, Boston, 1981.

[8] Cowen, C. C., The commutant of an analytic Toeplitz operator, Trans. AMS, 239(1978), 1–31.

[9] Cowen, C. C., On equivalence of Toeplitz operators, J. Operator Theory 7(1982), 167–172.

[10] Cowen, C. C. and J. J. long, Some subnormal Toeplitz operators, J. Reine angew. Math., 351(1984), 216–220.

[11] Cowen, C.C., More subnormal Toeplitz operators, J. Reine angew. Math. 367(1986), 215–219.

[12] Cowen, C. C., Hyponormal and subnormal Toeplitz operators, Pitman Research Notes in Mathematics Series 171(1988), 155–168.

[13] Cowen, M. J. and R. G. Douglas, Complex geometry and operator theory, Acta Math., 141(1978), 187–261.

[14] Douglas, R. G., Banach algebra techniques in operator theory, Academic press, New Youk, 1972.

[15] Douglas, R. G. and V.I. Paulsen, Hilbert modules over function algebra, Pitman Research Notes in Mathematics Series 217, 1989

[16] Halmos, P. R., A Hilbert space problem book, Springer-Verlag, New York, 1974.

[17] Halmos, P. R., Ten problem in Hilbert space, Bull. AMS., 76(1979), 887—933.

[18] Ma Jipu and Zhou Shaojie, A sufficient and necessary condition for subnormal operator (in Chinese), J. Nanjing Univ. Math. Biquarterly, No. 2(1985), 190–195.

[19] Ma Jipu, A new proof of the lemma by C. Cowen &. J. Long, J. Nanjing Univ. Math. Biquarterly, No. 1(1990), 1–2.

[20] Ma Jipu and Wang Gongbao, Near subnormal operators and subnormal operators, Science in China (Series A), 33(1) (1990), 1–9.

[21] Sun Shunhua, Bergman shift is not unitarily equivalent to a Toeplitz operator, Kexue Tongbao, No. 1(1983), 1–3 (in Chinese first, and the English version appeared in Kexue Tongbao 28(1983), 1027–1030.)

[22] Sun Shunhua, On hyponormal weighted shift, Chin. Ann. of Math., 5B(1)(1984), 101–108.

[23] Sun Shunhua, On hyponormal weighted shift (I), Chin. Ann. of Math., 6B(3)(1985), 359–361.

[24] Sun Shunhua, On Toeplitz' operators in the θ-class, Scientia Sinica (Series A) No. 9(1984), 806–811 (in Chinese first, and the English version appeared in Seientia Sinica (Series A) 28:3(1985), 235–241).

[25] Sun Shunhua, Some questions on unitary equivalence of Toeplitz operator, Kexue Tongbao, No. 8(1985) 570–572 (in Chinese first, the English version appeared in kexue Tongbao 30:10(1985)1292–1295).

[26] Sun Shunhua, On unitary equivalence of multiplication operators no Bergman space, Northeastern Math. J., 1(2)(1985). 213–222.

[27] Yu Dahai and Sun Shunhua, On unitary equivalence of multiplication operators on Bergman space (I), Northeastern Math. J., 4(2)(1988) 169–179 (in Chinese with English abstract).

[28] Sun Shunhua and Yu Dahai, Super-isometrically dilatable opearator, Science in China (Series A), No. 6(1989), 580–587 (in Chinese first, and the Enghish version appeared in Science in China (Seres A), 32(12)(1989), 1447–1457.)

[29] Sun Shunhua, Duality and Hankel operators, Operator Theory: Advances and Applications, Vol.48, Birkhauser Verlag Basel, 1990,p.p.373–385.

[30] Yu Dahai, The symbol characterization of the weighted shift Toeplitz opertors, Northeastern Math. J., 6(1)(1990), 75–84.

[31] Yu Dahai, Sun Shunhua and Dai Zhengqui, The Fredholm theory of Toeplitz operators on Bergman space, Northeastern Math. J., 4(4)(1988), 405–411.

[32] Yu Gaoliang, On reducibity and complete unitaty invariance of certain operator class, Kexue Tongbao, No. 9(1985), 1597 (in Chinese).

[33] Zheng Dechao, On hyponormal weighted unilateral shift of multiplicity $k(0 < k < \infty)$, Acta Mathematica Sinica, New series, 3(1)(1987), 77–81.

[33] Zhou Shaojie, Chen Fudong and Ma Jipu, An example of example of a hyponormal weighted unilateral shift which is not the trace class perturbation of subnormal werghted shift, Northeastern Math. J., 6(4) (1990), 489–496.

[34] Zhou Shaojie, Chen Fudong and Ma Jipu, An example of a hyponormal weighted unilateral shift which is not the trace class perturbation of subnormal weighted shift, Northeastern Math. J., 6(4)(1990), 489–496.

[35] Sz.-Nagty B. and C. Foias, Harmonic analysis of operators on Hilbert space, 1970.

[36] Pellev V. V., A description of Hankel operators of class S_p for $p > 0$, an investigation of

Shunhua Sun

the rate of ratinal application, and other applications, Math. USSR Sbornik, 50(1985), 465–494.

[37] Power, S. C., Hankel operators on Hilbert space, Pitman Research Notes in Math. Series 64, 1984.

Properties of Some Nonlinear Operators

Shengwang Wang

Department of Mathematics, Nanjing University, Nanjing

Abstract. This paper gives a brief survey of the author's work in the area of nonlinear analysis. It includes the properties of Uryson integral operators and Nemyckii operators, the extreme values of some nonlinear functionals and their applications to the solution of some nonlinear operator equations.

1. Properties of Uryson integral and Nemyckii operators

1.1. Complete continuity of Uryson integral operators

Let G be a closed bounded subset of the Euclidean space R^n, and assume function $K(s,t,u)$ to be defined on $(s,t,u) \in G \times G \times R$. What we are considering is the complete continuity and related topics of the Uryson integral operator $\mathcal{K}$ defined by

$$\mathcal{K}\varphi(s) = \int_G K(s,t,\varphi(t))dt$$

on the space $C(G)$ of real-valued continuous functions on G. We assume that $K(s,t,u)$ satisfies: for every s in G, $K(s,t,u)$ is continuous with respect to u for almost all t in G, and for every (s,u) in $G \times (-\infty, \infty)$, $K(s,t,u)$ is Lebesgue measurable with respect to t in G.

Theorem 1.1. Operator $\mathcal{K}$ is bounded on every bounded subset of $C(G)$, if and only if $\mathcal{K}$ is sequentially weakly continuous, and if and only if for every $\alpha > 0$ there exists $\beta > 0$ such that

$$\int_G \sup_{|u| \le \alpha} |K(s,t,u)|dt \le \beta \tag{1.1}$$

for all s in G.

Theorem 1.2. Operator $\mathcal{K}$ is compact, that is, $\mathcal{K}$ sends every bounded subset of $C(G)$ into compact subset of $C(G)$, if and only if

1991 Mathematics Subject Classification: 47H

(i) $K(s, t, u)$ satisfies (1.1);

(ii) for every $\alpha > 0, \varepsilon > 0$ and every s in G there exists $\delta > 0$ such that

$$\int_G \sup_{|u| \leq \alpha} |K(s + h, t, u) - K(s, t, u)| dt < \varepsilon,$$

whenever $\|h\| < \delta$, where $s + h$ is in G.

Moreover, if $\mathcal{K}$ is compact then it is completely continuous. Hence $\mathcal{K}$ is completely continuous if and only if (i) and (ii) hold.

As a special case of the Uryson integral operator, we consider the following Hammerstein integral operator

$$\mathcal{K}\mathfrak{F}\varphi(s) = \int_G K(s, t) f(t, \varphi(t)) dt,$$

where $K(s, t)$, defined on $(s, t) \in G \times G$, is Lebesgue measurable with respect to t for every s, $f(t, u)$, defined on $G \times (-\infty, \infty)$, is Lebesgue measurable with respect to t for every u and continuous with respect to u for almost every t.

Theorem 1.3. For a given $\alpha > 0$, let

$$a_\alpha(t) = \sup_{|u| \leq \alpha} |f(t, u)|.$$

If for every $\alpha > 0$, there exist $a_1 \geq a_2 > 0$ such that $a_1 \geq a_\alpha(t) \geq a_2$ for all t in G, then the operator $\mathcal{K}\mathfrak{F}$ is completely continuous if and only if $K(s, t)$ satisfies

(iii) for every s in G,

$$\int_G |K(s, t)| dt < \infty;$$

(iv) for every s in G and a given $\varepsilon > 0$, there exists $\delta > 0$ such that

$$\int_G |K(s + h, t) - K(s, t)| dt < \varepsilon$$

whenever $\|h\| < \delta$.

From Theorem 1.3, the linear integral operator $\mathcal{K}$ defined by

$$\mathcal{K}\varphi(s) = \int_G K(s, t) \varphi(t) dt$$

is completely continuous if and only if (iii) and (iv) hold.

1.2. Properties of Nemyckii operators

Let $M(\cdot)$ be a Young function defined on $(-\infty, \infty)$, that is, $M(\cdot)$ satisfies

(i) $M(0) = 0$, $M(\cdot)$ is even, nondecreasing on $[0, \infty)$ and $\lim_{u \to \infty} M(u) = \infty$. $M(u)$ may be infinite for sufficiently large $|u|$;

(ii) $M(u)$ is continuous and convex for those u's where $M(u)$ is finite.

Let $M_1(\cdot)$ be another Young function. If for every $\nu > 0$,

$$\overline{\lim_{u \to \infty}} \frac{M_1(\nu u)}{M(u)} < \infty,$$

then we say that $M(\cdot)$ is essentially stronger than $M_1(\cdot)$. Here we make the convention $\frac{\infty}{\infty} = 1$.

Assume that G is a Lebesgue measurable subset of R^n with finite measure. Let

$$L_M^* = \left\{ u(\cdot) : \text{there exists } \beta > 0 \text{ such that } \int_G M(\beta u(t))dt < \infty \right\};$$

$$E_M = \left\{ u(\cdot) : \text{for all } \beta > 0, \int_G M(\beta u(t))dt < \infty \right\}.$$

Endowed with the norm

$$\|u\|_{(M)} = \inf \left\{ \frac{1}{\beta} : \int_G M(\beta u(t))dt \le 1 \right\},$$

both L_M^* and E_M are Banach spaces.

Assume that $f(t, u)$ is a real-valued function of (t, u) in $G \times (-\infty, \infty)$ and satisfies: for every u in $(-\infty, \infty)$, $f(t, u)$ is Lebesgue measurable with respect to t and for almost all t in G, $f(t, u)$ is continuous with respect to u. Let $\mathfrak{F}$ be the operator defined by

$$\mathfrak{F}\varphi(t) = f(t, \varphi(t)).$$

Theorem 1.4. Assume that $\mathfrak{F}$ is defined on $S_r = \{u(\cdot) \in L_M^* : \|u\|_{(M)} \le r\}$ with its range in $L_{M_1}^*$. Then $\mathfrak{F}$ is Frechét differentiable at zero if and only if $f(t, u)$ satisfies

(i) there exists a measurable function $g(\cdot)$ on G such that the operator $\mathfrak{G}$ defined by

$$\mathfrak{G}\varphi(t) = g(t)\varphi(t)$$

is a bounded linear operator from L_M^* into $L_{M_1}^*$, and $\frac{f(\cdot, u) - f(\cdot, 0)}{u}$ converges to $g(\cdot)$ in measure as $u \to 0$;

(ii) for a given $\varepsilon > 0$, there exists $\mu_0 \ge \frac{1}{r}$ such that for all $\mu \ge \mu_0$,

$$M_1 \left(\frac{\mu}{\varepsilon} [f(t, u) - f(t, 0) - g(t)u] \right) \le M(\mu u) + a_{\mu,\varepsilon}(t),$$

where $a_{\mu,\varepsilon}(\cdot)$ is a nonnegative Lebesgue measurable function satisfying

$$\lim_{\mu \to \infty} \int_G a_{\mu,\varepsilon}(t)dt = 0.$$

When (i) and (ii) are satisfied, $\mathfrak{G}$ is the Frechét derivative of $\mathfrak{F}$ at zero.

Corollary 1.5. If $f(t,u) - f(t,0)$ is nonlinear with respect to u for t in a subset of G with positive measure then, under the conditions (i), (ii) in Theorem 1.4, $M(\cdot)$ is essentially stronger than $M_1(\cdot)$, and $g(\cdot)$ is in $L^*_{\tilde{M}}$, where $\tilde{M}(\cdot)$ is the Young function defined by

$$\tilde{M}(u) = \sup_{v \geq 0}\{M_l(uv) - M(v)\}.$$

Corollary 1.6. Assume that $\mathfrak{F}$ is defined on L^{p_1} with its range in $L^{p_2}(p_1, p_2 \geq 1)$. If $f(t,u) - f(t,0)$ is nonlinear with respect to u for t in a subset of G with positive measure, then $\mathfrak{F}$ is Frechet differentiable at zero if and only if

(i) for $p_1 > p_2$, there exists a Lebesgue measurable function $g(\cdot)$ such that the operator $\mathfrak{G}$ defined by

$$\mathfrak{G}\varphi(t) = g(t)\varphi(t)$$

is a bounded linear operator from L^{p_1} into $L^{p'_1}$, where p'_1 satisfies $\frac{1}{p_1} + \frac{1}{p'_1} = \frac{1}{p_2}$, $\frac{f(t,u) - f(t,0)}{u}$ converges to $g(t)$ in measure as $u \to 0$;

(ii) for a sufficiently large $v > 0$, there exists nonnegative $b_\nu(\cdot) \in L^{p_2}$ such that

$$|f(t,u) - f(t,0) - g(t)u| \leq b_\nu(t) + \nu|u|$$

and $\left(\displaystyle\int_G b_\nu(t)^{p_2}\,dt\right)^{1/p_2} = o\left(\frac{1}{\nu^{p_1/p_2}}\right)$.

When (i), (ii) are satisfied, $\mathfrak{G}$ is the Frechet derivative of $\mathfrak{F}$ at zero.

1.3. Further properties of $\mathfrak{F}$

In this last part we study the uniform continuity and α-Lipschitz continuity $(0 < \alpha \leq 1)$ of $\mathfrak{F}$. We assume that $\mathfrak{F}$ is defined on $S_{r+a}(a > 0)$ with range in $L^*_{M_1}$.

Theorem 1.7. The operator $\mathfrak{F}$ is uniformly continuous on S_r if and only if for given $\mu > 0$, there exists $\rho > 0$ such that

$$M_1(\mu[f(t, u+v) - f(t,u)])$$

$$\leq M\left(\tfrac{1}{r}u\right) + M(\rho v) + a_\mu(t),$$

where $a_\mu(\cdot)$ is a nonegative Lebesgue measurable function satisfying

$$\int_G a_\mu(t)\,dt \leq 1.$$

Theorem 1.8. $\mathfrak{F}$ is α-Lipschitz continuous $(0 < \alpha \leq 1)$ on S_r if and only if there exists $B > 0$ such that for a sufficiently large $\mu \left(\mu > \frac{1}{a} \right)$,

$$M_1 \left(\frac{\mu^{\alpha}}{B} [f(t, u + v) - f(t, u)] \right)$$

$$M \left(\frac{1}{r} u \right) + M(\mu v) + a_{\mu}(t),$$

where $a_{\mu}(t)$ is a nonnegative Lebesgue measurable function satisfying

$$\int_G a_{\mu}(t) dt \leq 1.$$

2. Extreme values of nonlinear functionals and their applications

In this section we study the existence and uniqueness of extreme points of some nonlinear functionals and their applications to the solution of the following equation

$$P(u) = \omega, \tag{2.1}$$

where u, ω are elements in a Hilbert space H. The inner product of H is denoted by $\langle \cdot, \cdot \rangle$, $P(\cdot)$ is defined on a dense subspace H_0 of H and it is the gradient of a functional Φ. Thus the existence and uniqueness of solutions of (2.1) can be reduced to those of extreme points of Φ.

Proposition 2.1. Suppose that the functional Φ defined on H_0 satisfies
 (i) $\Phi(\cdot)$ is lower semicontinuous at $u = 0$;
 (ii) there exists a nonnegative function $\chi(\cdot)$ on $[0, \infty)$ satisfying

  $a.$ $\chi(0) = 0, \quad \chi(t) > 0$ for $t > 0$;

  $b.$ $\chi(\cdot)$ increases in $[0, \delta_0]$ for some $\delta_0 > 0$,

and

$$\frac{1}{2} \Phi(u_1) + \frac{1}{2} \Phi(u_2) - \Phi \left(\frac{u_1 + u_2}{2} \right) \geq \chi(\|u_1 - u_2\|) \tag{2.2}$$

for u_1, u_2 in H_0 satisfying $\|u_1 - u_2\| \leq \delta_0$.
  Then $\Phi(\cdot)$ is bounded from below and there exists a unique u_0 in H such that every minimizing sequence $\{u_n\}$ of $\Phi(\cdot)$ converges to u_0 in the norm topology of H.

We now assume that $P(\cdot)$ in (2.1) is the gradient of a functional $\Psi(\cdot)$ in H_0 satisfying (2.2). Then

$$\Phi(u) = \Psi(u) - \langle \omega, u \rangle$$

satisfies all conditions in Proposition 2.1 and hence there exists a unique u_0 in H such that every minimizing sequence $\{u_n\}$ of $\Phi(\cdot)$ converges to u_0 as $n \to \infty$; u_0 is called a generalized solution to equation (2.1).

In the following, we assume that $\Phi(\cdot)$ satisfies some more conditions. By definition, if u_0 is a minimal point of $\Phi(\cdot)$ then it is called a weak solution of (2.1).

Let H_0' be another dense subspace of H satisfying $H_0 \subseteq H_0' \subseteq H$. Assume that there is another inner product $[\cdot,\cdot]$ on H_0' satisfying

$$[u, u] \geq \beta_0^2 \langle u, u \rangle$$

for some $\beta_0 > 0$ and all u in H_0'. Denote $|u| = [u, u]^{1/2}$. We assume that H_0' is complete with respect to the norm $|\cdot|$.

Proposition 2.2. Assume that $\Phi(\cdot)$ is a functional on H_0' satisfying

(i) $\Phi(\cdot)$ is lower semicontinuous in the norm $|\cdot|$;

(ii) there exists $\chi(\cdot)$ satisfying conditions (ii, a, b) in Proposition 2.1 and

$$\frac{1}{2}\Phi(u_1) + \frac{1}{2}\Phi(u_2) - \Phi\left(\frac{u_1 + u_2}{2}\right) \geq \chi(|u_1 - u_2|). \tag{2.3}$$

Then $\Phi(\cdot)$ has a unique minimal point u_0 in H_0' and every minimizing sequence $\{u_n\}$ converges to u_0 in the norm $|\cdot|$.

Now assume that $\Psi(\cdot)$ is defined on H_0' and satisfies (2.3). Let $P(\cdot)$ be the gradient of $\Psi(\cdot)$ on H_0. Clearly, the functional $\Phi(\cdot)$ defined by

$$\Phi(u) = \Psi(u) - \langle \omega, u \rangle$$

satisfies (2.3). Hence the equation (2.1) has a unique weak solution u_0 in H_0'.

Applying Propositions 2.1, 2.2, we may study the existence and uniqueness of solutions to some nonlinear partial differential equations. We omit the details here.

3. Solutions of Hammerstein operator equations

In this section, we are devoted to the study of the following Hammerstein operator equation

$$x = \mathcal{K}\mathfrak{F}(x), \tag{3.1}$$

where $\mathcal{K}$ is a bounded selfadjoint operator on a real Hilbert space H, $\mathfrak{F}$ is a potential operator on H. Previous works on this topic were concentrated on the case of $\mathcal{K}$ being positive or quasi-positive. We shall assume that $\mathcal{K}$ is an arbitrary bounded or compact selfadjoint operator on H. We say that λ is an eigenvalue of $\mathcal{K}$, if $x = \lambda \mathcal{K}x$ has nontrivial solutions.

Theorem 3.1. Assume that $\mathcal{K}$ is a compact and selfadjoint operator with eigenvalues

$$\cdots < \lambda_{-2} < \lambda_{-1} < 0 < \lambda_0 < \lambda_1 < \cdots,$$

$\mathfrak{F}$ is a continuous potential operator with potential f: grad $f = \mathfrak{F}$. If there exist an integer N and constants μ_N, μ_{N+1} such that

$$\lambda_N < \mu_N < \mu_{N+1} < \lambda_{N+1},$$

and for x, h in H,

$$\mu_N \|x\|^2 - c_1 \leq 2f(x) \leq \mu_{N+1}\|x\|^2 + c_2;$$

$$f(x + h) - f(x) < \lambda_{N+1}\|h\|^2 \quad (h \neq 0)$$

where c_1, c_2 are constants, then (3.1) has solutions in H. If we further assume that

$$f(x + h) - f(x) > \lambda_N \|h\|^2 \quad (h \neq 0),$$

then the solutions are unique.

Theorem 3.2. Assume that $\mathcal{K}$ is bounded and selfadjoint. $(\lambda_1, \lambda_2) \subseteq \rho(\chi)$ and $0 \notin (\lambda_1, \lambda_2)$. $\mathfrak{F}$ is a potential operator and bounded on every bounded subset of H. If there exist μ_1, μ_2 satisfying

$$\lambda_1 < \mu_1 < \mu_2 < \lambda_2$$

such that

$$\frac{1}{\mu_2}\|h\|^2 \leq \langle f(x + h) - f(x), h \rangle \leq \frac{1}{\mu_1}\|h\|^2,$$

then equation (3.1) has a unique solution in H.

References

[1] C. L. Dolph, Nonlinear integrai equations of the Hammerstein type, Trans. Amer. Math. Soc., 66(1949), 289–307.

[2] M. Golomb, Zur Theorie der Schwingungen, Monatshefte für Mathematik, 42(1935), 7–36.

[3] A. Hammerstein, Nichtlinear Integralglerchungen nebst Anwendungen, Acta Math. 54(1930), 117–176.

[4] R. Iglish, Existence und Eindeutigkeit Satze bei nichtlinear Integralgleichungen, Math. Ann., 108(1933), 161–189.

[5] M. A. Krasnosel'skii, Topological method in the theory of nonlinear integral equations, Oxford, Pergamon Press, 1964.

[6] L. A. Ladzhenskii, Conditions for complete continuity of P. S. Uryson's integral operator in the space of continuous functions, Dokl. Adad. Nauk SSSR 97(1954), 1105–1108.

[7] A. Langenbach, Applications of variational principle to some nonlinear differential equations, Dolk. Akad. Nauk SSSR 121(1958), 214–217.

[8] S. G. Mikhlin, Variational method in mathematical physics, Gostekhzdat, Moscow 1957.

[9] I. V. Shragin, On a nonlinear operator, Nuchn. gokl. vesch. sch., F.-M. N. no. 2(1958), 103–108.

[10] I. V. Shragin, On Hammerstein operators and equations, Nuchn. gokl. vesch. sch., F.-M. N. no. 3(1958), 107–110.

[11] M. M. Vainberg, Variational method in the theory of nonlinear integral equations, Gostekhzdat, Moscow 1956.

[12] S. W. Wang, A note on Marcinkewicz-Orlicz spaces, Bull. Acad. Polon. Sci., 7(1959), 707–710.

[13] S. W. Wang, On the solutions of Hammerstein integral equations ibid. 8(1960), 339–342.

[14] S. W. Wang, Convex functions of several variables and vector-valued Orlicz spaces, idib. 11(1963), 279–284.

[15] S. W. Wang, On the product of Orlicz spaces, idib. 11(1963), 19–22, Shuxue Jinzhan 7(1964), 282–294.

[16] S. W. Wang, The complete continuity and the strengthening continuity of Uryson integral operators, Dolk. Acad. Nauk SSSR, 151(1963), 1010–1013, Acta Math. Sinica 13(1963), 254–261.

[17] S. W. Wang, A note on the solutions of some nonlinear partial differential equations, idib. 150(1963), 967–970.

[18] S. W. Wang, On the differentiablity of Nemyckii operators, ibid. 150(1963), 1198–1201.

[19] S. W. Wang, On the solutions of Hammerstein integral equations II, ibid. 154(1964), 9–12.

[20] S. W. Wang, On the solutions of Hammerstein integral equations III, Acta Math. Sinica, 15(1965), 63–73.

[21] S. W. Wang, A kind of nonlinear operators in a Hilbert space and its applications, Shuxue Jinzhan, 9(1966), 27–32.

[22] S. W. Wang, Some properties of Nemyckii operators, Koxue Tonboa, Math. Issue (1980), 39–42.

Embedding Properties of
Locally Convex Riesz Spaces

Yau-chuen Wong

Department of Mathematics, The Chinese University of Hong Kong, Hong Kong

1. Introduction

For any locally convex Riesz space $(X, X_+, \mathcal{P})$, it is well-known that its strong dual $(X', X'_+, \beta(X', X))$, equipped with the dual cone X'_+, is a Dedekind complete locally convex Riesz space, so that its bidual (X'', X''_+) is a Dedekind complete Riesz space. It is also clear that (X, X_+) can be embedded as a Riesz subspace of (X'', X''_+) under the evaluation map. As any Riesz space X is a band in itself X, any band in a Riesz space is an order closed ideal, a subspace is an ideal if and only if it is an order-convex Riesz subspace, and the canonical embedding from an ideal always preserves artitrary supremum and infimum, it is natural to ask the following questions:

(a) What condition is necessary and sufficient for the embedding to preserve arbitrary supremum and infimum?

This also suggests the following three intimately related questions.

(b) What condition on X is necessary and sufficient for X to be an ideal in (X'', X''_+)?

(c) What condition on X (or X') is necessary and sufficient for X to be a band in (X'', X''_+)?

(d) What condition on X (or X') is necessary and sufficient for the semi-reflexivity of X?

These questions have been studied by many authors, in particular, for the first two questions in the case of Banach lattices. This note is mainly devoted to give such informations as well as some applications.

2. Locally convex Riesz spaces

We shall assume throughout this paper that the scalar field for vector spaces is the field $\mathbb{R}$ of real numbers, and that all topological vector spaces will be Hausdorff.

1991 Mathematics Subject Classification: 46A03

The author died in 1994

Let (X, X_+) be a Riesz space (i.e. vector lattice) with positive cone X_+. For any $x \in X$, the positive part, negative part and absolutely value of x are defined respectively by

$$x^+ = \sup\{x, o\} = x \lor o; \qquad x^- = (-x) \lor o \qquad \text{and} \quad |x| = x \lor (-x).$$

The index set $\{x_i : i \in \Lambda\}$ (abbreviated to $[x_i]$) of (X, X_+) is directed upwards (in symbols $x_i \uparrow$) if for any $i_1, i_2 \in \Lambda$ there is $i \in \Lambda$ such that

$$x_{i_1} \leq x_i \text{ and } x_{i_2} \leq x_i.$$

The symbol $x_i \downarrow$ denotes directed downwards and its definition is analogous. If $x_i \uparrow$, then the index set Λ equipped with the ordering $\leq$, defined by

$$i_1 \leq i_2 \text{ if } x_{i_1} \leq x_{i_2},$$

is a directed set, hence $\{x_i, i \in \Lambda\}$ is a net with $x_{i_1} \leq x_{i_2}$ whenever $i_1 \leq i_2$; in this case, $[x_i]$ is also called an increasing net. Decreasing nets are defined similarly. If $x_i \uparrow$ and $x = \sup\limits_{i \in \Lambda} x_i$ exists, then we write $x_i \uparrow x$. Recall that (X, X_+) is Dedekind complete (resp. σ-Dedekind complete) if $x_i \uparrow\leq x$ (resp. $x_n \uparrow\leq x$) then $\sup x_i$ (resp. $\sup x_n$) exists in X.

A net $\{x_i, i \in \Lambda\}$ in a Riesz space (X, X_+) is said to be order convergent to u (in symbols $x_i \overset{(o)}{\to} u$ or $u = (o) - \lim_i x_i$) if there exists a net $\{u_i, i \in \Lambda\}$ with the same indexed set Λ such that

$$|x_i - u| \leq u_i \downarrow o \text{ (i.e. } |x_i - u| \leq u_i \text{ and } u_i \downarrow o);$$

in this case, u is referred to as the order limit of $[x_i]$. It is easily seen that the order limit (when it exists) is unique.

It should be noted that our terminology for order convergence differs slightly from that of Peressini [22, p.43] in that we require an order convergent net to be order bounded.

A subset B of (X, X_+) is said to be order closed (resp. σ-order closed) if

$$u = (o) - \lim_i x_i \text{ with } x_i \in B \text{ (resp.} u = (o) - \lim_n x_n \text{ with } x_n \in B) \Rightarrow u \in B.$$

One can show that a solid set V in (X, X_+) is order closed (resp. σ-order closed) if and only if for any net $[u_i]$ (resp. sequence $[x_n]$) in V,

$$o \leq u_i \uparrow u \text{ (resp. } o \leq x_n \uparrow u) \Rightarrow u \in B.$$

[Let $[x_i]$ be a net in V with $x = (o) - \lim_i x_i$, and let $[y_i]$ be a net in X such that $|x_i - x| \leq y_i \downarrow o$. It is not hard to see that

$$(|x| - y_i)^+ = |x| \lor y_i - y_i \leq |x_i| \text{ and } (|x| - y_i)^+ \downarrow |x| \text{(with } x_i \in V),$$

the sufficiency then follows.]

Recall that a subspace M of (X, X_+) is a (or an):

(a) Riesz subspace if $x^+ \in M$ (for all $x \in M$);

(b) ideal (or ℓ-ideal) if it is solid;

(c) band or normal subspace (resp. σ-ideal) if it is solid as well as order closed (resp. σ-order closed).

Consequently an ideal M in (X, X_+) is a band (resp. σ-ideal) if and only if for any net $[u_i]$ (sequence $[x_n]$) in M,

$$u_i \uparrow x \text{ or } u_i \downarrow x \text{ (resp. } x_n \uparrow x \text{ or } x_n \downarrow x) \Rightarrow x \in M.$$

An ideal in a Dedekind complete Riesz space must be Dedekind complete.

Let (X, X_+) and (Y, Y_+) be Riesz spaces. A ℓ-homomorphism $T : (X, X_+) \to (Y, Y_+)$ (i.e. $Tx^+ = (Tx)^+$ for all $x \in X$) is a normal integral or order continuous (resp. σ-order continuous or integral) if

$$u = (o) - \lim_i x_i \text{ (resp. } u = (o) - \lim_n x_n) \text{ in } (X, X_+)$$

$$\Rightarrow Tu = (o) - \lim_i Tx_i \text{ (resp. } Tu = (o) - \lim_n Tx_n) \text{ in } (Y, Y_+)$$

or, equivalently,

$$u_i \downarrow o \text{ (resp. } x_n \downarrow o) \text{ in } (X, X_+) \Rightarrow Tu_i \downarrow o \text{ (resp. } Tx_n \downarrow o) \text{ in } (Y, Y_+),$$

and this is the case, if and only if

$$T(\sup B) = \sup TB \text{ (resp. } T(\sup\{z_n : n \geq 1\} = \sup_{n \geq 1} Tz_n)$$

whenever $\sup B$ (resp. $\sup_n z_n$) exists in X.

Moreover, a surjective ℓ-homomorphism $T : (X, X_+) \to (Y, Y_+)$ is a normal integral if and only if $\operatorname{Ker} T$ is a band in X (see [27, (I.3.5)]). In particular, if M is an ideal in (X, X_+), then the canonical embedding $j_M : M \to (X, X_+)$ is always a normal integral.

Let (X, X_+) be a Riesz space. We denote by X^* the algebraic dual of X, by X_+^* the dual cone of X_+, i.e.

$$X_+^* = \{f \epsilon X^* : f(u) \geq 0 \text{ for all } u \in X_+\},$$

and by X^b the order-bound dual of X (which is the vector space consisting of all order-bounded linear functionals on X). Then

$$X^b = X_+^* - X_+^*,$$

and (X_b, X_+^*) is a Dedekind complete Riesz space. The set consisting of all normal integrals (resp. integrals) order-bounded linear functionals on (X, X_+), denoted by X_n^b (resp. X_c^b), is a band in (X^b, X_+^*) with

$$X_n^b \subset X_c^b.$$

We also write

$$X_{nn} = (X_n^b)_n^b.$$

If X_n^b is total over X, then the evaluation map $K_X^{(n)} : X \to X_{nn}$, defined by

$$< f, K_X^{(n)} x > = < x, f > (= f(x)) \qquad \text{(for all } f \epsilon X_n^b),$$

is an ℓ-isomorphism (i.e. injective ℓ-homomorphism) from X into X_{nn}.

By a topological Riesz space (or topological vector lattice) we shall mean a Riesz space (X, X_+) equipped with a vector topology $\mathcal{P}$ which admits a local base of 0 consisting of solid sets in (X, X_+) ($\mathcal{P}$ is called a lattice topology). In addition, if $\mathcal{P}$ is locally convex, then $(X, X_+, \mathcal{P})$ is called a locally convex Riesz space (abbreviated by LCRS), and in this case, $\mathcal{P}$ is referred to as a locally solid topology on X.

It is clear that a locally convex topology $\mathcal{P}$ on (X, X_+) is locally solid if and only if $\mathcal{P}$ admits a local base of 0 consisting of convex and solid sets in X or, equivalently, $\mathcal{P}$ is determined by a family $\{p_\alpha : \alpha \epsilon \Gamma\}$ of Riesz seminorms in the sense that

$$|x| \leq |y| \Rightarrow p_\alpha(x) \leq p_\alpha(y),$$

and this is the case, if and only if $\mathcal{P}$ is locally o-convex (i.e. $\mathcal{P}$ admits a local base of 0 consisting of o-convex (i.e. order-convex and convex) and circled sets as well as locally decomposable (i.e. $\mathcal{P}$ admits a local base of 0 consisting of decomposable and convex sets in X)-Nachbin's result [18, p.89] (a generalization was given in [35, Theorem 3.1]).

Let $\mathcal{T}$ be a vector topology on a Riesz space (X, X_+). We say that $(X, X_+, \mathcal{T})$ or $\mathcal{T}$ is:

(a) boundedly Dedekind complete (Levi property in the terminology of Fremlin [8, p.49]) if each $\mathcal{T}$-bounded subset of X_+ which is directed upwards has a supremum;

(b) order regular (Fatou property in the terminology of Fremlin [8]) if $\mathcal{T}$ admits a local base of 0 consisting of order closed and circled sets in X.

It is clear that any order closed solid set B in a Dedekind complete Riesz space (X, X_+) must be Dedekind complete in the sense that

$$x_i \uparrow \leq u \text{ with } x_i \epsilon B \Rightarrow \sup x_i \text{ exists and } x \epsilon B;$$

and that any boundedly Dedekind complete topological Riesz space must be Dedekind complete. A topological Riesz space $(X, X_+, \mathcal{P})$ is order regular and Dedekind complete if and only if $(X, X_+, \mathcal{P})$ is locally Dedekind complete (in the terminology of Nakano [20]) in the sense that $\mathcal{P}$ adimits a local base of 0 consisting of solid and Dedekind complete sets in X.

The following theorem is one of the deepest results in the theory of topological Riesz spaces.

(2.1) **Nakano's theorem.** Let $(X, X_+, \mathcal{P})$ be a Dedekind complete, order regular (i.e. locally Dedekind complete) topological Riesz space. Then each order-interval [a,b] in X is $\mathcal{P}$-complete.

If, in addition, $(X, X_+, \mathcal{P})$ is boundedly Dedekind complete, then $(X, \mathcal{P})$ is complete.

Nakano's proof of his theorem contained a gap (see [20, Theorem 3.3]), filled later by Schaefer [23, Theorem 1], Aliprantis and Burkinshaw [2, Theorem 3.2], [3, Theorem] give another proof of the first part of Theorem 2.1, they also generalize the result concerned the completeness of $(X, \mathcal{P})$ under the assumptions that $(X, X_+, \mathcal{P})$ is boundedly Dedekind complete as well as each order interval in X is $\mathcal{P}$-complete (see [2, Theorem 5.4]). The proof of Theorem 2.1 can also be found in [22, (IV. 1.3), (IV.1.5)] or [39, (11.13) and (11.14)].

Let $(X, X_+, \mathcal{P})$ be a LCRS. The topological dual of $(X, X_+, \mathcal{P})$, denoted by X', is an ideal in (X^b, X_+^*), hence (X', X_+') is a Dedekind complete Riesz space, where

$$X_+' = X' \cap X_+^*.$$

It is also clear that the polar of any solid subset of X is solid and order closed in (X', X_+'), and that the solid hull of any $\mathcal{P}$-bounded subset of X is P-bounded, it follows that the strong topology $\beta(X', X)$ on X' is a locally solid topology with a local base of 0 consisting of order closed, solid, convex sets in X' (i.e. $\mathcal{P}$ is order regular), and hence that the bidual of $(X, X_+, \mathcal{P})$, denoted by (X'', X_+''), is a Dedekind complete Riesz space. Moreover, we have:

(2.2) Theorem. Let $(X, X_+, \mathcal{P})$ be a LCRS with the topological dual X'. Then $(X', X_+', \beta(X', X))$ is a Dedekind complete LCRS which is also order regular (i.e. locally Dedekind complete). Moreover, the following assertions hold:

(a) The evaluation map $K_X : (X, X_+) \to (X'', X_+'')$, defined by

$$< f, K_X x > = < x, f > \quad \text{(for all } x \epsilon X \text{ and } f \epsilon X'),$$

is an ℓ-isomorphism from (X, X_+) into (X'', X_+''), hence we identify X with a Riesz subspace of (X'', X_+'').

(b) ([30, Prop. 3]). If $(X, X_+, \mathcal{P})$ is infrabarrelled, then $(X', X_+', \beta(X', X))$ is boundedly Dedekind complete and a fortiori complete.

(c) ([34, Theorem]). If X' is a band in (X^b, X_+'), then X' is complete for $\beta(X', X)$.

The proof can be found in [37, (9.12)].

Schaefer [23, Theorem 3] and Kawai [11, p.290] proved part (b) for barrelled spaces.

As locally solid topologies are locally o-convex, and the converse is, in general, not true, it is natural to associate a locally solid topology from a given locally o-convex topology. To do this, let us recall that if V is a subset of a Riesz space (X, X_+), then the solid kernel of V, defined by

$$sK(V) = \{x \epsilon X : S_x = [-|x|, |x|] \subset V\} (= \cup \{S_x : S_x \subset V\}),$$

is either empty or the largest solid subset of X contained in V. Suppose now that $\mathcal{T}$ is a locally o-convex topology on (X, X_+) and that $\mathcal{U}$ is a local base of 0 consisting of o-convex and circled sets in X. Then the faimily

$$\mathcal{U}_s = \{sK(V) : V \in \mathcal{U}\}$$

determines a unique locally solid topology $\mathcal{T}_s$, called the locally solid topology associated with $\mathcal{T}$, such that $\mathcal{T}_s$ is the greatest lower bound of all locally solid topologies which are finer than $\mathcal{T}$ (see [29, (3,2)]). Moreover

$$(X, X_+, \mathcal{T}_s)' = \text{ the ideal generated by } (X, X_+, \mathcal{T})'$$

[see [29, (3.9)]], this is a generalization of a result of Kaplan [10, Prop. 10.1, p.349] also Gordon [9, Theorem 14]. In particular, let Y be an ideal in (X^b, X_+^*) (or more general, a decomposable subspace of (X^b, X_+^*)) which is total over X. Then $\sigma(X, Y)$ (the weak topology) is always a locally o-convex topology (see [21]), hence the locally solid topology associated with $\sigma(X, Y)$, denoted by $\sigma_s(X, Y)$, is consistent with the dual pair $< X, Y >$ (resp.

$$(X, X_+, \sigma_s(X, Y))' = \text{ the ideal in } (X^b, X_+^*) \text{ generated by } Y), \text{ moreover,}$$

$$\sigma_s(X, Y) = \text{ the topology of uniform convergence on}$$

$$\{[-f, f] : f \in Y \cap X_+^*\}$$

$$= |\sigma|(X, Y)(\text{the absolute weak topology on} X),$$

where $|\sigma|(X, Y)$ is determined by $\{p_f : f \epsilon Y \cap X_+^*\}$ and

$$p_f(x) = f(|x|) \qquad (\text{for all } x \epsilon X)$$

(see [36, (1.3.15)(b)] or [29, (3.10)]). This associated locally solid topology $\mathcal{T}_s$ has some remarkable properties (see [29]), in particular, we have the following:

(2.3) Proposition. Let $(X, X_+, \mathcal{P})$ be a LCRS with the topological dual X'.

(a) ([30, Prop. 2]). $\sigma_s(X', X)$ is order regular (in fact, locally Dedekind complete). i.e. it admits a local base of 0 consisting of order closed (in fact Dedekind complete), solid and convex sets in (X', X_+').

(b) ([32, Prop. 3.1] or [30, Prop. 4]). If $(X, X_+, \mathcal{P})$ is order-infrabarrelled(i.e. each $\sigma_s(X', X)$-bounded subset of X' is $\mathcal{P}$-equicontinuous), then $\sigma_s(X', X)$ is boundedly Dedekind complete (hence X' is complete for $\sigma_s(X', X)$ and also for $\beta(X', X)$), and X' is a band in (X^b, X^*).

3. Embedding of a locally convex Riesz space into its bidual

Let $(X, X_+, \mathcal{P})$ be a locally convex Riesz space with the bidual (X'', X''_+). It is known from (2.2)(a) that (X, X_+) can be embedded as a Riesz subspace of (X'', X''_+) (or $(X')^b_n$) under the evaluation map $K_X : (X, X_+) \to (X'', X''_+)$. Therefore it is natural to ask the following questions:

(a) What condition is necessary and sufficient for the embedding to preserve the supremum and infimum for infinite subsets of X (i.e. K_X is order continuous)?

(b) (resp. (c)) What condition on X (or X') is necessary and sufficient for X to be an ideal (resp. a band) in (X'', X''_+)?

(d) What condtion on X (or X') is necessary and sufficient for X to be semi-reflexive?

The answer of the first question is related to the order continuity of the locally solid topology as shown by the following result.

(3.1) Theorem. For a LCRS $(X, X_+, \mathcal{P})$, the following statements are equivalent:

(a) The evaluation map $K_X : (X, X_+) \to (X'', X''_+)$ is order continuous, i.e. preserves arbitrary supremum and infimum $((X, X_+)$ is called a regular Riesz subspace of $(X'', X''_+))$.

(b) $\mathcal{P}$ is order continuous (Lebesque in the terminology of Fremlin [8]) in the sense that

$$0 \le u_i \downarrow 0 \Rightarrow 0 = \mathcal{P} - \lim_i u_i.$$

(c) $X' \subset X^b_n$ (i.e. continuous linear functionals are order continuous).

(d) The ideal in (X'', X''_+) generated by X is a Dedekind completion of (X, X_+).

(e) Each order-dense ideal in (X, X_+) is $\mathcal{P}$-dense in X.

(f) Each $\mathcal{P}$-closed ideal in X is order closed (i.e. a band in X).

(g) For any $f \epsilon X'$, the ideal, defined by

$$N_f = \{x \epsilon X : |f|(|x|) = 0\},$$

is a band in X.

(h) X is an order dense Riesz subspace of $(X')^b_n$.

(i) Every order closed ideal (i.e. band) in (X', X'_+) is $\sigma(X', X)$-closed.

(j) $\mathcal{P}$ is order regular (i.e. it admits a local base of 0 consisting of order closed, solid and convex sets in X) and every order-bounded increasing net in X_+ is a $\mathcal{P}$-Cauchy net (called monotone Cauchy property or Pre-Lebesque property in the terminology of Fremlin [8]).

(k) $\sigma_s(X, X')$ is order regular.

(l) $\sigma_s(X, X')$ is order continuous.

If $(X, \mathcal{P})$ is complete, then (a) is equivalent to the following:

(m) $\mathcal{P}$ has the monotone Cauchy property (i.e. Pre-Lebesque property).

The proof can be found in [39,(13.1)] for (a)-(g), and in [4, (9.1), (10.3), (11.6), (11.7)] for (h)-(m).

Remarks. (i) Fremlin[8] shows that for any LCRS $(X, X_+, \mathcal{P})$ the monotone Cauchy property of $\mathcal{P}$ is equivalent to say that any order bounded disjoint sequence in X_+ is a $\mathcal{P}$-null sequence.

(ii) For any LCRS $(X, X_+, \mathcal{P})$, the absolute weak topology $\sigma_s(X, X')$ always has the monotone Cauchy property $\big($hence the implication (k) $\Rightarrow$ (l) in (3.1) follows from (b) $\Leftrightarrow$ (j) in (3.1) $\big)$.

(iii) As an immediate consequence of (3.1) (j), any complete LCRS $(X, X_+, \mathcal{P})$ for which $\mathcal{P}$ is order continuous, is always Dedekind complete.

(iv) By a similar argument given in the proofs of (a) $\Leftrightarrow$ (b)$\Leftrightarrow$(c)$\Leftrightarrow$(g) in (3.1), one can show that the following statements are equivalent for a LCRS $(X, X_+, \mathcal{P})$:

(a) The evaluation map $K_X : (X, X_+) \to (X'', X_+'')$ is σ-order continuous.

(b) $\mathcal{P}$ is σ-order continuous (σ-Lebesque in the terminology of Fremlin[8]) in the sense that

$$o \leq u_n \downarrow 0 \Rightarrow 0 = \mathcal{P} - \lim_n u_n.$$

(c) $X' \subset X_c^b$.

(d) For any $f \epsilon X'$, the ideal N_f is a σ-ideal.

The equivalence (a)$\Leftrightarrow$(b) is due to Nakano [20], the equivalence (b)$\Leftrightarrow$(g) is due to Ando $\big($see [17, (47.3)]$\big)$, the equivalence (a)$\Leftrightarrow$(c)$\Leftrightarrow$(d)$\Leftrightarrow$(e)$\Leftrightarrow$(f)$\Leftrightarrow$(h) was proved in [17, Note XIV, Theorem 47.8] for normal Riesz spaces, while statements (i) through (m) can be found in [4, (11.6), (11.7), (10.3)].

The following result, due to Luxemberg and Zaanen [14, Note X (33.8) and Note XVI, (61.1) (iii)] for the normed case, can be found in [4, (17.8)].

(3.1)* Theorem. For a metrizable LCRS $(X, X_+, \mathcal{P})$, the following statements are equivalent.

(b) $\mathcal{P}$ is order continuous.

(n) (i) $\mathcal{P}$ has the monotone Cauchy (i.e. Pre-Lebesque) property;

(ii) if $0 \leq u_n \downarrow 0$ and $[u_n]$ is $\mathcal{P}$-Cauchy, then $0 = \mathcal{P} - \lim_n u_n$ (called quasi σ-order continuous or pseudo σ-Lebesque property in the terminology of [4, (17.1)].

(l) (i) Same as (n)(i).

(ii) $\mathcal{P}$ is σ-order continuous (σ-Lebesque in the terminology of Fremlin [8]) in the sense that

$$0 \leq u_n \downarrow 0 \Rightarrow 0 = \mathcal{P} - \lim_n u_n.$$

It should be noted that the metrizability condition in the preceding result is essential, Aliprantis [1] (also see [4, (8.6)] gives an example to show that the preceding result is false for non-metrizable case.

The following result answers the second question posed at the beginning of this section, namely: What condition is necessary and sufficient for X to be an ideal in (X'', X''_+)?

(3.2) Theorem. The following statements are equivalent for a LCRS $(X, X_+, \mathcal{P})$.

(a) X is an ideal in (X'', X''_+).

(b) X is an ideal in $(X')_n^b$.

(c) $X' \subset X_n^b$ and (X, X_+) is Dedekind complete.

(d) $\mathcal{P}$ is order continuous and (X, X_+) is Dedekind complete.

(e) Every order-bounded increasing net in X_+ is $\mathcal{P}$-convergent (i.e. $\mathcal{P}$ has the monotone convergent property or Bolzano property).

(f) $\sigma_s(X', X)$ is consistent with the dual pair $< X, X' >$.

(g) Each order-interval in X is $\sigma(X, X')$-compact.

(h) Each order-bounded disjoint sequence in X is a $\mathcal{P}$-null sequence and each order-interval in X is $\mathcal{P}$-complete.

(i) $\mathcal{P}$ has the monotone Cauchy property and each order interval is $\mathcal{P}$-complete.

(j) X is an order dense ideal in $(X')_n^b$.

Of course, the order continuity of $\mathcal{P}$ $\big($in (d)$\big)$ can be replaced by any one of the statements in (3.1).

The proof can be found in [37, (10.5)].

Remark. The equivalence of (c), (d) and (e) are true for a locally o-convex Riesz space (see [29, (3.13) and (3.14)]).

The equivalence between (a), (b), (c), (d) and (f) were proved by Kawai [11, theorem 5.1] in 1957, while the equivalence between (a), (c), (d), (e) and (g) were verified by Wong [29, (3.13), (3.14) and (3.9)] for the case of locally o-convex Riesz spaces during 1969. In 1977, Burkinshaw and Dodds [5, Prop. 5.1] only give the equivalence of (a) and (h), not all equivalent statements in (3.3) are due to Burkinshaw and Dodds in [5, Prop 5.1] (Please compare with the information given by Aliprantis and Burkinshaw in [4, Note p.160]).

As a consequence of (3.2) and (3.1), we obtain the following:

(3.2.1) Corollary. Let $(X, X_+, \mathcal{P})$ be a LCRS, and suppose that $\mathcal{P}$ is order continuous. Then the following statements are equivalent.

(a) X is order-convex in (X'', X''_+).

(b) Each order-interval in X is $\mathcal{P}$-complete.

(c) The Mackey topology $\tau(X', X)$ is locally decomposable.

Proof. (a) ⇒ (b): It is clear that a subspace is an ideal if and only if it is order-convex and a Riesz subspace, hence the implication follows from (3.2)(h).

(b) ⇒ (a): As $\mathcal{P}$ is order continuous, it follows from (3.1)(j) that $\mathcal{P}$ is order regular and has the monotone Cauchy property, and hence from (3.2)(i), together with (b), that X is an ideal, and surely order-convex in (X'', X''_+).

(a) ⇔ (c): As the cone X'_+ is generating, it follows from a result of [33, Proposition 1] or [38, (24.16)(b)] that these two statements are equivalent.

It is natural to ask whether the implication (b) ⇒ (a) in (3.2.1) holds without the assumption on the order continuity of $\mathcal{P}$.

Let $(X, X_+, \mathcal{P})$ and $(Y, Y_+, \mathcal{T})$ be two LCRSs. We say that X is embedded in Y (or Y contains X) if there exists a Riesz subspace M of Y and a linear map T such that T is an ℓ-isomorphism and homeomorphism form X onto M.

As an immediate consequence of (3.2) ((exactly parts (a) and (d) of (3.2)) and Remark (iv) of Theorem 3.1, we obtain the following:

(3.2.2) Corollary. For a complete LCRS $(X, X_+, \mathcal{P})$, the following statements are equivalent:

(a) $\mathcal{P}$ is order continuous.

(a)* Any statement in (3.1).

(b) X is an ideal in (X'', X''_+).

(b)* Any statements in (3.2).

(c) l^∞ is not embedded in $(X, X_+, \mathcal{P})$ and (X, X_+) is σ-Dedekind complete.

Proof. We have to verify the equivalence of (a) and (c).

(a) ⇒ (c): By Remark (iv) of (3.1), (X, X_+) is Dedekind complete and a fortiori σ-Dedekind complete. On the other hand, $\mathcal{P}$ has the monotone Cauchy (i.e. Pre-Lebesque) property, it then follows from [4, (10.7)] that l^∞ is not embedded in $(X, X_+, \mathcal{P})$.

(c) ⇒ (a): By [4, (10.7)] again, $\mathcal{P}$ has the montone Cauchy property, it then follows from Theorem 3.1(m) that $\mathcal{P}$ is order continuous.

The equivalence of (a) and (b) in the preceding result was proved by many authors for the case of Banach lattices (see [27, §3 of Chapter II]); while the equivalence of (a) and (c) in Corollary 3.2.2, due to Lozanovskij for the case of Banach Lattices, is one of the deepest results in the theory of Banach Lattices (see [27, (II.3.7), p.78]).

As the topological dual (X', X'_+) of any LCRS $(X, X_+, \mathcal{P}0$ is Dedekind complete, and the topologies $\beta(X', X)$ and $\sigma_s(X', X)$ are order regular (see (2.2) and (2.3)(a)), the following result is an immediate consequence of (3.2) and (3.1):

(3.2.3) Corollary. For a LCRS $(X, X_+, \mathcal{P})$ with the topological dual X', the following statements are equivalent:

(a) $\beta(X',, X)$ (resp. $\sigma_s(X', X)$) is order continuous.

(b) $\beta(X', X)$ (resp. $\sigma_s(X', X)$) has the monotone Cauchy property.

(c) Any order-bounded disjoint sequence in X'_+ is a $\beta(X', X)$- (resp. $\sigma_s(X', X)$-) null sequence.

(d) $\beta(X', X)$ (resp. $\sigma_s(X', X)$) has the monotone convergent property.

(e) Each order-interval in X' is $\sigma(X', X'')$-compact (resp. $\sigma(X', l(X))$-compact, where $l(X)$ is the ideal in X'' generated by X).

Consequently, if $\beta(X', X)$ is order continuous then so does $\sigma_s(X', X)$.

The following result answers the third question posed at the beginning of this section, namely: What condition is necessary and sufficient for X to be a band in X''?

(3.3) Theorem (Wong [32, Thm 2, Cor. 5] or [39, (13.9), (13.12)]). For a LCRS $(X, X_+, \mathcal{P})$, the following statements are equivalent.

(a) X is a band in (X'', X''_+).

(b) $\mathcal{P}$ is both order continuous and boundedly Dedekind complete.

(c) $(X, X_+, \sigma_s(X, X')$ is both locally Dedekind complete (or order regular) and boundedly Dedekind complete (called a Nakano space).

(d) X is $\sigma_s(x, X')$-complete.

(e) Each $\sigma(X, X')$-bounded increasing net in X_+ is $\sigma(X, X')$-convergent (it is said that $\sigma(X, X')$ has the boundedly convergent property).

(f) $\mathcal{P}$ has the boundedly convergent property.

(g) X is a band in $(X')^b_n$.

(h) $X = (X')^b_n$.

(i) X is $\mathcal{P}$-complete and $\mathcal{P}$ has the boundedly Cauchy property (i.e. every $\mathcal{P}$-bounded increasing net in X_+ is $\mathcal{P}$-Cauchy).

(j) X is $\mathcal{P}$-complete, and each $\mathcal{P}$-bounded increasing sequence in X_+ is $\mathcal{P}$-Cauchy.

A Banach lattice satisfying any one of the statements in (3.3) is called a KB-space by Vulikh (see [25, p.92]).

The proof can be found in [32, Theorem 2, Corollary 5] or [39, (13.9), (13.12)] or [37, (10.6)].

Remark. Any one of the statements in (3.3) implies that X is complete for $\beta(X', X)$ (see [34, Corollary 1]).

In 1969, Wong showed that statements (b) through (f) in (3.3) are mutually equivalent for a locally o-convex Riesz space (see [32, Theorem 2] or [39, (13.9)]), and that statements (a), (d) and (g) are mutually equivalent for a LCRS $\big($see [32, Corollary 5] or [39, (13.12)]$\big)$.

It is strange that Aliprantis and Burkinshaw said in their book [4, Note, p.160] that (3.3) is due to Burkinshaw and Dodds [5, Prop. 5.2] in an article written in 1977; in fact, in that article [5, Prop. 5.2] they only give the following:

(3.3)* Theorem. Let $(X, X_+, \mathcal{P})$ be a LCRS. Then the following statements are equivalent.
   (a) X is a band in (X'', X''_+).
   (k) (i) X is $\mathcal{P}$-complete;
       (ii) each disjoint sequence in X_+ with $\mathcal{P}$-bounded partial sum is a $\mathcal{P}$-null sequence.

Other criteria for a Banach lattice to be a KB-space can be found in [27, §6.7 of Chapter II]. Moreover, Lozanovskji shows (see [27, (II.6.5), p.89]) that a Banach lattice is a KB-space if and only if it does not contain c_o, similar to the analogous statement concerning Banach lattices with order continuous norm which has been extended to the case of complete LCRS (see(3.2.2)).

The following is a dual result of (3.3):

(3.3.1) Corollary ([32, Corollary 4], [31,(4.1)]). For a LCRS $(X, X_+, \mathcal{P})$, the following statements are equivalent:
   (a) X' is a band in (X^b, X^*_+).
   (b) $(X', X'_+, \sigma(X', X))$ is boundedly Dedekind complete.
   (c) $(X', X'_+, \sigma_s(X', X))$ is a Nakano space.
   (d) X' is $\sigma_s(X', X)$-complete.
   (e) $\sigma(X', X)$ has the boundedly convergent property.
   If, in addition, $(X, X_+, \mathcal{P})$ is infrabarrelled, then (a) is equivalent to:
   (f) $(X, X_+, \mathcal{P})$ is order-infrabarrelled (i.e. each barrel in $(X, \mathcal{P})$ which absorbs all order-bounded subsets of X is a $\mathcal{P}$-neighbourhood of 0).

Recall that a Riesz space (X, X_+) is perfect if X^b_n is total over X and $X = X_{nn}$. Using (3.3), we are able to give some criteria of perfectness in terms of the topological completeness.

(3.3.2) Corollary. Let (X, X_+) be a Riesz space such that X^b_n is total over X. Then the following statements are equivalent:
   (a) (X, X_+) is perfect.
   (b) X is a band in X_{nn}.
   (c) X is complete for $\sigma_s(X, X^b_n)$.
   (d) Each $\sigma(X, X^b_n)$-bounded increasing net in X_+ has a $\sigma(X, X^b_n)$-limit.
   (e) $(X, X_+, \sigma(X, X^b_n))$ is boundedly Dedekind complete.
   (f) $(X, X_+, \sigma_s(X, X^n_n))$ is both locally and boundedly Dedekind complete.

Proof. Given X the topology $\sigma_s(X, X^b_n)$, then the result follows from (3.3).

The equivalence of (a) and (e) in (3.3.2) was given by Nakano [19]; while the equivalence of (c) and (e) is due to Schaefer [23, Theorem 4].

As a consequence of (3.3) and (3.2), we have the following:

(3.3.3) Corollary. Let $(X, X_+, \mathcal{P})$ be a LCRS. Suppose that $\mathcal{P}$ is order continuous and that (X, X_+) is Dedekind complete. Then X is order closed in (X'', X''_+) if and only if each $\mathcal{P}$-bounded increasing net in X_+ has an upper bound (called boundedly majorant property).

Proof. Under the hypothesis, X is an ideal in X'', hence the necessity follows from (3.3)(f). To prove the sufficiency, let $[u_i]$ be a $\mathcal{P}$-bounded increasing net in X_+. Then there exists an $u \epsilon X$ such that $0 \leq u_i \uparrow \leq u$, hence there exists an $x \epsilon X$ such that $x = \mathcal{P} - \lim_i u_i$ (by (3.2)(e)) since X is an ideal in X'', thus X is order closed in X'' (by (3.3)(f)).

It is natural to ask whether the preceding result is true without the assumptions that $\mathcal{P}$ is order continuous and X is Dedekind complete.

The following result answers the final question posed at the beginning of this section, namely: What condition is necessary and sufficient for the semi-reflexivity of X?

(3.4) Theorem (Wong [32, Theorem 6]). For a LCRS $(X, X_+, \mathcal{P})$, the following statements are equivalent:

(a) X is semi-reflexive.

(b) $(X, X_+, \sigma_s(X, X'))$ is a Nakano space (i.e. locally and boundedly Dedekind complete), and $X'' \subset (X')^b_n$.

(c) X is $\sigma_s(X, X')$-complete, and $X'' \subset (X')^b_n$.

(d) $\mathcal{P}$ is both order continuous and boundedly Dedekind complete, and $X'' \subset (X')^b_n$.

(e) $\sigma(X, X')$ has the boundedly convergent property (i.e. each $\sigma(X, X')$-bounded increasing net in X_+ is $\sigma(X, X')$-convergent), and $X'' \subset (X')^b_n$.

(f) $\mathcal{P}$ has the boundedly convergent property, and $X'' \subset (X')^b_n$.

(g) X is a band in (X'', X''_+), and $X'' \subset (X')^b_n$.

(h) $X = (X')^b_n$, and $X'' \subset (X')^b_n$.

(i) X is $\mathcal{P}$-complete, $\mathcal{P}$ has the boundedly Cauchy property (or equivalently, each $\mathcal{P}$-boundedly increasing sequence in X_+ is $\mathcal{P}$-Cauchy), and $X'' \subset (X')^b_n$.

(j) X is a band in $(X')^b_n$, and $X'' \subset (X')^b_n$.

Proof. In view of (3.3), statements (b) through (j) are mutually equivalent, we have only to verify the implications (a) $\Rightarrow$ (b) and (j) $\Rightarrow$ (a).

(a) $\Rightarrow$ (b): Observe that X is always a Riesz subspace of $(X')^b_n$, it then follows from the semi-reflexivity of X that $X'' = X \subset (X')^b_n$.

On the other hand, the semi-inflexivity of X implies that $X = X'' = (X', X'_+, \beta(X', X))'$ and $(X', \beta(X', X))$ is barrelled, it then follows that $(X, X_+, \sigma_s(X, X'))$ is locally Dedekind complete (by (2.3(a)) as well as boundedly Dedekind complete (by (2.3)(b)).

(j) $\Rightarrow$ (a): If suffices to show that the band in $(X')_n^b$ generated by X'', denoted by $b(X'')$, is contained in $b(X)$ (the band in $(X')_n^b$ generated by X), and this is the case if and only if the set, defined by

$$X_\perp = \{h \epsilon X' : h(x) = 0 \qquad \text{(for all } x \epsilon X)\},$$

is contained in the set

$$(X'')_\perp = \{f \epsilon X' : \psi(f) = 0 \qquad \text{(for all } \psi \epsilon X'')\}$$

(see [13, (31.6)(i), p.375]). It is easily seen that $X_\perp = \{0\}$, and hence that $X_\perp \subset (X'')_\perp$.

Remark. It is easily verified that (a) and (h) in (3.4) are equivalent.

As semi-reflexivity and reflexivity are equivalent for normed spaces, the preceding result contains the classical result of Ogasawara (see Luxemburg and Zaanen [15, Theorem 40.1]). Other criteria for reflexivity of Banach lattices can be found in [27, § 7 of Chapter II].

It is strange again that Aliprantis and Burkinshaw claimed in their book [4, Note. p.163] that the equivalence of (a) and (d) in (3.4) is due to Burkinshaw and Dodds [5, Prop. 5.4] in an article written in 1977; in fact, in that article [5, (5.4) and (5.2)], they give the following:

(3.4)* Theorem. For a LCRS $(X, X_+, \mathcal{P})$, the following statements are equivalent:
(a) X is semi-reflexive.
(k) (i) X is complete for $\mathcal{P}$;
   (ii) each disjoint sequence in X_+ which is $\mathcal{P}$-bounded partial sum is a a $\mathcal{P}$-null sequence;
   (iii) each $\mathcal{P}$-bounded disjoint sequence in X_+ is a $\sigma(X, X')$-null sequence.
(l) (i)* Same as (b)(i);
   (ii)* Same as (b)(ii);
   (iii)* each order bounded disjoint sequence in X'_+ is a $\beta(X', X)$-null sequence.

As an immediate consequence of (3.4)(d) and Remark (i) of (3.1), we have the following:

(3.4.1) Corollary. A LCRS $(X, X_+, \mathcal{P})$ is semi-reflexive if and only if it satisfies the following three conditions:
   (i) $(X, X_+, \mathcal{P})$ is a Nakano space (i.e. locally and boundedly Dedekind complete);
   (ii) every order-bounded disjoint sequence in X_+ is a $\mathcal{P}$-null sequence;
   (iii) every order-bounded disjoint sequence in X'_+ is a $\beta(X', X)$-null sequence.

Proof. The result follows from the following equivalent statements:

X is semi-reflexive.

$\Leftrightarrow \mathcal{P}$ is both order continuous and boundedly Dedekind complete, and

$$X'' \subset (X')_n^b$$

$\Leftrightarrow \mathcal{P}$ is order regular and has the monotone Cauchy property, $(X, X_+, \mathcal{P})$ is boundedly Dedekind complete, and $\beta(X', X)$ is order continuous.

$\Leftrightarrow$ (i), (ii) and (iii) hold (since $\beta(X', X)$ is always order regular).

References

[1] Aliprantis, C.D., Some order and topological properties of locally solid linear topological Riesz spaces, Proc Amer Math. Soc., **40**(1973), 443–447.

[2] Aliprantis, C.D. and O. Burkinshaw, A new proof of Nakano's theorem in locally solid Riesz space, Math. Z., **27**(1975), 666–678.

[3] ——,——, Nakano's theorem revisited, Michigan Math. J., **23**(1976), 173–176.

[4] ——,——, Locally solid Riesz spaces, Academic Press, 1978.

[5] Burkinshaw, O. and P.G. Dodds, Disjoint sequences, compactness and semi-reflexivity in locally convex Riesz space, Illinois J. Math., **21**(1977), 759–775.

[6] Buskes, G. and I. Labuda, On Levi-like properties and some of their applications in Riesz space theory, Canad. Math. Bull., **31**(1988), no.4, 477–486.

[7] Dodds, P.G. and D.H. Fremlin, Compact operators in Banach lattices, Israel J Math., **34**(1979), 287–320.

[8] Fremlin, D.H., Topological Riesz spaces and measure theory, Cambridge Univ. Press, (1974).

[9] Gordon, H., Topologies and projections on Riesz spaces, Trans. Amer. Math. Soc., **94**(1960), 529–551.

[10] Kaplan, S., On the second dual of the space of continuous functions, I., Trans, Amer. Math. Soc., **86**(1957), 70–90.

[11] Kawao, I., Locally convex lattices, J. Math. Soc. Japan, **9**(1957), 281–314.

[12] Labuda, I., On boundedly order-complete locally solid Riesz spaces, Studia Math., **81**(1985), no.3, 245–258.

[13] Luxemburg, W.A.J. and A.C. Zaanen, Notes on Banach function spaces, IX, Proc. Acad. Sci. Amsterdam, **A67**(1964), 104–119.

[14] ——,——, Notes on Banach functions spaces, X, ibid., **A67**(1964), 493–509.

[15] ——,——, Notes on Banach functions spaces, XIII, ibid., **A67**(1964), 530–543.

[16] Luxemburg, W.A.J. and A.C. Zaanen, Riesz spaces I., North-Holland (1971), Amsterdam.

[17] Luxemburg, W.A.J., Notes on Banach function spaces, XIV, Proc. Acad. Sci. Amsterdam, **A68**(1965), 229–248.

[18] Nachbin, L., Topology and order, Van Nostrand, (1965), New York.

[19] Nakano, H., Modulared semi-ordered linear spaces, Maruzen Co. Tokyo (1950).

[20] ——, Linear topologies on semi-ordered linear spaces, J. Fac. Sci. Hokkaido Univ., **12**(1953), 87–104.

[21] Namioka, I., Partially ordered linear topological spaces, Memoirs Amer. Math. Soc., **24**(1957).

[22] Peressini, L., Ordered topological vector spaces, Harper and Row (1967), New York.

[23] Schaefer, H.H. On the completeness of topological vector lattices, Mich. Math. Soc., (1960), 303–309.

[24] ——, Topological vector spaces, Springer-Verlag (1971).

[25] ——, Banach lattices and positive operators, Springer-Verlag (1974).

[26] ——, Aspects of Banach lattices, In: MAA Stud. Math. (R.C. Bartle, ed.) (Math. Asoc. of Amer., Washington)(1980), 158–221.

[27] Schwarz, Hans-Ulrich, Banach lattices and operators, Band 71, Teubner-Texte Bur Mathematik (1984).

[28] Wnuk, W., Full Riesz subspaces in some locally solid Riesz spaces, Comment. Math. Prace Math., **29**(1990), no.2, 325–329.

[29] Wong Yau-chuen, Locally o-convex Riesz spaces, Proc. London Math. Soc., **19**(1969), 289–309.

[30] ——, Local Dedekind-completeness and bounded Dedekind-completeness in topological Riesz spaces, J. London Math. Soc., **1**(1969), 207–212.

[31] ——, Order-infrabarrelled Riesz spaces, Math. Ann., **183**(1969), 17–32.

[32] Wong Yau-chuen, Reflexivity of locally convex Riesz spaces, J. London Math. Soc., **1**(1969), 725–732.

[33] ——, Relationship between order completeness and topological completeness, Math. Ann., **199**(1972), 73–82.

[34] ——, A note on completeness of locally convex Riesz spaces, J. London Math. Soc., **(2)6**(1973), 417–418.

[35] ——, Open decompositions on ordered convex spaces, Proc. Cambridge Phil Soc., **74**(1973), 49–59.

[36] ——, The topology of uniform convergence on order-bounded sets, Lecture Notes in Math. 531, Springer-Verlag (1976).

[37] ——, An intorduction to ordered vector spaces, Lecture delivered at Institute of Mathematics, Academia Sinica, Taiwan (1980).

[38] ——, Introductory theory of topological vector spaces, Marcel Dekker (1992).

[39] Wong Yau-Chuen and Kung-Fu Ng, Partially ordered topological vector spaces, Oxford Math. Monographs, Clarendon Press, Oxford (1973).

[40] Zaanen. A.C., Riesz spaces, II. North-Holland Amsterdam (1983).

Advances of Research on
Orlicz Spaces in Harbin

Congxin Wu

Department of Mathematics, Harbin Institute of Technology, Harbin

Tingfu Wang

Harbin University of Science and Technology, Harbin

Shutao Chen

Department of Mathematics, Harbin Normal University, Harbin

This article makes a general survey of the contributions of the mathematicians in Harbin to Orlicz space theory and its applications. It consists of three parts — weak topology and rotundity of the spaces, geometric property of the spaces, generalizations of the spaces and applications of the space theory.

1. Weak topology and rotundity

1.1. Introduction

Suppose that $M(R \to R^+)$ is an even, convex and continuous function. We call M an N-function, if it satisfies that $\lim_{u \to \alpha} M(u)/u = \alpha \in \{0, \infty\}$ and that $M(u) = 0$ iff $u = 0$; and its complementary N-function N on R is defined by $N(v) = \max_{u \in R}\{uv - M(u)\}$. We always denote by M, N such a pair of N-functions and by p, q their right hand side derivatives, respectively. The function M is said to be strictly convex on R (resp., on $[a, b]$), or simply, $M \in SC$ (resp., $SC_{[a,b]}$) if

$$M((u + v)/2) < (M(u) + M(v))/2 \quad \text{for all} \quad u \neq v \text{ in } R \text{ (resp., in}[a, b]).$$

We also say that the complement N is smooth on R (resp., on $[a, b]$), or simply, $N \in S$ (resp., $S_{[a,b]}$) whenever $M \in SC$ (resp., $SC_{[a,b]}$).

1991 Mathematics Subject Classification:46E30, 46B25, 46B20, 46B99

For a measure space (G, M, μ) and a measurable function x on G, $\rho_M(x) \equiv \int_G M(x(t))d\mu$ is called the modulus of x, and the Orlicz space $L_M(G)$ and its subspace $E_M(G)$ are defined by

$$L_M(G) \equiv \{x : \rho_M(\lambda x) < \infty \text{ for some } \lambda > 0\} \text{ and}$$

$$E_M(G) \equiv \{x : \rho_M(\lambda x) < \infty \text{ for all } \lambda > 0\},$$

and the Orlicz norm $\| \cdot \|_M$ and the Luxemburg norm $\| \cdot \|_{(M)}$ on $L_M(G)$ by

$$\|x\|_M = \inf_{k>0} k^{-1}(1 + \rho_M(kx)) \text{ and } \|x\|_{(M)} = \inf\{\lambda > 0 : \rho_M(x/\lambda) \leq 1\}.$$

We only focus on the following two cases: i) $G \subset R^m$ is a closed bounded set with Lebesgue measure μ and ii) $G = \mathbb{N}$ (the set of positive integers) with $\mu(i) = 1$ for all i in $\mathbb{N}$. We also write in the first case

$$(L_M(G), \| \cdot \|_M) = L_M, \quad (L_M(G), \| \cdot \|_{(M)}) = L_{(M)},$$

$$(E_M(G), \| \cdot \|_M) = E_M, \quad (E_M(G), \| \cdot \|_{(M)}) = E_{(M)};$$

and in the second case

$$(L_M(\mathbb{N}), \| \cdot \|_M) = l_M, \quad (L_M(\mathbb{N}), \| \cdot \|_{(M)}) = l_{(M)},$$

$$(E_M(\mathbb{N}), \| \cdot \|_M) = h_M, \quad (E_M(\mathbb{N}), \| \cdot \|_{(M)}) = h_{(M)}.$$

Moreover, $M \in \Delta_2$ means in the first case that there are $u_0 > 0$ and $K \geq 2$ such that $M(2u) \leq KM(u)$ for all $u \geq u_0$, and in the second case, the inequality above holds for all $u \in [0, u_0]$. We also write $M \in \nabla_2$ if $N \in \Delta_2$. (The reader may consult [1] and [2] for details).

1.2. Weak topology and isomorphic subspaces

For a given Banach space X, we denote by $B(X)$ and $S(X)$ respectively, the unit ball and the unit sphere of X, and by X^* the dual of X.

Theorem 1.1[3-6]. Let $x \in L_M(G)$, $k_x^* = \inf\{k > 0 : \rho_N(p(k|x|)) \geq 1\}$ and $K_x^{**} = \sup\{k > 0 : \rho_N(p(k|x|)) \leq 1\}$, and let $\theta(x) = \inf\{\lambda > 0 : \rho(\lambda^{-1}x) < \infty\}$, $x_n = x|_{G_n}$, the restriction of x to $G_n \equiv \{t \in G : |x(t)| \leq n\}$ for $G \subset R^m$ and, $\equiv \{i \in \mathbb{N} : i \leq n\}$ for $G = \mathbb{N}$. Then
  i) $\|x\|_M = k^{-1}(1 + \rho_M(kx))$ iff $k \in K(x) \equiv [k_x^*, k_x^{**}]$;
  ii) $\inf\{k \in K(x) : \|x\|_M = 1\} > 1$ iff $M \in \Delta_2$;
  iii) $\sup\{k \in K(x) : \|x\|_M = 1\} < \infty$ iff $M \in \nabla_2$;
  iv) $\theta(x) = \lim_n \|x - x_n\|_M = \lim_n \|x - x_n\|_{(M)}$
      $= \inf\{\|x - \omega\|_M : \omega \in E_M(G)\} = \inf\{\|x - \omega\|_{(M)} : \omega \in E_{(M)}(G)\}.$

Theorem 1.2[7,8]. i) $L_M(G)$ is weakly sequentially complete iff $M \in \Delta_2$;

ii) $A \subset L_M(G)$ is weakly compact iff $\lim\limits_{\lambda \to 0} \sup\limits_{x \in A} \lambda^{-1} \rho(\lambda x) = 0$ and $\lim\limits_{m} \theta(\min\limits_{k \leq m} |x_k - x|) = 0$

for all sequences $\{x_k\}$ in A with $\lim\limits_{k} \int_E [x_k(t) - x(t)]dt = 0$ for all $E \in \Sigma$;

iii) $\{x_n\}$ is weakly convergent to 0 in $L_M(G)$ iff $\lim\limits_{n} \int_E x_n dt = 0$, $\lim\limits_{\lambda \to 0} \sup\limits_{n} \lambda^{-1} \rho_M (\lambda x_n) = 0$ and $\lim\limits_{m} \theta\left(\min\limits_{k \leq m} |x_{n_k} - x|\right) = 0$ for all $E \in \Sigma$ and subsequences $\{x_{n_k}\}$ of $\{x_n\}$.

Each $f \in L_M^*(l_M^*)$ can be uniquely decomposed into $f = v + \phi$, where $v \in L_N(l_N)$ and ϕ is a singular functional, i.e., $\phi(E_M) = \{0\}(\phi(h_M) = \{0\})$. We write $f|_E(x) \equiv f(x|_E)$ for any functional f, set E and $x \in L_M(G)$, and f is called positive if $f(x) \geq 0$ whenever $x(t) \geq 0$. Every singular functional ϕ can be represented as $\phi = \phi^+ - \phi^-$, the difference between the positive singular functionals ϕ^+ and ϕ^- [9].

Theorem 1.3[6,10]. Let $\phi \neq 0$ be singular on $L_M(G)$. Then

i) $\forall \varepsilon > 0, \exists E \in \Sigma$ such that $\|\phi^+|_E\| < \varepsilon$ and $\|\phi^-|_{G \backslash E}\| < \varepsilon$;

ii) ϕ is not norm attaining on $B(L_M)(B(l_M))$;

iii) ϕ is norm attaining on $B(L_M)(B(l_M))$ iff $\exists E \in \Sigma$ such that $\|\phi^+|_E\| = 0$ and $\|\phi^-|_{G \backslash E}\| = 0$;

iv) ϕ is an extreme point of $B(L_{(M)}^*)(B(l_{(M)}^*))$ iff $\|\phi\| = 1$ and either $\phi|_E = 0$ or $\phi|_{G \backslash E} = 0$ $\forall E \in \Sigma$.

[11] presents the criteria for functionals which are norm attaining and support functionals on $L_M, l_M, L_{(M)}$ and $l_{(M)}$.

Let X, Y be two Banach spaces. Y is called an almost isometrically complemented copy of X provided $\forall \varepsilon > 0, \exists$ a subspace of X and an isomorphism T from the subspace to Y with $\|T\|\|T^{-1}\| < 1 + \varepsilon$.

Theorem 1.4[12]. i) L_M has no subspace isometric to c_0;

ii) $M \overline{\in} \Delta_2$ iff $L_{(M)}$ copies c_0 iff $L_{(M)}$ isometrically copies l^∞ iff $L_M(L_{(M)}, l_M, l_{(M)})$ almost isometrically complementarily copies l^∞ iff $E_M(E_{(M)}, h_M, h_{(M)})$ almost isometrically complementarily copies c_0.

iii) $M \overline{\in} \nabla_2$ iff $L_M(E_M, l_M, h_M)$ complementarily copies l^1 iff $L_M(L_{(M)}, l_{(M)}, l_M)$ complementarily copies l^1 iff $L_M(L_{(M)}, l_M, l_{(M)}, E_M, E_{(M)}, h_M, h_{(M)})$ almost isometrically complementarily copies l^1;

iv) $L_M(l_M)$ copies l^1 iff it is nonreflexive.

1.3. Rotundity and smoothness

We denote by $UKR, WUR, LUKR, LWUR, MPLUR, R, KR, URED$ and KC, successively, the uniformly k-rotund, the weakly uniformly rotund, the locally uniformly k-rotund, the locally weakly uniformly rotund, the mid-point locally uniformly rotund, the rotund, the k-rotund, the uniformly rotund in every direction and the fully k-convex Banach spaces, and by SS, VS, S, SC and UC, successively, the strongly smooth, the very smooth, the smooth, the strictly convex and the uniformly convex Banach spaces.

The criteria of the rotundity and smoothness for Orlicz spaces are summarized in the following table.

	$L_{(M)}$	L_M	$l_{(M)}$	l_M
UKR	[27] $M \in \Delta_2$, $M \in UC$	[28] $M \in \Delta_2$, $M \in UC$	[29,30] $M \in \Delta_2$, $M \in UC[0, M^{-1}(1/(k+1))]$	[31] $M \in \Delta_2$, $UC[0, \pi_M(1)]$
WUR	[32] $M \in \Delta_2$ ∇_2, SC	[4] as above	[33] $M \in \Delta_2, \nabla_2$, $UC[0, M^{-1}(1/2)]$ or $SC[0, M^{-1}(1)]$	[34] as above
LUKR	[35, 36] $M \in \Delta_2$, SC	[37, 36] $M \in \Delta_2$, ∇_2, SC	[37, 38] $M \in \Delta_2, SC[0, M^{-1}(1/(k+1))]$, ∇_2 or $SC[M^{-1}(1/(k+1)), M^{-1}(1/k)]$	[34, 38] $M \in \Delta_2, \nabla_2$, $SC[0, \pi_M(1/k)]$
LWUR	[35] as above	[4] as above	[37] $M \in \Delta_2, SC[0, M^{-1}(1/2)]$, ∇_2 or $SC[M^{-1}(1/2), M^{-1}(1)]$	[34] $M \in \Delta_2, \nabla_2$, $SC[0, \pi_M(1)]$
MPLUR	[16, 39] as above	[16, 39] $M \in \Delta_2$, SC	[16, 39] $M \in \Delta_2$, $SC[0, M^{-1}(1/2)]$	[16, 39] $M \in \Delta_2$, $SC[0, \pi_M(1)]$
R	[3] as above	[3] $M \in SC$	[4] as above	[15] $M \in SC[0, \pi_M(1)]$
KR	[27] as above	[28] as above	[40] $M \in \Delta_2$, $SC[0, M^{-1}(1/(k+1))]$	[40] $M \in SC[0, \pi_M(1/k)]$
URED	[35] as above	[41] $M \in (*)$, SC	[37] as above	[42] $M \in (**)$, $SC[0, \pi_M(1)]$
KC	[43] $M \in \Delta_2$, ∇_2, SC	[44] $M \in \Delta_2, \nabla_2$, SC	[43] $M \in \Delta_2, \nabla_2$, $SC[0, M^{-1}(1/2)]$	[44] $M \in \Delta_2, \nabla_2$, $SC[0, \pi_M(1)]$
SS	[23] $M \in \Delta_2$, ∇_2, S	[6] $M \in \Delta_2$, ∇_2, S	[24] $M \in \Delta_2, \nabla_2$, $S[0, M^{-1}(1)]$	[24] $M \in \Delta_2, \nabla_2$ $S[0, \pi_M(1/2)]$
VS	[23] as above	[6] as above	[24] as above	[24] as above
S	[23] $M \in \Delta_2$, S	[6] $M \in \Delta_2$, S	[24] $M \in \Delta_2$, $S[0, M^{-1}(1)]$	[24] $M \in \Delta_2$, $S[0, \pi_M(1/2)]$

where

$$\pi_M(\alpha) = \inf\{s > 0 : N(p(s)) \geq \alpha\},$$

(*): for any $u' > 0, \varepsilon > 0, \varepsilon' > 0$, there exist $\tau > 0, D > 0$, such that for any $u > u'$,

$$p((1 + \varepsilon)u) \leq (1 + \tau)p(u) \implies p(u) \leq Dp(\varepsilon' u), \tag{3}$$

(**) : for any $u' > 0, \varepsilon > 0, \varepsilon' > 0$, there exist $\tau > 0, D > 0$, such that (3) holds for all $u \in (0, u')$.

Remark. A few results in this paper are also obtained independently by some mathematicians out of Harbin, which are mentioned in [2].

Characterizations have been given for Orlicz spaces $E_M, E_{(M)}, h_M$ and $h_{(M)}$ to be Asplund and weak Asplund spaces, and for the spaces $L_M, L_{(M)}, l_M$ and $l_{(M)}$ to be Gateaux differentiabity spaces in [6, 20, 23–26, 45].

For localization of convexity and smoothness, characterizations have also been obtained for points in the spaces $L_M, L_{(M)}, l_M$ and $l_{(M)}$, which are extreme points, strongly extreme points, weakly uniformly rotund points and smooth points in [13–19]. For instance

Theorem 1.5[18,19]. i) If $M \bar{\in} \Delta_2$, then $B(L_{(M)})$ and $B(L_M)$ have no WURP; ii) If $M \in \Delta_2$, then x is a URP of $B(L_{(M)})(B(L_M))$ iff x is a WURP of $B(L_{(M)})(B(l_{(M)}))$ iff $\mu\{t \in G : kx(t) \in I\} = 0$ for any $k \in K(x)$, whenever p is a constant on the interval I, and either $\rho_N(p(k|x|)) = 1$ or $\mu\{t \in G : k|x(t)| = b\} = 0$ for any $SAI[a, b]$ of $M(\{i \in \mathbb{N} : x(i) \neq 0\}$ which is a singleton, or $k \in K(x), i \in \mathbb{N} \Longrightarrow kx(i) \in S_M; k|x(j)| = b \Longrightarrow \sum_{i \neq j} N(p(k|x(i)|)) +$

$N(p_-(k|x(j)|)) < 1$; and $k|x(j)| = a \Longrightarrow \sum_{i \neq j} N(p_-(k|x(i)|)) + N(p(k|x(j)|))) > 1$ for every $SAI[a, b]$ of M, where WURP means weakly uniformly rotund point (a point $x \in B(X)$ is called a WURP if $\{x_n\} \subset B(X), \|x + x_n\| \to 2 \Longrightarrow x_n \xrightarrow{w} x$) and SAI means structural affine interval of the function $M(M$ is affine on the interval but not affine on any interval containing properly the interval).

Theorem 1.6[20−−22]. i) $S_s(L_{(M)}) = \{x \neq 0 : \theta(x/\|x\|) < 1, \mu\{t \in G : p_-(|x(t)|/ \|x\|) < p(|x(t)|/\|x\|)\} = 0\}$;

ii) $S_s(L_M) = \{x \neq 0 : \rho_N(p_-(k|x|)) = 1, k = \min\{K(x)\}$, or $\theta(kx) < 1$ and $\rho_N(p(k|x|)) = 1\}$;

iii) $S_s(l_{(M)}) = \{x \neq 0 : x$ has only one nonzero coordinate or otherwise with $\theta(x/\|x\|) < 1\}$ and M is smooth at each point $|x(i)|/\|x\|$;

iv) $S_s(l_M) = \{x \neq 0 : \rho_M(p_-(k|x|)) = 1$, or $\theta(kx) < 1$ but, $\rho_N(p(k|x|)) = 1$ or $J \equiv \{j \in \mathbb{N} : p_-(k|x(j)|) < p(k|x(j)|), k = \min\{K(x)\}\}$ is unique or empty $\}$;

v) The smooth point set of $X \in \{L_{(M)}, L_M, l_{(M)}, l_M\}$ is dense in X.

2. Geometric property

2.1. Nonsquareness and nonsquare points

In the sense of Schäffer, $x \in S(X)$ is called a nonsquare point or *S-NP* (uniformly nonsquare point or *S-UNP*) provided $\max\{\|x+y\|\}, \|x-y\|\} > 1(1+\delta_x$ for some $\delta_x > 0)$ for all y in $S(X)$; we say that X is nonsquare (locally uniformly nonsquare) or *S-N* (*S-LUN*) if each point of $S(X)$ is an *S-NP* (*S-UNP*); and X is called uniformly nonsquare (*S-UN*) if δ_x mentioned above can be chosen so that it is independent of $x \in S(X)$.

In the sense of James, we can obtain the notion of *J-NP, J- UNP, J-LUN* space, *J-N* space and *J-UN* space by substituting min, $<$ and 2 for, successively, max, $>$ and 1 in the sense of Schäffer.

Results on the nonsquareness and nonsquare points[52--57] are listed below in Tables 2.1 and 2.2.

Table 2.1

$\|x\| = 1$	S-NP	S-UNP	J-NP	J-UNP
L_M, l_m	always	always	always	$M \in \nabla_2$
$L_{(M)}, l_{(M)}$	$R_M(x) = 1$	$M \in \Delta_2$	$\theta(x) < 1$	$\theta(x) < 1$

Table 2.2

	S-N	S-LUN	S-UN	J-N	J-LUN	J - UN
L_M, l_M	always	always	$M \in \Delta_2 \cap \nabla_2$	always	$M \in \nabla_2$	$M \in \Delta_2 \cap \nabla_2$
$L_{(M)}, l_{(M)}$	$M \in \Delta_2$	$M \in \Delta_2$	$M \in \Delta_2 \cap \nabla_2$	$M \in \Delta_2$	$M \in \Delta_2$	$M \in \Delta_2 \cap \nabla_2$

2.2. Normal and uniformly normal structures

X is said to have the normal (weakly normal) structure if for every bounded closed (weakly compact) convex set $C(\subset X)$ containing at least two elements, there exists $p \in C$ such that $\sup\{\|p - x\| : x \in C\} < \text{diam}(C)$; X is said to have the uniformly normal structure if for each such closed bounded convex C there are $p \in C$ and $h < 1$ such that $\sup\{\|p - x\| : x \in C\} < h\text{diam}(C)$.

In 1984, T. Landes (Trans. Amer. Math. Soc., 285(1984), 523–533) gave some criteria for the normal and the weakly normal structures of $l_{(M)}$, and analogous criteria for that of $L_{(M)}$ would easily follow in the same way. Other criteria listed in Table 2.3 were given[58−61] by mathematicians in Harbin.

Table 2.3 ($[a_i, b_i]$ denotes an SAI of M)

	W-structure	N-Structure	U-Structure
L_M	always	$\lim\sup_{b_i \to \infty} (b_i/a_i) < \infty$	$M \in \Delta_2 \cap \nabla_2$
l_M	always	$\lim\sup_{b_i \to \infty} (b_i/a_i) < \infty$	$M \in \Delta_2 \cap \nabla_2$
$L_{(M)}, l_{(M)}$	$M \in \Delta_2$	$M \in \Delta_2$	$M \in \Delta_2 \cap \nabla_2$

2.3. *H*-property

A point $x \in S(X)$ is called an *H*-point if for $\{x_n\} \subset S(X)$, $x_n \xrightarrow{\omega} x$ implies $\|x_n - x\| \to 0$. If all points in $S(X)$ are *H*-points, then we say that X has the *H*-property.

Table 2.4 [63, 65, 66]

	$x \in S(X)$ being *H*-point	*H*-property
$L_M, L_{(M)}$	Extreme point $+M \in \Delta_2$	$M \in \Delta_2 \cap SC$
$l_M, l_{(M)}$	$M \in \Delta_2$	$M \in \Delta_2$

In [62, 68, 72–76, 92, 93, 95], characterizations and criteria have been given for L_M, $L_{(M)}$, l_M and $l_{(M)}$ to have the H^*, the W^*PC, the PC, the denting, the W^*-denting and the Mazur intersection properties, and for points of the spaces to be H^*, W^*PC, PC, denting, W^*-denting points.

2.4. Girth and flatness

Let C be a symmetric with respect to the origin, closed and rectifiable curve lying in $S(X)$, i.e., $C \equiv \{g^s \in S(X) : 0 \le s \le \lambda(C)\}$, where g^s is a continuous mapping from $[0, \lambda(C)]$ to $S(X)$ with $g^{\lambda(C)/2+s} = -g^s$ and with $\lambda(C)$, the length of the curve C. We denote by Γ the family of all such curves, and define the girth of $B(X)$ by girth $B(X) \equiv \inf_{C \in \Gamma} \lambda(C)$. We call the space X flat if $\lambda(C) = 4$ for some $C \subset \Gamma$.

Table 2.5[69–71]

	Flat	Girth = 4 and nonattaining	Girth > 4
$L_{(M)}, l_{(M)}$	$M \bar{\in} \Delta_2$	$M \in \Delta_2 \backslash \nabla_2$	$M \in \Delta_2 \cap \nabla_2$
L_M, l_M	impossible	$M \bar{\in} \Delta_2 \cap \nabla_2$	$M \in \Delta_2 \cap \nabla_2$

2.5. λ-property and stability

Let $x \in B(X)$ and let

$$\lambda(x) = \sup\{\lambda \in [0,1] : x \in \lambda e + (1-\lambda)y, y \in B(X) \text{ and } e \in \text{Ext } B(X)\}$$

If $\lambda(x) > 0$, then we call x a λ-point of $B(X)$; and X is said to have the λ-property (uniform λ-property) if $\lambda(x) > 0$ for all $x \in B(X)$ ($\inf\{\lambda(x) : x \in B(X)\} > 0$).

Table 2.6[77–78] ($[a_i, b_i]$ denotes an SAI of M)

	$L_{(M)}$	L_M	$l_{(M)}$	l_M
λ-property	always	always	always	always
U-λ-property	$M \in SC$	sup$\{b_i/a_i :$ $b_i > 1\} < \infty$	$M \in SC[0, \eta]$ for some $\eta > 0$	sup$\{b_i/a_i :$ $0 < b_i \le 1\} < \infty$

x in $B(X)$ is called a stable point if the set-valued mapping $u \in S(X) \to \{(y, z) : y+z = 2u\} \in B(X) \times B(X)$ is lower semicontinuous at x. X is said to be stable if every x in $B(X)$ is stable. A. S-Granero (Bull. Pol. Acad. Sci. Math., 37 (1989), 7–12) and M. Wisia (Arch. Math. 54(1990), 1–9) have discussed the stability property of $L_{(M)}$ and $l_{(M)}$, and also discussed that of other spaces in [78].

Table 2.7[78]

	Stable point $x \in S(X)$	Stable space X
$L_{(M)}, l_{(M)}$	$R_M(x) = 1$	$M \in \Delta_2$
L_M, l_M	$K(x)$ is singleton	mapping $x \to K(x)$ is single valued

2.6. WM-property

Let $x \in S(X)$ and let $\nabla_x = \{f \in S(X^*) : f(x) = 1\}$. X is said to have the WM property if $\forall x, x_n \in S(X), \|x + x_n\| \to 2$ implies $\exists f \in \nabla_x$ with $f(x_n) \to 1$.

Table 2.8[80−81]

	$L_{(M)}$	L_M	$l_{(M)}$	l_M
WMP	i) $M \in \Delta_2$; ii) $M \in \nabla_2 USC$	i) $M \in \Delta_2 \cap \nabla_2$ ii) p is conti. at end points of SAIs of M.	i) $M \in \Delta_2$; ii) $M \in \nabla_2$ or $M \in SC[0, M^{-1}(1)]$	i) $M \in \Delta_2 \cap \nabla_2$ ii) p is conti. at a and b whenever $2N(p(a)) + N(p(b)) \le 1$

2.7. Geometric constants

We recall the packing coefficient Λ_X of a space X:

$$\Lambda_X \equiv \sup\{r > 0 : \exists \{x_i\} \subset (1 - r)B(X), \|x_i - x_j\| \ge 2r \text{ for } i, j \in \mathbb{N}, i \ne j\},$$

and let $D_X = \sup\{\inf[\|x_n - x_m\| : m \ne n \in \mathbb{N}] : \{x_n\}_{n=1}^\infty \subset S(X)\}$. Then $\Lambda_X = D_X/(2 + D_X)$.

Theorem 2.2[84−−88]. Let $M \in \Delta_2, x \circ x = (x_1, x_1, x_2, x_2, \cdots)$ for $x = (x_i)$ and let $D_n = \sup \inf[\|x^{(1)} \pm x^{(2)} \pm \cdots \pm x^{(n)}\| : x^{(i)} \in X, i = 1, \cdots, n]$. Then
  i) $D(l_{(M)}) = \sup\{c_x : R_M(x/c_x) = \frac{1}{2}, x \in S(l_{(M)})\}$,

ii) $D(l_M) = \sup\limits_{\|x\|_M=1} \inf\limits_{k>1} \left\{ c_{x,k} : R_m(k_x/c_{x,k}) = \dfrac{k-1}{2} \right\}$,

iii) $D(l) = \sup\{\|x \circ x\| : x \in S(l)\}$, where $l \in \{l_M, l_{(M)}\}$,

iv) $D_n(l_M) = \sup \left\{ c_x : R_M(x/c_x) = \dfrac{1}{n} \right\}$.

The rotund coefficient coef (X) of a Banach space X is defined by coef $(X) = \sup\{\varepsilon \in (0,2) : \delta_X(\varepsilon) = 1\}$ where $\delta_X(\varepsilon) \equiv \inf\{1 - \|x + y\|/2 : x, y \in S(X)$ with $\|x - y\| \geq \varepsilon\}$, the convexity modulus of X. Let $0 < \varepsilon < 1$ and let $V(\varepsilon) = \{(u,a) : u \geq 0, a \in [0,1], 2M\left(u\left(\dfrac{1+a}{2}\right)\right) \geq (1-\varepsilon)(M(u) + M(au))\}$ and $C(\varepsilon) = \sup\left\{c > 0 : \sum\limits_{i=1}^{n} \lambda_i M(u_i(1 - a_i)/c) \geq \sum\limits_{i=1}^{n} M(u_i(1 + a_i)/c, n \in \mathbb{N}, (u_i, a_i) \in V(\varepsilon), \{\lambda_i\} \subset R^+$ with $\sum\limits_{1}^{n} \lambda_i \leq \mu(G)$ and $1 - \varepsilon < \sum\limits_{1}^{n} \lambda_i(M(u_i) + M(a_i u_i))/2 \leq 1\}$. With the notations above, coef $(L_{(M)})$ and coef (L_M) are obtained in [90,91]. We only give coef$(L_{(M)})$ here.

Theorem 2.3. coef$(L_{(M)}) = \inf\{C(\varepsilon) : \varepsilon > 0\}$.

The weakly convergently sequential coefficients for the spaces l_M and $l_{(M)}$ and the Jame and the Schäffer uniformly nonsquare coefficients have been obtained in [92, 93].

3. Generalizations of Orlicz spaces and applications

3.1. Geometry of Musielad-Orlicz (sequence) space

First, recall some definitions and notations.

Let X be a complex Banach space, (T, Σ, μ) an atomless measure space. And let $\Phi(t,s) : T \times R^+ \to R^+$ with

i) $\Phi(t,0) = 0$, $\lim\limits_{s\to\infty} \Phi(t,s) = \infty$ and $\Phi(t,s_0) < \infty$ μ-a.e. for some $s_0 > 0$;

ii) $\Phi(t,\cdot)$ is convex on R^+ (μ-a.e. on T) and

iii) $\Phi(\cdot,s)$ is μ-measurable for every $s \in R^+$.

Further, if $\exists$ a constant $K \geq 2$ and a nonnegative μ-measurable function $\delta(t)$ on T with $\displaystyle\int_T \Phi(t,\delta(t))d\mu < \infty$ such that

$$\Phi(t,2s) \leq K\Phi(t,s) \text{ whenever } s \geq \delta(t) \ (\mu\text{-a.e.}),$$

then we denote $\Phi \in \Delta$. Let

$$E(t) = \sup\{s > 0; \Phi(t,s) < \infty\} \text{ and } e(t) = \sup\{s > 0 : \Phi(t,s) = 0\}.$$

Let $\Phi = \{\Phi_n\} : X \times \mathbb{N} \to R^+$ satisfy the following conditions:

i) $\Phi_n(\theta) = 0$ and $\Phi_n(x_n) < \infty$ for all $n \in \mathbb{N}$ and for some $x_n \in X$;

ii) $\Phi_n(x)$ is convex for all $n \in \mathbb{N}$;

iii) $\Phi_n(e^{it}x) = \Phi_n(x)$ for $t \in R, n \in \mathbb{N}$ and $x \in X$;

iv) $\Phi_n(tx) : (0, \infty) \to [0, \infty]$ is, for any fixed $x \in X$, left continuous in t.

Moreover, if $\exists \lambda > 1, K > 1, q > 0, \{c_n\} \in l^1$ and $n_0 \in \mathbb{N}$ such that

$$\Phi_n(\lambda x) \leq K\Phi_n(x) + |c_n| \text{ whenever } \Phi_n(x) < a, n \geq n_0,$$

then we also denote $\Phi \in \Delta$.

Let X_T be the collection of all strongly μ-measurable functions $x(t) : T \to X$. For each $x \in X_T (x = \{x_n\} \in X^{\mathbb{N}})$, define

$$e_\Phi(x) = \int_T \Phi(t, \|x(t)\|)d\mu \left(e_\Phi(x) = \sum_{n=1}^{\infty} \Phi_n(x_n) \right),$$

and call

$$L_{(\Phi)} \equiv \{x(t) \in X_T : \exists k > 0 \text{ s.t. } e_\Phi(kx) < \infty\}$$

$$(l_{(\Phi)} \equiv \{x \in X^{\mathbb{N}} : \exists k > 0 \text{ s.t. } e_\Phi(kx) < \infty)$$

the Musielak-Orlicz space (Musielak-Orlicz sequence space) with the Luxemburg norm $\|x\|_{(\Phi)} = \inf\{k > 0 : e_\Phi(x/k) \leq 1\}$.

x in a convex set $D \subset X$ is called a complex extreme point of D, if $x + \lambda y \in D$ with $|\lambda| \leq 1$ and $y \in X$ implies $y = \theta$; and X is said to be complex strictly convex (CSR) if each $x \in S(X)$ is a complex extreme point of $B(X)$, and X complex uniformly convex (CUR) if $\forall \varepsilon > 0, \exists \delta > 0$ such that $\|x + \lambda y\|$ with $x, y \in X, \|y\| \geq \varepsilon$ and with $|\lambda| \leq 1$ implies $\|x\| < 1 - \delta$. We say that $x_n \in X$ is a complex strictly convex point of Φ_n, if $2\Phi_n(x_n) = \Phi(x_n + \lambda y) + \Phi(x_n - \lambda y)$ with $|\lambda| \leq 1$ implies $y = \theta$.

In [98–102], Wu Congxin, Chen Shutao and Sun Huiying obtained the criteria for extreme points, complex extreme points, SR, CSR and CUR of the spaces $L_{(\Phi)}$ and $l_{(\Phi)}$, and that for UR of the spaces were given by H. Hudzik (Bull. Acad. Polon. Sci. Math., 32 (1984), 303–313) and A. Kaminska (J. Approx. Theory, 47(1986), 302–322).

For brevity, we denote $\Phi \in (P)$ if Φ has a property (P). Now, we define some properties for Φ as follows:

(i) $\sup\{k : \Phi_n(kx) < \infty\} > 1$ for all $x \neq 0$ in X with $\Phi_n(x) < 1$ and $n \in \mathbb{N}$;

(ii) $(C(ii))$ for $n \neq m$ in $\mathbb{N}$ and $x, y \in X, \Phi_n(x) + \Phi_m(y) \leq 1$ implies that either x is a strictly (complex strictly) convex point of Φ_n or y is that of Φ_m;

$(S_{iii})((iii))\Phi_n(x)$ is not a constant on $\{y + \lambda z : \lambda \in [0,1]\}(\{y + \lambda z : |\lambda| \leq 1\})$ for any $y \neq 0, z$ in X with $\Phi_n(y) \leq 1$ and $n \in \mathbb{N}$;

(iv) $\forall \varepsilon > 0, \exists \delta > 0$ such that $\|x\|_{(\Phi)} > \varepsilon$ implies $e_\Phi(x) > \delta$;

(v) $\forall \varepsilon > 0, \exists \delta > 0$ such that $e_\Phi(x) \leq 1 - \varepsilon$ implies $\|x\|_{(\Phi)} \leq 1 - \delta$.

Theorem 3.1[98]. x in $B(L_{(\Phi)})$ is a complex extreme point of $B(L_{(\Phi)})$ iff the following conditions hold:

i) $\lim_{k \to 1^-} e_{\Phi}(kx) = 1$ or $\|x(t)\| = E(t)$ (μ-a.e.),

ii) For any $y \neq \theta$ in $L_{(\Phi)}, \mu G_{x,y} = 0$, where $G_{x,y} = \{t \in T : 2\Phi_0(t, \|x(t)\|) = \Phi_0(t, \|x(t) + \lambda y(t)\|) + \Phi_0(t, \|x(t) - \lambda y(t)\|), |\lambda| \leq 1\}$ and $\Phi_0(t, s) = \lim_{r \to E(t)^-} \Phi(t, r)$ if $s = E(t), = \Phi(t, s)$ otherwise.

For the properties of SR, CSR, UR and CUR, we would also like to express the criteria for them by the following table.

	$L_{(\Phi)}$	$l_{(\Phi)}$
SR	$\Phi \in \triangle^{[96]}$ $X \in SR$ $\Phi \in SC$	$\Phi \in \triangle, (i)^{[97]}$ $\Phi \in (ii), (S_{iii})$
CSR	$\Phi \in \triangle^{[98]}$ $X \in CSR$ $\mu\{t \in T : e(t) > 0\} = 0$	$\Phi \in \triangle, (i)^{[100]}$ $\Phi \in (C_{ii}), (iii)$
UR	$\Phi \in \triangle$ (Hudzik) $X \in UR$ $\Phi \in UC$	$\Phi \in (iv), (v)$ (Kaminska) $\Phi \in (S_{vi})$
CUR	$\Phi \in \triangle^{[99]}$ $X \in CUR$ $\mu(t \in T : e(t) > 0\} = 0$	$\Phi \in (iv), (v)^{[100]}$ $\Phi(vi)$

Now we turn to discussing the Orlicz norm for the Musielak-Prlicz sequence spaces. For this purpose, let X be the complex number field. In [103], Wu Congxin and Sun Huiying gave a calculating formula for Orlicz norm and by this formula obtained some criteria for complex extreme points and CSR for Orlicz norm. Ye Yining[104,105] investigated the differentiability property of Musielad-Orlicz sequence spaces. In [106], Wu Congxin and Duan Yangzheng extended $L_{(\Phi)}$ further to X being locally convex spaces.

3.2. Geometry of vector-valued Orlicz spaces

Suppose that $T \times R^n \to R^+$ fulfils i) For each $t \in T, M(t, \cdot)$ is even, continuous and convex on R^n, and for every $u \in R^n, M(\cdot u)$ is measurable; ii) For any $t \in T, M(t, u) = \theta$ iff $u = \theta$ and $\lim_{|u| \to \infty} M(t, u)/u = \infty$; iii) $\exists K \geq 1$ such that $M(t, u) \leq KM(t, v)$ for all $t \in T$ and $|u| \leq |v|$. It is easy to see that the corresponding vector-valued Orlicz space $L_{(M)}$ with the Luxemburg norm is not a special case of Musielak-Orlicz space $L_{(\Phi)}$. Chen Shutao[107,109] gave some characterizations of the extreme points, UR and SR of $L_{(M)}$. He also pointed out that proofs of the two results on $L_{(M)}$ in M. S. Skaff's work (Pacific J. Math., 28(1969), 193–206, 414–430) had some mistakes and corrected them.

3.3. Embedding theorm

In 1958, Wu Congxin introduced the Banach spaces

$$D_\Phi^h = \{x(t) \in L_{(\Phi)} : h(t)x(t) \in L_{(\Phi)}\}, \|x\|_{D_\Phi^h} = \|x\|_{(\Phi)} + \|hx\|_{(\Phi)}$$

where $h(t)$ is a fixed unbounded measurable function on R^n with $\text{mes}\triangle_m \equiv \text{mes}\{t \in R^n : |h(t)| \le m\} < \infty, m \in \mathbb{N}$. Then he improved and extended Ding Xiaxi's results (Science Record, New Ser., 1(1957), 287–290), by giving the necessary and sufficient conditions for the embedding, and also generalized the Sobolev embedding theorem, that is,

Theorem 3.2[111]. $D_{\Phi_2}^h \subset L_{(\Phi_1)}$ iff i) $\exists a > 0, s_0 > 0$ such that $\Phi_1(as) \le \Phi_2(s)$ whenever $s \ge s_0$ and ii) for each $x \in L_{(\Phi_1)}, \exists b > 0$ such that $\int_{R^n\backslash\triangle_1} \Phi_1(bx(t)/h(t))dt < \infty$.

Later, Wu proved that D_Φ^h and $L_{(\Phi)}$ are topologically equivalent. See Wu's works [110–113] for details.

3.4. Lebesgue-Orlicz points

Theorem 3.3[116]. Let $M(t)$ be an N-function. Then

i) For each $x \in L_M(0,1)$ and $t_0 \in (0,1), t_0$ being an LB point of $x(t)$ implies that t_0 is an LO point of $x(t)$ iff $M \in \triangle'$ (namely, $\exists L > 0$ and $s_0 \ge 0$ such that $M(st) \le LM(s)M(t)$, whenever $s, t \ge s_0$);

ii) For each $x \in L_M(0,1)$ and $t_0 \in (0,1), t_0$ being an LO point of $x(t)$ implies that t_0 is an LB point of $x(t)$ iff $M \in \bigtriangledown'$ (the conjugate function $N \in \triangle^1$). (t_0 is called an LO or LB point of $x(t)$, if

$$\lim_{\triangle t \to 0} \left\| M^{-1}\left(\frac{1}{2\triangle t}\right)[x(t) - x(t_0)]\chi_{[t_0-\triangle t, t_0+\triangle t]} \right\|_M = 0$$

or

$$\lim_{\triangle t \to 0} \|x(t_0 + \triangle t) - x(t_0)\|_M = 0, \text{ respectively})$$

Note that Salehov (Vorone2 Gos. Uaiv. Trudy Sem. Funct. Anal., 10(1968), 114–121) only obtained the sfficiency of the theorem above with an additional condition $M \in \bigtriangledown_2$ for i).

3.5. Best approximation

For a Banach space X and a closed convex set C in X, if for every $x \in X$, there exists a unique $y \in C$ such that $\|x - y\| = \inf_{z \in C} \|x - z\|$, then y is called the best approximation of x in C, which we denote by $y = \pi(xC)$, and the operator $\pi(\cdot|C) : X \to X$ is called a best approximation operator.

Assume that $M \in \Delta_2$, its right derivative p is continuous and strictly increasing, and (T, Σ, μ) is a complete, nonatomic and finite measure space. For Orlicz spaces L_M with Luxemberg norm or Orlicz norm, Wang Yuwen and Chen Shutao [117] gave some necessary and sufficient conditions for $y \in C$ to be a best approximation of $x \in L_M$ in C. Recently, Duan Yanzheng and Chen Shutao [118] further obtained some sufficient conditions for S : $L_M \to L_M$ to be a best approximation operator $\pi(\cdot|L_M(\Sigma'))$ for some σ-Lattice $\Sigma' \subset \Sigma$, and this generalized the main theorems of Landerssand-Rogge (see Proc. Amer. Math. Soc., 76(1979), 307–309) and Dykstra (see Ann. Math. Statist., 41 (1970), 698–701).

3.6. Control theory

For the optimal boundary control, Wang Yuwen and Chen Shutao [119] considered the distributed parameter system

$$(\lambda - \triangle)y = x(t \in \Omega), \frac{\partial y}{\partial n} = u(t \in \Gamma_1), y = v(t \in \Gamma_0),$$

where Ω is a bounded open domain in R^m with smooth boundary $\Gamma = \Gamma_1 \cup \Gamma_0, \lambda > 0, u \in L^2(\Gamma_1), v \in L^2(\Gamma_0), x \in L^2(\Omega), \triangle$ is the Laplacian operator, and using Orlicz spaces they have solved the optimal control problem which guarantees the uniqueness of control.

Wang Tingfu and Wang Yuwen [120] considered the distributed parameter system:

$$x(t) = \int_0^t k(t, x)bu(s)ds$$

where $b, x(t)$ are in a Banach space $X(0 \le t \le T), K(t, s) \in L(X \to X)(0 \le s \le t \le T), u(t) \in L_M(0, T)$, and under certain assumptions, proved that these exists in the system a unique minimum Orlicz norm control.

3.7. Other applications

In [121], Wu Congxin investigated the decomposition of linear operators in Orlicz spaces, and in [122], Wu Congxin and Duan Yanzheng considered the continuity and complete continuity of a kind of nonlinear operators in Orlicz spaces. Wu Congxin [123] gave some properties for generalized absolutely continuous functions with Orlicz metric.

Recently, Wu Congxin and Fu Yongqiang [124, 125] proved that under certain conditions a class of quasi-linear partial differential equations has a weak solution in Sobolev-Orlicz spaces $W^2 L_M(\Omega)$.

References

[1] Wu Congxin, Wang Tingfu, Orlicz Spaces and Applications, Heilongjiang Sci. & Tech. Press, 1983.

[2] Wu Congxin, Wang Tingfu, Chen Shutao, Wang Yuwen, Theory of Geometry of Orlicz Spaces, Harbin Inst. of Tech. Press, 1986.

[3] Wu Congxin, Zhao Shanzhong, Chen Junao, On calculation of Orlicz norm and rotundity of Orlicz spaces, J. Harbin Inst. of Tech., (2)(1978)1–12.

[4] Chen Shutao, Some rotundities of Orlicz spaces with Orlicz norm, Bull. Pol. Acad. Sci. Math., 34(9–10)(1986)585–596.

[5] Chen Shutao, Wang Tingfu, Uniformly rotund points of Orlicz spaces, J. Harbin Normal Univ. 8(3) (1992) 5–10.

[6] Chen Shutao, Smoothness of Orlicz spaces, Comment. Math., 27(1987) 49–58.

[7] Wang Yuwen, Weakly sequential completeness of Orlicz spaces, Chin. Northeast Math., 1(1985) 241–246.

[8] Chen Shutao, Sun Huiying, Weak convergence and weak compactness in Orlicz spaces (to appear).

[9] Chen Shutao, Sun Huiying, Weak convergence and weak compactness in abstract M spaces (to appear).

[10] Chen Shutao, Sun Huiying, Wu Congxin, λ-property of Orlicz spaces, Bull. Pol. Acad. Sci. Math., 39(1)(1991) 63–69.

[11] Chen Shutao, Hudzik H., Kaminska A., Support functionals and smooth points in Orlicz spaces equipped with the Orlicz norm (to appear).

[12] Chen Shutao, Hudzik H., Sun Huiying, Complemented copies of 1, in Orlicz spaces, Math. Nachr., 158(1992) (in press).

[13] Lao Bingyuan, Zhu Xiping, Extreme points of Orlicz spaces, J. Zhongshan Univ., (2)(1983) 97–103.

[14] Wang Zuoqiang, Extreme points of Orlicz sequence spaces, J. Daqing Oil College., (1)(19983) 112–121.

[15] Cheng Shutao, Shen Yaquan, Extreme points and rotundity of Orlicz sequence spaces, J. Harbin Normal Univ., (2)(1985) 1–6.

[16] Cui Yunan, Wang Tingfu, Strongly extreme points of Orlicz spaces, Chin. J. Math., 7(4) (1987) 335–340.

[17] Wang Tingfu, Cui Yunan, Li Yanhong, Packing constants and strongly extreme points in Orlicz spaces, Chin. Adv. Math., 15(2)(1986) 217–218.

[18] Wang Tingfu, Ren Zhongdao, Zhang Yunfeng, Uniformly rotund points in Orlicz spaces, Chin. J. Math., (3) (1993).

[19] Wang Tingfu, Li Yanhong, Zhang Yonglin, UR points and WUR points of sequence Orlicz spaces, SEA. Bull. Math. (in press).

[20] Chen Shutao, Yu Xintai, Smooth points of Orlicz spaces, Comment. Math., 31(1991) 39-47.

[21] Chen Shutao, Smooth points of Orlicz spaces with Orlicz norm, J. Harbin Normal Univ., 2 (1989) 1-4.

[22] Wang Baoxiang, Zhang Yunfeng, Smooth points of Orlicz sequence spaces, J. Harbin Normal Univ., 7(3) (1991) 18-22.

[23] Wang Tingfu, Chen Shutao, Smoothness and differentiability of Orlicz spaces, Chin. J. Engin. Math., 1(3) (1987) 113-115.

[24] Tao Liangde, Some rotundities and smoothness of Orlicz sequence spaces, J. Harbin Normal Univ., 4(2) (1988).

[25] Wang Tingfu, Shi Zhongrui, Criteria for KUC, NUC and UKK of Orlicz spaces, Northeastern Math., 3(1987) 160-172.

[26] Chen Shutao, Duan Yanzheng, Convex functions or Orlicz spaces, Collected Youth Science Articles, Heilongjiang, (1990) 1-2.

[27] Wang Tingfu, Chen Shutao, K-rotundity of Orlicz spaces, J. Harbin Normal Univ. (4) (1985) 11-15.

[28] Shi Zhongrui, K-uniform rotundity of Orlicz spaces, J. Heilongjiang Univ., (2) (1987) 41-44.

[29] Wang Tingfu, Chen Shutao, K-rotundity of Orlicz sequence spaces, Canad. Math. Bull. 34(1)(1991) 128-135.

[30] Wang Tingfu, Uniformly convex condition of spaces, J. Harbin Univ. of Sci. & Tech., (2) (1983) 1-8.

[31] Wang Tingfu, Shi Zhongrui, KUR of spaces, SEA. Bull. Math., 14(1) (1990) 33-44.

[32] Wang Tingfu, Wang Yuwen, Li Yanhong, Weakly uniform convexity of Orlicz spaces, Chin. J. Math., 6(1986) 209-214.

[33] Li Yanhong, Weakly uniformly rotundity of Orlicz sequence spaces, Chin. Nature J., 9(1986) 471-472.

[34] Tao Liangde, Rotundity of Orlicz sequence spaces, J. Harbin Normal Univ., (1) (1986) 11-15.

[35] Chen Shutao, Wang Yuwen, Locally uniform rotundity of Orlicz spaces, Chin. J. Math., 5(1985) 9-14.

[36] Shi Zhongrui, Fan Ying, Locally uniform k-rotundity of Orlicz spaces (to appear).

[37] Chen Shutao, Shen Yaquan, Locally uniform rotundity of Orlicz spaces, J. Harbin Normal Univ. add., (2)(1985) 1-5.

[38] Cui Yunan, Locally uniform k-rotundity of Orlicz sequence spaces (to appear).

[39] Cui Yunan, Midpoint locally uniform rotundity of Orlicz spaces, Chin. Nature J. 9(1986) 230-231.

[40] He Miaohong, K-rotundity of Orlicz sequence spaces J. Qiqihar Normal College (to appear).

[41] Wang Tingfu, Shi Zhongrui, Cui Yunan, Uniform rotundity in every direction of Orlicz spaces, Comment. Math. (to appear).

[42] Wang Tingfu, Shi Zhongrui, Cui Yunan, Uniform rotundity in every direction of Orlicz sequence spaces, Acta Szeged Math. (to appear).

[43] Chen Shutao, Lin B., Yu Xintai, Rotund reflexive Orlicz spaces are fully convex, AMS, Contemporary Math., 85(1989) 79–86.

[44] Wang Tingfu, Zhang Yunfeng, Wang Baoxiang, Fully k-convexity of Orlicz spaces, J. Harbin Normal Univ. 5(3)(1989) 19–21.

[45] Liang Yanhang, On the set of smooth points of Orlicz sequence spaces l_M, J. Harbin Normal Univ., 7(3) (1991) 8–15.

[46] Chen Shutao, Hudzik H., On some convexities of Orlicz and Orlicz-Bochner spaces, Comment. Math. Caro. Univ., 29(1988) 13–29.

[47] Cui Yunan, Wang Tinfu, Convexity of Orlicz spaces, Comment. Math., 31(2)(1991) 49–57.

[48] Li Ronglu, General representation theorem of bounded linear fuctionals in Orlicz spaces, J. Harbin Inst. of Tech., (3)(1980))) 91–94.

[49] Wang Tingfu, Wang Baoxiang, Strongly and very smooth points of Orlicz spaces, Chin. Northeastern Math., 8(2) (1992) 223–230.

[50] Wang Tingfu, Zhang Yunfeng, l_N-weak compactness of Orlicz sequence spaces, J. Harbin Univ. of Sci. & Tech., 16(3) (1992).

[51] Wu Yanping, Sequential compactness and weak convergence of Orlicz spaces, Chin, Nature J., 5(1982) 234.

[52] Tingfu Wang, Zhongrui Shi, Yanhong Li, On uniformly nonsquare points and nonsquare points of Orlicz spaces, Comm. Math. Univ. Carolinae, 33:3(1992), 477–484.

[53] Tingfu Wang, On uniform non-l_n, Sci. Researr., 5:1(1985), 125–126.

[54] Shutao Chen, Nonsquareness of Orlicz spaces, Math. Ann., 6A(1958), 607–613, (Chinese).

[55] Yuwen Wang, Shutao Chen, On the definition of nonsquareness in normed spaces, ibid, 9A(1988), 69–73.

[56] Yuwen Wang, Shutao Chen, Nonsquareness, flatness and B-property of Orlicz Comm. Math. (Prace Mat.), 28(1988).

[57] Shutao Chen, Locally uniform nonsquareness of Orlicz sequence spaces, J. Harbin Inst. of Tech., 1(1987), 1–5, (Chinese).

[58] Tingfu Wang, Baoxiang Wang, Normal structure and Lami-Dozo property of Orlicz spaces, Math. Resear. Expo., 12(1992), 477–478.

[59] Shutao Chen, Yangzheng Duan, Uniform and weakly uniform normal structure of Orlicz spaces, Comm. Math. Univ. Carolinae, 32(1991), 219–225.

[60] Tingfu Wang, Zhongrui Shi, On uniformly Normal structure of Orlicz spaces with Orlicz norm, ibid, 34(1993).

[61] Shutao Chen, Huiying Sun, On uniformly Normal structure of Orlicz spaces studia Math., (to appear).

[62] Yunan Cui, Tingfu Wang, The roughness of the normal on Orlicz spaces, Comm. Math., (Prace Mat.) 31(1991), 49–57.

[63] R. Pluciennik, Tingfu Wang, Yonglin Zhang, *H* points and denting points of Orlicz spaces, Com. Math., (Prace Mat.) 33(1993).

[64] Tingfu Wang, (*G*) and (*K*) properties of Orlicz spaces, Comm. Math. Univ. Carolinae, 31(1990), 307–313.

[65] Shutao Chen, Yuwen Wang, *H* property of Orlicz spaces, Math. Ann. 8A(1987), 61–67, (Chinese).

[66] Congxin Wu, Shutao Chen, Yuwen Wang, *H* Property of Orlicz sequence spaces, J. Harbin Inst. of Tech., (1985) (Math.), 6–11.

[67] Tingfu Wang, Yunan Cui, Donghai Ji, Mazer's intersection property of Orlicz spaces, (to appear).

[68] Tingfu Wang, Zhongrui Shi, *KUR, NUR* and *UKK* of Orlicz spaces, Math. Northeast China, 3(1987), 160–172. (Chinese).

[69] Tingfu Wang, Girth and Reflexivity of Orlicz sequence spaces, Math. Ann., 6A(1985), 567–574.

[70] Yuwen Wang, Nonsquareness and flatness of Orlicz spaces, Math. Resear. Expo., 4(1984), 94 (English summary).

[71] Congxin Wu, Shutao Chen, Yuwen Wang, Criteria of reflexivity and flatness of Orlicz sequence spaces, Math. *N-E*-China, 2(1986), 49–57.

[72] Tingfu Wang, *P*-Rotundity of Orlicz sequence spaces, Math. Quar., 7(1992), 18–21. (Chinese)

[73] Yining Ye, Miaohong He, R. Plucinnik, *P*-Convexity and reflexivity of Orlicz sequence spaces, Comm. Math., (Prace Mat.), 31(1991), 203–216.

[74] BaoXiang Wang, Exposed point of Orlicz spaces, J. Baoji Norm. Univ. 12(1989), no2, 43–49. (Chinese)

[75] Tingfu Wang, Zhongrui Shi, Donghai Ji, The Criteria of Strongly exposed points in Orlicz spaces, (to appear).

[76] Shutao Chen, Geometry of Orlicz spaces, (to appear).

[77] Shutao Chen, Huiying Sun, On λ-property of Orlicz spaces, Bull. Pol. Acad. Aci. Math., 39(1991), 63–69.

[78] Congxin Wu, Huiying Sun, On λ-property of Orlicz spaces L_M, Comm. Math. Univ. Carolinae, 31(1991), 731–741.

[79] Huiying Sun, Shutao Chen, Stable points of Orlicz spaces with Orlicz norm, (to appear)

[80] Shutao Chen, Yanming Li, Baoxiang Wang, WM prroperty and LKR of Orlicz sequence spaces, Fasciculi Math., 4(1991), 19–25.

[81] Shutao Chen, Yanzheng Duan, WM property of Orlicz Spaces, Math. Northeast China, 8:3(1992).

[82] Tingfu Wang, Donghai Ji, U-spacial property of Orlicz spaces (to appear)

[83] Tingfu Wang, Some geomtric properties of Orlicz spaces, Teubner Texte Math., 120(1991), 49–53.

[84] Yining Ye, Packing constant of Orlicz sequence spaces, Math. Ann., 4A(1983), 487–493. (Chinese)

[85] Tingfu Wang, Packing constant of Orlicz sequence spaces with Orlicz norm, Math. Ann., 8A(1987), 508–513. (Chinese)

[86] Tingfu Wang, Youming Liu, Packing constant of a type of sequence spaces, Comm. Math., (Prace Mat.) 30:1(1990), 197–203.

[87] Tingfu Wang, Yunan Cui, Yuanhong Li, Packing constant and strong extreme point of Orlicz sequence spaces, Adv. Math., 15(1986), 217–218 (Chinese)

[88] Tingfu Wang, Geometric Constant and nonsquenceness, Teubner Texte Math., 103(1988), 37–40.

[89] Tingfu Wang, On d_n of Orlicz spaces, Pure. Appl. Math., (1987), n 0.3, 38–41.

[90] Tingfu Wang, Yonglin Zhang, Geometric coefficient of Orlicz spaces, J. Harbin Univ. of Sci. and Tech., 15:4 (1991), 70–78. (Chinese)

[91] H. Hudzik, Tingfu Wang, Baoxiang Wang, On the convexity characterstic of Orlicz spaces, Japonica Math., 37:4(1992), 691–699.

[92] Tingfu Wang, Yunan Cui, Weakly convergent sequence coefficient for Orlicz sequence space, (to appear).

[93] Donghai Ji, Tingfu Wang, Nonsquare coefficients of Orlicz spaces and L^P, (to appear).

[94] Tingfu Wang, Quandi Wang, Some notations on K_x of Orlicz spaces, (to appear).

[95] Tingfu Wang, Yunan Cui, Quandi Wang, Rough coefficient of Orlicz spaces (to appear).

Research on Some Topics of Banach Spaces
and
Topological Vector Spaces in Harbin

Congxin Wu

Department of Mathematics, Harbin Institute of Technology, Harbin

This paper presents a survey of the contributions of mathematicians in Harbin to some topics of Banach spaces and topological vector spaces, especially, to Köthe sequence spaces and the infinite matrix operator algebras on them, Banach spaces containing no copy of c_0, abstract functions and integrals, differentiability of convex functions and abstract duality.

Owing to the limitation of space, we will only state a few results for each topic mentioned above, and the reader may refer to the references listed at the end of this paper for more information.

1. Locally convex sequence spaces and matrix algebras

In 1957, at the suggestion of Prof. Jiang Zejian, Prof. Wu Congxin examined the Köthe theory of sequence spaces, and established the theory of perfect matrix algebra $\Sigma(\lambda)$ as a class of locally convex algebras [1–11].

Suppose that ω is the space of all sequences of complex numbers. Then each linear subspace $\lambda \subset \omega$ is called a sequence space and

$$\lambda^* = \{\overline{u} = (u_k) \in \omega : [\overline{u}\overline{x}] \triangleq \sum_{k=1}^{\infty} |u_k x_k| < \infty \text{ for all } \overline{x} = (x_k) \in \lambda\}$$

its Köthe dual. Clearly, $\lambda \subset \lambda^{**}$. If $\lambda^{**} = \lambda$, then λ is called (in this section, λ will always be) a perfect space. Thus, (λ, λ^*) is a dual pair with $\overline{u}\overline{x} \triangleq \sum_{k=1}^{\infty} u_k x_k$. We denote by $\sigma(\lambda, \lambda^*), \tau(\lambda, \lambda^*)$ and $\beta(\lambda, \lambda^*)$, respectively, the weak, the Mackey and the strong topologies,

1991 Mathematics Subject Classification:46B25, 46A05, 46A45, 46A20, 46B20, 46B99, 16A99, 26A51, 49A51, 26B45

and by $n(\lambda, \lambda^*)$ the family of seminorms $\left\{\|\overline{x}\|_{\overline{u}} \triangleq [\overline{ux}] = \sum_{k=1}^{\infty} |u_k x_k| : \overline{u} \in \lambda^*\right\}$, the normal topology on λ. Clearly, $\beta(\lambda, \lambda^*) \geq \tau(\lambda, \lambda^*) \geq n(\lambda, \lambda^*) \geq \sigma(\lambda, \lambda^*)$.

Wu [7] proved a converse version of a Köthe's metrization theorem and [8] gave a characterization for $(\lambda, n(\lambda, \lambda^*))$ to be normalizable. He [7] (in 1965, but the publication time was delayed) showed independently the Grothendiech-Pietsch's nuclearity theorem for perfect spaces, and recently, $n(\lambda, \lambda^*) = \sigma(\lambda, \lambda^*)$ iff $\lambda = \omega$. He (1964, [5,6], the publication time was also delayed) first introduced the topologies for $\Sigma(\lambda)$, the perfect matrix ring or the set of all infinite matrices $A = (a_{lk})$ mapping λ into λ, such that $\Sigma(\lambda)$ becomes a locally convex algebra, i.e., the strong, the weak and the Mackey topologies, denoted by, respectively, T_s, T_w and T_k, are generated by the kinds of neighbourhood system of zero $\{A \in \Sigma(\lambda) : \sup[|\overline{u}A\overline{x}| < \varepsilon : \overline{u} \in N, \overline{x} \in M]\}, \varepsilon > 0, \overline{u}A\overline{x} = \overline{u}(A\overline{x})$, both $N \subset \lambda^*$ and $M \subset \lambda$ being bounded, finite and weakly compact.

Theorem 1[6]. The multiplication in $\Sigma(\lambda)$ is continuous with respect to T_s iff $\exists$ a bounded set $N_0 \subset \lambda^*$ s.t. $\forall$ bounded set $N \subset \lambda^*, \exists \alpha > 0$ with $N \subset \alpha N_0$ iff $(\lambda, \beta(\lambda, \lambda^*))$ is a Banach space iff $(\Sigma(\lambda, T_s)$ is a Banach algebra iff $(\Sigma(\lambda), T_s)$ is an m-convex algebra iff the multiplication in $\Sigma(\lambda^*)$ is T_s-continuous. (See [2] for T_w in $\Sigma(\lambda)$).

For approximating of $A \in \Sigma(\lambda)$ by finite matrix $A_{mn}(a_{lk} = 0, l > m, k > n)$, row-finite matrix $A_{m\infty}(a_{lk} = 0, l > m)$ and column-finite matrix $A_{\infty n}(a_{lk} = 0, k > n)$, we have

Theorem 2[4,1]. $\forall A \in \Sigma(\lambda), M_{mn} \in \{A_{mn}, A_{m\infty}, A_{\infty m}\}$ T_s-converges to A iff for every bounded set B in $\lambda(\lambda^*), (B, \beta(\lambda, \lambda^*))((B, \beta(\lambda^*, \lambda))$ is relatively sequentially compact iff the sequential convergences in $(\Sigma(\lambda), T_s)$ and $(\Sigma(\lambda), T_w)$ are equivalent.

We say that $A \in \Sigma(\lambda)$ is $T \in \{T_s, T_w, T_k\}$ completely continuous if it maps any bounded set in λ to a relatively $T_\lambda \in \{\beta(\lambda, \lambda^*), \sigma(\lambda, \lambda^*), \tau(\lambda, \lambda^*)\}$ compact set and denote $C_T(\lambda) = \{A \in \Sigma(\lambda) : A$ is T completely continuous $\}$. For approximating of $A \in C_{T_s}$ by $A_{mn}, A_{m\infty}$ and $A_{\infty n}$ we also have

Theorem 3[4]. For each $A \in C_{T_s}(\lambda), \{a_{\infty n}\}(a_{m\infty}, a_{mn})$ T_s-converges to A iff $\lambda^*(\lambda$, both λ and $\lambda^*)$ is (are) sequentially $\beta(\lambda^*, \lambda)(\beta(\lambda, \lambda^*)$, correspondingly both $\beta(\lambda, \lambda^*)$ and $\beta(\lambda^*, \lambda))$ compact.

Recently, Wu obtained a generalization of a well-known theorem for l_2, i.e.,

Theorem 4. $C_s(\lambda)$ is the minimal non-trivial T_s-closed ideal of $\Sigma(\lambda)$ iff λ and λ^* are sequentially $\beta(\lambda, \lambda^*)$ and $\beta(\lambda^*, \lambda)$ separable.

Also the T_k-closed ideal of $\Sigma(\lambda)$ was studied in [26].

In [1,3,5,7], Wu also considered the criteria of bounded set and convergence sequences in $\Sigma(\lambda)$ for some concrete sequence spaces, e.g., nuclear spaces, convergence-free spaces and so on. These generalize many results of some mathematicians.

After 1984, they further presented the matrix algebra $\Sigma(\lambda, \mu)$ between two sequence spaces λ and μ, which is different from $\Sigma(\lambda \to \mu)$ in Köthe's sense. In this case, they proved that $\Sigma(\lambda, \mu)$ is multiplicatively T_s-continuous iff $\Sigma(\lambda, \mu)$ is an m-convex algebra. But the m-convex algebra $\Sigma(\omega, \phi)$ is not a B_0-algebra where $\phi = \omega^*$, and so the discussion about the multiplicative T_s-continuity of $\Sigma(\lambda, \mu)$ is more complicate than that of $\Sigma(\lambda)$ (see [8–11]).

Let X and Y be two vector spaces over complex field such that (X, Y) forms a dual pair. For a perfect space λ, let

$$\lambda[X] = \{\overline{x} = (x_k) \in X^{\mathrm{N}} : (\langle x_k, y \rangle) \in \lambda \text{ for } y \in Y\}.$$

Define its Köthe dual with respect to the dual pair (X, Y) as follows:

$$\lambda[X]^* = \left\{\overline{y} = (y_k) \in Y^{\mathrm{N}} : \sum_{k=1}^{\infty} |\langle x_k, y_k \rangle| < \infty \text{ for each } \overline{x} \in \lambda[X]\right\}.$$

Then $(\lambda[X], \lambda[X]^*)$ form a dual pair with the bilinear functional

$$\langle \overline{x}, \overline{y} \rangle = \sum_{k=1}^{\infty} \langle x_k, y_k \rangle, \quad \overline{x} = (x_k) \in \lambda[X], \quad \overline{y} = (y_k) \in \lambda[X]^*.$$

In [15], a locally convex topology $\mathcal{J}$ on $\lambda[X]$ is called a GAK-topology if for each $\overline{x} = (x_k) \in \lambda[X]$, $\mathcal{J} - \lim_n \overline{x}(k > n) = 0$, where $\overline{x}(k > n) = (0, \cdots, 0, x_{n+1}, x_{n+2}, \cdots)$. And $(\lambda[X], \mathcal{J})$ is called a GAK-space if $\mathcal{J}$ on $\lambda[X]$ is a GAK-topology.

Theorem 5[15]. Each compatible topology on $\lambda[X]$ with respect to the dual pair $(\lambda[X], \lambda[X]^*)$ is a GAK-topology. Consequently, $\sigma(\lambda[X], \lambda[X]^*)$ and $\tau(\lambda[X], \lambda[X]^*)$ on $\lambda[X]$ are GAK-topologies.

Theorem 6[15]. Regarding the following statements (a)-(f) and (1)-(2):

(a) the strong topology $\beta(\lambda[X], \lambda[X]^*)$ on $\lambda[X]$ is a GAK-topology;

(b) in $\lambda[X]^*$, $\sigma(\lambda[X]^*, \lambda[X])$ convergent sequences coincide with $\sigma(\lambda[X]^*, \lambda[X])$ bounded and coordinate $\sigma(Y, X)$ convergent sequences;

(c) $(\lambda[X], \beta(\lambda[X], \lambda[X]^*))$ is a sequentially separable space;

(d) each $\sigma(\lambda[X]^*, \lambda[X])$ bounded subset of $\lambda[X]^*$ is relatively $\sigma(\lambda[X]^*, \lambda[X])$ compact;

(e) $(\lambda[X], \tau(\lambda[X], \lambda[X]^*))$ is a barrelled space;

(f) $(\lambda[X], \beta(\lambda[X], \lambda[X]^*))' = \lambda[X]^*$;

(1) $(X, \beta(X, Y))$ is a sequentially separable space;

(2) $(X, \tau(X, Y))$ is a barrelled space,

we have that (a) $\Longleftrightarrow$ (b), (c) $\Longleftrightarrow$ (a) + (1) and (d) $\Longleftrightarrow$ (e) $\Longleftrightarrow$ (f) $\Longleftrightarrow$ (a)+ (2).

They also discussed the hereditary properties on $\lambda[X]$ from λ and X, such as the boundedness, compactness, sequential completeness, normedness, metrizability etc..

For information about the sequence spaces $l_p[X], l_p(X)$ and their applications, see [14, 17, 18, 20].

2. On Banach spaces containing no copy of c_0

As early as 1964, Jiang Zejian and Zou Chengzu [27] had made a study of Banach spaces containing no copy of c_0 in the view of spectral operator theory. Twenty years after, Chinese mathematicians paid attention to the c_0-absence once again and obtained a series of results [14, 21, 28, 29].

Theorem 1[27]. A dual space X^* of a Banach space X contains no copy of c_0 if and only if the conjugate operators of spectral operators on X are spectral operators on X^*.

See also [27] for spaces of c_0-absence characterized by Boolean algebra.

Theorem 2[28]. A Banach space X contains no copy of c_0 iff every continuous linear operator $T : c_0 \to X$ is compact.

Theorem 3[14]. A Banach space X contains no copy of c_0 iff $l_1(X)$ is a GAK-space iff each operator $T_{(x_i)} : X^* \to l_1, T_{(x_i)} f = \{f(x_i)\} (f \in X^*)$ is compact.

Theorem 4[29]. The space $X(X^*)$ contains no copy of c_0 iff every operator $T_{(x_i)} : X^* \to l_1$, where $\sum_{i=1}^{\infty} x_i$ is a weakly unconditional Cauchy series $(T : X \to l_1)$, is compact.

See [21] for characterizations of locally convex spaces containing no copy of c_0.

3. Abstract functions of bounded variation

As is well known, in 1938, I. M. Gelfand first introduced the abstract functions of bounded variation from $[a, b]$ to a Banach space. After Gelfand's work, many mathematicians investigated various properties and extensions of this kind of abstract functions, and also paid attention to the abstract functions of absolute continuity. Wu Congxin, Liu Tiefu, Xue Xiaoping et. al also did good jobs on these topics [30–46]. We state a few results here, and for the definitions of symbols and terminology appearing in the following, the reader is referred to [30–32, 39].

Theorem 1[31]. The following statements are equivalent: (i) $x \in SV_k[a,b](k \geq 2)$; (ii) for any (or for some) $r \in [2,k]$, $x^{k-r} \in SV_r[a,b]$; (iii) $D_+x^{k-2}, D_-x^{k-2} \in SV[a,b]$ and $x^{k-2} \in SAC[a,b]$ (or $\forall 2 \leq r \leq k, x^{k-r}(t) = (B) \int_a^t x^{k-r+1}(s)ds + x^{k-r}(a)$); (iv) for any (or for some) $r \in [2,k]$, there is $g \in SV_{r-1}[a,b]$ such that $x^{k-r}(t) = (B) \int_a^t g(s)ds + x^{k-r}(a)$; (v) for any (or for some) $r \in [2,k]$, $x^{k-r}(t)$ is strong rth convexly expressible; (vi) there is a dense subset D of U such that for every $f \in D$, 1) $f(x(t))$ is an ordinary function of bounded kth variation, 2) $\exists h \in V[a,b], V_{t_1 k}^{t_2}[f(x)] \leq V_{t_1}^{t_2}(h)$ when $[t_1,t_2] \subset [a,b]$.

Theorem 2[31]. If E is a normed space which is weakly complete, then the following statements are equivalent: (i) $x \in WV_k[a,b](k \geq 2)$; (ii) for any (or for some) $r \in [2,k]$, the weak $(k-r)$th derivative $wx^{k-r} \in WV_r[a,b]$; (iii) $WD_+[wx^{k-2}], WD_-[wx^{k-2}] \in WV[a,b]$ and $wx^{k-2} \in WAC[a,b]$ (or $\forall 2 \leq r \leq k, wx^{k-r}(t) = (B) \int_a^t wx^{k-r+1}(s)ds + wx^{k-r}(a)$); (iv) for any (or for some) $r \in [2,k]$, there is $g \in WV_{r-1}[a,b]$ such that $wx^{k-r}(t) = (B) \int_a^t g(s)ds + wx^{k-r}(a)$; (v) for any (or for some) $r \in [2,k]$, $wx^{k-r}(t)$ is weak rth convexly expressible; (vi) for any (or for some) $r \in [2,k]$, there is $M > 0$ such that

$$\sup_x \left\| \sum_{i=0}^{n-k} \varepsilon_i [Q_{r-1}(x^{k-r}; t_{i+1}, \cdots, t_{i+k}) - Q_{r-1}(x^{k-r}; t_i, \cdots, t_{i+k-1})] \right\| < M, \quad \varepsilon_i = \pm 1.$$

Theorem 3[31]. The following statements are equivalent: (i) $x \in WV_k[a,b](k \geq 2)$; (ii) 1) (vi) 1) of Theorem 1, 2) $\exists M > 0$ such that $\forall f \in D, V_{ak}^b[f(x)] < M$; (iii) (vi) (or (v)) of Theorem 2 when $r = k$; (iv) $\forall f \in E^*$, there is $g_f \in V_{k-1}[a,b]$ such that $f[x(t)] = \int_a^t g_f(s)ds + f[x(a)]$.

Theorem 4[32]. Suppose that the locally convex space X is metrizable. Then for the abstract function from $[a,b]$ to X, the H-bounded variation and strongly bounded variation are equivalent iff X is barreled.

Theorem 5[32]. Suppose that the locally convex space X is metrizable. Then for the abstract function from $[a,b]$ to X, the H-bounded variation and weakly bounded variation are equivalent iff X is semi-nuclear.

Theorem 6[33]. The following statements are true.

(1) $X(t)$ is a function of bounded variation on Köthe sequence space λ iff $x_k(t)(k = 1, 2, \cdots)$ are ordinary functions of bounded variation and $\{\vee_a^b(x_k)\} \in \lambda^{**}$.

(2) If λ is perfect (namely $\lambda^{**} = \lambda$), then $X(t)$ is a function of bounded variation iff $X(t)$ can be expressed by the difference of two increasing functions $X^{(1)}(t) = \{x_k^{(1)}(t)\}$ and $X^{(2)}(t) = \{x_k^{(2)}(t)\}$:

$$X(t) = X^{(1)}(t) - X^{(2)}(t) \quad (x_k(t) = x_k^{(1)}(t) - x_k^{(2)}(t), k = 1, 2, \cdots).$$

Theorem 7[34]. For any pair $X(t), Y(t)$ of functions of bounded variation on λ, the product $X(t)Y(t) = \{x_k(t)y_k(t)\}$ is also a function of bounded variation on λ iff for each function of bounded variation $z(t)$ on λ, we have $\lambda^*_{z(t)} \subset \lambda^*$, where $\lambda^*_{z(t)} = \{U^{z(t)} : U \in \lambda^*\}$ and $u_k^{z(t)} = (|z_k(a)| + \vee_a^b(z_k))u_k(k = 1, 2, \cdots)$.

Theorem 8[35]. If λ is perfect, then for any $\{X^{(m)}(t)\} \subset V([a, b], \lambda), \{X^{(m)}(t) : t \in [a, b], m = 1, 2, \cdots\}$ and $\left\{\left\{\vee_a^b(X_k^{(m)})\right\}_k : m = 1, 2, \cdots\right\}$ being bounded sets of λ implies that there are a subsequence $\{X^{(m_l)}(t)\}$ and a function $X(t)$ in $V([a, b], \lambda)$ such that $\{X^{(m_l)}(t)\}$ weakly (strongly) converges to $X(t)$ for every $t \in [a, b]$ iff λ is sequentially srongly separable (correspondingly, all bounded sets of λ are sequentially relative strongly compact).

[34, 35] also considered the continuity and differentiability in $V([a, b], \lambda)$, and the relations among many kinds of functions of bounded variation from $[a, b]$ to λ. [36, 37] further discussed the functions of 2nd bounded variation from $[a, b]$ to λ.

Theorem 9[39]. The following statements are equivalent: (i) $x \in NAC_k[a, b](k \geq 2)$; (ii) for any (or for some) $r \in [1, k]$, $x^{k-r} \in NAC_r[a, b]$; (iii) $x^{k-1} \in NAC[a, b]$ and $\forall 2 \leq r \leq k, x^{k-r}(t) = (B)\int_a^t x^{k-r+1}(s)ds + x^{k-r}(a)$; (iv) for any (or for some) $r \in [1, k]$, there is $g \in NAC_{r-1}[a, b]$ such that $x^{k-r}(t) = (B)\int_a^t g(s)ds + x^{k-r}(a)$; (v) for any (or for some) $r \in [1, k]$, $\vee_{ar}^l(x^{k-r}) \in AC[a, b]$; (vi) for any (or for some) $r \in [1, k]$, there is a dense subset D of unit sphere U of E^* such that $\forall f \in D$, 1) $f(x^{k-r}(t)) \in NAC_r[a, b]$, 2) $\exists h \in AC[a, b]$, such that $\vee_{t_1 r}^{t_2}[f(x^{k-r})] < \vee_{t_1}^{t_2}(h)$ when $[t_1, t_2] \subset [a, b]$.

[36, 42] also discussed the case where E is a locally convex space or a Köthe sequence space.

Theorem 10[42]. Let E be a reflexive Banach space, $x(t)$ be a function from $[a, b]$ to E. Then $x(t)$ can be expressed by $x(t) - x(a) = (P)\int_a^t y(s)ds$ and $WDx(t) = y(t)$ a.e. iff $x(t) \in NAC[a, b]$ and $x(t)$ satisfies the locally Lipschitzian condition almost everywhere.

[43] defined the Riemann-Stieltjes integral for an abstract function from $[a, b]$ to a locally convex algebra, and by means of the function of strongly bounded variation, gave an existence theorem and a convergence theorem for this kind of (RS) integral.

Theorem 11[44]. Every linear continuous operator T from $D[a, b]$ to E may be expressed in the form

$$T(g) = \int_a^b g(t)df_l + \sum_{\substack{i=0 \\ b_i > a}} g(b_i - 0)(\psi_{b_i} - \psi_{b_i, -0}) + \sum_{\substack{i=0 \\ b_i < b}} g(b_i + 0)(\psi_{b_i + 0} - \psi_{b_i})$$
$$+ \sum_{\substack{i=0 \\ c_i > a}} [g(c_i) - g(c_i - 0)]\phi_{c_i},$$

where $\{c_i\}$ denotes the points of discontinuity of $g(t)$, $f_t + \psi_t \in V_E[a, b]$ (the space of all functions with bounded variation in E), $f_t \in VC_E[a, b]$ (the space of all continuous functions in $V_E[a, b]$), ψ_t whose points of discontinuity is $\{b_i\}$ is the jump function of $f_t + \psi_t$, ϕ_t vanishes except at a denumerable set of points and the sum of the absolute values of $\phi(x)$ at these points is finite.

Finally, [4] presented the functions of bounded variation which take values in a Banach lattice, and discussed the relations with respect to the abstract L-space and the generalized abstract L-space, and [4] also investigated the properties of set-valued functions of bounded variation.

4. Differentiability of convex functions and some geometric topics

We Congxin and Cheng Lixin [71] extended the deep Preiss differentiability theorem: "Every locally Lipschitz function on a nonempty open convex set D of an Asplund space is densely Frechet differentiable in D", to locally lower (upper) semi Lips. functions, and also extended it in other ways. For instance

Theorem 1[71]. For every lower semicontinuous, lower bounded and nowhere Frechet differentiable function on a nonempty open subset D of an Asplund space, there exists a dense G_δ subset of D such that for each x in the subset we have

$$\liminf_{\substack{t \to 0+ \\ y \in B}} \frac{f(x + ty) - f(x)}{t} = -\infty$$

where B is the unit ball of the space.

Wu and Cheng also showed the following

Theorem 2[47]. Suppose that f is a continuous convex function on an open convex subset D of a separable Banach space E. Then for each Gateaux differentiability point $x \in D$ of f, there is a closed convex set $C \subset D$ with $x \in C$ and with $C_x \equiv \bigcup_{\lambda > 0} \lambda(C - x)$ being dense in E such that x is a Frechet differentiability point of f_C, the restriction of f to C.

Theorem 3[78]. Every complemented subspace of a weak Asplund space is again a weak Asplund space.

Theorem 4[8]. With f and D as above and with $\dim E \geq 2$, then f is Frechet differentiable at $x_0 \in D$ with $f(x_0) > \inf_D f$ if and only if

i) $\lim_{t \to 0+} \frac{f(x_0 + tx_0) - f(x_0)}{t} = \lim_{t \to 0-} \frac{f(x_0 + tx_0) - f(x_0)}{t}$

and

ii) $\lim_{r \to 0+} u, v \in S \cap U(x_0, r) \dfrac{f(x_0) - f(\frac{u+v}{2})}{\|u - v\|}$

where S denotes the level set $\{x \in D; f(x) = f(x_0)\}$ and $U(x_0, r)$ stands for the open ball centered at x_0 with radius r.

As an application of Theorem 4, Wu and Cheng [10] also gave the characterizations of Frechet [Gateaux] differentiability points of the original norms on $c_0(\Gamma)$ and $l_\infty(\Gamma)$ and Day's norm on $c_0(\Gamma)$, and they also considered the differentiability properties of convex functions on locally convex spaces and so on in [48, 71–80].

S. Kaijser and Q. Guo have generalized the important Dvorelzky theorem to nonsymmetric case.

Theorem 5[70]. For each $\varepsilon > 0$, there exists $\delta > 0$ with the following property. Let $B \subset R^N$ be a compact convex body and let b_0 be any interior point of B. Then for each integer $1 \leq n < \delta \log N$, there is an n-dimensional affine subspace F of R^N passing through b_0 such that $B \cap F$ is $(1 + \varepsilon)$-Euclidean.

Suppose that E_0 is a closed proper subspace of a Banach space E, and suppose that $x_0^* \in E_0^*$. Let S_ε denote all those functionals on E which are the norm-preserving extensions of x_0^*. For each fixed $\varepsilon > 0$, we set $S_\varepsilon = \{x^* \in E^*, x^* = x_0^* \text{ in } E_0 \text{ with } \|x^*\| \leq \|x_0^*\|_{E_0} + \varepsilon\}$. The following question was asked by Lin Borluh.

Is $S_\varepsilon \backslash S_0$ nonempty?

Cheng Lixin, Cheng Lianchang and Wei Wenzhan [49], [50] gave the question an affirmative answer.

Theorem 6[49]. With E, E_0 and x_0^* as above, letting $I = [\|\|x_0^*\|\|_{E_0}, \infty)$, for each $\alpha \in I$, there exists $x_\alpha^* \in E^*$ such that $x_\alpha^* = x_0^*$ in E_0 and $\|x_\alpha^*\| = \alpha$.

A point $x \in B$ [the closed unit ball of the space E] is called a λ-point provided there exist an extreme point e of B, y in B and $\lambda(0, 1]$ such that $x = \lambda e + (1 - \lambda)y$; the space E is said to have the λ-property if each point in B is a λ-point.

Chen Shutao, Sun Huiying and Wu Congxin have shown the following

Theorem 7[51]. The Orlicz space L_M with Luxemburg norm has the λ-property; moreover, the dual space L_M^* has no λ-property if M does not satisfy condition $\triangle_2$.

Hence, it gives a negative answer to the Aron-Lohman's question: Does the dual space E^* have the λ-property if the Banach space E has ?

5. Abstract duality

Let G be an abelian topological group and E a nonempty set. For every nonempty $F \subseteq G^E$ we call the pair (E, F) an abstract duality pair with respect to G, or simply, a G-valued mapping pair.

Subseries convergence is a central problem in dual pair theory even now. Kalton (Math. Ann., 208 (1974), 267–278), Thomas (Ann. Inst. Fourier, 20 (1970), 55–191), Swartz (Publ. DE'INSTITUTE Math., Tome 26 (40) (1977), 288–292; SEA Bull. Math. (1) 12 (1988), 31–38) and L Ronglu [81] have discussed the subseries convergence problem in other special mapping pairs such as $(\Omega, C(\Omega, X))$ and $(X, L(X, Y))$, etc. But these results are special cases of a very general result below.

For a G-valued mapping pair (E, F), let $\sigma(E, F)$ be the topology on E of pointwise convergence by F, and $\overline{F}^s$ the sequential $\sigma(G^E, E)$ closure of F in G^E.

Theorem 1. If a sequence $\{x_j\}$ in E is subseries $\sigma(E, F)$ convergent, i.e., for every nonempty $\triangle \subset \mathbb{N}$ there is an $x_\triangle \in E$ such that $\sum_{j \in \triangle} f(x_j) = f(x_\triangle)$ for all $f \in F$, then $\{x_j\}$ is subseries convergent in the topology of uniform convergence on the family of conditionally $\sigma(\overline{F}^s, E)$-sequentially compact subsets of $\overline{F}^s$.

Many important results in analysis, for instance, the Orlicz-Pettis-Bennet-Kalton theorm, the Thomas-Swartz theorem, etc, are special cases of the above general result, and also can be improved by the result (see [84] for details, and see [83], also [84] for imformation about the continuity of limit functions on toplogical groups G and for the Schur summability of function matrices).

References

[1] Wu Congxin, Perfect spaces and perfect matrix rings (I), Science Record, 3(1959), 95–102.

[2] ——, Perfect spaces and perfect matrix rings (II), Science Record, 3(1959), 103–106.

[3] ——, Perfect spaces and perfect matrix rings (III), Acta Math. Sinica, 14(1964), 319–327.

[4] ——, On the complete continuity of matrix operators in perfect spaces, J. Jilin Univ., (1962), no. 1, 61–66.

[5] ——, Ideals of completely continuous matrix operators in perfect spaces, J. Harbin Inst. of Tech. (1977), no. 3, 32–38.

[6] ——, Perfect matrix algebras (I), Acta Math. Sinica, 21(1978), 161–170.

[7] ——, Some problems of nuclear perfect spaces, Acta Math. Sinica, 22(1979), 653–666.

[8] ——, Characterizations for normedness and metrizability of sequence spaces, J. Harbin Inst. of Tech., (1993), no. 4.

[9] Wu Congxin, Wang Hongtao, Matrix algebra $\Sigma(\lambda, \mu)$ and its topologies, J. Harbin Inst. of Tech., Math. issue (1984), 1–5.

[10] ——, ——, Multiplicative continuity under strong topology for $\Sigma(\lambda, \mu)$, Science Bulletin, 30(1985), 157–158.

[11] ——, ——, Multiplicative continuity under strong topology for $\Sigma(\lambda, \mu)$ on convergence free spaces, J. Harbin Inst. of Tech., Math. issue (1985), 12–15.

[12] Wu Congxin, Liu Lei, Matrix transformations on some vectorvalued sequence spaces, SEA Bull. Math., 17(1993), no. 1.

[13] Liu Lei, Wu Congxin, The topology on sequence spaces with values in Banach spaces, J. Harbin Inst. of Tech., Math. issue (1991), 8–9.

[14] Wu Congxin, Bu Qingying, The vector-valued sequence space $l_p[X]$ and Banach spaces not containing a copy of c_0, A Friendly Collection of Mathematical Papers I, Jilin Univ. Press, (1990), 9–16.

[15] ——, ——, Vector-valued sequence space $\Lambda[X]$ and its Köthe dual (I), J. Northeastern Math., 8(1992), 275–282.

[16] Wu Congxin, Bu Qingying, Characterizations of CMC (X) being GAK-space, J. Harbin Inst. of Tech., (1993), no. 1, 93–96.

[17] ——, ——, Köthe dual of Banach sequence spaces $l_p[X]$ and Grothendieck space, Comment. Math. Univ. Caroline, 34(1993), no. 2.

[18] ——, ——, The sequential completeness of operator spaces $L(l_p, X)$ and $K(l_p, X)$, J. Math. Res. & Expos., 12(1992), 366.

[19] ——, ——, Unconditionally convergent series of operators on Banach spaces, J.Math. Anal. Appl., to appear.

[20] ——, ——, Banach sequence spaces $l_p[X]$ and their properties, SEA Bull. Math., to appear.

[21] Li Ronglu, Bu Qingying, Locally convex spaces containing no copy of c_0, J. Math. Anal. Appl., 172(1993), 205–211.

[22] Bu Qingying, The locally convex space X for which $\lambda(X) \equiv \lambda[X]$, J. Harbin Inst. of Tech., Math. issue (1991), 142–144.

[23] ——, Barrelledness of vector-valued sequence space $c_0(X)$, J. Congcheng Shuxue Xuebao, 9(1992), 64–72.

[24] ——, Sequential representation of compact operator space $K(l_p, X)$, Hebei Jidian Xueyuan Xuebao, 10(1993), no. 2, 62–68.

[25] ——, Barrelledness of vector-valued sequence space $\Lambda(X)$, J. Harbin Inst. of Tech., (1993), no. 4.

[26] ——, Ideal of infinite matrix operators on perfect sequence spaces, Functiones of Approximation, 22(1993).

[27] Jiang Zejian and Zou Chengzu, On spectral operators, J. of Jilin Univ., 1(1964), 65–74.

[28] Li Ronglu, A characterization of Banach spaces containing no copy of c_0, Bull. Chin. Sci., 7(1984), 444.

[29] Wu Congxin and Xue Xiaoping, Bounded linear operators from Banach spaces not containing c_0 into l_1, J. of Math. (PRC), vol. 12, 4(1992), 430–434.

[30] Wu Congxin and Liu Tiefu, Abstract bounded second variation functions, Northeastern Math. J., 1(1985) 41–53.

[31] Wu Congxin and Liu Tiefu, Abstract kth bounded variation functions, Science Bulletin, 31(1986), 931–932.

[32] Wu Congxin and Xue Xiaoping, Abstract bounded variation functions on locally convex space, Acta Math. Sinica, 33(1990), 107–112.

[33] Wu Congxin, Abstract bounded variation functions on sequence space (I), J. Harbin Inst. of Tech., (1959), no.2, 93–100.

[34] Wu Congxin, Abstract bounded variation functions on sequence space (II), Acta Math. Sinica, 13(1963), 548–557.

[35] Wu Congxin, Abstract bounded variation functions on sequence space, Scientia Sinica, 13(1964), 1359–1380.

[36] Wu Congxin and Zao Linsheng, Abstract 2nd bounded variation functions on sequence space (I), J. Math. Res. Exp., 2(1982), no. 4, 143–150.

[37] Wu Congxin and Zao Linsheng, Abstract 2nd bounded variation functions on sequence space (I), J. Math. Res. Exp., 4(1982), no. 1, 97–106.

[38] Wu Congxin and Liu Tiefu, Some notes of the abstract functions of 2nd absolute continuity, Science Bulletin, 31(1986), 646–647, Northeastern Math. J., 2(1986), 371–378.

[39] Wu Congxin and Liu Tiefu, Abstract functions of absolute kth continuity, J. Harbin Inst. of Tech., (1986) no. 1, 123–124.

[40] Wu Congxin and Xue Xiaoping, Abstract functions of absolute continuity on locally convex space, Chin. Annals of Math., 12A(1990), Supplement, 84–86.

[41] Wu Congxin and Zao Linsheng and Liu Tiefu, Bounded variation functions and their generalizations and applications, Heilongjiang Scientific & Technique Pub. House, 1988.

[42] Wu Congxin and Xue Xiaoping, A remark of Pettis integral, Science Bulletin, 34(1989), 1836.

[43] Wu Congxin and Zhang Bo, Riemann-Stieltjes integral on abstract functions, J. Harbin Inst. of Tech., (1990) no. 2, 1–7.

[44] Liu Tiefu, Linear oprator between $D[a, b]$ and a Banach space E, J. Math., 1(1988) no. 2, 105–112.

[45] Wu Congxin and Ma Ming, Bounded variation of abstract function whose value is in a Banach lattice, Northeastern Math. J., 8(1992), 293–298.

[46] Xue Xiaoping and Zhang Bo, Properties of set-valued function with bounded variation in Banach space, J. Harbin Inst. of Tech., (1991) no. 3, 102–105.

[47] Wu Congxin and Cheng Lixin, A note on the differentiability of convex functions, Proc. Amer, Math. Soc. 121(1994), 1057–1062.

[48] Wu Congxin and Cheng Lixin, Characterizations of the differentiability points of the norms on $c_0(\Gamma)$ and $l_\infty(\Gamma)$, Northeastern Math. J. (to appear).

[49] Cheng Lixin and Wei Wenzhan, A Generalized Hahn-Banach extension theorm, J. Guanxi Teacher's College, No. 4(1988).

[50] Cheng Lixin and Chen Lianchang, The final answer for an open problem, J. Jianghan Petro. Inst. (3) 10(1988), 136–138.

[51] Chen Shoutao, Sun Huiying and Wu Congxin, λ-Property of Orlicz spaces, Bull. Polish Acad. Sci. Math. 39(1991), 63–69.

[52] Cheng Lixin, Orthogonalities of Banach spaces, J. Jianghan Petrol. Inst. (1) 9(1987), 1–5.

[53] S. Kaijser and Q. Guo, An estimate of the Minkowski distance between convex bodies. Uppsala Univ. Dept. Math Report 1(1992).

[54] J. Zhu, Topics in Banach space theory. Doctorial thesis of Lancaster Univ, Britain.

[55] Chen Shoutao and Wang Yuwein. On definition of non-squane normed spaces, Chin. Ann. of Math. 9A(1988), 330–334.

[56] Wu Congxin and Sun Huiying, On complex uniform convexity of Musielak-Orlicz spaces. Northeastern Math. J. 4(1988), 389–396.

[57] Wu Congxin and Guo Qi, On uniform convexity of locally convex spaces, Chin. Ann. of Math. 11A(1990), 351–354.

[58] Guo Qi and Wu Congxin, Strict convexity and smoothness in locally convex spaces, Northeastern Math. J., 5(1989), 465–472.

[59] Wu Congxin and Guo Qi, Uniform convexity and strict convexity in metric linear space, J. Liaoning Univ., (1989) No.3, 1–5.

[60] Wu Congxin and Li Yongjing; Extreme points and linear bounded operators, Northeastern Math. J., 8(1992), 475–476.

[61] Li Yongjin, Almost uniform convexity and reflexivity, J. Harbin Inst. of Tech., 9(1991) Supplement, 145–147.

[62] Li Yongjin, Complex convexity and complex smoothness, J. of Math., 12(1992).

[63] Li Yongjin, A note of WLUC points, J. Univ., (1992) No. 1.

[64] Cheng Lixin, Chen Lianchang and Cheng Wei. Orthogonalities of Banach spaces and Hilbert space. J. Daqing Petro. Inst. (4) 13(1989), 75–77.

[65] Cheng Lixin and Chen Lianchang, Comment:"L^p–orthogonality of Banach space", J. Math. Res. Exp. (1) 7(1987), 175–176.

[66] Cheng Lixin, Wang Tinfu and Chen Lianchang, A new class of characteristic functions for Banach spaces. J. Nature (8) (1988), 633–634; J. Harbin Univ. of Sci and Tech., (3)(1988), 93–97.

[67] Chen Lixin, Subinner-product and suborthogonality in Banach spaces, J. Jianghan Petrol. Inst. (1) 12(1990), 80–89.

[68] Cheng Lixin, On the characteristic functions of Banach spaces, J. Jianghan Petro. Inst. (3) 15(1993).

[69] Cheng Lixin, Cheng Lianchang and Wei Weinzhan, The claracteristic functions and the moduli of convexity and smoothness of Banach spaces, J. of Math., 10(1990), 309–314.

[70] S. Kaijser and Q. Guo, A Dvoretzky theorem for general convex bodies, Uppsala Univ. Dept. of Math. Report 1(1992).

[71] Wu Congxin and Cheng Lixin, Extensions of the Preiss Differentiability, Theorem, J. Funct. Anal. 124(1994), 112–118.

[72] Cheng Lixin and Nan Chaoxun, A sufficiency and necessity condition for Gateaux and Frechet differentiability of continuous gauges on Banach spaces, Bull. Sci., 34(1989), 795.

[73] Cheng Lixin, Li Jianhua and Nan Chaoxun, Gateaux and Frechet differentiability of continuous gauges on Banach spaces, Adv. in Math., 20(1991), 326–333.

[74] Cheng Lixin, Two notes on the smoothensss of Banch spaces, J. Math. Res. Exp. 9(1989), 315–316.

[75] Cheng Lixin, Smoothness and strong smoothness of Banach spaces, J. Jianghan Petro. Inst. (1) 11(1989), 102–107.

[76] Cheng Lixin and Wei Wenzhan, Some differentiability properties of continuous convex Functions on Banach spaces, J. Guanxi Univ. (1) 16(1991), 65–70.

[77] Cheng Lixin, Differentiability of convex functions and Asplund spaces, Acta. Math. Sci. (1) 15(1995).

[78] Wu Congxin and Cheng Lixin, On weak Asplund spaces, to appear.

[79] Wu Congxin and Cheng Lixin, Differentiability of convex functions on locally convex spaces, J. Harbin Inst. of Tech., E-1, 1(1994), 7-12.

[80] ——, ——, Approximation of functions on metric spaces and its application to differentiability of convex functions on meager sets, Wuhan Univ. Press, 1995.

[81] Li Ronglu and C. Swartz, K-convergence and the Orlicz-Pettis theorem, publ. De'Inst. Math., Tome 49(63), 1991, 117-122.

[82] ——, ——, Spaces for which the uniform convergence principle holds, Studia Sci. Math. Hungarica, 27(1992), 373-384.

[83] ——, ——, A nonlinear Schur theorem, Acta Sci. Math., to appear.

[84] ——, —— and Cho Min-Hyung, Abstract duality, to appear.

Some Results on Operator Algebra

Liangsen Wu

Department of Mathematics, East China Normal University, Shanghai

Abstract. This is a brief summary of my work on Operator Algebra, which contains the following:

1. Type classification of Von Neumann algebras by the use of their automorphism groups.
2. Tensor products of preclosed operators on C^*-algebras.
3. Continuum of ideals on $R(\Phi_2) \otimes_{\max} R'(\Phi_2)$.
4. C^*-algebraic isomorphism determined by completely positive maps.
5. When can the stable algebra determine the structure of a C^*-algebra?
6. Pure completely positive maps as a dual object of C^*-algebras.
7. Semi-simple H^*-algebra of Hilbert-Schmidt operators.

1. Type classification of Von Neumann algebras by the use of their automorphism groups

J. Von Neumann initiated the classification of Von Neumann algebras by the use of their traces. Afterwards Kaplansky developed the projection lattice classification of Von Neumann algebras. The advent of Tomita- Takesaki theory changed the scope of the theory of Von Neumann algebras greatly and provided powerful tools for analysing Type III Von Neumann algebras. It is an interesting problem how to classify Von Neumann algebras in the framework of Tomita-Takesaki theory.

In 1973 Pederson and Takesaki proved that a Von Neumann algebra $\mathcal{M}$ is semi-finite if and only if there exists a faithful normal semi-finite weight on $\mathcal{M}$ whose modular automorphism group is implemented by a strongly continuous one-parameter unitary group in $\mathcal{M}$.

In 1987 the author proved a theorem as follows

Theorem.[1] p328. If $\mathcal{M}$ is a Von Neumann algebra, and the modular automorphism group $\{\sigma_t^\phi\}$ of a faithful semi-finite normal weight φ on $\mathcal{M}$ is globally outer, then there is no non-zero finite projection in $\mathcal{M}$.

1991 Mathematics Subject Classification: 46L05

This work is supported by the Chinese Science Foundation

Now we can classify all Von Neumann algebras by the use of their automorphism groups.

Definition. 1. We say that a Von Neumann algebra $\mathcal{M}$ is semi-finite if the modular automorphism group $\{\sigma_t^\varphi\}$ of a faithful semi-finite normal weight φ on $\mathcal{M}$ is implemented by a strongly continuous one-parameter unitary group $\{u(t)\}$ in $\mathcal{M}$.

      2. We say that a Von Neumann algebra $\mathcal{M}$ is purely infinite if it has no non-trivial semi-finite reduced algebra, or equivalently, the modular automorphism group is globally outer.

      3. We say that a semi-finite Von Neumann algebra (with trace τ) is properly τ-infinite if $\tau(z) = +\infty$ for every non-zero $z \in \mathrm{Proj}(\mathcal{M} \cap \mathcal{M}')$.

In this way the type classification of Von Neumann algebras is entirely possible within the frame of Tomita-Takesaki theory with the help of Arveson's spectral theory of one-parameter automorphism groups.

2. Tensor products of preclosed operators on C^*-algebras

Takesaki found out that the C^*-norm on an algebraic tensor product is not unique. Then mysterious properties of the C^*-norm received much attention of many specialists. Okayasu showed that the minimal C^*-norm is not a uniform cross norm. It was known that even if σ, τ are bounded linear operators on C^*-algebras A_1, A_2, respectively, $\sigma \otimes \tau$ could be unbounded on $A_1 \otimes_{\min} A_2$. Naturally, we will ask the question: What kind of properties of the operators σ and τ are preserved under the tensor product operation?

    First of all, we prove the following result.

Theorem 1[2]. Let A_1 and A_2 be C^*-algebras and T_1 and T_2 be densely defined preclosed operators on A_1 and A_2. Then $T_1 \otimes T_2$ is preclosed on $A_1 \otimes_{\min} A_2$.

Sketch of proof. The main thing we have to prove is that if A_1, A_2 are C^*-algebras and A_1^*, A_2^* are conjugate spaces of A_1, A_2, respectively, then the algebraic tensor product

$$F = A_1^* \otimes A_2^*$$

is $\sigma((A_1 \otimes_{\min} A_2)^*, A_1 \otimes_{\min} A_2)-$ dense in $(A_1 \otimes_{\min} A_2)^*$. Equivalently, we will prove that F is total in $(A_1 \otimes_{\min} A_2)^*$.

    Suppose that $x \in A_1 \otimes_{\min} A_2$ and $\langle x, f \rangle = 0, \forall f \in F$. We shall show that $x = 0$. Let $\sigma(A_1)$ and $\sigma(A_2)$ be the state spaces of A_1 and A_2, $\omega_1 \in \sigma(A_1)$ and $\omega_2 \in \sigma(A_2)$.

Let $\prod_{\omega_1}$ and $\prod_{\omega_2}$ be the cyclic representations corresponding to ω_1 and ω_2 with representation Hilbert spaces $\mathcal{H}_1$ and $\mathcal{H}_2$; we construct

$$\prod_\omega = \prod_{\omega_1} \otimes \prod_{\omega_2}.$$

Therefore, $\forall \xi_1, \eta_1 \in \mathcal{H}_1$ and $\forall \xi_2, \eta_2 \in \mathcal{H}_2$,

$$\left(\prod_\omega (x)(\xi_1 \otimes \xi_2) \mid \eta_1 \otimes \eta_2 \right) = \langle x, f \otimes g \rangle = 0,$$

where $\langle x_1, f \rangle = \left(\prod_{\omega_1}(x_1)\xi_1 \mid \eta_1 \right)$, $\langle x_2, g \rangle = \left(\prod_{\omega_2}(x_2)\xi_2 \mid \eta_2 \right)$.

It follows that $\prod_\omega(x) = 0$. This implies that $\|x\|_{min} = 0$. Therefore x=0.

For the Banach space case, we have even proved theorems as follows:

Theorem 2[2]. If β is a reflexive norm on $E_1 \otimes E_2$,($\beta \geq \lambda$) and T_1, T_2 are densely defined preclosed operators on Banach spaces E_1 and E_2, then $T_1 \otimes T_2$ is preclosed on $E_1 \otimes_\beta E_2$.

An interesting property of the injective C^*-norm is the following

Theorem 3[2]. If A_1 and A_2 are C^*-algebras, the injective C^*-cross norm on the algebraic tensor product $A_1 \otimes A_2$ is reflexive.

Combining Theorem 2 and Theorem 3, we get another proof of Theorem 1.

3. Continuum of ideals in $R(\Phi_2) \otimes_{\max} R'(\Phi_2)$

Let Φ_2 be the free group on two generators. Let $\mathcal{K} = \mathcal{L}^2(\Phi_2)$ be the Hilbert space of all complex valued functions $f(g)$ on Φ_2 such that

$$\sum_{g \in \Phi_2} |f(g)|^2 < \infty.$$

For each $g_1 \in \Phi_2$ we define the unitary operator $U(g_1)$ on $\mathcal{K}$ by

$$(U(g_1)f)(g) = f(g_1^{-1}g) \quad \forall f \in \mathcal{K}.$$

A suprising result concerning $R(\Phi_2) \otimes_{\max} R'(\Phi_2)$ is the existence of a continuum of ideals in $R(\Phi_2) \otimes_{\max} R'(\Phi_2)$.

Theorem [3]. There is a continuum of ideals in $R(\Phi_2) \otimes_{\max} R'(\Phi_2)$.

Sketch of proof.

Step 1. By the use of the universal property of the projective C^*-tensor product, we can construct a homomorphism η such that

$$R \otimes_{\max} R' \xrightarrow{\eta} C^*(R, R'),$$

$$R \otimes_{\max} R'/I \cong C^*(R, R'),$$

in which I is $Ker(\eta)$.

Step 2. I is a proper ideal of $R \otimes_{\max} R'$.

Connes's characterization : Let N be a factor of Type II_1 with a separable predual acting in $\mathcal{K} = L^2(N, \tau)$. Then the following conditions are equivalent for $\theta \in \mathrm{Aut}(N)$,

(a) $\theta \in \overline{\mathrm{Int}(N)}$,

(b) There exists an automorphism of the C^*-algebra generated by N and N' in $\mathcal{K}$ which is θ on N and an identity on N'.

Using this result, we can show that I is a proper ideal of $R \otimes_{\max} R'$.

Step 3. With the help of Connes's result, we have the following : If $\alpha, \beta \in \mathrm{Aut}(R(\Phi_2))$, then $(\alpha \otimes Id)(I) = (\beta \otimes Id)(I)$ if and only if $\alpha^{-1}\beta \in Int(R(\Phi_2))$.

Step 4. There is a group of outer automorphisms $\alpha_\lambda, \lambda \in [a, b]$ with continuous parameter in $Aut(R(\Phi_2))$. This is a consequence of Behncke's results.

Step 5. Setting $I_\lambda = (\alpha_\lambda \otimes Id)(I)$, it follows that $I_\lambda \neq I_\mu$ for $\lambda \neq \mu$. So $\{I_\lambda\}$ is a continuum of ideals in $R \otimes_{\max} R'$.

4. C^*-Algebraic isomorphism determined by completely positive maps

The problem whether the state space of a C^*-algebra can determine the structure of the C^*-algebra has received considerable attention of many specialists.

Let A and B be C^*-algebras, $\sigma(A)$ and $\sigma(B)$ state spaces. In 1965 Kadison proved that the affine homeomorphism between $\sigma(A)$ and $\sigma(B)$ can determine the Jordan isomorphism between A and B. In 1980 Alfsen, Olson and Shultz gave a geometric characterization of the state spaces of unital C^*-algebras among compact convex sets. They defined the notion of an orientation of the state space and showed that the state space as a compact convex set with orientation completely determines the C^*-algebra up to an isomorphism. We consider some sets of completely positive maps related to the C^*-algebra A instead of $\sigma(A)$ and find out some conditions under which they can determine the C^*-isomorphic classes of the algebras.

1. $M_n \to A$ case

Let A, B be unital C^*-algebras,

$\mathcal{K}_A = \{\varphi \mid \varphi$ are completely positive maps from $M_n(C) \to A$ with $\|a(\varphi)\| \leq 1\}$,

$$a(\varphi) = \begin{pmatrix} \varphi(e_{11}) & \cdots & \varphi(e_{1n}) \\ \vdots & \ddots & \vdots \\ \varphi(e_{n1}) & \cdots & \varphi(e_{nn}) \end{pmatrix},$$

where $\{e_{ij}\}$ is the matrix unit of $M_n(C)$.

Suppose that $SU(n)$ is the set of all $n \times n$ unimodular unitary matrices and α is the automorphism group on $M_n(C)$ defined by

$$\alpha_g(x) = gxg^{-1}, \qquad x \in M_n(C), g \in SU(n).$$

Using $(\alpha_g\varphi)(x) = \varphi(\alpha_g^{-1}(x))$, $\varphi \in \mathcal{K}_A, x \in M_n(C)$, α induces an action on $\mathcal{K}_A$.

Theorem 1 [4]. Let A, B be C^*-algebras. For $n \geq 3$, if Φ is an α-invariant affine isomorphism from $\mathcal{K}_A$ to $\mathcal{K}_B$, $\Phi(0) = 0$, (α-invariance means that $\alpha\Phi = \Phi\alpha$), then A and B are *-isomorphic.

For $n = 2$, we are provided with a counterexample.

2. $A \to M_n(C)$ case

$\mathcal{D}_A = \{\varphi \mid \varphi$ are completely positive maps from $A \to M_n(C)$ with $Tr\varphi(I) \leq 1\}$.

By the use of

$$(\alpha_g\varphi)(x) = \alpha_g(\varphi(x)), \qquad \varphi \in \mathcal{D}_A, x \in A, g \in SU(n),$$

we can define an action on $\mathcal{D}_A$.

Theorem 2[5]. Let A and B be unital C^*-algebras. If Φ is an α-invariant affine homeomorphism from $\mathcal{D}_A$ to $\mathcal{D}_B$ ($n \geq 3$), $\Phi(0) = 0$, then A is *-isomorphic to B.

The results obtained can be viewed as non-commutative Kadison-Shultz theorems.

5. When can the stable algebra determine the structure of a C^*-algebra?

For any C^*-algebra A, we define the elements of $K_0(A)$ as the formal difference of equivalent classes of projections in matrix algebras over A. It is an interesting problem whether $K_0(A)$ can determine the structure of A. In a particular case, Elliot proved that if A and B are A_F-algebras and $\sigma: K_0(A) \to K_0(B)$ is an isomorphism of scaled order groups, then there is an isomorphism $\phi: A \to B$. In the general case, we can not expect so good a result, but we can raise a question: When can the stable algebra $\mathcal{K} \otimes A$ determine the structure of a C^*-algebra A?

Let $\mathcal{K}$ be a set of a compact operators on a separable Hilbert space. $\mathcal{K}$ can be viewed as the closure of the union of $M_n(C)$, that is, $\mathcal{K} = \overline{\cup_{n=1}^{\infty}(M_n(C))}$, with an embedding $M_n \to M_{n+1}$ by $a \to diag(a, 0)$.

$$\mathcal{K}_{\infty}(A) = \cup_{n=1}^{\infty}(M_n \otimes A).$$

$\mathcal{K} \otimes A$ is the closure of $\mathcal{K}_{\infty}(A)$.

$SU(\infty)$ is the infinite unimodular unitary group defined by

$$SU(\infty) = \cup_{n=1}^{\infty} SU(n)$$

with embeddings from $SU(n)$ to $SU(n + 1)$ by $a \to \mathrm{diag}(a, 1)$.

Using $SU(\infty)$, we can define an action α on $\mathcal{K} \otimes A$,

$$\alpha_u(x) = uxu^*, \quad x \in \mathcal{K}_{\infty}(A), u \in SU(\infty).$$

Theorem 1[7]. Let A and B be C^*-algebras. If Ψ is an α-invariant affine isomorphism between $(\mathcal{K} \otimes A)_+^1$ and $(\mathcal{K} \otimes B)_+^1$, $\Psi(0) = 0$, then A is *−isomorphic to B.

Definition. Suppose that A and B are C^*-algebras. If $\mathcal{K} \otimes A \cong \mathcal{K} \otimes B$, then A is said to be stably isomorphic to B.

Theorem 2[7]. Assume that A and B are C^*-algebras. If A is α-invariant stably isomorphic to B, then there is a *−isomorphism between A and B.

Let $\mathcal{O}_n$ be Cuntz algebra.

Theorem 3[7]. Let A and B be C^*-algebras. If Ψ is an α−invariant affine isomorphism between $(\mathcal{O}_n \otimes \mathcal{K} \otimes A)_+^1$ and $(\mathcal{O}_n \otimes \mathcal{K} \otimes B)_+^1$ with $\Psi(0) = 0$, then $A \cong B(3 \leq n < \infty)$.

6. Pure completely positive maps as a dual object of C^*-algebra

The starting point of this work is Shultz's paper in which he proved

Theorem. Let A and B be C^*-algebras. Suppose that $P(A)$ and $P(B)$ are pure state spaces of A and B, $\psi \colon P(B) \cup \{0\} \to P(A) \cup \{0\}$ is a bijection with $\psi(0) = 0$. Then ψ is induced by a *-isomorphism of A onto B iff ψ and ψ^{-1} are uniformly continuous and ψ preserves orientation and transition probabilities.

Let A and B be unital C^*-algebras, n be a fixed integer, $n \geq 3$. Set

$$CP(A, M_n) = \{\phi | \phi \text{ is a pure completely positive map from } A \text{ to } M_n \text{ with } Tr\phi(I) = 1\}.$$

First of all, we consider the non-commutative version of the above theorem, that is, we consider $CP(A, M_n)$ in place of $P(A)$. Suppose that α is a natural action induced by $SU(n)$ on $CP(A, M_n)$. Namely, we define

$$\alpha_g \phi(x) = g\phi(x)g^{-1}, \ \forall x \in A, g \in SU(n), \phi \in CP(A, M_n).$$

Theorem 1 [8]. If $\psi \colon CP(B, M_n) \cup \{0\} \to CP(A, M_n) \cup \{0\}$, $(n \geq 3)$, a bijection with $\psi(0) = 0$, is α-invariant, preserves transition probabilities and ψ and ψ^{-1} are uniformly continuous, then ψ gives rise to a *-isomorphism between A and B.

The motivation for the above theorem is as follows. Alfsen and Shultz defined the notion orientation of the state spaces of a C^*-algebra, and proved that the state space with orientation can determine the structure of the C^*-algebra. We consider the matrix algebra of a C^*-algebra instead of the state space, and define the notion of α-invariance with which the matrix algebra can determine the structure of the C^*-algebra. In the theorem we use the α-invariance in place of the orientation and set $CP(A, M_n)$ as the non-commutative version of the pure state space. The theorem obtained can be regarded as a non-commutative Shultz theorem.

Recently the theory of pure completely bounded and completely positive maps is developing rapidly. This is another motivation for our paper.

If we consider M_{n^∞} in place of M_n (where M_{n^∞} is a UHF-algebra of type n^∞), we can get a generalization of the above theorem. We give some notations. Let A be a unital C^*-algebra.

$$CP(A, M_{n^\infty}) = \{\phi | \phi \text{ is a completely positive map from } A \text{ to } M_{n^\infty} \text{ with } Tr\phi(I) = 1\}$$

(Tr is the trace in M_{n^∞})

$$M_{n^\infty} = \overline{\cup_{k=1}^{\infty} \varphi_k(M_{n^k})},$$

where $\{\varphi_k\}$ are embeddings from M_{n^k} to $M_{n\infty}$.

$$SSU(\infty) = \bigcup_{k=1}^{\infty} \varphi_k(SU(n^k)).$$

By the use of $SSU(\infty)$, we can define a natural action on $CP(A, M_{n\infty})$ as follows:

$$\alpha_g\phi(x) = g\phi(x)g^{-1}, \; \forall x \in A, g \in SSU(\infty), \phi \in CP(A, M_{n\infty}),$$

which is denoted by α.

When we consider $M_{n\infty}$ in place of M_n, we can obtain an order isomorphism between $L(A, M_{n\infty})$ and $(M_{n\infty} \otimes A)^*$ (with respect to $L(A, M_{n\infty})^{\oplus}$). The restriction of this order isomorphism to $CP(A, M_{n\infty})$ can be viewed as a map from $CP(A, M_{n\infty})$ to $P(A \otimes M_{n\infty})$ (the pure state space of $A \otimes M_{n\infty}$). In the same way we can define the transition probabilities between the elements in $CP(A, M_{n\infty})$. Then we can get the following theorem.

Theorem 2 [8]. Suppose that Ψ is a bijection,

$$\Psi: CP(B, M_{n\infty}) \cup \{0\} \rightarrow CP(A, M_{n\infty}) \cup \{0\}$$

$$\Psi(0) = 0.$$

If Ψ is α-invariant, preserves transition probabilities and Ψ and Ψ^{-1} are uniformly continuous, then Ψ induces a *-isomorphism between A and B.

7. Triangular model of semi-simple L^*-algrbra of H-S operators

In this section, we will consider the triangular model of some kind of infinite dimensional Lie algebras composed of Hilbert- Schmidt operators.

Definition. An L^*−algebra is defined as a Lie algebra L over the complex field such that the vector space of L is a Hilbert space and for each $x \in L$ there is an x^* in L with

$$([x, y], z) = (y, [x^*, z]) \quad \forall y, z \in L.$$

L will be called semi-simple if and only if $L = [L, L]$ and this is equivalent to the mapping from x to D_x being one-one.

Let L_H be the semi-simple L^*−algebra composed of Hilbert-Schmidt operators. The Lie product is given in the usual way, $[A, B] = AB - BA$, $\forall A, B \in L_H$, and the inner product is defined by $(A, B) = \text{trace}(B^*A)$.

Theorem 1[9]. If e_α is a nonzero root vector of L_H, then e_α is a finite rank operator.

Definition of Triangular Model. We denote by L_2^r $(1 \leq r \leq +\infty)$ the Hilbert space $L_2^r[0,1]$. Thus, an element $f \in L_2^r$ is an r-dimensional vector function $f = \{f_\nu(t)\}_1^r$ with measurable components $f_\nu(t)(0 \leq t \leq 1)$ such that

$$|f|^2 = \int_0^1 \sum_{\nu=1}^r |f_\nu|^2 \, dt \leq \infty.$$

For the scalar product of the elements $f, g \in L_2^r$, we have

$$(f,g) = \int_0^1 g^*(t)f(t)dt = \int_0^1 \sum_{\nu=1}^r f_\nu(t)\overline{g_\nu(t)}dt.$$

Let $p(t)$ $(0 \leq t \leq 1)$ be the truncation projector-function defined by the conditions: $p(0) = 0$, $p(1) = 1$ and

$$(p(s)f)(t) = \begin{cases} f(t), & 0 \leq t < s, \\ 0, & s < t \leq 1. \end{cases}$$

Let A be some (abstract) Voltera operator acting in a Hilbert space H_1. A (concrete) Voltera operator $\mathcal{A}$, acting on L_2^r and having $p(t)$ as an eigen-projector function, is called a triangular model of A, if $\mathcal{A}$ is unitary equivalent to A or to an inessential extension of A.

Theorrem 2 [9]. Let L_H be a semi-simple L^*−algebra of H-S operators on H_1, α a nonzero root and e_α the corresponding root vector. Then e_α has a triangular model

$$(\mathcal{A}f)(t) = \int_t^1 \mathcal{A}(t,s)f(s)ds,$$

$$\mathcal{A}(t,s) = \|a_{\mu\nu}(t,s)\|_1^r, \quad (0 \leq t \leq s \leq 1)$$

$$\int_0^1 \int_t^1 \sum_{\mu,\nu=1}^r |a_{\mu\nu}(t,s)|^2 \, dsdt < \infty,$$

$$r \leq \left[\frac{1}{\alpha(h)}\right] + 1.$$

Definition. The algebra L consisting of linear bounded operators on a Hilbert space H_1 is called Voltera, if every operator in L is a Voltera operator.

Theorem 3 [9]. If L is a semi-simple L^*−algebra, then L is not a Voltera algebra.

References

[1] L.S.Wu, Type Classification of Von Neumann Algebras in the Framework of Tomita-Takesaki Theory and Arvesoon's Spectral Theory, J. Operator Theory., 17(1987), 327–333.

[2] L.S.Wu, Tensor Products of Preclosed Operators on C^*-algebras, Proceedings of AMS., 88(1983), 265–269.

[3] L.S.Wu, Continuum of Ideals in $R(\Phi_2) \otimes R'(\Phi_2)$, Proceedings of Japan Academy, 88(1985), 265–269.

[4] L.S.Wu, Completely Positive Maps and *-Isomorphism of C^*-Algebras, Chin Ann of Math., 9B(1988), 27–31.

[5] L.S.Wu, Completely Positive Maps and *-Isomorphism of C^*-algebras II, Acta Mathematica Sinica., 8(1992), 406–412.

[6] L.S.Wu, C^*-Algebraic Isomorphism Determined by Completely Positive and Completely Bounded Maps, Proceedings of the Satelite Conference of ICM-90 —Current Topics in Operator Algebras—World Scientific, (1991), 178–184.

[7] L.S.Wu, When Can the Stable Algebra Determine the Structure of a C^*-Algebra?, Chin Ann of Math., 15B(1994), 153–156.

[8] L.S.Wu, Pure Completely Positive Maps as a Dual Object of C^*-Algebras, Nihonkai Math Journal, 5(1994), 139–147.

[9] L.S.Wu, The Triangular Model of Semi-Simple L^*-Algebra of H-S Operators, Chin Ann of Math., 5B(1984), 306–310.

Operator Theory in Indefinite Inner Product Space

Shaozong Yan and Yusheng Tong

Institute of Mathematics, Fudan University, Shanghai

Operator theory in indefinite inner product space is not a logical generalization of operator theory in Hilbert space. It has a solid foundation. For example, the time-space in relativity theory is just the indefinite inner product space. And the mathematical form reflecting the changes in the observer-the Lotentz group is the group of unitary operators in these spaces.

In the past fifteen years, a group of mathematicians headed by Daoxing Xia and Shaozong Yan in Fudan University have done much research in the field of spectral theory of operators in indefinite inner product space. The geometric structure of these spaces are described. The spectral properties of the operators are thoroughly discussed. Many sharp theorems and important counterexamples are given at the same time. We list a part of them as follows.

1. Geometry of indefinite inner product space

Suppose that Π is a linear space, $(\cdot, \cdot)$ is a bilinear Hermite functional, i.e.

$$(x, y) = \overline{(y, x)}, \text{ for any } x, y \in \Pi,$$

$$(\alpha x + \beta y, z) = \alpha(x, z) + \beta(y, z) \text{ for any } \alpha, \beta \in C, \ x, y, z \in \Pi$$

and there is not any $x \neq 0$ such that $(x, y) = 0$ for all $y \in \Pi$. Then $(\Pi, (\cdot, \cdot))$ is called an indefinite inner product space.

Suppose that $(\Pi, (\cdot, \cdot))$ is an indefinite inner product space, $x \in \Pi$, x is called a positive (or semi-positive) vector if $(x, x) > 0$ (or ≥ 0). Similarly, x is called a negative (or semi-negative) vector if $(x, x) < 0$ (or $\leq$). If $(x, x) = 0$, then x is called a neutral vector or isotropic vector.

1991 Mathematics Subject Classification: 47B50

Suppose that $(\Pi, (\cdot, \cdot))$ is an indefinite inner product space, $x, y \in \Pi$. We say x is orthogonal to y if $(x, y) = 0$, and denote it by $x \perp y$. Suppose that L_1 and L_2 are two linear subspaces. If $L_1 \perp L_2$ (i.e. $x \perp$ for any $x \in L_1, y \in L_2$) and $L_1 \cap L_2 = \{0\}$, then we denote the linear sum $L_1 + L_2$ by $L_1 \oplus L_2$.

Suppose that $\Pi, (\cdot, \cdot))$ is an indefinite inner product space. If there exist a positive subspace H_+ and a negative subspace H_- such that $\Pi = H_- \oplus H_+$, and $(H_\pm(\cdot, \cdot))$ are Hilbert spaces, then $\Pi, (\cdot, \cdot))$ is called a Krein space. In this case, $\Pi = H_- \oplus H_+$ is called a regular decomposition of Π. In particular, $(\Pi(\cdot, \cdot))$ is called a Pontrjagin space and denote by Π_K if $\dim H_- = k < +\infty$.

Under a regular decomposition $\Pi = H_- \oplus H_+$, a positive-definite inner product can be derived as follows:

$$[x_- + x_+, y_- + y_+] = -(x_-, y_-) + (x_+, y_+)$$

where $x_\pm, y_\pm \in H_\pm$. Then $(\Pi, [\cdot, \cdot])$ is a Hilbert space, and the norm corresponding to $[\cdot, \cdot]$ is denoted by $\| \cdot \|$. It can be proved that topologies derived from different regular decompositions of a Krein space are equivalent.

In the operator theory of indefinite inner product space, "the standard decomposition" introduced by Shaozong Yan is more valuable than the regular decomposition.

Let $(\Pi(\cdot, \cdot))$ be a Krein space, and N, P, Z and Z^* be closed negative, positive, neutral and another neutral subspaces of Π respectively. Suppose $\Pi = N \oplus \{Z + Z^*\} \oplus P$. Then this decomposition is called a standard decomposition of Π.

It can be proved that if $\Pi = N \oplus \{Z + Z^*\} \oplus P$ is a standard decomposition, then

i) N, P and $Z + Z^*$ are complete subspaces;

ii) there exists a regular decomposition $\Pi = H_- \oplus H_+$ such that $H_- \supset N, H_+ \supset P$;

iii) $N \oplus Z$ and $N \oplus Z^*$ are maximal semi-negative subspaces, $P \oplus Z$ and $P \oplus Z^*$ are maximal semi-positive subspaces.

There are abundant results on the geometry of Pontrjagin space and Krein space; refer to [7, 8, 12].

2. Preliminary for operators on indefinite inner product space

Suppose that $\Pi(\cdot, \cdot))$ and $\Pi', (\cdot, \cdot)')$ are Krein spaces and T is a densely defined operator from Π to Π'. Then there exists a unique linear operator $T^* : \Pi' \to \Pi$,

$$\mathfrak{D}(T^*) = \{y| \text{ there exists } z \text{ such that } (Tx, y)' = (x, z) \text{ for } x \in \mathfrak{D}(T)\}$$

$$T^* : y \to z$$

T^* is called the adjoint operator of T.

T is called a symmetric operator if $T \subset T^*$. T is called a selfadjoint operator if $T = T^*$. T is called a normal operator if $TT^* = T^*T$.

An example is given to show that in a separable Krein space, there may be a selfadjoint operator A such that $\sigma_P(A) = \mathbb{C}$.

Theorem 2.1[16][33]. Let T be a closed densely defined operator on $(\Pi_K, (\cdot, \cdot))$. Then both T^*T and TT^* are selfadjoint operators, and the numbers of the negative eigenvalues of T^*T and TT^* do not exceed $2k$.

Suppose that $(\Pi, (\cdot, \cdot))$, $\Pi', (\cdot, \cdot)'$ are indefinite inner product spaces, V is a linear operator from Π to Π'. V is called an isometry if $(Vx, Vy)' = (x, y)$ holds for any $x, y \in (\mathfrak{V})$. V is called a unitary operator if it is an isometry and $\mathfrak{D}(V) = \Pi$, $\mathfrak{R}(V) = \Pi'$.

There are great differences between isometries defined on Hilbert space and those on Krein space. For an isometry V defined on a Krein space, V may be not continuous if $\mathfrak{R}(V)$ is not closed (even if non-degenerate). But we have the following results:

Theorem 2.2[33]. Let Π, Π' be two Krein spaces, and V be an isometry from Π to Π'. If $\mathfrak{R}(V)$ is a complete subspace of Π', then V is continuous.

For Π_K space, the condition in the above theorem can be improved.

Theorem 2.3[33]. Let V be an isometry from Π_K to $\Pi_{K'}$. If $\overline{\mathfrak{R}(V)}$ is non-degenerate, then V is continuous.

Suppose that the complex number $\bar{\xi}(\mathrm{Im}\xi \neq 0)$ is not an eigenvalue of the selfadjoint operator A. The Cayley transformation of A is defined by

$$V = (A - \bar{\xi}I)(A - \xi I)^{-1}.$$

Even for a bounded selfadjoint operator, its Cayley transformation may be not unitary. Moreover we can give an example to show that there is a bounded selfadjoint operator with an unbounded Cayley transformation.

Suppose that P is a linear operator on Krein space Π, $\mathfrak{D}(P) = \Pi$. P is called a (orthogonal) projection if $P^2 = P$ $P^* = P$ and $P\Pi$ is called a projective subspace. Suppose that P_1, P_2 are two projection on Π. We say that P_1 is not less than P_2 if $P_1\Pi \supset P_2\Pi$, and denote it by $P_1 \geq P_2$.

There is an example showing that for two projections P_1, P_2, there may not exist a projection P such that $P\Pi = (P_1\Pi) \cap (P_2\Pi)$.

Lemma. Suppose that $_n(n = 1, 2, \cdots)$ are projections, $P_1 \leq P_2 \leq \cdots, \leq P_n \cdots$ (or $P_1 \geq P_2 \geq \cdots \geq P_n \leq \cdots$) and $s - \lim\limits_{n \to \infty} P_n = P$. Then P is a projection

Some examples show that in the above lemma, the assumption "$s - \lim\limits_{n \to \infty} P_n = P$ cannot be abandoned.

3. Spectral theory of operators in Π_K space

3.1. Selfadjoint and unitary operators

The triangular models of unitary operator and selfadjoint operator given by Shaozong Yao are very useful to establish the spectral theory of these operators.

Theorem 3.1[5]. For selfadjoint operator A on Π_k, there is an upper triangular model, i.e. there is a standard decomposition $\Pi_K = N \oplus \{Z + Z^*\} \oplus P$ together with six linear operators $\{S, A_N, A_P, F, G, Q\}$ such that

$$
A = \begin{pmatrix} S & F & G & Q \\ & A_N & & -F^* \\ & & A_P & G^* \\ & & & S^* \end{pmatrix} \begin{matrix} Z \\ N \\ P \\ Z \end{matrix}
$$

where $A_N^* = A_N, A_P^* = A_P, Q = Q^*$. Moreover

$$
\sigma(A) = \sigma(S) \cup \sigma(S^*) \cup \sigma(A_N) \cup \sigma(A_P)
$$

$$
\sigma_P(A) = \sigma(S) \cup \sigma(S^*) \cup \sigma(A_N) \cup \sigma_P(A_P)
$$

The triangular model is based on the existence of invariant semi-positive subspace. In this direction we should mention here that early in 1963, Wuchan Su, whose adviser was Daoxing Xia, and M. A. Naimark got an important result at the same time. They proved that for any family of commutative unitary operators on Π_K, there must be a common invariant semi-positive subspace.

Using the triangular model, Shaozong Yan obtains a series of important results.

Theorem 3.2[5]. Let A be a selfadjoint operator on Π_K. Then A is a bounded operator if and only if $\sigma(A)$ is a bounded set.

Theorem 3.3[5]. Let A be a selfadjoint operator on a separable Π_K space, $\Pi_K = H_- \oplus H_+$ be a regular decomposition. Then for any $\varepsilon > 0$, there must exist a selfadjoint Hilbert-Schmidt operator A_1 such that $\|A_1\| < \varepsilon$ and all eigenvectors of $A + A_1$ span the whole space.

From the triangular model of selfadjoint operator A, we can easily get the triangular model of A^m. Then we can prove that any quasi-nilpotent selfadjoint operator on Π_K is in fact nilpotent.

For a selfadjoint operator A on Π_K, there may not exist any selfadjoint square root even if A is bounded and $(Ax, x) \geq 0$ for all $x \in \Pi_K$. A sufficient and necessary condition for the existence of the square root of a non-negative operator in Π_K space is given. We omit the details here.

An important concept for the spectral theory is the spectral system of selfadjoint operators.

Suppose that $\{E_t\}$ is a projection-valued function defined on $(-\infty, \infty)$ except for a finite number of points $\lambda_1, \lambda_2, \cdots, \lambda_n$ and $\{E_t\}$ satisfies the following conditions
i) $E_t \leq E_{t'}$ if $t < t'$;
ii) $s - \lim_{t \to -\infty} E_t = 0, s - \lim_{t \to \infty} E_t = I$
iii) $s - \lim_{t \to t_0 t_0} E_t = E_{t_0}$, if $t_0 \neq \lambda_1, \lambda_2, \cdots, \lambda_\ell$.
    Then $\{E_t\}$ is called a spectral system. If for some λ_i and intervals $\Delta = (t, t']$ containing λ_i,

$$\overline{\lim}_{\Delta \to \lambda_i} \|E_\Delta\| = \infty,$$

where $E_\Delta = E_{t'} - E_t$, then λ_i is called a critical point of $\{E_t\}$.

Suppose that A is a selfadjoint operator on Π_K, λ is a real eigenvalue of A, and the principal subspace $\Phi_\lambda(A)$ is degenerate. Then λ is called a critical point of A.

The set of all critical points of $\{E_t\}$ and A are denoted by $C(E_t)$ and $C(A)$.

"Critical point" forms an essential difference between the spectral theory on Hilbert space and that on indefinite inner product space.

Theorem 3.4[16]. Let A be a selfadjoint operator on Π_K, $\sigma(A) \subset (-\infty, \infty), C(A) = \{\lambda_1, \lambda_2, \cdots, \lambda_\ell\}$. Then there exists a spectral system $\{E_t\}$ such that
i) E_t is defined for all $t \bar{\in} C(A)$;
ii) For any $(\alpha, \beta], \alpha, \beta \bar{\in} \sigma(A), E_\Delta \Pi_K$ is an invariant subspace of A, $\sigma(A_\Delta) \subset \bar{\Delta}$ where $A_\Delta = A|_{E_\Delta \Pi_K}$;
iii) $E_\Delta \Pi_K$ is a Hilbert space if $C(A) \cap \Delta = \emptyset$;

Theorem 3.5[16]. Let A be given as in the above theorem, and corresponding to the standard decomposition $\Pi_K = W \oplus \{Z + Z^*\} \oplus P, A = \{S, A_N, A_P, F, G, Q\}$. Then $C(A) \subset \sigma(S)$. Moreover, if $\lambda \in \sigma(S)$, then the following propositions are equivalent:

i) $\sup\limits_{\Delta}\{\|E_\Delta\| \mid \Delta = (\alpha, \beta], \Delta \cap \sigma(S) = \{\lambda\}\} < \infty.$

ii) both $s - \lim\limits_{\alpha \to \lambda} E_\Delta$ and $s - \lim\limits_{\beta \to \lambda} E_\Delta$ exist.

iii) $\Phi_\lambda(A)$ is non-degenerate.

3.2. Normal operators

In terms of the triangular model of selfadjoint operators, we can study systematically the spectrum properties of normal operators on Π_K space.

Theorem 3.6[30]. Let $N = A_1 + iA_2$ be a normal operator with the property that $\sigma(A_i) \subset \mathbb{R}$, $i = 1, 2$. Then

i) $\sigma(N) = \sigma_a(N)$;

ii) if $\lambda = \alpha + i\beta \in \sigma(N)$, then there exists a sequence of unit vectors $\{x_n\}$ such that

$$(A_1 - \alpha)x_n \to 0, (A_2 - \beta)x_n \to 0;$$

iii) if $\lambda = \alpha + i\beta \in \sigma_P(N)$, then there exists a vector $x \neq 0$ such that

$$(A_1 - \alpha)x = 0, (A_2 - \beta)x = 0;$$

iv) if $\alpha \in \sigma(A_1)$, then there exists β such that $\alpha + i\beta \in \sigma(N)$;

v) $\Phi_\lambda(N) = \Phi_{\bar\lambda}(N^*)$, and $\Phi_\lambda(N) \perp \Phi_\mu(N)$ if $\lambda \neq \mu$.

vi) if $\lambda = \alpha + i\beta$ is a critical point of N, i.e., $\Phi_\lambda(N)$ is a non-positive subspace, then $\alpha \in C(A_1)$ and $\beta \in C(A_2)$.

Let N be the same as in the above theorem, and E_t^1, E_t^2 be the spectral functions of A_1 and A_2. We define

$$E(\Delta_1 \times \Delta_2) = E^1(\Delta_1)E^2(\Delta_2).$$

This projection-valued function can be extended to the spectral measure of N which is defined in the set

$$E = \{O \mid \partial O \text{ doesn't contain any critical points}\}$$

To discuss the functional calculus of normal operator N on Π_K space, we introduce a special family of functions.

Suppose that $\mathcal{F}$ is a sub-family of bounded Borel functions in the neighborhood of $\sigma(T)$: $F \in \mathcal{F}$ iff there exist a neighborhood O_F of O, an analytic function f in O such that

$$|F(z) - f(z)| \leq M|z|^{2n+1}, \ \forall z \in O_F$$

where M is a constant and n is an integer determined by N.

Let N be the same as in the above theorem, $C(N) = \{0\}$. E_z is the spectral function of N, $F \in \mathcal{F}$. We define $F(N)$ as follows:

$$F(N) = \int_{\mathbb{C} \backslash O_F} F(z) dE_z + \int_{O_F} (F(z) - f(z)) dE_z + f(NE(O_F)).$$

$F(N)$ is well defined. When F is an analytic function, our definition meets the original definition given by Riesz.

Suppose that $F_1, F_2 \in \mathcal{F}$. Then

$$
\begin{aligned}
(\alpha F_1 + \beta F_2)(N) &= \alpha F_1(N) + \beta F_2(N) \\
(F_1 F_2)(N) &= F_1(N) F_2(N) \\
(F_1(N))^* &= F_1^*(N).
\end{aligned}
$$

Theorem 3.7[30]. Let N and F be given as above. Then

$$\sigma(F(N)) = F(\sigma(N)),$$

where $F(\sigma(N))$ denotes the essential range of F in $\sigma(N)$.

3.3. Contractions

A linear bounded operator T is called a contraction if $(Tx, Tx) \leq (x, x)$ for any $x \in \Pi_K$. There is also a triangle model for contractions on Π_K although the invariant maximal semi-negative subspace may be not reducible.

Theorem 3.8[19]. Let T be a contraction on Π_K. Then there is a standard decomposition of Π_K such that $\Pi_K = N \oplus \{Z + Z^*\} \oplus P$ and

$$
T = \begin{pmatrix} S & F & G & B \\ & T_N & T_1 & C \\ & & T_P & D \\ & & & S^{*-1} \end{pmatrix}
\begin{matrix} Z \\ N \\ P \\ Z^* \end{matrix}
$$

where S is an injection on Z; $B = \frac{1}{2}S(C^*C - D^*D + 2Q), Q + Q^* \leq 0, \{T_N, T_1, T_P\}$ forms a contraction on $N \oplus P$.

Theorem 3.9[19]. Let T be given as in the above theorem. Then
i) $\sigma(T) = \sigma(S) \cup \sigma(S^{*-1}) \cup \sigma(T_P) \cup \sigma(T_N)$;
ii) $\sigma_e(T) = \sigma_e(T_P)$ where $\sigma_e(T)$ is the essential spectrum and ind $(T - \lambda I) = $ ind $(T_P - \lambda I)$ if $\lambda \in \rho_e(T)$;
iii) $\sigma(S) \cup \sigma'(S^{*-1} \cup \sigma(T_N) \subset \sigma_P(T)$ where $\sigma'(S^{*-1}) = \{\lambda | \lambda \in \sigma(s^{*-1}), |\lambda| \geq 1\}$;

iv) if $\lambda \in \sigma(S* - 1)\backslash\sigma'(S^{*-1})$, then $\lambda \in \sigma_e(T)$ or $\lambda \bar{\in} \sigma_e(T)$ but $\lambda \in \sigma_P(T) \cup \overline{\sigma_P(T^*)}$.

It is also proved that any weakly continuous semi-group of contractions on Π_K is in fact strongly continuous and there exists its J-unitary dilation in the sense of Nagy. B. S, Foias. C.

4. Operators on Krein space

It is more difficult to study operators on Krein space than on Pontrjagin space. Generally speaking, some restriction is necessary. In fact, although many authors noticed this topic several years ago, only few results have been given. In this field Shaozong Yan has obtained some interesting results.

4.1. Unitary and selfadjoint operators

Theorem 4.1[9][10][11]. Let U be a unitary operator on Krein space. Suppose that there exists a non-constant analytic function f defined on a simply connected region Ω containing $\sigma(U)$ and a finite number of disjoint connected closed sets $\sigma_1, \cdots, \sigma_n$ such that $\sigma(f(U)) = \cup_{i=1}^n \sigma_o$. Then there must be a decomposition $\Pi = \Pi^1 \dot{+} \Pi^2 \dot{+} \cdots \dot{+} \Pi^n$, where Π^i are invariant subspaces of U such that $\sigma(f(U)|_{\Pi^i} \subset \sigma_i$ $(i = 1, 2, \cdots, n)$. Denote the unit circle by C_0, Then Π^i must be a complete space if $\sigma(U|_{\Pi^i}) \cap C_0 \neq \emptyset$. Besides, for each i, there is i', such that $\sigma(U(\Pi^{i'}|) = \left\{ |\frac{1}{\lambda} \in \sigma(U|_{\Pi^i} \right\}$, and $\Pi^i \dot{+} \Pi^{i'}$ is a complete space.

Theorem 4.2[9][10][11]. Let U be a unitary operator on a Krein space. Suppose there exists a non-zero polynomial P such that $P(U)$ is a quasi-nilpotent operator. Then

i) there is a polynomial $q(t)$ with the lowest degree such that $q(U)$ is a quasi-nilpotent operator; if the above $q(t) = \Pi_{i=1}^n (t - a_i)^{m_i}$ the $\sigma(U) = \{a_1, a_2, \cdots, a_n\}$

ii) there exists a decomposition $\Pi = \Pi' \oplus \cdots \oplus \Pi''$ where Π^i are invariant subspaces of U, $\sigma(U|_{\Pi^i})$ is either a singleton $\{a_i\}$ with $|a_i| = 1$ or $\{a_i, \frac{1}{a_i}\}$ with $|a_i| \neq 1$, $\Pi^i = \Pi_1^i \dot{+} \Pi_2^i$ where Π_1^i, Π_2^i are neutral subspaces and are invariant for $U, (\sigma U|_{\Pi_1^i}) = \{a_i\}, \sigma(U|_{\Pi_2^i}) = \{\frac{1}{a_i}\}$.

There are also some similar results for selfadjoint operators. We omit the details here.

We have constructed a unitary operator on Krein space which hasn't any eigenvalue but its spectrum is a singleton. We have also given a selfadjoint operator A satisfying $\rho(LA) = \mathbb{C}$.

4.2. Contractions

For contractions on Krein space, we have found out an important class.

Suppose that T is a densely defined closed contraction on Krein space Π, and there exists a regular decomposition $\Pi = H_- \oplus H_+$ such that $_- \subset \mathfrak{D}(T)$ and $\Pi = \overline{TH_-} \oplus (TH_-)^\perp$ is also a regular decomposition. Then T is called a regular contraction.

A densely defined contraction on Π_K must be a regular contraction. Any regular contraction must be continuous.

Theorem 4.3[14]. Let T be a regular contraction on Krein space Π. Then for any regular decomposition $\Pi = H_- \oplus H_+, \Pi = TH_- \oplus (TH_-)^\perp$ holds, and it is also a regular decomposition.

Theorem 4.4[14]. T is a regular contraction if and only if T^* is also a regular contraction.

Lastly we should mention that Daoxing Xia has studied the scattering problem and got some enlightening results on Krein space. Some others in Fudan University have got a series of interesting results on reduction and perturbation for operators, they have also worked on operator algebras on Pontrjagin space. The reader may refer to [26, 27, 33].

References

[1] Yan Shaozong, On unitary operators on Π_k, Fudan Xuebao, 16(1977), 38–49.

[2] Yan Shaozong, On selfadjoint operators on Π_k, Fudan Xuebao, 18(1979), 22–31.

[3] Yan Shaozong, Unitary dilation of operators, Science Bulletin (Kexue Tongbao), 25(1980), 289–291.

[4] Yan Shaozong, On contractions on Π_k, Fudan Xuebao, 19(1980), 441–451.

[5] Yan Shaozong, Unitary operators and selfadjoint operators on indefinite inner product space, Sci. Sinica, 24, 1615–1625.

[6] Yan Shaozong, Square roots of selfadjoint cperators on Π_k, Chin. Ann. of Math., 3(1982), 19–29.

[7] Yan Shaozong, The structure of Π space I, Chin. Ann. of Math., 5B(1984), 91–100.

[8] Yan Shaozong, The structure of Π space II, Chin. Ann. of Math., 5B(1984), 265–275.

[9] Yan Shaozong, Unitary operators on Π space I, Fudan Xuebao, 20(1981), 424–432.

[10] Yan Shaozong, Unitary operators on Π space II, Chin. Ann. of Math., 4A(1983), 207–216.

[11] Yan Shaozong, Unitary operators on Π space III, Acta Math. Sinica, 27(1984), 749–759.

[12] Yan Shaozong, Some questions on the space with an indefinite metric, Acta Math. Sinica New series, 1(1985), 285–293.

[13] Yan Shaozong, Unitary operators on space Π, Acta Math. Sinica, New series, 2(1986), 1–13.

[14] Yan Shaozong, Contractions on space Π, Chin. Ann. of Math., 7B(1986), 75–89.

[15] Yan Shaozong, Maximal semi-negative common invariant subspace, Northeastern Math., 1(1985), 121–125.

[16] Yan Shaozong, Tong yusun, Unbounded selfadjoint operators on Π_k, Chin. Ann. of Math., 2B(1981), 157–180.

[17] Yan Shaozong, Tong Yusun, On reduction to indefinite metric, Sci. Sinica, 29(1986), 241–253.

[18] Yan Shaozong, Chen Xiaoman, Isometric operators on Π spaces (to appear).

[19] Yan Shaozong, Chen Xiaoman, The spectral theory of contractions on Π_k spaces I, Acta Math. Sinica, New Series, 8(1992), 309–318.

[20] Tong Yusun, On minimal J-unitary dilation, Fudan Xuebao, 19(1980), 181–188.

[21] Tong Yusun, Dissipative operator in space with an indefinite metric, Fudan Xuebao, 20(1981), 233–237.

[22] Tong Yusun, Three theorems on conditional positive-definite functional, Fudan Xuebao, 21(1982), 94–101.

[23] Tong Yusun, Tensor product spaces with indefinite inner products and linear operators on these spaces, Fudan Xuebao, 21(1982), 341–350.

[24] Tong Yusun, Double commutates and generalized spectral decomposition for selfadjoint operators on Π_k space, Fudan Xuebao, 21(1982), 433–438.

[25] Tong Yusun, K-quasi-triangular operators on indefinite inner product spaces, Chin. Ann. of Math., 5A(1984), 551–558.

[26] Tong Yusun, Perturbations of definitizable operators in a space with an indefinite metric, Chin. Ann. of Math., 7B(1986), 435–451.

[27] Tong Yusun, Reduction of operator on indefinite inner product space, Acta Math. Sinica, 29(1986), 658–660.

[28] Chen Xiaoman, The spectrum of contractive operators on Π_k (to appear).

[29] Chen Xiaoman, Unitary dilations of semi-groups of contractions on Π space (to appear).

[30] Chen Xiaoman, Huang Chaocheng, Normal operators on Π_k space, Northeastern Math. J., 1(1985), 247–252.

[31] Zhang Yinan, The symmetric representation of the conditional positive-definite, functional, Fudan Xuebao, 20(1981), 61–66.

[32] Li Shaokuan, Semi-hyponormal operators on indefinite inner product space, Fudan Xuebao, 22(1983), 95–98.

[33] Xia Daoxing, Yan Shaozong, Spectral theory for linear operators II-Operator theory on indefinite inner product space, Academic Press, China, 1987.

Random Metric Theory in China

Zhaoyong You

Department of Mathematics, Xi'an Jiaotong University, Xi'an

Tiexin Guo

Department of Mathematics, Xiamen University, Xiamen

Xi Lin

Jimei Navigation College, Xiamen

Linhu Zhu

Department of Mathematics, Xi'an Jiaotong University, Xi'an

1. Introduction

The theory of random metric (briefly, RM) spaces was initiated by A. Špaček [1, 2, 3]. An RM space is a space in which the distance between a pair of points in the space is a nonnegative real-valued random variable, instead of a nonnegative real number. In earlier years, Chinese scholars were principally concerned in the development of the theory of E-spaces and their applications [4, 5, 6]. E-spaces [3, Chap. 9, pp. 143, Def. 9.1.1] are a typical and wide class of random metric spaces, which are used to provide a practicable framework only for random elements with values in a separable metric space or strongly measurable functions with values in a nonseparable metric space.

A genuine and systematic development of random metric theory is only a matter of recent years when more and more Chinese scholars are devoted to the studies of random metric theory and its applications led by the first author of the present paper. It is because of the following three facts that the related authors have been turning their attention at present from the theory of probabilistic metric (briefly, PM) spaces to random metric theory: (1) Example 1 of Mushtari's paper [7] has illustrated that even for random normed (briefly, RN) spaces of random variables, the probabilistic normed (briefly, PN) spaces determined by the RN spaces can't reflect some important properties possessed by the RN spaces

1991 Mathematics Subject Classification: 46A16, 46A22, 46E40, 46H25, 60H25

themselves, for instance, the property "convergence in measure is equivalent to almost sure convergence", and thus the contents of random metric theory will not be properly covered by those of probabilistic metric theory, hence the study of random metric theory necessarily has some new interest in itself; (2) The theory of RM spaces can't be completely replaced by that of E-spaces, for RM spaces are able to work with wider objectives in random functional analysis than E-spaces, for example not every set of random variables (in the sense of O. Hanš [8]) from a probability space to a nonseparable metric space forms an E-space with target the metric space, but it is always an RM space by a natural construction [9, 10, 11]. Another disadvantage of E-spaces is that they aren't selfcontained, for instance, even for E-norm spaces [3, Chap. 15], the sets of almost surely bounded random linear functionals on the E-norm spaces aren't again E-norm spaces [9]; (3) Following the further development of random functional analysis [12–16], we were forced not only to depict the deep properties that random metric theory itself possesses but also to put forward the more general frameworks motivated by the idea of introducing RM spaces, such as random inner product spaces (briefly, RIP spaces) introduced by Lin & Guo [17], random pseudonormedly determined spaces, espestially random seminormedly determined spaces introduced by You & Guo [16], and random normed modules (particularly random inner product modules) introduced by Guo [13].

The main purpose of this paper is to give a survey of recent developments of random metric theory in China. Limited by the length of the present paper, we only collect the above-mentioned results that are fundamental and motivating for the future development of random metric theory. This paper includes four sections. Section 1 is just the introduction; Section 2 is devoted to mapping generated RM spacess and H. Sherwood's open problem; Section 3 is devoted to RN spaces; Section 4 is devoted to pointwise best approximation in random normed modules and applications to best approximation in classical L^p-spaces.

Throughout the whole paper, (Ω, σ, u) is a given probability space; K is the field of real numbers or complex numbers; $\overline{K} = [-\infty, +\infty]$ or the set of generalized complex numbers; $L(\Omega, K) = \{\xi : \Omega \to \overline{K} | \xi$ is measurable and $|\xi(\omega)| < +\infty$ a.s (almost surely) $\}$, where functions equal a.s are identified. Obiously $L(\Omega, K)$ is not only a ring under ordinary pointwise addition and multiplication but also a module over itself. Especially $L(\Omega, R)$ is also a complete linear lattice under ordinary pointwise ordering; $L^+(\Omega) = \{\xi \in L(\Omega, R) | \xi(\omega) \geq 0$ a.s $\}$. $D^+ = \{F : [0, +\infty) \to [0, 1] | F$ is left continuous, $\inf F = 0, \sup F = 1\}$.

Finally every RM space (S, χ) with base (Ω, σ, u) determines a PM space (S, F) under t-norm $T(a, b) = \text{Max}(a+b-1, 0)$, $\forall a, b \in [0, 1]$, and the probabilistic metric $F : S \times S \to D^+$ defined by $F_{pq}(t) = u\{\omega \in \Omega | X_{pq}(\Omega) < t\}$, $\forall t \geq 0, p, q \in S$ induces a metrizable uniformity structure with base $\{(p, q) \in S \times S | F_{pq}(\varepsilon) > 1 - \lambda\}, \varepsilon > 0, 0 < \lambda < 1$. If the uniformity structure is complete, then we call the RM space (S, χ) complete.

2. Mapping generated RM spaces and H. Sherwood's open problem

lemma 2.1[9]. Let $\xi, \eta : (\Omega, \sigma, u) \to \overline{R} = [-\infty, +\infty]$ satisfy the condition that $\{\omega \in \Omega | \xi(\omega) < +\infty\}$ contains a measurable set of probability one, and so does η. $A \subset L(\Omega, R)$, if $\{\xi(\omega) \leq a(\omega) \leq \eta(\omega)\}$ contains a measurable set of probability one for each $a \in A$, then so do $\wedge A$ and $\vee A$.

Remark 2.1. For any two functions ξ and $\eta : (\Omega, \sigma, u) \to \overline{R}$, that we write $\xi(\omega) \leq \eta(\omega)$ a.s means $\{\xi(\omega) \leq \eta(\omega)\}$ contains a measurable set of probability one. Such a statement consistently works in the following of this paper.

Theorem 2.1[10]. Let (M, d) be a metric space, S be a set of some mappings from (Ω, σ, u) to (M, d) and satisfy the condition that for any p and q in S there exists $\xi \in L^+(\Omega)$ such that $d(p(\omega), q(\omega)) \leq \xi(\omega)$ a.s. If the mapping $\chi : S \times S \to L^+(\Omega)$ is introduced by $X_{pq} = \wedge \{\xi \in L^+(\omega) | d(p(\omega), q(\omega)) \leq \xi(\omega) \text{a.s}\}$, then (S, χ) is an RM space with base (Ω, σ, u). (Functions in S equal a.s are identified).

Remark 2.2. In [10], the above (S, χ) is called a mapping generated RM space with base (Ω, ρ, u) and target (M, d). An RM space with base $(\Omega, \sigma.u)$ is an ordered pair (S, χ) and $\chi : S \times S \to L^+(\Omega)$ satisfies the following conditions: (I) $X_{pq}(\omega) = 0$ a.s iff $p = q$; (II) $X_{pq}(\omega) = \chi_{qp}(\omega)$ a.s $\forall p, q \in S$; (III) $X_{pq}(\omega) \leq X_{pq}(\omega) + X_{qr}(\omega)$a.s$\forall p, q, r \in S$.

Definition 2.1. An ordered pair (S, χ) is called a random normed linear space (RN space) over the field K with base (Ω, σ, u), where S is a linear space over the field K, if $\chi : S \to L^+(\Omega)$ is a mapping satisfying the following conditions:
   (I) $X_{pq}(\omega) = 0$ a.s iff $p = \theta$ (the zero of S);
   (II) $X_{\alpha p}(\omega) = |\alpha| X_p(\omega)$ a.s $\forall p \in S, \alpha \in K$;
   (III) $X_{p+q}(\omega) \leq X_p(\omega) + X_q(\omega)$ a.s $\forall p, q \in S$.

Theorem 2.2[10]. Let $(B, \|\cdot\|)$ be a normed linear space over the field K, S be a linear space over the field K consisting of some mapping from (Ω, σ, u) to $(B, \|\cdot\|)$ such that for any p in S there exists $\xi \in L^+(\Omega)$ with $\|p(\omega)\| \leq \xi(\omega)$ a.s. If a mapping $\chi : S \to L^+(\omega)$ is introduced by $X_p = \wedge \{\xi \in L^+(\Omega) | \|p(\omega)\| \leq \xi(\omega) \text{a.s}\}$, then (S, χ) is an RN space with base (Ω, σ, u).

Remark 2.3. The above (S, χ) is called a mapping generated RN space over the field K with base (Ω, σ, u) and target $(B, \|\cdot\|)$ [10]. Let $S = \{p : (\Omega, \sigma, u) \to B|$ there exists $\xi \in L^+(\Omega)$ such that $\|p(\omega)\| \leq \xi(\omega) \text{a.s}\}$. Clearly S is a linear space, and thus (S, χ) is a mapping generated RN space. Since each B-valued random element (in the sense of Hanš [8]) belongs to S, we denote by $L(\Omega, B)$ the linear closure of the set of all B-valued random elements in (S, χ). We can prove $L(\Omega, B)$ is complete if B is complete (see [10]).

Definition 2.2[10]. Let S be a set of some mappings from (Ω, σ, u) to a metric space (M, d). S is called closed under the pointwise limitation of sequence if there exists a measurable set Ω' of probability one for any sequence $\{p_n\}$ in S such that whenever $d(p_n(\omega), p(\omega)) \to 0 \ \forall \omega \in \Omega'$ for some p from Ω to M, we have $p \in S$.

Theorem 2.3[10]. Let (S, χ) be a mapping generated RM space with base (Ω, σ, u) and target (M, d). If S is closed under pointwise limitation of sequence, then (S, χ) is complete. Particularly, all M-valued random element generated RM spaces are complete if (M, d) is a complete metric space.

Theorem 2.4[11]. Let (S, χ) be a complete RM space with base (Ω, σ, u), T be a selfmapping on S such that for some positive $\alpha < 1$, $X_{TpTq}(\omega) \leq \alpha X_{pq}(\omega)$ a.s $\forall p, q \in S$. Then T has a unique fixed point.

Theorem 2.5[21]. There exists a complete RM space of some real-valued random variables such that there is a probabilistic metric contraction mapping on PM space determined by this RM space, but it has no fixed point.

Remark 2.4. Theorem 2.4 has given H. Sherwood's open problem [3] a negative answer, which has also illustrated the contraction condition of Theorem 2.3 can't be relaxed unconditionally. For detailed disucussions, we refer the reader to [29].

Remark 2.5. Ekeland's variational principle and Caristi's fixed point theorems on ordinary metric spaces have been successfully generalized to RM spaces [5, 6, 23].

3. Random normed spaces

Definition 3.1[16]. Let $\triangle$ be a directed set. An ordered pair $(S, \{\chi^d\}_{d \in \triangle})$ is called a random pseudonormedly determined space over K with base (Ω, σ, u) (briefly, RPN determined space) if S is a linear space over K and the mappings $\chi^d : S \to L^+(\Omega)$ satisfy the following conditions:

(I) $X_p^d(\omega) = 0$ a.s $\forall d \in \triangle$ iff $p = \theta$ (the zero of S);
(II) for d and β in $\triangle$ with $d \leq \beta$, then $X_p^d(\omega) \leq X_p^\beta(\omega)$ a.s $\forall p \in S$;
(III) $X_{\alpha p}^d(\omega) = |\alpha| X_q^d(\omega)$ a.s $\forall p \in S, \alpha \in K, d \in \triangle$;

(IV) for any $d \in \Delta$ there exists $\beta \in \Delta$ such that $X_{p+q}^d(\omega) \leq X_p^\beta(\omega) + X_q^\beta(\omega)$ a.s $\forall p, q \in S$.

Furthermore if the condition $X_{p+q}^d(\omega) \leq X_q^d(\omega) + X_q^d(\omega)$ a.s $\forall p, q \in S, d \in \Delta$ is satisfied, then the random pseudonormedly determined space is called a random seminormedly determined space (briefly, RSN determined space).

Lemma 3.1[16]. Let $(S, \{\chi^d\}_{d \in \Delta})$ be an RPN determined space with base (Ω, σ, u) over K. Then $(S, \{\chi^d\}_{d \in \Delta})$, endowed with the linear topology with neighbor base $\{u_\theta(d, \varepsilon, \lambda) | d \in \Delta, \varepsilon > 0, 0 < \lambda < 1\}$, where $u_\theta(d, \varepsilon, \lambda) = \{p \in S | u\{\omega \in \Omega | X_p^d(\omega) < \varepsilon\} > 1 - \lambda\}$, is a Housdorff linear topological space.

Remark 3.1. From now on, the topological structure of $(S, \{\chi^d\}_{d \in \Delta})$ in this paper is always assumed to be the above-stated linear topological one without special statement. Obviously a net $\{p_\gamma | \gamma \in \Gamma\}$ in S converges to some point p in S iff $X_{p_\gamma p}^d(\omega)$ converges in probability measure u to 0 for each $d \in \Delta$. Random normed spaces are special cases of RPN determined spaces, and $L(\Omega, K)$ is a simpler RN space with random norm $\chi : L(\Omega, K) \to L^+(\Omega)$ defined by $X_p(\omega) = |p(\omega)|$ $\forall p \in L(\Omega, K)$.

Definition 3.2[13]. Let $(S, \{X^d\}_{d \in \Delta})$ be an RPN determined space over the field K with base (Ω, σ, u). A continuous linear mapping from S to $L(\Omega, K)$ is called a continuous random linear functional on S. A continuous random linear functional f on S is called almost surely (a.s) bounded if there exist $c \in L^+(\Omega)$ and some $d \in \Delta$ such that $|f(p)|(\omega) \leq c(\omega) X_p^d(\omega)$ a.s $\forall p \in S$.

Definition 3.3[13]. An ordered triple $(S, \chi, *)$ is called a random normed module (RN module) over the field K with base (Ω, σ, u) if the following conditions are satisfied for the mappings $\chi : S \to L^+(\Omega)$ and $* : L(\Omega, K) \times S \to S$:

(I) (S, χ) is an RN space over the field K with base (Ω, σ, u);

(II) $(S, *)$ is a module over the ring $L(\Omega, K)$;

(III) $X_{\xi * p}(\omega) = |\xi(\omega)| X_p(\omega)$ a.s $\forall p \in S, \xi \in L(\Omega, K)$;

(IV) if $\xi(\omega) = \alpha$ (a constant in K) a.s, then $\xi * p = \alpha \cdot p$ $\forall p \in S$ is implied.

Remark 3.2. In Definition 3.3, if the module's condition, $\xi * (\eta * p) = (\xi \cdot \eta) * p$ $\forall \xi, \eta \in L(\Omega,), p \in S$, where $(\xi \cdot \eta)(\omega) = \xi(\omega) \cdot \eta(\omega)$, is removed, then $(S, \chi, *)$ is just the so-called strong RN space which is first studied by the authors of [15, 25]. However we found in our later works [13, 20] the strong RN spaces too weak to guarantee many important theorems to hold. Owing to (IV) of Def. 3.3, we can abbreviate $(S, \chi, *)$ to (S, χ), $\xi * p$ to $\xi \cdot p$ whenever $*$ is given.

Theorem 3.1[15]. Let S be a linear space over the field $K, \tilde{S} \subset S$ a subspace of $S, f :$ $\tilde{S} \to L(\Omega, K)$ a liner mapping, $\xi : S \to L^+(\Omega)$ a random seminorm on S, and $|f(p)|(\omega) \leq$ $X_p(\Omega)$ a.s $\forall p \in \tilde{S}$. Then there exists a linear mapping $g : S \to L(\Omega, K)$ such that the following hold:

(1) $f(p) = g(p)$ $\forall p \in \tilde{S}$;

(2) $|g(p)|(\omega) \leq X_p(\omega)$ a.s $\forall p \in S$.

Definition 3.4[15]. Let (S, χ) be a RN space with base (Ω, σ, u) over the field $K, f :$ $S \to L(\Omega, K)$ an a.s bounded random linear functional. $X_f^* = \wedge\{c \in L^+(\Omega)| \ |f(p)|(\omega) \leq$ $c(\omega)X_p(\omega)$ a.s $\forall p \in S\}$ is called the random norm of f. Denoting by S^* the set of all a.s bounded random linear functionals on S, and defining $\chi^* : S^* \to L^+(\Omega)$ by $X^*(f) = X_f^*$, then it is proved that (S^*, χ^*) is also an RN space with base (Ω, σ, u) over the field K, which is called the random conjugate space of (S, χ).

Corollary 3.1[9]. Let (S, χ) be the same as in Def. 3.4, M be a subspace of S, and $f : M \to L(\Omega, K)$ an a.s bounded random linear functional on M. Then there exists $\tilde{f} \in S^*$ such that the following hold:

(1) $\tilde{f}(p) = f(p)$ $\forall p \in M$;

(2) $X_{\tilde{f}}^* = \wedge\{c \in L^+(\Omega)| \ |f(p)|(\omega) \leq c(\omega)X_p(\omega)$ a.s $\forall p \in M\}$.

Remark 3.3. In Corollary 3.1, a case for real RN spaces is due essentially to Zhu [5]. Corollary 3.1 guarantees that there exist sufficiently many a.s bounded, and hence also, continuous random linear functionals on an RN space.

Theorem 3.2[13]. Let $(S_1, \chi^{(1)})$ and $(S_2, \chi^{(2)})$ be two RN modules with the same base (Ω, σ, u) over the field $K, T : S_1 \to S_2$ a linear operator. Then T is a.s bounded i.e., there exists $c \in L^+(\Omega)$ such that $X_{Tp}^{(2)}(\Omega) \leq c(\omega)X_p^{(1)}(\omega)$ a.s $\forall p \in S_1$ iff T is a continuous module homomorphism.

Theorem 3.3[13]. Let $(S, \chi, *)$ be a strong RN space with base (Ω, σ, u) over the field K. If the proposition that f is a.s bounded iff f is continuous and $f(\xi * p) = \xi \cdot f(p)$ $\forall \xi \in$ $L(\Omega, K), p \in S$ always holds for any random linear functional f on S, then $(S, X, *)$ must be an RN module.

Remark 3.4. In fact, the random conjugate space (S^*, X^*) of an RN apace (S, X) is an RN module with $* : L(\Omega, K) \times S^* \to S^*$ defined via $(\xi * f)(p) = \xi \cdot f(p)$ $\forall \xi \in L(\Omega, K), f \in$ $S^*, p \in S$, and hence so is (X^{**}, χ^{**}). Define $J : S \to S^{**}$ by $J(p)(f) = f(p)$ $\forall p \in S, f \in S^*$. Then from Cor. 3.1 we see J preserves random norm, i.e., $X_{Jp}^{**} = X_p$ $\forall p \in S$. We denote by $M(S)$ the least closed submodule of closed submodules containing JS of S^{**}, where JS

stands for the image of S under J.

Theorem 3.4[13]. Let $(S_1, X^{(1)})$ and $(S_2, X^{(2)})$ be two RN spaces with the same base (Ω, σ, u) over K, and $(S_2, X^{(2)})$ be complete. Then the RN space $(B(S_1, S_2), \chi)$ of all a.s bounded linear operators from S_1 to S_2, where $\chi : B(S_1, S_2) \to L^+(\Omega)$ is introduced by $\chi_T = \wedge\{c \in L^+(\Omega)|\chi_{Tp}^{(2)}(\omega) \le c(\omega)X_p^{(1)}(\omega) \text{ a.s } \forall p \in S_1\}\forall T \in B(S_1, S_2)$, is random-norm-preserving isometrically isomorphic to a closed subspace of $(B(M(S_1), M(S_2)), \chi)$. Particularly for an RN space (S, χ) we have that (S^*, X^*) is a random-norm-preserving module isomorphic to $((M(S))^*, \chi^{***})$ [15].

Theorem 3.5[13]. Let $(S_1, \chi^{(1)})$ and $(S_2, \chi^{(2)})$ be the same as in Theorem 3.4. Then $(B(S_1, S_2), \chi)$ is complete, particularly (S^*, χ^*) is complete for an RN space (S, χ) [15].

Definition 3.5[13]. Let $(S, \{\chi^d\}_{d \in \Delta})$ be an RPN determined space with base (Ω, σ, u). $A \subset S$ is called probabilistically bounded if A is linearly topologically bounded, and is called a.s bounded if $\vee_{p \in A} X_p^d \in L^+(\Omega)$ for each $d \in \Delta$.

Theorem 3.6[13]. Let $(S_i, \chi^{(i)})$ be two RN modules with base $(\Omega, \sigma, u)(i = 1, 2)$. $\{T_\gamma | \gamma \in \Gamma\}$ is a family of a.s bounded linear operators from S_1 to S_2. If S_1 is complete, then the following hold:

(1) $\{T_\gamma\}$ is probabilistically bounded in $B(S_1, S_2)$ iff $\{T_\gamma p\}$ is probabilistically bounded in S_2 for any $p \in S_1$;

(2) $\{T_\gamma\}$ is a.s bounded in $(B(S_1, S_2), \chi)$ iff $\{T_\gamma p\}$ is a.s bounded in S_2 for any $p \in S_1$;

Theorem 3.7[13]. Let $(S, \{\chi^d\}_{d \in \Delta})$ be an RSN determined space with base (Ω, σ, u), and $A \subset S$ be a subset. Then the following hold:

(1) A is probabilistically bounded in S iff $f(A)$ is probabilistically bounded in $L(\Omega, K)$ for any $f \in S^*$;

(2) A is a.s bounded in S iff $f(A)$ is a.s bounded in $L(\Omega, K)$ for any $f \in S^*$, where S^* is the set of all a.s bounded random linear functionals on S.

4. Pointwise best approximation in RN modules

Definition 4.1[13]. An ordered triple $(S, \chi, *)$ is called a random inner product module with base (Ω, σ, u) over the field K (briefly, RIP module) if the mappings $\chi : S \times S \to L(\Omega, K)$ and $* : L(\Omega, K) \times S \to S$ satisfy the following conditions:

(I) (S, χ) is a random inner product space with base (Ω, σ, u) over the field K (briefly, RIP space), namely, the following four conditions hold: (1) $X_{p,p} \in L^+(\Omega)\forall p \in S$, and

$X_{p,p}(\omega) = 0$ a.s iff $p = \theta$ (the zero of S); (2) $X_{p,q}(\omega) = \overline{X_{q,p}(\omega)}$ (the conjugate number of $X_{q,p}(\omega)$) a.s $\forall p, q \in S$; (3) $X_{\alpha p,q}(\omega) = \alpha X_{p,q}(\omega)$ a.s $\forall \alpha \in K, p, q \in S$; (4) $X_{p,q+r}(\omega) = X_{p,q}(\omega) + X_{q,r}(\omega)$ a.s $\forall p, q, r \in S$;

   (II) $(S, *)$ is a module over the ring $L(\Omega, K)$;

   (III) if $\xi(\omega) = \alpha$ a.s (α is some constant in K) then $\xi * p = \alpha \cdot p \forall p \in S$ is implied.

Remark 4.1. Let (S, χ) be an RIP space with base (Ω, σ, u). In [17], Lin and Guo proved $(S, \tilde{\chi})$ is an RN space with $\tilde{\chi}^2_{p+q}(\omega) + \tilde{\chi}^2_{p-q}(\omega) = 2\tilde{\chi}^2_p(\omega) + 2\tilde{\chi}^2_q(\omega)$ a.s $\forall p, q \in S$, and conversely, where $\tilde{\chi} : S \to L^+(\Omega)$ is defined by $\tilde{\chi}_p(\omega) = \sqrt{\chi_{p,p}(\omega)} \forall p \in S$. For RIP modules we have also completely similar conclusions.

Definition 4.2[20]. Let (S, χ) be an RN space with base $(\Omega, \sigma, u), M \subset S$ be a subset, and $p \in S$. Some point p_0 of M is called a pointwise best approximantion of p to M if $X_{p-p_0}(\omega) \leq X_{p-q}(\omega)$ a.s $\forall q \in M$, i.e., $X_{p-p_0} = \wedge_{q \in M} X_{p-q}$. If any point p in S has a pointwise best approximantion in M, then we say that M is pointwise proximal in S.

   Let (S, χ) be an RN module with base (Ω, σ, u) over K, p be a positive number. Denote by $L^p(S)$ the set $\left\{ q \in S | \int_\Omega (X_q(\omega))^p du < +\infty \right\}$. When $p \geq 1, L^p(S)$ is a normed linear space over K with the norm $\| \cdot \|_p : L^p(S) \to R^+$ introduced by $\|q\|_p = \left(\int_\Omega [X_q(\omega)]^p du \right)^{1/p} \forall q \in L^p(S)$. When $p < 1, L^p(S)$ is a quasinormed space over K with the quasinorm $\| \cdot \|_p : L^p(S) \to R^+$ introduced by $\|q\|_p = \int_\Omega [X_q(\omega)]^p du \forall p \in L^p(S)$. For a set A of $(L^p(S), \| \cdot \|_p)$, we say that A is proximal in $L^p(S)$ if for any $q \in L^p(S)$ there exists some point q_0 of A such that $\|q - q_0\|_p = \inf_{a \in A} \|q - a\|_p$.

Theorem 4.1[20]. Let (S, χ) be a complete RN module with base $(\Omega, \sigma, u), M \subset S$ be a closed submodule, and $p > 0$ any fixed number. Then the following are equivalent:

   (I) M is pointwise proximal in S;

   (II) $L^p(M)$ is proximal in $L^p(S)$.

   Let Y be a closed subspace of a Banach space $(X, \| \cdot \|)$ over K. Denote by $S(\Omega, X)$ and $S(\Omega, Y)$ respectively the sets of all X-valued and Y-valued strongly measurable functions (see [20]) on (Ω, σ, u). Then $S(\Omega, X)$ will be a complete RN module with base (Ω, σ, u) by introducing a natural random norm $\chi : S(\Omega, X) \to L^+(\Omega)$ as $X_p(\omega) = \|p(\omega)\|$, and a module operation $* : L(\Omega, K) \times S(\Omega, X) \to S(\Omega, X)$ as $(\xi * p)(\omega) = \xi(\omega) \cdot p(\omega)$. Accordingly $S(\Omega, Y)$ will be a closed submodule of $S(\Omega, X)$. In $S(\Omega, X)$, functions equal a.s are identified. Clearly $L^p(S(\Omega, X)) = L^P(u, X), L^p(S(\Omega, Y)) = L^P(u, Y)$ for a fixed positive number p.

Theorem 4.2[20]. Let $X, Y, p > 0$ be the same as stated above. Then the following are equivalent:

(I) $S(\Omega, Y)$ is pointwise proximal in $S(\Omega, X)$;

(II) $L^p(u, Y)$ is poximal in $L^p(u, X)$.

Theorem 4.3[20]. In Theorem 4.3, if Y is also separable, then the following are equivalent:

(I) $L^1(u, Y)$ is proximal in $L^1(u, X)$;

(II) Y is proximal in X.

Theorem 4.4[20]. Let (S, χ) be a complete RIP module, $M \subset S$ a subspace of S. Then M is pointwise proximal in S iff M is a closed submodule of S.

Theorem 4.5[18]. Let (S, χ) be an RIP space with base (Ω, σ, u), $M \subset S$ be a subspace, and $p \in S, p_0 \in M$. Then p_0 is a pointwise best approximantion of p to M iff $X_{p-p_0,q}(\omega) = 0$ a.s $\forall q \in M$.

Corollary 4.1. Let (S, χ) be a complete RIP module, $M \subset S$ be a subspace. Then M has a pointwise orthoprojective complement subspace $M^\perp$, namely $S = M + M^\perp, X_{p+q}(\omega) = 0$ a.s $\forall p \in M, q \in M^\perp$ iff M is a closed submodule.

Remark 4.2. From Corollary 4.1 we can see that classical projection decomposition theorem in Hilbert spaces can't be generalized to complete RIP spaces until the RIP spaces and related subspaces which are RIP modules. Thus the present paper has also modified the corresponding principal results of Liu and Gong's paper [19].

Theorem 4.6. Let (S, χ) be an RIP module over K with base $(\Omega, \sigma, u), f : S \to L(\Omega, K)$ be an a.s bounded random linear functional on S and $\{f(p)|p \in S\}$ be a closed subspace of $L(\Omega, K)$. Then there exists a unique point q in S such that $f(p)(\omega) = X_{p,q}(\omega)$ a.s $\forall p \in S$.

Remark 4.3. Theorem 4.6 was first considered by Liu and Gong [19], and Liu and Gong also gave a generalization of classical Lax-Milgram theorem in Hilbert spaces to selfconjugate RIP spaces. However, they neglected the fact that their results don't hold until they founded the corresponding RIP modules.

References

[1] A. Špaček, Note on K. Menger's probabilistic geometry, Czechoslovak Math. J. 6(81), 1956, 72–74.

[2] A. Špaček, Random metric spaces, Trans Second prague Conf. Information Theory, statist. Decision Functions and Random Processes, 1960, 627–638.

[3] B. Schweizer & A. Skler, Probabilistic Metric spaces, North-Holland, 1983.

[4] LinXi, A class of probabillstic normed spaces and random operators, Chinese Science Bulletin, 4(1983), 199–201.

[5] Zhu Lin-Hu, Metric properties of probabilistic metric spaces and their applications, Ph. D thesis, Xi'an Jiaotong University, 1988.

[6] You Zhao-Yong & Zhu Lin-Hu, Ekeland's varational principle on E-spaces, Chinese J. of Engg. Math., 3(1988), 1–7. (In Chinese)

[7] D. kh. Mushtari, On almost sure convergence in linear spaces of random varables, Prob. Theory Appl. 15(1970), 337–342.

[8] O. Hanš, Random operator equations, Forth Berkerley symp. 1960, pp. 185–202.

[9] Guo Tie-Xin, The theory of RM spaces with applications to random functional analysis, Master's thesis, Xi'an Jiaotong Univ., 1989.

[10] Guo Tie-Xin, Random metric theory and its applications, Ph. D thesis, Xi'an Jiaotong Univ., 1992.

[11] Lin Xi & Guo Tie-Xin, Random metric spaces and their applications, Chinese J. of Math. Res. & Expo., 12(4), 1992, 499–504.

[12] Guo Tie-Xin & Zhu Lin-Hu, Random metric space-system, Chinese J. of Engineering Mathematics, 8(2), 1991, 208–212. (In Chinese)

[13] Guo Tie-Xin, A new approach to random functional analysis, Proceedings of the first China postdoctoral academic congress (II), Chinese national defence and industry Press, ISBN 7–118–01173–8/N. 1, 1993, 1150–1154. (In Chinese), also see: Chinese J. of Northeast Mathematics, to appear in English.

[14] Guo Tie-Xin, Metrics, Probabilistic Metrics and Random Metrics, Chinese J. of System Science & Mathematics, to appear in English.

[15] You Zhao-Yong, Zhu Lin-Hu and Guo Tie-Xin, Random conjugate spaces for random normed spaces, J. of Xi'an Jiaotong Univ., 25(3), 1991, 133–134. (Abstract in Chinese)

[16] You Zhao-Yong & Guo Tei-Xin, Random linear topological spaces, ibid, 25(6), 1991, 331–334. (Abstract in Chinese)

[17] Lin Xi & Guo Tei-Xin, Random inner product spaces, Chinese Science Bulletin, 35(22), 1990, 1707–1709. (In Chinese)

[18] Lin Xi & Li Chuan-Mu, Orthoprojection theorems in RIP spaces, Chinese J. of Engg. Math., 8(2), 1991. (In Chinese)

[19] Liu Qing-Rong & Gong Fu-Zhou, Orthoprojection theorems in a class of RIP apaces and applications, Chinese Anal. of Math., A series, 13(3), 1992, 296–305.

[20] You Zhao-Yong & Guo Tie-Xin, Pointwise best approximation with applications to best approximation in $L^p(u, X)$, J. of Approximation Theory, 78(3), 1994, 314–320.

[21] Gong Huai-Yun & Xiang Shu-Huang, A note on complete E-spaces, manuscript, 1992.

[22] Li Tian, Several basic problems on E-spaces, Master's thesis, Xi'an Jiaotong Univ., 1989.

[23] Bai Yian-Qin and Xong Dao-Tong, Drop theorem and Petal theorem in RM spaces, Chinese J. of Engg. Math., 1990. (In Chinese)

[24] Guo Tie-Xin, Extension theorems of continuous random linear operators on random domains, J. Math. Anal. Appl., 192(2), 1995.

[25] Gong Fu-Zhou, The theory of Prob. metric spaces and its applications to random functional analysis.

[26] Wang Zi-Kun, An Introduction to random functional analysis, Chinese Math. advance, 5(1), 1962, 45-71. (In Chinese)

[27] Li Chuan-Mu, Some basic problems in random functional analysis, Master's thesis, Xi'an Jiaotong Univ., 1991.

[28] Chen Guo-Hua, Some basic results in probabilistic metic spaces, Master's thesis, Xi'an Jiaotong Univ., 1988.

[29] Guo Tie-Xin & Lin Xi, Complete W-spaces, J. of Xi'an Jiaotong Univ., 26(4), 1992, 13-18, 52. (In Chinese)

[30] Guo Tie-Xin, A general theorem of solutions of setvalued random operator equations on separable metric spaces, ibid, 26(2), 1992, 106. (Abstract in Chinese)

Some Results on the Joint Spectrum
for N-Tuple of Linear Operators

Dianzhou Zhang

Department of Mathematics, East China Normal University, Shanghai

If $A_1, A_2, \cdots, A_n$ are mutually commuting linear operators on Hilbert space H, then the joint spectrum $\mathrm{Sp}(A)$ for n-tuple $A = (A_1, A_2, \cdots, A_n)$ can be defined in terms of the Kaszul complex by J. L. Taylor [1]. Since 1982, we have tried to generalize the results on the spectrum properties of a single operator to the case of joint spectrum for n-tuple of linear operators [2]. This article is a ten years survey for this subject.

1. An n-tuple of hypo-normal operators

If $A \in B(H)$ and $A^*A - AA^* \geq 0$, we say A is hyponormal. An n-tuple of operators will be said to be double commuting, if $A_i A_j = A_j A_i$ and $A_i A_j^* = A_j^* A_i$, $i \neq j$, $i, j = 1, 2, \cdots, n$. We shall say that a point $z = (z_1, z_2, \cdots, z_n)$ of C^n belongs to the joint approxiamate spectrum $\sigma_\pi(A)$, if there exists a sequence $x_k \in H, \|x_k\| = 1$, such that

$$\|(A_i - z_i)x_k\| \longrightarrow 0, \quad k \to \infty.$$

Theorem 1.1. If $A = (A_1, A_2, \cdots, A_n)$ is a double commuting n-tuple of hyponormal operators, then its joint spectrum has a Cartesian decomposition, i.e.

$$\mathrm{Re}(\mathrm{Sp}(A)) = \mathrm{Sp}(\mathrm{Re}A), \qquad \mathrm{Im}(\mathrm{Sp}(A)) = \mathrm{Sp}(\mathrm{Im}A),$$

and

$$\mathrm{Re}(\sigma_\pi(A)) = \sigma_\pi(\mathrm{Re}A), \qquad \mathrm{Im}(\sigma_\pi(A)) = \sigma_\pi(\mathrm{Im}A).$$

We denote the product spectral measure of $\mathrm{Re}\, A = (\mathrm{Re}\, A_1, \mathrm{Re}\, A_2, \cdots, \mathrm{Re}\, A_n)$ by $E = \prod_{i=1}^{n} E_i$. Let $\Delta = \prod_{i=1}^{\infty} \Delta_i$, where Δ_i is an interval in $\mathbb{R}$, $D_\Delta = \{(x_1 + iy_1, x_2 + iy_2, \cdots, x_n + iy_1)\}$

1991 Mathematics Subject Classification: 47A13

$iy_n) \in \mathbb{C}^n$, $(x_1, x_2, \cdots, x_n) \in \Delta\}$; $H_\Delta = E(\Delta)H$; $A_\Delta = (A_{1\Delta}, A_{2\Delta}, \cdots, A_{n\Delta})$; $A_{k\Delta} = E(\Delta)A_k|_{H(\Delta)}$, $k = 1, 2, \cdots, n$.

Theorem 1.2. If $A = (A_1, A_2, \cdots, A_n)$ is a double commuting n-tuple of hyponormal operators, then

1. $\sigma_p(A_\Delta) = \sigma_p(A) \cap D_\Delta$;
2. $\text{Sp}(A_\Delta) \subset \overline{D}_\Delta$;
3. $\sigma_\pi(A_\Delta) \cap D_\Delta = \sigma_\pi(A) \cap D_\Delta$;
4. $\sigma_r(A_\Delta) \cap D_\Delta = \sigma_r(A) \cap D_\Delta$, $\quad (\sigma_r(A) = \sigma_\pi(A^*))$;
5. $\text{Sp}(A_\Delta) \cap D_\Delta = \text{Sp}(A) \cap D_\Delta$.

The proof of the above theorems can be found in [3].

According to Xia's work [4], any hyponormal operator has a symbol operator A:

$$A_\pm = s - \lim_{t \to \pm\infty} e^{itX} A e^{-itX}, \quad \text{where} \quad A = X + iY.$$

If $A = (A_1, A_2, \cdots, A_n)$ is a double commuting n-tuple of hyponormal operators, we will call $A^K = (A_1^K, A_2^K, \cdots, A_n^K)$, where $K = (k_1, k_2, \cdots, k_n)$, $0 \le k_i \le 1$, $A_i^K = k_i A_i^+ + (1 - k_i)A_i^-$ the generalized symbol operator of n-tuple A. Then we have

$$\text{Sp}(A) = \bigcup_K \text{Sp}(A^K).$$

We also consider the joint spectrum for an n-tuple of subnormal operators. If $S = (S_1, S_2, \cdots, S_n)$ is an n-tuple of subnormal operators, and there is a minimum commuting normal extension $N = (N_1, N_2, \cdots, N_n)$ which acts on $K \supset H$, then

$$\text{Sp}(N, K) \subset \text{Sp}(S, H).$$

2. Functional model [5]

Let $T = (T_1, T_2, \cdots, T_n) = (X_1 + iY_1, X_2 + iY_2, \cdots, X_n + iY_n)$ be an n-tuple of linear operators, $\Delta_j = \sigma(X_j)$, $\Delta = \Delta_1 \times \Delta_2 \times \cdots \times \Delta_n$, $\Omega = (\Delta, \mathcal{B}_\Delta, \mu)$, and D be a Hilbert space. If R is a map from Δ to $L^2(D)$, then we can define an operator $\hat{R}$ on $L^2(\Omega) \otimes D$:

$$\hat{R}f(x) = R(x) \cdot f(x), \quad x \in \Delta.$$

We consider $\tilde{H} = \hat{R}(L^2(\Omega) \otimes D)$. If the j-th component of measure μ is the Lebesgue measure, then we can define an operator P_j on H, $x = (x_1, x_2, \cdots, x_n) \in \Delta$:

$$(P_j f)(x) = \lim_{\varepsilon \to 0} \frac{1}{2\pi i} \int_{-\infty}^{\infty} \frac{f(x_1, \cdots, s_j, \cdots, x_n)}{x_j - (s_j + i\varepsilon)} ds_j,$$

$\hat{x}_j$ being also an operator on $\tilde{H}$: $(\hat{x_j}f)(x) = x_j \cdot f(x)$, $j = 1, 2, \cdots, n$.

Theorem 2.1. T is a double commuting n-tuple of hyponormal operators if and only if

1. There exist a decomposition $H = \oplus H_\sigma$ and a Hilbert space $\tilde{H} = \oplus \tilde{H}_\sigma, \tilde{H}_\sigma = \hat{R}_\sigma(L^2(\Omega_\sigma) \otimes D_\sigma), \Omega_\sigma = (\Delta, \mathcal{B}_\Delta), \mu_\sigma)$. The j-th component of μ_σ is the Lebesgue measure if $\sigma \in N_K^+ = \{(\varepsilon_1, \varepsilon_2, \cdots, \varepsilon_n)|\varepsilon_j = 1\}$. In addition, there is a unitary operator $W = \oplus W_\sigma$ from H onto $\tilde{H}$.

2. There exists a measurable function $\alpha_j^\sigma(\cdot)$ with unitary bounded operator-value. $\alpha_j^\sigma(\cdot) = 0$ if $\sigma \in N_j^- = \{\varepsilon_1, \varepsilon, \cdots, \varepsilon_n)|\varepsilon_j = -1\}$; $\alpha_j^\sigma(\cdot)$ is independent of x_k if $\sigma \in N_j^+(k \neq j)$.

Let $\alpha_j(\cdot) = \oplus \alpha_j^\sigma(\cdot)$, $\beta_j(\cdot) = \oplus \beta_j^\sigma(\cdot)$, $R(\cdot) = \oplus R_\sigma(\cdot)$, $D = \oplus D_\sigma$. We have

$$R(\cdot)\alpha_j(\cdot) = \alpha_j(\cdot)R(\cdot) = \alpha_j(\cdot),$$

$$R(\cdot)\beta_j(\cdot) = \beta_j(\cdot)R(\cdot) = \beta_j(\cdot),$$

$$\alpha_j(\cdot)\alpha_k(\cdot) = \alpha_k(\cdot)\alpha_j(\cdot),$$

$$\beta_j(\cdot)\beta_k(\cdot) = \beta_k(\cdot)\beta_j(\cdot),$$

$$\alpha_j(\cdot)\beta_k(\cdot) = \beta_k(\cdot)\alpha_j(\cdot), \quad (j \neq k),$$

$$\alpha_j(\cdot) \geq 0, \qquad \beta_j(\cdot) \geq 0,$$

such that

$$WT_jW^{-1} = \hat{x}_j + i(\hat{\alpha}_j P \hat{\alpha}_j + \hat{\beta}_j), \quad j = 1, 2, \cdots, n.$$

3. Mosiac functions and principal functions [5]

Theorem 3.1. If $T = (T_1, T_2, \cdots, T_n)$ is a double commuting n-tuple of hyponormal operators, then there is a bounded operator-valued measurable function $B(x, y)$ (Mosiac function of T), $x = (x_1, x_2, \cdots, x_n)$, $y = (y_1, y_2, \cdots, y_n)$, which has a compact support, and $0 \leq B(x, y) \leq I$. For any $z_i \notin \sigma(\hat{\beta}_j), j = 1, 2, \cdots, n$, we have

$$\ln(1 + \alpha_j(x)(\beta_j(x) - z_j)^{-1}\alpha_j(x)) = \int \frac{B(x, y)}{\prod\limits_{j=1}^{n}(y_j - z_j)} dy.$$

In addition, for any bounded continuous function $\psi(y)$ on Sp (Y), we have

$$\int \psi(y) B(x, y) dy = \alpha(x) \int \psi(\beta_1(x) + k_1 \alpha_1^2(x), \cdots, \beta_n(x) + k_n \alpha_n^2(x)) dk \cdot \alpha(x),$$

where $I^n = [0.1] \times \cdots \times [0, 1]$, $dk = dk_1 \cdots dk_n$, $\alpha(x) = \alpha_1(x) \cdots \alpha_n(x)$.

Theorem 3.2. If $T = (T_1, T_2, \cdots, T_n)$ is a double commuting n-tuple of hyponormal operators, the essential support of the Mosiac function is $D(T)$, then we have $D(T) \subset \text{Sp}(T)$; moreover, if T is also a complete non-normal n-tuple of operators, then

$$D(T) = \text{Sp}(T).$$

Theorem 3.3. If T is a double commuting n-tuple of hyponormal operators, then

$$\left\| \prod_{j=1}^{n} [T_j^*, T_j] \right\| \leq \left(\frac{1}{\pi}\right)^n m(\text{Sp}(T)),$$

where $[T_j^*, T_j] = T_j^* T_j - T_j T_j^*, j = 1, 2, \cdots, n$, and m is the Lebesgue measure.

This is a generalized Putnam inequality in the case of n-tuple of operators.

Now we can define the principal function of a double commuting n-tuple of hyponormal operators:

$$g(x, y) = tr B(x, y),$$

where $B(x, y)$ is the Mosiac function.

Theorem 3.4. If $T = (T_1, T_2, \cdots, T_n)$ is a double commuting n-tuple of hyponormal operators and T is joint approximate normal (i.e. $[T_j^*, T_j]$ is an operator of trace class), then we get

$$tr \prod_{j=1}^{n} [P_j, Q_j] = \left(\frac{1}{2\pi i}\right)^n \int J(P_1, Q_1, \cdots, P_n, Q_n) g(x, y) dx dy$$

for any polynomials in two elements $P_1, Q_1, \cdots, P_n, Q_n$.

Finally, we can get similar results in the case of n-tuple of double commuting semi-hyponormal operators [5].

4. Joint essential spectrum and index [11]

Let $A = (A_1, A_2, \cdots, A_n)$ be a commuting n-tuple of linear operators on a Banach space X, $E(X, A)$ be the Koszul complex introduced by A, its boundary operators being d_k, $k = 1, 2, \cdots, n$. If for every k, Im d_k is closed, and $\dim(\text{Ker}d_{2k}/\text{Im } d_{2k-1}) < \infty$ or $\dim(\text{Ker}d_{2k+1}/\text{Im } d_{2k}) < \infty, k = 1, 2, \cdots, n$, then we call A an n-tuple of semi-Fredholm operators, and $\sum(-1)^{k+1}\dim(\text{Ker}d_k/\text{Im } d_{k-1})$ the index of A. In particular, if $\dim(\text{Ker}d_k/\text{Im } d_{k-1}) < \infty$, $k = 1, 2, \cdots, n$, we say that A is an n-tuple of Fredholm operators.

We denote all Fredholm operators by $\mathcal{F}$, and the joint essential spectrum of an n-tuple of linear operators by $\mathrm{Sp}_e(A)$:

$$\mathrm{Sp}_e(A) = \{(z_1, z_2, \cdots, z_n) \in \mathbb{C}^n : (z_1 - A_1), \cdots, (z_n - A_n) \notin \mathcal{F}\}.$$

In this paragraph, we will discuss the joint essential spectrum for tensor product of several n-tuples of operators.

Theorem 4.1. Let X, Y be Banach spaces, $A = (A_1, A_2, \cdots, A_n)$ be a commuting n-tuple of linear operators, and

$$\hat{A}_i = A_i \otimes I \in L(X \hat{\otimes} Y), \qquad \hat{A} = (\hat{A}_1, \cdots, \hat{A}_n).$$

Then we have

$$\mathrm{Sp}(\hat{A}, X \hat{\otimes} Y) = \mathrm{Sp}(A).$$

In general, if X_i are Banach spaces, $T_i \in L(X_i)$, $i = 1, 2, \cdots, n$, $X = X_1 \otimes X_2 \otimes \cdots \otimes X_n$, $\hat{T}_i = I \otimes \cdots \otimes I \otimes T_i \otimes \cdots \otimes I, \hat{T} = (\hat{T}_1, \cdots, \hat{T}_n)$, then

$$\mathrm{Sp}(\hat{T}, X) = \mathrm{Sp}(T_1, X_1) \times \cdots \times \mathrm{Sp}(T_n, X_n).$$

Furthermore, we have

$$\mathrm{Sp}_e(\hat{T}, X) = \bigcup_{j=1}^{\infty} \mathrm{Sp}(T_1, X_1) \times \cdots \times \mathrm{Sp}_e(T_j, X_j) \times \cdots \times \mathrm{Sp}(T_n, X_n).$$

In the case of Hilbert space, we get

Theorem 4.2. If $A = (A_1, A_2, \cdots, A_n)$ and $B = (B_1, B_2, \cdots, B_n)$ are the commuting n-tulpes of linear operators acting on Hilbert spaces H and K respectively, then

$$\mathrm{Sp}(A \otimes I, I \otimes B) = \mathrm{Sp}(A) \times \mathrm{Sp}(B),$$

$$\mathrm{Sp}_e(A \otimes I, I \otimes B) = (\mathrm{Sp}_e(A) \times \mathrm{Sp}(B)) \bigcup (\mathrm{Sp}(A) \times \mathrm{Sp}_e(B)),$$

and if A and B are Frodholm n-tuples, we will have

$$\mathrm{Ind}(A \otimes I, I \otimes B) = -\mathrm{Ind}A \cdot \mathrm{Ind}B.$$

We can prove this theorem by using Curto's matrix [16].

This result can be generalized to the case of a resolvable algebra [2] [5].

5. Unbounded operators

The joint spectrum for an n-tuple of unbounded operators has been defined by Eschmier [6] in the case of $\rho(A) \neq \emptyset$, i.e. every operator A has a non-empty resolvent set. We will give another definition of joint spectrum for an n-tuple of unbounded operators without the condition $\rho(T) \neq \emptyset$ [7].

Let $A = (A_1, A_2, \cdots, A_n)$ be an n-tuple of dense defined closed operators on Hilbert space H. We set $D_0 = H$, and $D_{j_1 \cdots j_p} = \left\{ x \in \bigcap_{i=1}^{p} D(A_{j_i}), A_{j_i}(x) \in D_{j_1 \cdots \hat{j_i} \cdots j_p} \right\}$, $j_1 < j_2 < \cdots < j_p$. Similarly, we can define $D^*_{j_1 \cdots j_p}$ for $A^* = (A_1^*, \cdots, A_n^*)$.

Now, we will say A is commuting, if

(1) $A_i A_j x = A_j A_i x$, for $x \in D(A_i A_j) \bigcap D(A_j A_i)$,

(2) $\overline{D_{1,2,\cdots,n} \cap D^*_{1,2,\cdots,n}} = H$.

Let $E_p^n(A, H) = \oplus_{j_1, \cdots, j_p} D_{j_1, \cdots, j_p} \otimes (S_{j_1} \wedge S_{j_2} \wedge \cdots \wedge S_{j_p})$, $E_p^n(A, H)$ be dense in $E_p^n(H)$,

$$d_p(A)(x S_{j_1} \bigwedge S_{j_2} \bigwedge \cdots \bigwedge S_{j_p}) = \sum_{i=1}^{n} (-1)^{i-1} A_{j_i} x S_{j_1} \bigwedge \cdots \bigwedge S_{j_i} \bigwedge \cdots \bigwedge S_{j_p}.$$

Then we will call $\{E_p^n(A, H), d_p(A)\}$ a Koszul chain complex introduced by the n-tuple A.

Definition. If $\{D(\overline{d}_p), \overline{d}_p\}$ is an exact chain complex, then we will say $A = (A_1, \cdots, A_n)$ is regular. For any $z = (z_1, \cdots, z_n)$, if $z - A = (z_1 - A_1, \cdots, z_n - A_n)$ is not regular, we say z is a point of the joint spectrum of $A : z \in \mathrm{Sp}(A)$.

Theorem 5.1. We have

(1) A is regular if and only if A^* is regular.

(2) A is regular if and only if for any p, the operator $d_p^* \overline{d}_p + \overline{d}_{p+1} d_{p+1}^*$ has a bounded inverse in $E_p^n(H)$.

(3) $\mathrm{Sp}(A)$ is a closed set in $\mathbb{C}^n$.

(4)   If $A = (A_1, \cdots, A_n)$ is an n-tuple of unbounded normal operators, the spectral measures $\{E_i\}$ are commutative, i.e. $E_i E_j = E_j E_i, i \neq j$. Then

$$\mathrm{Sp}(A) = \mathrm{Supp} E, \quad \text{where} \quad E = \prod_{i=1}^{n} E_i.$$

As is well known, the Banach algebra and C^* algebra play a great role in the spectral theory of linear bounded operators. However, for an unbounded operator, we have to be aided by the GB^* algebra and EC# algebra [9] [10].

Let $A = (A_1, \cdots, A_n)$ be a commuting n-tuple of unbounded operators with spectral measure $E = \prod_{i=1}^{n} E_i$, and the family $\mathcal{F} = \{f(z) : f \text{ is a continuous function on Supp } E\}$, the operator $E(f) = \int_{\mathrm{Supp} f} f(z) \cdot dE(z)$ have its domain $D(E(f))$, $D = \bigcap_{f \in \mathcal{F}} D(E(f))$.

Theorem 5.2. We have [8]

(1) D is dense in H;

(2) D is an invariant subspace of every $E(f), f \in \mathcal{F}$;

(3) $\mathcal{U} = \{E(f)|_D : f \in \mathcal{F}\}$ is a closed commuting EC# algebra;

(4) $\mathcal{A} = \{E(f), f \in \mathcal{F}\}$ is a separable (with unit element) and commuting GB^* algebra, if we introduce a matrix ρ by

$$\rho(T_1, T_2) = \sum_{n=1}^{\infty} \frac{1}{2^n} \frac{\|E(\Delta_n)(T_1 - T_2)\|}{1 + \|E(\Delta_n)(T_1 - T_2)\|},$$

where $T_1, T_2 \in \mathcal{A}, \{\Delta_n\}$ is a compact subset sequence which is monotonically increasing and convergent to $\mathbb{C}^n$.

If $\mathcal{A}$ is a GB^* algebra introduced by an n-tuple A, $\mathcal{A}_0$ is a subset consisting of bounded elements of $\mathcal{A}$, M_0 is the total of non-zero multiplicative linear functionals on $\mathcal{A}_0$, then we can get a theorem about the extension of multiplicative functions.

Theorem 5.3. If $\mathcal{A}$ is a commuting GB^* algebra and with a unit element, then for any multiplicative functional $\varphi \in M_0$, there is an extension φ' on $\mathcal{A}$ such that

(1) φ' is an extension of φ,

(2) φ' is a partial homomorphism, i.e.

   (a) $\varphi'(\lambda x) = \lambda \varphi'(x)$ $(\lambda \in \mathbb{C}, x \in \mathcal{A}), 0 \cdot \infty = 0$;

   (b) $\varphi'(x_1 + x_2) = \varphi'(x_1) + \varphi'(x_2)$ $(x_1, x_2 \in \mathcal{A}$, and $\varphi'(x_1), \varphi'(x_2)$ do not assume ∞ simultaneously;

   (c) $\varphi'(x_1 \cdot x_2) = \varphi'(x_1) \cdot \varphi'(x_2)$ $(x_1, x_2 \in \mathcal{A}$, and do not appear in the case $0 \cdot \infty$ or $\infty \cdot 0)$;

   (d) $\varphi'(x^*) = \overline{\varphi'(x)}$ (letting $\overline{\infty} = \infty$).

Theorem 5.4. If $\mathcal{A}$ is a commuting GB^* algebra introduced by an n-tuple $T = (T_1, \cdots, T_n)$, φ' is the extension of a multiplicative functional φ from M_0 to $\mathcal{A}$, then we have

$$\mathrm{Sp}_E(T) = \{\varphi'(T_1), \cdots, \varphi'(T_k), \cdots, \varphi'(T_n) : \varphi \in M_0\}.$$

Using the EC# algebra introduced by A , we can get an estimate of the joint resolvent of a normal n-tuple:

$$\mathrm{dist}(z, \mathrm{Sp}_E(A)) = \|(A \overset{\sim}{-} z)^{-1}\|^{-1},$$

where $(A \overset{\sim}{-} z)$ is the closure of operator $(A \overset{\frown}{-} z)_D$ on $H \otimes \mathbb{C}^{2n-1}$, and $(A \overset{\frown}{-} z)_D$ is the Curto matrix of $(A - z)$, whose elements are restricted on D [10].

Other results we obtained are related to the elementary operators and operator equations [12], an n-tuple of contractive operators and $A\aleph_0$ algebra [5], compact perturbation of joint spectrum and the joint Weyl theorem [13].

References

[1] Taylor, J. L., A joint spectrum for several commuting operators, J. Funct. Anal., 19(1975), 390–421.

[2] Zhang Dianzhou et al., Joint spectrum for n-tuple of linear operators (in Chinese), East China Normal University Press, Shanghai, (1992), 1–262.

[3] Zhang Dianzhou and Huang Danren, On the joint spectrum for n-tuple of hyponomal operators, Chinese Annals of Mathematics, 7b, 1(1986), 14–23.

[4] Xia, D., Spectral Theory of Linear Operator (I), (in Chinese), Academy Press, Beijing, 1983.

[5] Hu Shanwen, Theory of Joint Spectrum, Dissertation, Fudan University, Shanghai, (1988).

[6] Eschmeier, J., Spektralzerlegungen und Functionakakule für vertauschende tupel stetiger und abgeschlossener Operator in Banachraumen, Schriftenreine des Mathematika, Instituts der Universität Münster, 2. serie, helf 20, Juli, (1981), 1045.

[7] Zhang Dianzhou and Wang Zongyao, Talor joint spectrum for n-tuple of closed operators on Hilbert space, Scientia Scinica, Series A, 6(1985).

[8] Huang Danren and Zhang Dianzhou, Joint spectrum and unbounded operator algebras, Acta Matimatica, 3, 2(1986).

[9] Inoue, A., On a class of unbounded operator algebras, Pacific J. Math., 65(1976), 77–95.

[10] Allan. G. R., On a class of locally convex algebras, Proc. London Math. Soc., 3, 17(1967), 91–114.

[11] Hu Shanwen, Joint essential spectrum and index of tensor product of linear operators in Banach spaces, Kexue Tongbao, 11, 34(1989), 885–888.

[12] Huang Danren, On the joint spectrum for n-tuple of linear operators and its applications, Dissertation (Master degree), East China Normal University, (1984).

[13] Cai Jun, A classification of joint spctrum and perbutations, Dissertation (Master degree), East China Normal University, (1984).

[14] Huang Danren, Joint numerical ranges for unbounded normal operators, Proc. Edinburgh Math. Soc., 28(1985), 225–232.

[15] Hu Shanwen, Commuting n-tuple of closed operators which posses spectral capacity, Chinese Annals of Mathematics, 2, 8B, (1987), 156–159.

[16] Curto, R. E., Fredholm and invertible n-tuples of operators, Tran. Math. Soc., 226(1981), 129–159.

Some Structural Properties of Banach Spaces

Junfeng Zhao

Department of Mathematics, Wuhan University, Wuhan

In this paper, we describe briefly some structural properties of Banach space. We organize this paper into three parts. First, we investigate several long James-type Banach spaces and get representation theorems of long James-type Banach spaces and their duals by using transfinite basis. Second, we use Edgar ordering to consider sums of Banach spaces, sequence Banach spaces, function Banach spaces and compare them with c_0, l_p, James space, long James space and so on. We find out the long James space $J(\omega_1)$ is a predecessor of the continuous function space $C[0, \omega_1]$ in this ordering,however $J(\omega_1)$ and $C[0, \omega_1]$ both are not predecessors of l_∞.So we find out the existence of new ordering chains on the class of Banach spaces. We also describe the ordering structure about vector valued long James spaces $J(\eta, X), J(\eta, l_p)(1 < p < \infty)$ and we get $J(\eta, l_p) = J(\eta)(1 < p < \infty)$ and $J(\eta, l_p) < J(\eta, l_\infty)$ and some others. Finally, we introduce and use some geometric constants of Banach space,for example $R_\alpha(X)$, to investigate the geometric properties of Banach spaces.

1.

We will use the following definitions for transfinite series and basis in a Banach space X. Let η be an ordinal, and let $x_\alpha \in X$ be given for each $\alpha < \eta$.The value (when it exists) of the series

$$\sum_{\alpha < \gamma} x_\alpha$$

is defined recursively as follows. If $\gamma = 0$, then

$$\sum_{\alpha < 0} x_\alpha = 0.$$

If $\gamma = \beta + 1$ is a successor,then

$$\sum_{\alpha < \gamma} x_\alpha = \sum_{\alpha < \beta} x_\alpha + x_\beta,$$

1991 Mathematical Subject Classification: 46B20

provided the series on the right-hand side converges. If γ is a limit, then

$$\sum_{\alpha<\gamma} x_\alpha = \lim_{\beta<\gamma}(\sum_{\alpha<\beta} x_\alpha),$$

where the limit is taken in the norm topology of X.

A transfinite sequence $(x_\alpha)_{\alpha<\eta}$ of vectors is called a basis for X if and only if for each $y \in X$, there is a unique sequence $(c_\alpha)_{\alpha<\eta}$ of scalars such that

$$y = \sum_{\alpha<\eta} c_\alpha x_\alpha.$$

Let η be an ordinal, and let $f : [0, \eta] \to R$ be a function. The square variation of f is

$$\sup(\sum_{i=1}^{n} |f(\alpha_i) - f(\alpha_{i-1})|^2)^{1/2} \qquad *$$

where the sup is taken over all finite sequences $\alpha_0 < \alpha_1 < \cdots < \alpha_n$ in $[0, \eta]$. Let $J(\eta)$ be the set of all continuous functions f on $[0, \eta]$ with finite square variation and $f(0) = 0$. Then $J(\eta)$ is a Banach space with the norm (*). $[0, \eta]$ is a well-ordered set with the ordering topology.

Let ω_1 be the first uncountable infinite ordinal number. The bidual $J(\omega_1)^{**}$ of $J(\omega_1)$ can be identified with the set of all functions g on $[0, \omega_1]$ satisfying $g(0) = 0$ and

$$\sup(\sum_{i=1}^{n} |g(\alpha_i) - g(\alpha_{i-1})|^2 < \infty).$$

Theorem 1[5]. Suppose $g \in J(\omega_1)^{**}$. Then $lim_{\beta<\gamma}g(\beta)$ exists for any limit ordinal $\gamma \in [0, \omega_1]$.

Theorem 2[5]. Suppose that $g \in J(\omega_1)^{**}, \gamma, \beta$ are ordinal numbers in $[0, \omega_1]$, d_β is a real number, $d_0 = 0$ and

$$d_\beta = g(\beta) - \begin{cases} g(\beta - 1), & \text{if } \beta \text{ is an isolated point,} \\ \lim_{\gamma<\beta} g(\gamma), & \text{if } \beta \text{ is a limit point.} \end{cases}$$

Then

$$\sum_{0<\beta<\gamma} d_\beta = \begin{cases} g(\gamma - 1), & \text{if } \gamma \text{ is an isolated point,} \\ \lim g(\beta), & \text{if } \gamma \text{ is a limit point.} \end{cases}$$

Theorem 3[5]. Suppose that $H_\alpha = \chi_{[\alpha,\omega_1]}$, $\chi_{[\alpha,\omega_1]}$ is a characteristic function of $[\alpha, \omega_1]$,

$$d_\alpha = \begin{cases} g(\alpha) - g(\alpha - 1), & \text{if } \alpha \text{ is an isolated point,} \\ g(\alpha) - \lim_{\beta<\alpha} g(\beta), & \text{if } \alpha \text{ is a limit point.} \end{cases}$$

Then

$$g = \sum_{0 < \alpha < \omega_1} d_\alpha H_\alpha,$$

where $g \in J(\omega_1)^{**}$.

Theorem 4[5]. The transfinite sequence $\{H_\alpha\}_{\alpha \in (0,\omega_1]}$ is a basis of the Banach space $J(\omega_1)^{**}$.

Theorem 5[5]. Suppose $\gamma \in (0, \omega_1 + 1]$ and $g \in J(\omega_1)^{**}$. Suppose $Q_\gamma g = g_\gamma$, where $g_\gamma = \sum_{\alpha < \gamma} d_\alpha H_\alpha$, and

$$d_\alpha = g(\alpha) - \begin{cases} g(\alpha - 1), & \text{if } \alpha \text{ is an isolated point,} \\ \lim_{\beta < \alpha} g(\beta), & \text{if } \alpha \text{ is a limit point.} \end{cases}$$

Then for any ordinal number $\gamma \in (0, \omega - 1]$, Q_γ is a projection on $J(\omega - 1)^{**}$, and Q_{ω_1+1} is an identity operator on $J(\omega - 1)^{**}$.

Definition 1. A Banach space X is said to have the approximation property (A.P. for short) if, for every compact set K in X and every $\epsilon > 0$, there is an operator $T : X \twoheadrightarrow X$ of finite rank (i.e. $Tx = \sum_{i=1}^{n} x_i^*(x)x_i$, for some $\{x_i\}_{i=1}^{n} \subset X$ and $\{x_i^*\}_{i=1}^{n} \subset X^*$) such that $\|Tx - x\| \leq \epsilon$, for every $x \in K$.

Theorem 6[5]. Suppose γ is a finite integer. Then $J(\omega + \gamma)$ has the approximation property.

Let us consider the long James-type vector valued space $J(\eta, l_2)$, $J(\eta, l_p)$. We only mention $J(\eta, l_2)$ here. Banach space $J(\eta, l_2)$ consists of the vector valued functions F defined on $[0, \eta]$ with values in l_2 which satisfy the following conditions:

(i) $F(0) = 0$;

(ii) $|F| = \sup(\sum_{i=1}^{n} \|F(\alpha_i) - F(\alpha_{i-1})\|_{l_2}^2)^{1/2}$

where the sup is taken over all finite sequences $\alpha_0 < \alpha_1 < \cdots < \alpha_n$ in $[0, \eta]$;

(iii) F is continuous in the norm $\| \cdot \|_{l_2}$.

The Banach space l_2 consists of the sequences $a = (a_0, a_1, \ldots, a_n, \ldots)$ with norm $\|a\| = (\sum_{n=0}^{\infty} a_n^2)^{1/2}$. The unit vector $e_n (n = 0, 1, 2, \ldots)$ is the vector that has 1 in its $(n + 1)$th slot and zeros elsewhere.

Let us consider functions $(\phi_{\alpha,i})$ defined by

$$\phi_{\alpha,i} = \begin{cases} e_i, & \text{if } \gamma \in (\alpha, \eta], \\ 0, & \text{if } \gamma \notin (\alpha, \eta], \end{cases}$$

where $\alpha \in [0, \eta], \gamma \in [0, \eta], i \in [0, \omega), e_i = (0, 0, \cdots, 0, 1, 0, 0, \cdots) \in l_2$ and ω is the first infinite ordinal number.

We know that $\phi_{\alpha,i} \in J(\eta, l_2)$ and $\|\phi_{\alpha,i}\| = 1$ for any $\alpha \in [0, \eta]$ and for any $i \in [0, \omega)$.

Suppose $F \in J(\eta_1, l_2)$, $F = (F(\alpha))_{\alpha \in [0,\eta]}$, where $F(\alpha) = (F_{\alpha,1}, F_{\alpha,2}, \cdots, F_{\alpha,i}, \cdots) \in l_2$.

Theorem 7[4]. $((\phi_{\alpha,i})_{i \in [0,\eta)})_{\alpha \in [0,\eta)}$ is a transfinite basis of the Banach space $J(\eta, l_2)$. Furthermore, for any $F \in J(\eta, l_2)$ we have

$$F = \sum_{\alpha \in [0,\eta]} \sum_{i \in [0,\omega]} C_{\alpha,i} \phi_{\alpha,i}$$

where $C_{\alpha,i} = F_{\alpha+1,i} - F_{\alpha,i}$.

Theorem 8[4]. $((\phi_{\alpha,i})_{i \in [0,\omega)})_{\alpha \in [0,\eta)}$ is a transfinite basis of the l_p valued long James type Banach space $J(\eta, l_p)(1 \leq p < \infty)$ and for $F \in J(\eta, l_p)$,

$$F = \sum_{\alpha \in [0,\eta]} \sum_{i \in [0,\omega]} C_{\alpha,i} \phi_{\alpha,i}$$

where $C_{\alpha,i} = F_{\alpha+1,i} - F_{\alpha,i}$,

$$\phi_{\alpha,i}(\gamma) = \begin{cases} e_i, & \text{if } \gamma \in (\alpha, \eta], \\ 0, & \text{if } \gamma \notin (\alpha, \eta], \end{cases}$$

e_i being a unit vector in l_p that has 1 in its (i+1)th slot and zeros elsewhere.

Long James-type X-valued Banach space $J(\eta, X)$ consists of functions F on $[0, \eta]$ with values in Banach space X which satisfy the following conditions:

(i) $F(0) = 0$;

(ii) $|F| = \sup(\sum_{i=1}^{n} \|F(\alpha_i) - F(\alpha_{i-1})\|_X^2)^{1/2} < \infty$, where the sup is taken over all finite sequences $\alpha_0 < \alpha_1 < \cdots < \alpha_n$ in $[0, \eta]$;

(iii) F is X-norm continuous on $[0, \eta]$.

Theorem 9[4]. Suppose X is a Banach space with a Schauder basis $\{x_i\}_{i=0}^{\infty}$, and

$$\phi_{\alpha,i}(\gamma) = \begin{cases} x_i, & \text{if } \omega \in (\alpha, \eta], \\ 0, & \text{if } \gamma \notin (\alpha, \eta], \end{cases}$$

where $i \in [0, \omega)$, $\alpha \in [0, \eta)$. Then $((\psi_{\alpha,i})_{i \in [0,\omega)})_{\alpha \in [0,\eta)}$ is a transfinite basis of the X-valued long James-type Banach space $J(\eta, X)$, and for any $F \in (\eta, X)$ we have

$$F = \sum_{\alpha \in [0,\eta]} \sum_{i \in [0,\omega]} C_{\alpha,i} \psi_{\alpha,i}$$

where $C_{\alpha,i}(F) = F_{\alpha+1,i} - F_{\alpha,i}$.

Let us consider the associated transfinite sequence of coefficient functionals $((C_{\alpha,i})_{i \in [0,\omega)})_{\alpha \in [0,\eta)}$. $((C_{\alpha,i})_{i \in [0,\omega)})_{\alpha \in [0,\eta)}$ is a transfinite biorthogonal system.

Theorem 10[4]. The biorthogonal functionals $((C_{\alpha,i})_{i\in[0,\omega)})_{\alpha\in[0,\eta)}$, associated with $((\phi_{\alpha,i})_{i\in[0,\omega)})_{\alpha\in[0,\eta)}$ which is a transfinite basis of the Banach space $J(\eta, l_2)$, form a transfinite basis of the dual space $J(\eta, l_2)^*$, and we have the representation:

$$l = \sum_{\alpha\in[0,\eta]} \sum_{i\in[0,\omega]} C_{\alpha,i} l(\phi_{\alpha,i})$$

for any $l \in J(\eta, l_2)^*$.

Theorem 11[4]. The transfinite sequence of coefficient functionals, associated with the transfinite basis $((\phi_{\alpha,i})_{i\in[0,\omega)})_{\alpha\in[0,\eta)}$ of the space $J(\eta, l_p)(1 < p < \infty)$, forms a transfinite basis of the dual $J(\eta, l_p)^*(1 < p < \infty)$, and for any $l \in J(\eta, l_p)^*$ we have

$$l = \sum_{\alpha\in[0,\eta]} \sum_{i\in[0,\omega]} C_{\alpha,i} l(\phi_{\alpha,i}).$$

Theorem 12[4]. The Banach spaces $J(\eta, l_p)(1 \leq\, < \infty)$ and $J(\eta, l_2)^*(1 < p < \infty)$ have the approximation property for any ordinal number η. If a Banach space X has a Schauder basis, then $J(\eta, X)$ has the approximation property for any ordinal number η.

Let us now consider the dual $J(\eta, l_2)^*$ of the Banach space $J(\eta, l_2)$. For any $F \in J(\eta, l_2), \alpha \in [0, \eta], i \in [0, \omega)$,

$$g_{\alpha,i}(F) = F_{\alpha,i}$$

is a linear bounded functional on $J(\eta, l_2)$, i.e.

$$((g_{\alpha,i})_{i\in[0,\omega)})_{\alpha\in[0,\eta]} \in J(\eta, l_2)^*.$$

Theorem 13[8]. $((\phi_{\alpha,i}, g_{\alpha,i}), i \in [0, \omega)), \alpha \in [0, \eta)$ is not a biorthogonal system.

Theorem 14[8]. $[\phi_{\alpha,n+1}, \phi_{\alpha,n+2}, \cdots]$ is a proper subset of $[g_{\alpha,n+1}, g_{\alpha,n+2}, \cdots]_\perp$ for any $\alpha \in [0, \eta)$ and for any $\eta \in [0, \omega)$, where the notation $[x_1, x_2, \cdots]$ indicates the closure of linear span of the set $x_1, x_2, \cdots$. $[g_{\alpha,n+1}, g_{\alpha,n+2}, \cdots]_\perp = \{\phi \in J(\eta, l_2) :< \phi, g >= 0$ for any $g \in [g_{\alpha,n+1}, g_{\alpha,n+2}, \cdots]\}$.

Theorem 15[8]. For any $l \in J(\eta, l_2)^*$,

$$l = \sum_{\alpha\in[0,\eta]} \sum_{i\in[0,\omega]} (U_{\alpha,i} - U_{\alpha+1,i}) g(\phi_{\alpha,i})$$

holds in topology $[J(\eta, l_2)^*, J(\eta, l_2)]$ in the Banach space $J(\eta, l_2)^*$ where

$$U_{\alpha,i} = \begin{cases} l(\phi_{\alpha-1,i}), & \text{if } \alpha \text{ is a non-limit ordinal,} \\ \lim_{\beta<\alpha} l(\phi_{\beta,i}), & \text{if } \alpha \text{ is a limit ordinal,} \end{cases}$$

$\alpha \in [0, \eta), i \in [0, \omega)$.

Theorem 16[8]. Suppose $\beta \in [0, \eta]$ is a limit ordinal number in the topological space $[0, \eta]$, $\{\gamma_n\}_{n=1}^{\infty}$ is a sequence of ordinal numbers in $[0, \eta]$ and $\eta_1 < \eta_2 < \ldots < \eta_n < \ldots \uparrow \beta$ when $n \to \infty$. Let

$$V_1 = \{l \in J(\eta, l_2)^* \mid \|l\|_n = \|l\mid_{((\phi_{\alpha,i})_{i \in [0,\omega)})_{\alpha \in [\gamma_n, \beta]}} \| \to 0, \text{ as } n \to \infty\},$$

$$V_2 = \{ \quad l \in J(\eta, l_2)^* \mid < l, f_m > \to, \text{ as } m \to \infty \text{ for all sequences}$$
$$k_1 < k_2 < \cdots, f_m \in ((\phi_{\alpha,i})_{i \in [0,\omega)})_{\alpha \in [\gamma_{k_m}, \gamma_{k_{m+1}})}\}.$$

Then $V_1 = V_2$.

Theorem 17[8]. The bidual $J(\eta, l_2)^{**}$ can be identified with the Banach space of all transfinite matrices of scalars $((b_{\alpha,i})_{i \in [0,\omega)})_{\alpha \in [0,\eta)}$ such that for any $\gamma \in [0, \eta)$ the series

$$\sum_{\alpha \in [0,\eta]} \sum_{i \in [0,\omega]} b_{\alpha,i} \phi_{\alpha,i}$$

is w^* convergent and

$$\sup_{\gamma \in [0,\eta]} \| \sum_{\alpha \in [0,\eta]} \sum_{i \in [0,\omega]} b_{\alpha,i} \phi_{\alpha,i} \|_{J(\eta,l_2)^{**}} < \infty.$$

This correspondence is given by

$$x^{**} \leftrightarrow ((x^{**}(C_{\alpha,i})_{i \in [0,\omega)})_{\alpha \in [0,\eta)}$$

for $x^{**} \in J(\eta, l_2)^{**}$. The norm of x^{**} is equal to

$$\sup_{\gamma \in [0,\eta]} \| \sum_{\alpha \in [0,\eta]} \sum_{i \in [0,\omega]} x^{**}((C_{\alpha,i})\phi_{\alpha,i} \|_{J(\eta,l_2)^{**}}.$$

Furthermore, for any $i \in [0, \omega), \alpha \in [0, \eta)$ we have $\mid x^{**}(C_{\alpha,i}) \mid \leq \|x^{**}\|$ where $\phi_{\alpha,i}, C_{\alpha,i}$ are defined as before.

2.

Definition 2. Let X and Y be Banach spaces. Then $X < Y$ means $X = \cap T^{**-1}[Y]$, where the intersection is taken over all bounded linear operators $T : X \to Y$.

264   *Junfeng Zhao*

This definition can be rephrased as follows: $X < Y$ if and only if any $\alpha \in X^{**}$, satisfying $T^{**}(\alpha) \in Y$ for all bounded linear operators $T : X \to Y$ must be in X.

Theorem 18[1]. Let Γ be any abstract set. Then $c_0(\Gamma) < c_0$.

A Banach space is said to have the property (X) if and only if $\alpha \in X^{**}$ satisfying $\alpha(\sum f_n) = \sum \alpha(f_n)$ for every sequence $(f_n) \subset X^*$ with $\sum | f_n(x) | < \infty$ for all $x \in X$ must be in X (the sum $\sum f_n$ is taken in the weak* topology of X^*).

Theorem 19[1]. Let Γ be a set with $l_1(\Gamma) < l_1$. For each $\gamma \in \Gamma$, let X_γ be a Banach space and let X_γ have property (X). Then

$$(\oplus_{\gamma \in \Gamma} X_\gamma)_1 < (\oplus_{\gamma \in \Gamma} X_\gamma)_0.$$

Theorem 20[1]. Let Γ be an abstract set and for each $\gamma \in \Gamma$, X_γ be a Banach space with $X_\gamma < l_1(\Gamma)$. Then

$$(\oplus_{\gamma \in \Gamma} X_\gamma)_1 < l_1(\Gamma).$$

Theorem 21[1]. Let Γ be an abstract set for each $\gamma \in \Gamma$, let X_γ, Y_γ be Banach spaces such that $X_\gamma < Y_\gamma$. Then $X < Y$ where $X = (\oplus_{\gamma \in \Gamma} X_\gamma)_0, Y = (\oplus_{\gamma \in \Gamma} X_\gamma)_0$.

Theorem 22[1]. Let Γ be an abstract set for each $\gamma \in \Gamma$, let X_γ be a Banach space such that $X_\gamma < l_\infty$. Then

$$X = (\oplus_{\gamma \in \Gamma} X_\gamma)_0 < l_\infty.$$

Space $C[0, \omega_1]$ consists of all continuous functions on $[0, \omega_1]$ with max norm. The dual of $C[0, \omega_1]$ consists of all functions f on $[0, \omega_1]$ satisfying $|f| = \sum_{\gamma \in [0,\omega_1]} |f(\gamma)| < \infty$, written as $C[0, \omega_1]^* = l_1[0, \omega_1]$. The bidual of $C[0, \omega_1]$ consists of all functions α on $[0, \omega_1]$ satisfying $\|\alpha\| = \sup_{\gamma \in [0,\omega_1]} | \alpha(\gamma) | < \infty$, written as $C[0, \omega_1]^{**} = l_{\infty[0,\omega_1]}$.

Theorem 23[1]. $J(\omega_1) < C[0, \omega_1]$.

Lemma 1[1]. Suppose A is a weak* separable bounded subset in $l_1[0, \omega_1]$. Then there is a $\beta \in [0, \omega_1)$ such that $f = 0$ on (β, ω_1) for all $f \in A$.

Theorem 24[1]. Suppose A is a weak* separable bounded subset in $l_{1[0,\omega_1]}$. Suppose a net $f_\delta \subset A$ and $f \in A$. If $f_\delta \longrightarrow_\delta f$, then $\lim_\delta f_\delta(\omega_1) = f(\omega_1)$.

Theorem 25[1]. $C[0, \omega_1] \neq l_\infty$.

Theorem 26[1]. Suppose Γ is an abstract set. Then $l_\infty(\Gamma) < l_\infty$ if and only if $\bar{\bar{\Gamma}} \leq \aleph_0$.

For any $x \in J(\eta, l_2)$,

$$x = \sum_{\alpha \in [o, \eta)} \sum_{i \in [o, \omega)} C_{\alpha, i} \phi_{\alpha, i}, \quad C_{\alpha, i} = x_{\alpha+1, i} - x_{\alpha, i},$$

and let $\langle Z_{\gamma, i}, x \rangle = x(\gamma)_i$.

Theorem 27[7]. For any $\gamma \in [0, \eta]$ and $i \in [0, \omega)$, then $\sum_{\alpha < \gamma} C_{\alpha, i} \in J(\eta, l_2)^*$ and $\sum_{\alpha < \gamma} C_{\alpha, \gamma} = Z_{\gamma, i}$.

Theorem 28[7]. Suppose $J(\eta, l_2)^{**}$ is a bidual of $J(\eta, l_2)$, $x^{**} \in J(\eta, l_2)^{**}$. For any $\gamma \in [0, \eta]$ define

$$x^{**}(\gamma) = (\langle x^{**}, \sum_{\alpha < \gamma} C_{\alpha, i} \rangle)_{i \in [0, \omega)}.$$

Then $x^{**}(\gamma) \in l_2$ and

$$\|x^{**}\| \geq \sup(\sum_{i=1}^{n} \|x^{**}(\gamma_i) - x^{**}(\gamma_{i-1})\|_{l_2}^2)^{1/2}.$$

Theorem 29[7]. $J(\eta, l_2)^{**}$ consists of all l_2-valued functions x^{**} on $[0, \eta]$ satisfying the following conditions :
  (i) $x^{**}(0) = 0$;
  (ii) $\sup(\sum_{i=1}^{n} \|x^{**}(\beta_i) - x^{**}(\beta_{i-1})\|_{l_2}^2)^{1/2} \leq \|x^{**}\|$,
where sup is taken over all finite sequences $\beta_0 < \beta_1 < \ldots < \beta_n$ in $[0, \eta]$.

Theorem 30[7]. Suppose

$$W_1 = x^{**} \in J(\eta, l_2)^{**}; x^{**} \text{ is } w^*\text{-continuous on } [0, \eta],$$

$$W_2 = x^{**} \in J(\eta, l_2)^{**}; x^{**} \text{ is } l_2\text{-norm continuous on } [0, \eta].$$

Then $W_1 = W_2$.

Theorem 31[7]. For any ordinal number η, $J(\eta, l_2) < J(\eta)$.

Theorem 32[7]. $J(\eta, l_2) = J(\eta)$.

Theorem 33[7]. $J(\eta, l_p) = J(\eta)(1 < p < \infty). J(\eta, l_p) < J(\eta, l_\infty)(1 < p < \infty)$.

Theorem 34[7]. $J(\omega_1) \neq c_0, c_0 \neq J(\eta)$.

Theorem 35[7]. $J(\eta, c_0)$ is equivalent to c_0 if and only if η is a countable ordinal number.

3.

The following is a joint work with Yin Hungsheng and Wang Chonghu.

Definition 3[17]. A real Banach space X is said to be flat if the girth of X is equal to A and is achieved by some centrally symmetric simple closed curve. Such a centrally symmetric simple closed curve is called a girth curve for X.

$$\text{girth}(X) = \inf\{l(c) : c \quad \text{is a centrally symmetric simple}$$
$$\text{closed curve lying in} S(X)\};$$

$$S(X) = x : \|x\| = 1;$$

$$R_\alpha(X) = \sup\{\lambda : S(X) \quad \text{contains a } \lambda\text{-separated subset}$$
$$\text{which is of cardinality} \alpha\}.$$

Theorem 36[17]. If X is a flat Banach space, then $R_{\aleph_0}(X) = 2, R_{\aleph_0}(x^*) = 2$, where x^* is the dual of X.

Theorem 37[17]. Suppose $\{B_i\}_{i=1}^{\infty}$ is a sequence of Banach spaces. If one of them is a flat Banach space, then the product space $(\oplus B_i)_p (1 < p < \infty)$ is also a flat Banach space.

Theorem 38[2]. $R_\alpha(X) = Q_\alpha(X)$, where the Kottman constant is defined by

$$Q_\alpha(X) = \sup\{\lambda : B(X) \text{contains } \lambda\text{-separated subset which is of cardinality } \alpha\}.$$

$$(B(X) = x \in X : \|x\| \leq 1).$$

Theorem 39[2].
$$Q_\alpha(X) = \frac{2P_\alpha(X)}{1 - P_\alpha(X)}$$
$$R_\alpha(X) = \frac{2P_\alpha(X)}{1 - P_\alpha(X)},$$

where $P_\alpha(X)$ is another Kottman constant defined by

$$P_\alpha(X) = \sup\{\lambda : B(X) \text{ contains } \alpha \text{ many disjoint open balls of radius } \lambda\}$$

Theorem 40[2]. For any positive integer $n > 0$, we have

$$\sqrt{2} - 1 \leq P_n(X) \leq 1/2.$$

Theorem 41[2]. Suppose Banach spaces X and Y are isomorphic, T is an isomorphism from X to Y. Then

$$\frac{P_\alpha(X)}{P_\alpha(Y)} \leq \|T\|\|T^{-1}\|.$$

Especially, if for some α, $R_\alpha(X) \neq R_\alpha(Y)$, then X is not almost isometric to Y.

References

[1] Edgar, G. A., Zhao, Jun Feng, The ordering structure on Banach spaces, Pacific J. Math., 116(1985)2, 255–263.

[2] Yang, Chang Sun, Zhao, Jun Feng, Some estimations of convexity modolus for Banach spaces, to appear.

[3] Yin, Hong Sheng, Zhao, Jun Feng, Wang, Chong Hu, On λ-separated subsets of unit spheres of normed linear spaces, Nanjing University (Mathematics Edition) (1980), 129–134.

[4] Zhao, Jun Feng, The vector valued long James type Banach spaces $J(\eta, l_p)(1 \leq p < \infty)$, System Sciences and Mathematics 2(1985)1, 52–62.

[5] Zhao, Jun Feng, The tranfinite basis of the bidual space of long James spaces, Acta Mathematica Scientia (English Ed.) 5(1985)3, 295–303.

[6] Zhao, Jun Feng, The transfinite bases of bidual space of long James space $J(\omega_1)$, Acta Mathematica Scientia 5(1985)3, 321–329.

[7] Zhao, Jun Feng, The ordering structure of Banach spaces. II, *J.* of Wuhan University (Natural Science Edition) (1985)3, 11–19.

[8] Zhao, Jun Feng, Some properties of $J(\eta, l_2)$, Chin. Ann. of Math., 7B(1986)4, 401–407.

[9] Zhao, Jun Feng, Some Problems on the development of the geometry of Banach spaces, Acta Mathematica Scientia 7(1987)4, 38–59.

[10] Zhao, Jun Feng, On structures of long James B-space and their ordering positions, Mathematical Methods (1988)1100–1123.

[11] Zhao, Jun Feng, The finite dimension perturbation of basic sequence and BHB extension, Acta Mathematica Scientia 10(1990)4, 404–408.

[12] Zhao, Jun Feng, Some problems on the structure of Banach spaces, SEA Bull. Math. Special Issue (1993) 185–192, World Scientific Publishing.

[13] Zhao, Jun Feng, Lu, Wan Zhong, A condition equivalent to uniform smoothness of Banach Spaces, Acta Mathematia Scientia 13(1993)2, 190–198.

[14] Zhao, Jun Feng, Some structural problems of infinite dimensional subspace of Banach spaces, Acta Mathematia Scientia 14(1994)2, 190–198.

[15] Zhao, Jun Feng, A remark about Pettis integrable property, Proceedings of the International Conference on Banach Space Theory–"Banach Space Theory and Its Applications", 1994.4.

[16] Zhao, Jun Feng, The structure theory on Banach spaces, 1991, Wuhan University Press.

[17] Zhao, Jun Feng, Wang, Chong Hu, Ying, Hong Sheng, On Some properties of flat Banach spaces, J. of Mathematics 1(1980)1, 96–101.

The Plurisubharmonic Dentability and the Analytic Radon-Nikodym Property for Bounded Subsets in Complex Banach Spaces

Shangquan Bu

Department of Applied Mathematics, Tsinghua University, Beijing

Let X be a complex Banach space. Following [2], X is said to have the analytic Radon-Nikodym property if every uniformly bounded analytic function from the open unit disk of $\mathbf{C}$ with values in X, $f : \mathbf{D} \to X$ has radial limits almost everywhere on the torus $\mathbf{T} = \{e^{i\theta} : \theta \in [0, 2\pi[\}$ in X, which means that for almost all $\theta \in [0, 2\pi[$, $\lim_{r \uparrow 1} f(re^{i\theta})$ exists. An upper semi-continuous function $\phi : X \to \mathbf{R}$ is plurisubharmonic (see [3]) if, for every $x, y \in X$, $f(x) \leq \int_0^{2\pi} f(x + e^{i\theta}y) \frac{d\theta}{2\pi}$. We shall denote by $PSH_0(X)$ the set of all Lipschitz plurisubharmonic functions f on X satisfying $f(0) = 0$. For $f, g \in PSH_0(X)$, define $d(f, g)$ as the Lipschitz constant $\|f - g\|_{Lip}$ of $f - g$. $PSH_0(X)$ equipped with the metric d becomes a complete metric space. A sequence of functions $(f_n)_{n \geq 0}$ in $L^1([0, 2\pi[^{\mathbf{N}}, X)$ is called an X-valued analytic martingale if $f_0 \equiv x_0 \in X$, $f_n \in L^1([0, 2\pi[^n, X)$ and for every $(\theta_1, \theta_2, \cdots, \theta_{n-1}) \in [0, 2\pi[^{n-1}$, $f_n(\theta_1, \theta_2, \cdots, \theta_n) = f_{n-1}(\theta_1, \theta_2, \cdots, \theta_{n-1}) + d_n(\theta_1, \theta_2, \cdots, \theta_{n-1})e^{i\theta_n}$. Among other known characterizations of the analytic Radon-Nikodym property, we have the following

Theorem A (see [3, 4]). Let X be a complex Banach space. Then the following are equivalent:

1. X has the analytic Radon-Nikodym property.

2. Every non-empty bounded subset C of X is plurisubharmonically dentable: for every subset D of C, for every $\epsilon > 0$, there exist a Lipschitz plurisubharmonic function ϕ on X and an $M \in \mathbf{R}$ such that $\{x \in C : \phi(x) > M\}$ is non-empty and has a diameter less than ϵ.

3. For every non-empty bounded subset C of X, for every function $f : C \to \mathbf{R}$ bounded above and $\epsilon > 0$, there exists a Lipschitz plurisubharmonic function ϕ such that $\{x \in C : f(x) + \phi(x) > 0\}$ is non-empty and has a diameter less than ϵ.

1991 Mathematical Subject Classification: 46B20

Supported by the Natural Science Foundation of China

The analytic Radon-Nikodym property has been localized for bounded subsets in complex Banach spaces in [1]. Recall that a non-empty bounded subset C of X is said to have the analytic Radon-Nikodym property if, for every function $f : C \to \mathbf{R}$ bounded above and $\epsilon > 0$, there exists a Lipschitz plurisubharmonic function ϕ on X so that $\{x \in C : f(x) + \phi(x) > 0\}$ is non-empty and has a diameter less than ϵ. So a complex Banach space X has the analytic Radon-Nikodym property if and only if every bounded subset C of X has the analytic Radon-Nikodym property. Comparing Theorem A with the corresponding results on the Radon-Nikodym property setting in [5], it is natural to ask whether the plurisubharmonic dentability characterizes the analytic Radon-Nikodym property of C. The purpose of this paper is to give a negative answer to this question. Precisely we shall show the following

Theorem 1. There exist an index set I and a non-empty closed bounded subset C of $c_0(I)$, such that C is plurisubharmonically dentable and C has not the analytic Radon-Nikodym property.

The subset C in the theorem above was first introduced and studied in [6], where the authors have shown the plurisubharmonic dentability of C, but with non-Lipschitz plurisubharmonic functions. We shall show that the non-Lipschitz plurisubharmonic functions used in [6] can be regularized to be Lipschitz and plurisubharmonic (see the lemma below). To prove that C has not the analytic Radon-Nikodym property, we shall use the following characterization of the analytic Radon-Nikodym property for bounded subsets in complex Banach spaces.

Theorem B (see [1]). Let C be a non-empty bounded subset of a complex Banach space X. Then the following are equivalent:

1. C has the analytic Radon-Nikodym property,

2. Every X-valued analytic martingale $(M_n)_{n \geq 0}$ satisfying the following conditions converges almost surely: There exist an increasing sequence of positive integers $(n_k)_{k \geq 1}, E \in L^1(\mathbf{T}^{\mathbf{N}}, X)$ and an adapted sequence of integrable functions $(E_n)_{n \geq 0}$ in $L^1(\mathbf{T}^{\mathbf{N}}, X)$ which converges to E in $L^1(\mathbf{T}^{\mathbf{N}}, X)$ and is such that, for every $k \in \mathbf{N}$, $M_{n_k}(\theta_1, \theta_2, \cdots, \theta_{n_k}) + E_{n_k}(\theta_1, \theta_2, \cdots, \theta_{n_k}) \in C$ for almost all $(\theta_1, \theta_2, \cdots, \theta_{n_k}) \in \mathbf{T}^{n_k}$.

If we suppose furthermore that C is closed, then the above conditions are equivalent to

3. For every bounded above upper semicontinuous function $f : C \to \mathbf{R}$, for every $\epsilon > 0$, there exists a Lipschitz plurisubharmonic function $\phi \in PSH_0(X)$ with $\|\phi\|_{Lip} \leq \epsilon$ such that $f + \phi$ strongly exposes C from above, which means that there exists $x_0 \in C$, such that $f(x_0) + \phi(x_0) = Sup_{x \in C}(f(x) + \phi(x))$ and $lim_{n \to \infty} diam\{x \in C : f(x) + \phi(x) > f(x_0) + \phi(x_0) - \frac{1}{n}\} = 0$. In fact, the set of such ϕ's is a dense G_δ-subset of $\{\phi \in PSH_0(X) : \|\phi\|_{Lip} \leq \epsilon\}$ with the metric d defined above.

The proof of the main theorem will use the following lemma.

Lemma. Let f be a plurisubharmonic function on a complex Banach space X, Lipschitz on every bounded subset of X. Then for every $M > 0$, there exists a Lipschitz plurisubharmonic function ϕ on X such that $f(x) = \phi(x)$ for every $\|x\| \leq M$.

Proof of Theorem 1. Let $(m_n)_{n\geq 1}$ be a sequence of integers, $m_n \geq 2$ and $\sum_{n\geq 1} m_n^{-1} < \infty$. Define the tree

$$I = \bigcup_{n=1}^{\infty} \prod_{j=1}^{n} \{1,\, 2,\, \cdots,\, m_j\},$$

a natural order structure can be equipped on I: if $(k_1, k_2, \cdots, k_n)$, $(h_1, h_2, \cdots, h_m) \in I$, $(k_1, k_2, \cdots, k_n) < (h_1, h_2, \cdots, h_m)$ if $n < m$ and if $h_i = k_i$ for $1 \leq i \leq n$. For $t = (k_1, k_2, \cdots, k_n) \in I$, denote $|t| = n$. An initial segment S of I will be a finite subset $S = \{t_j\}_{j=1}^{q}$ of I such that $|t_j| = j$ and $t_j < t_{j+1}$ for $1 \leq j < q$. To each initial segment $S = \{t_j\}_{j=1}^{q}$, we can find $k_j \in \{1, 2, \cdots, m_j\}, 1 \leq j \leq q$, such that $t_j = (k_1, k_2, \cdots, k_j)$. Denote $|S| = q$. Let $S = \{t_j\}_{j=1}^{q}$ be an initial segment, $t_j = (k_1, k_2, \cdots, k_j)$, $k_j \in \{1, 2, \cdots, m_j\}, 1 \leq j \leq q$. We associate an atom $a = a(S) \in c_0(I)$ with

$$a(t) = \begin{cases} e^{i2\pi k_{j+1}/m_{j+1}} & \text{if } t = t_j, 1 \leq j \leq q-1 \\ 0 & \text{otherwise .} \end{cases}$$

Denote by C the subset of $c_0(I)$ consisting of all atoms associated with the initial segments, C is a closed bounded subset of $c_0(I)$. By [6], for every subset D of C and $\epsilon > 0$, there exists a plurisubharmonic function ϕ on $c_0(I)$ Lipschitz on every bounded subset of $c_0(I)$ and there exists $M \in \mathbf{R}$, such that $\{x \in D : \phi(x) > M\}$ is non-empty and has a diameter less than ϵ. By the lemma, there exists a Lipschitz plurisubharmonic function φ on $c_0(I)$ such that for every $x \in C, \phi(x) = \varphi(x)$, so $\{x \in D : \phi(x) > M\} = \{x \in D : \varphi(x) > M\}$ is non-empty and has a diameter less than ϵ. This shows that C is plurisubharmonically dentable. We shall show that C has not the analytic Radon-Nikodym property. For each element $c \in c_0(I)$, the support of c will be the subset $\{i \in I : c(i) \neq 0\}$ of I.

Let $m \in \mathbf{N}, m \geq 2, \alpha \in \mathbf{C}, |\alpha| = 1$, define $k = l_m(\alpha) \in \{1, 2, \cdots, m\}$ such that if $\alpha = e^{ix}, x \in [0, 2\pi[$, then $x \in [2\pi(l_m(\alpha) - 1)/m, 2\pi l_m(\alpha)/m[$; one has in this case $|\alpha - e^{i2\pi l_m(\alpha)/m}| \leq 2\pi/m$.

Now, let $k_1 \in \{1, 2, \cdots, m_1\}$ and $t_1 = (k_1) \in I$. Let $c \in c_0(I)$ defined by

$$c(t) = \begin{cases} 1 & \text{if } t = t_1 \\ 0 & \text{otherwise,} \end{cases}$$

and let $F_1(\theta_1) = ce^{i\theta_1}$, $\theta_1 \in [0, 2\pi[$. For each $\theta_1 \in [0, 2\pi[$, we associate $t_2(\theta_1) = (k_1, l_{m_2}(\theta_1)) \in I$ with an initial segment $S_2(\theta_1) = \{t_1, t_2(\theta_1)\}$, define $G_1(\theta_1) = a(S_2(\theta_1))$. Then G_1 and F_1 have the same support $\{t_1\}$ and $|(G_1(\theta_1) - F_1(\theta_1))(t_1)| \leq 2\pi/m_2$. For each $\theta_1 \in [0, 2\pi[$, let

$c(\theta_1)$ be defined by

$$c(\theta_1)(t) = \begin{cases} 1 & \text{if } t = t_2(\theta_1) \\ 0 & \text{otherwise,} \end{cases}$$

and let $F_2(\theta_1, \theta_2) = F_1(\theta_1) + c(\theta_1)e^{i\theta_2}$, $\theta_2 \in [0, 2\pi[$. The support of F_2 is $\{t_1, t_2(\theta_1)\}$, $F_2(\theta_1, \theta_2)(t_1) = F_1(\theta_1)(t_1)$, $F_2(\theta_1, \theta_2)(t_2(\theta_1)) = e^{i\theta_2}$. Let $t_3(\theta_1, \theta_2) = (k_1, l_{m_2}(\theta_1), l_{m_3}(\theta_1, \theta_2)) \in I$ and let $S_3(\theta_1, \theta_2) = \{t_1, t_2(\theta_1), t_3(\theta_1, \theta_2)\}$ be an initial segment. Define $G_2(\theta_1, \theta_2) = a(S_3(\theta_1, \theta_2))$. The support of G_2 is $\{t_1, t_2(\theta_1)\}$,

$$G_2(\theta_1, \theta_2)(t_1) = G_1(\theta_1)(t_1),$$

$$G_2(\theta_1, \theta_2)(t_2(\theta_1)) = e^{i2\pi l_{m_3}(\theta_2)/m_3},$$

$$|(G_2 - F_2)(\theta_1, \theta_2)(t_2(\theta_1))| \le 2\pi/m_3.$$

Suppose now that F_1, F_2, $\cdots$, F_n, G_1, G_2, $\cdots$, G_n, t_1, $t_2(\theta_1)$, $\cdots$, $t_{n+1}(\theta_1, \cdots, \theta_n)$ have been obtained so that the support of F_i (resp. G_i) is $\{t_1, t_2(\theta_1), \cdots, t_i(\theta_1, \cdots, \theta_{i-1})\}$, $\{F_i\}_{i=1}^n$ is a $c_0(I)$-valued analytic martingale, G_j is C-valued, $F_i, G_i \in L^\infty([0, 2\pi[^i, c_0(I))$, $|(F_i - G_i)(\theta_1, \theta_2, \cdots, \theta_i)(t_i(\theta_1, \theta_2, \cdots, \theta_{i-1}))| \le 2\pi/m_{i+1}$ and for $1 \le j < i$,

$$F_i(\theta_1, \cdots, \theta_i)(t_j(\theta_1, \cdots, \theta_{j-1}))$$

$$= F_j(\theta_1, \cdots, \theta_j)(t_j(\theta_1, \cdots, \theta_{j-1})),$$

$$G_i(\theta_1, \cdots, \theta_i)(t_j(\theta_1, \cdots, \theta_{j-1}))$$

$$= G_j(\theta_1, \cdots, \theta_j)(t_j(\theta_1, \cdots, \theta_{j-1})).$$

Let $c(\theta_1, \cdots, \theta_n) \in c_0(I)$ be defined by

$$c(\theta_1, \cdots, \theta_n)(t) = \begin{cases} 1 & \text{if } t = t_{n+1}(\theta_1, \cdots, \theta_n), \\ 0 & \text{otherwise.} \end{cases}$$

Let $t_{n+2}(\theta_1, \cdots, \theta_{n+1}) = (k_1, l_{m_2}(\theta_1), \cdots, l_{m_{n+2}}(\theta_{n+1})) \in I$, and let $S_{n+2}(\theta_1, \cdots, \theta_{n+1}) = \{t_1, t_2(\theta_1), \cdots, t_{n+2}(\theta_1, \cdots, \theta_{n+1})\}$ be an initial segment for $\theta_{n+1} \in [0, 2\pi[$. Define

$$G_{n+1}(\theta_1, \cdots, \theta_{n+1}) = a(t_{n+2}(\theta_1, \cdots, \theta_{n+1})),$$

$$F_{n+1}(\theta_1, \cdots, \theta_{n+1}) = F_n(\theta_1, \cdots, \theta_n) + c(\theta_1, \cdots, \theta_n)e^{i\theta_{n+1}},$$

$\theta_{n+1} \in [0, 2\pi[$. Then the support of G_{n+1} (resp. F_{n+1}) is $\{t_1, t_2(\theta_1), \cdots, t_{n+1}(\theta_1, \cdots, \theta_n)\}$, $c(\theta_1, \cdots, \theta_n)e^{i\theta_{n+1}}$ is an analytic martingale difference, and furthermore, it holds true that

$$|(G_{n+1} - F_{n+1})(\theta_1, \cdots, \theta_{n+1})(t_{n+1}(\theta_1, \cdots, \theta_n))| \le 2\pi/m_{n+2}$$

and for $1 \le j < n + 1$,

$$G_{n+1}(\theta_1, \cdots, \theta_{n+1})(t_j(\theta_1, \cdots, \theta_{j-1}))$$

$$= G_j(\theta_1, \cdots, \theta_j)(t_j(\theta_1, \cdots, \theta_{j-1})),$$

$$F_{n+1}(\theta_1, \cdots, \theta_{n+1})(t_j(\theta_1, \cdots, \theta_{j-1}))$$

$$= F_j(\theta_1, \cdots, \theta_j)(t_j(\theta_1, \cdots, \theta_{j-1})).$$

$(F_n)_{n \geq 1}$ is a $c_0(I)$-valued analytic martingale, $(G_n)_{n \geq 1}$ is a sequence of C-valued adapted functions in $L^\infty([0, 2\pi[^{\mathbf{N}}, c_0(I))$. $(G_n - F_n)_{n \geq 1}$ is a convergent sequence in $L^1([0, 2\pi[^{\mathbf{N}}, c_0(I))$. Indeed, the support of $(G_n - F_n)(\theta_1, \cdots, \theta_n)$ is contained in $\{t_1, t_2(\theta_1), \cdots, t_k(\theta_1, \cdots, \theta_{k-1}), \cdots\}$. If $m \geq n$, then

$$(G_m - F_m)(\theta_1, \cdots, \theta_m)(t_n(\theta_1, \cdots, \theta_{n-1}))$$

$$= (G_n - F_n)(\theta_1, \cdots, \theta_n)(t_n(\theta_1, \cdots, \theta_{n-1})),$$

which means that the sequence $((G_n - F_n)(\theta_1, \cdots, \theta_n)(t_m(\theta_1, \cdots, \theta_{m-1})))_{n \geq 1}$ is a stationary sequence when n is big enough for a fixed $m \in \mathbf{N}$, so for every $(\theta_1, \theta_2, \cdots, \theta_n, \cdots) \in [0, 2\pi[^{\mathbf{N}}$, $(G_n - F_n)(\theta_1, \cdots, \theta_n)$ converges in $c_0(I)$, $G_n - F_n$ converges almost surely in $L^1([0, 2\pi[^{\mathbf{N}}, c_0(I))$. Because $(F_n - G_n)_{n \geq 1}$ is uniformly bounded, we have the convergence of $(F_n - G_n)_{n \geq 1}$ in $L^1([0, 2\pi[^{\mathbf{N}}, c_0(I))$. It is not hard to see from the construction that for every $n \in \mathbf{N}$, $\|F_n - G_n\|_{L^1([0,2\pi[^{\mathbf{N}}, c_(I))} = 1$, by Theorem B, C has not the analytic Radon-Nikodym property and this finishes the proof of Theorem 1.

Proof of Lemma. Without loss of generality, we can suppose that there exists $c > 0$ such that for every $x \in B(0, M)$, $f(x) > 0$, and $f(x) < c$ for every $x \in B(0, 2M)$. Let k, $N \in \mathbf{R}^+$ be such that $N/M > k > 2c/3M + 2N/3M$, and let $\phi(x) = k\|x\| - N$, $x \in X$. Consider the function

$$\varphi(x) = \begin{cases} f \vee \phi(x) & x \in B(0, 2M) \\ \phi(x) & x \notin B(0, 2M). \end{cases}$$

The fact that φ is Lipschitz can be easily seen from the definition of φ. The conditions on k, N ensure that $\phi(x) < 0$ for each $x \in B(0, M)$ and $\phi(x) > c$ for every $x \in B(0, 2M) \setminus B(0, 3M/2)$, so $\varphi = f$ on $B(0, M)$ and $\phi = \varphi$ on $X \setminus B(0, 3M/2)$. φ is plurisubharmonic on $B(0, 2M)$ and plurisubharmonic on $X \setminus B(0, 3M/2)$, which implies that φ is plurisubharmonic on X since the plurisubharmonicity is a local property: if h is an upper semi-continuous functon on a Banach space X, h is plurisubharmonic on X if and only if for every x, $y \in X$, there exists $\epsilon > 0$, such that for every $0 < r \leq \epsilon$, we have $f(x) \leq \int_0^{2\pi} f(x + rye^{i\theta}) \frac{d\theta}{2\pi}$. The lemma is proved.

The second result of this paper is the following remarks about the analytic Radon-Nikodym property for bounded subsets.

Theroem 2. Let X be a complex Banach space, and let C_1, C_2 $\cdots$ C_n be n bounded subsets of X with the analytic Radon-Nikodym property. Then the union $\cup_{i=1}^n C_i$ has also the analytic Radon-Nikodym property.

To avoid repetition, we shall use the following standard notation. If C is a bounded subset of a complex Banach space X, ϕ is a bounded above function on C and $\alpha > 0$, then

$$S(C, \phi, \alpha) = \{x \in C : \phi(x) > Sup_C \phi - \alpha\},$$

$diam(C)$ will denote the positive number $Sup\{\|x - y\| : x, y \in C\}$.

Proof of Theorem 2. It is clear that it will suffice to show the theorem for $n = 2$, and since a bounded subset of X has the analytic Radon-Nikodym property if and only if its closure in X has the analytic Radon-Nikodym property, without loss of generality, we can suppose furthermore that both C_1 and C_2 are closed. Assume that C_1 and C_2 have the analytic Radon-Nikodym property. We have to show that for every bounded above upper semicontinuous function $f : C_1 \cup C_2 \to \mathbf{R}$ and for every $\epsilon > 0$, there exists a Lipschitz plurisubharmonic function $h \in PSH_0(X)$ with $\|h\|_{Lip} \le 1$ and $\alpha > 0$, such that

$$\text{diam}\{S(C_1 \cup C_2, f + h, \alpha)\} \le \epsilon. \tag{$*$}$$

Fix then such a function f and $\epsilon > 0$.

If we consider the restrictions of f on C_1 and C_2, Theorem B gives two dense G_δ-subsets H_1 and H_2 of $PSH_0(X)$, such that for every $\phi \in H_i$, $f + \phi$ strongly exposes C_i from above for $i = 1, 2$, i.e.,

$$\lim_{n \to \infty} \text{diam}\{S(C_i, f + \phi, 1/n)\} = 0, \tag{$**$}$$

since C_1 and C_2 have the analytic Radon-Nikodym property. As $PSH_0(X)$ is complete with the metric d, the intersection $H_1 \cap H_2$ is also a dense G_δ-subset of $PSH_0(X)$, in particular, there exist $\phi \in H_1 \cap H_2$ and $x_i \in C_i$, such that for $i = 1, 2$, $(f + \phi)(x_i) = Sup_{C_i}(f + \phi)$ and $\|\phi\| \le 1/2$.

We shall distinguish the following three cases:

1. $(f + \phi)(x_1) \ne (f + \phi)(x_2)$. Without loss of generality, we can suppose that $(f + \phi)(x_1) > (f + \phi(x_2)$. Then it is clear that $S(C_1 \cup C_2, f + \phi, 1/n) \subset S(C_1, f + \phi, 1/n)$ for n big enough. This fact together with $(**)$ gives already $(*)$ with $h = \phi$ and $\alpha = 1/n$ for n big enough.

2. $x_1 = x_2$. In this case, $S(C_1 \cup C_2, f + \phi, 1/n) = S(C_1, f + \phi, 1/n) \cup S(C_2, f + \phi, 1/n)$, and

$$\text{diam}\{S(C_1 \cup C_2, f + \phi, 1/n)\}$$

$$\le \text{diam}\{S(C_1, f + \phi, 1/n)\} + diam\{S(C_2, f + \phi, 1/n)\},$$

where in the last line, we have used the hypothesis that $x_1 = x_2 \in S(C_1, f + \phi, 1/n) \cap S(C_2, f + \phi, 1/n)$ for every $n \in \mathbf{N}$. The above estimation together with $(**)$ gives again $(*)$ with $h = \phi$ and $\alpha = 1/n$ for n big enough.

3. $(f + \phi)(x_1) = (f + \phi)(x_2)$ and $x_1 \ne x_2$. We can suppose that $Sup_{y_1, y_2 \in C_1 \cup C_2}\|y_1 - y_2\| \le 1$. Fix an integer $n \in \mathbf{N}$ such that $diam\{S(C_i, f + \phi, 1/n)\} \le \epsilon$ for $i = 1, 2$. Set

$\delta = Inf\{1/8n, \|x_2 - x_1\|/16\}$, $\varphi(x) = \|x - x_1\|/8n - \|x_1\|/8n$ for $x \in X$. φ is then a Lipschitz plurisubharmonic function on X, $\varphi \in PSH_0(X)$ and $\|\varphi\|_{Lip} = 1/8n$. We can suppose furthermore that $\epsilon < \|x_2 - x_1\|/2$. We shall show that the following inclusion holds true:

$$S(C_1 \cup C_2, f + \phi + \varphi, \delta) \subset$$

$$S(C_1 \cup C_2, f + \phi, 1/n). \qquad (***)$$

Indeed, if we set $M = Sup_{C_1 \cup C_2}(f + \phi) = f(x_1) + \phi(x_1)$, $M' = Sup_{C_1 \cup C_2}(f + \phi + \varphi)$, and letting $x \in S(C_1 \cup C_2, f + \phi + \varphi, \delta)$, we have

$$(f + \phi)(x) > M' - \delta - \varphi(x)$$

$$\geq M' - 1/8n - 1/8n > M - 1/n,$$

which implies that $x \in S(C_1 \cup C_2, f + \phi, 1/n)$. We shall show that

$$S(C_1 \cup C_2, f + \phi + \varphi, \delta)$$

$$\subset S(C_2, f + \phi, 1/n). \qquad (****)$$

Let $x \in S(C_1 \cup C_2, f + \phi + \varphi, \delta)$ and assume that $x \in S(C_1, f + \phi, 1/n)$. Then

$$(f + \phi + \varphi)(x) \leq M + \|x - x_1\|/8n$$

$$\leq M + diam\{S(C_1, f + \phi, 1/n)\}/8n$$

$$\leq M + \epsilon/8n \leq M' - \|x_2 - x_1\|/8n + \|x_2 - x_1\|/16n$$

$$= M' - \|x_2 - x_1\|/16n \leq M' - \delta,$$

which is a contradiction. This fact together with $(***)$ gives $(****)$ since $S(C_1 \cup C_2, f + \phi, 1/n) = S(C_1, f + \phi, 1/n) \cup S(C_2, f + \phi, 1/n)$.

Finally, by $(****)$,

$$diam\{S(C_1 \cup C_2, f + \phi + \varphi, \delta)\}$$

$$\leq diam\{S(C_2, f + \phi, 1/n)\} \leq \epsilon,$$

which gives again $(*)$ with $h = \phi + \varphi$ and $\alpha = \delta$ since $(\phi + \varphi)(0) = 0$ and $\|\phi + \varphi\|_{Lip} < 1$. This finishes the proof of Theorem 2.

References

[1] S. Bu, On the analytic Radon-Nikodym property for bounded subsets in complex Banach spaces. J. London Math. Soc., (2), 47(1993), 484–496.

[2] A.V. Bukhvalov and A.A.Danilevich, Boundary properties of analytic and harmonic functions with values in Banach spaces, Math. Zametki, 31(1982), 203–214; Math. Notes, 31(1982), 104–110.

[3] N.Ghoussoub, B.Maurey and J.Lindenstrauss, Analytic martingales and pluri-subharmonic barriers in complex Banach spaces. Banach space theory Contemporary Mathematics, 85(1989), 111–130.

[4] N.Ghoussoub and B.Maurey, Plurisubharmonic martingales and barriers in complex quasi-Banach spaces. Ann. Inst. Fourier, 39(1989), 1007–1060.

[5] J.Bourgain: La propriété de Radon-Nikodym. Publication de Mathématiques de l'Uinversité Pierre et Marie Curie, 36(1979).

[6] N.Ghoussoub, B.Maurey and W.Schachermayer, Pluriharmonically dentable complex Banach spaces. J. für die reine und angewandte Mathematik, Band 402, 39(1989), 76–127.

Existence of Entire Solutions of
an Elliptic Equation on $\mathbb{R}^N$

Yanheng Ding and Shujie Li

Institute of Mathematics, Academia Sinica, Beijing

1. Introduction and main results

Let us consider the following semilinear elliptic partial differential equation on $\mathbb{R}^N$:

$$\begin{cases} -\Delta u + q(x)u = g(x, u) \\ u \in W^{1,2}(\mathbb{R}^N), \end{cases} \qquad (E)$$

where $q \in C(\mathbb{R}^N)$ and $g \in C(\mathbb{R}^N \times \mathbb{R})$. Equations like (E) arise in various branches of applied mathematics and physics, see [5, 12] and references therein. In recent years the existence and multiplicity of solutions to (E) have been studied extensively by using variational method, see, e.g., [1, 3, 5-7, 12], particularly, the literature of [6, 12].

In this paper we prove several results on the existence of solutions to (E) under some assumptions.

First, we assume that q and g satisfy

(Q_1) $q \in C(\mathbb{R}^N)$ and $q(x) \to \infty$ as $|x| \to \infty$;

(g_1) $g \in C(\mathbb{R}^N \times \mathbb{R})$ and there is a $\mu > 2$ such that

$$0 < \mu G(x, u) \equiv \mu \int_0^u g(x, t)dt \leq g(x, u)u, \quad \forall x \in R^N \text{ and } u \in \mathbb{R}\backslash\{0\};$$

(g_2) $|g(x, u)| = o(|u|)$ as $u \to 0$ uniformly in $x \in \mathbb{R}^N$;

(g_3) there are $a_1, a_2 > 0$ such that

$$|g(x, u)| \leq a_1 + a_2|u|^{\beta - 1}, \forall x \in \mathbb{R}^N \text{ and } u \in \mathbb{R},$$

where $\beta > 2$, and $\beta < \frac{2N}{N-2}$ if $N \geq 3$;

(g_4) $0 < \underline{b} \equiv \inf\limits_{y \in R^N, |z|=1} G(y, z)$ if $q(x) < 0$ for some $x \in \mathbb{R}^N$.

Assumptions like (g_1) – (g_3) have been used extensively in many papers for the study on (E). Our interest is to show that these assumptions, jointly with (Q_1), ensure the existence of nontrivial solutions of (E). Precisely, we have

1991 Mathematics Subject Classification: 35J60

Theorem 1.1. Let q and g satisfy (Q_1) and $(g_1) - (g_2) -(g_3) - (g_4)$. Then (E) has at least one nontrivial solution. Moreover, if g is also odd in $u \in \mathbb{R}$, then (E) has infinitely many solutions u_k such that

$$\int_{\mathbb{R}^N} (|\nabla u_k|^2 + q(x)u_k^2) \to \infty \quad \text{as} \quad k \to \infty.$$

If g has the particular form:

$$g(x, u) = p(x)|u|^{\mu-2}u + h(x), \tag{1.1}$$

then a multiplicity result still can be obtained. In fact, we have

Theorem 1.2. Let $N \geq 2, q$ satisfy (Q_1) and $q(x) > 0$ for all $x \in \mathbb{R}^N$, and g have the form (1.1). Assume $h \in L^{\mu'}(\mathbb{R}^N) \cap C(\mathbb{R}^N)$ and $p \in L^{\bar{r}}(\mathbb{R}^N)$ where $\mu' = \frac{\mu}{\mu-1}$ and $\bar{r} = \frac{2N}{2\mu-N\mu+2N}$, and moreover, one of the following is true:

(p_1) $p(x) > 0$ a.e. on $\mathbb{R}^N$ and $\mu \in (2, \frac{2N}{N-1})$, or

(p_2) $p(x) \geq 0$ on $\mathbb{R}^N$, and there is $h_0 > 0$ such that $p(x) \geq h_0$ for all $x \in$ supp h, and $\mu \in (2, \frac{2N-2}{N-2})$ if $N \geq 3, \mu \in (2, \infty)$ if $N = 2$.
Then (E) has infinitely many solutions.

Next, we consider the case where g is sublinear with respect to $u \in \mathbb{R}$. It seems that much less is known on the existence of solutions of (E) in such a case. We assume q and g satisfy

(Q_2) $q \in C(\mathbb{R}^N)$, and there exists $\alpha < 2$ such that $q(x)|x|^{\alpha-2} \to \infty$ as $|x| \to \infty$;

(g_5) there is $1 < \theta \in (\frac{2N}{2-\alpha+N}, 2)$ such that

$$0 < g(x, u)u \leq \theta G(x, u), \quad \forall x \in \mathbb{R}^N \quad \text{and} \quad u \in \mathbb{R}\backslash\{0\};$$

(g_6) there are $\nu > \max\{0, \frac{\alpha-2+N}{2-\alpha+N}\}$ and $a_1, a_2 > 0$ such that

$$G(x, u) \geq a_1|u|^\theta \quad \text{and} \quad |g(x, u)| \leq a_2|u|^\nu, \quad \forall x \in \mathbb{R}^N \quad \text{and} \quad |u| \leq 1;$$

(g_7) there are $1 < \bar{\theta} \in (\frac{2N}{2-\alpha+N}, \theta]$ and $b_1, b_2 > 0$ such that

$$G(x, u) \geq b_1|u|^{\bar{\theta}} \quad \text{and} \quad |g(x, u)| \leq b_2|u|, \forall x \in \mathbb{R}^N \quad \text{and} \quad |u| \geq 1.$$

One of our results reads as

Theorem 1.3. Let q satisfy (Q_2) and g satisfy (g_5)-(g_6)-(g_7). Then (E) has at least one nontrivial solution. Moreover, if g is also odd in $u \in \mathbb{R}$, then (E) has infinitely many solutions.

We remark that, since $\alpha < 2$, we always have $\frac{2N}{2-\alpha+N} < 2$. If further $\alpha < 2 - N$, then $\frac{2N}{2-\alpha+N} < 1$, and so we can deal with the case where g is bounded.

Theorem 1.4. Let q satisfy (Q_2) with $\alpha < 2 - N, q(x) \geq 0$ for all $x \in \mathbb{R}^N$, and g satisfy

(g_8) there is $M > 0$ such that $|g(x,u)| \leq M$ for all $x \in \mathbb{R}^N$ and $u \in \mathbb{R}$;

(g_9) there are $a, \omega > 0$ and $\theta \in (1,2)$ such that

$$G(x,u) \geq a|u|^\theta, \quad \forall x \in \mathbb{R}^N \quad \text{and} \quad |u| \leq \omega;$$

(g_{10}) $g(x,u) \to 0$ as $u \to 0$ uniformly in $x \in \mathbb{R}^N$.

Then (E) has at least one nontrivial solution. Moreover, if g is also odd in $u \in \mathbb{R}$, then (E) has infinitely many solutions.

Finally, we consider the case where q is bounded. There have been many known existence results concerning this case. However, most of these results are obtained under the assumption that $q(x) > 0$ a.e. on $\mathbb{R}^N$ so that the spectrum $\sigma(A)$ of the Schrödinger operator $A = -\Delta + q(x)$ is contained in the interval $(0,\infty)$, or the assumption that $q(x)$ is periodic and $0 \notin \sigma(A)$. These assumptions seem crucial in the proofs of the results. It is natural to ask that does (E) has nontrivial solutions if $q(x) < 0$ for x in some bounded domain of $\mathbb{R}^N$ and q is not periodic? We will give a result concerning this question.

Writting $x = (x_1, \cdots, x_N)$ for $x \in \mathbb{R}^N$, we assume:

(Q_3) i) there is a bounded open set $\Omega \subset \mathbb{R}^N$ such that $q(x) < 0$ for $x \in \Omega$;

ii) there is a $q_\infty \in C(R^N), q_\infty(x) > 0$ and being T_i periodic in $x_i (1 \leq i \leq N)$, such that $q(x) \leq q_\infty(x)$ and $q_\infty(x) - q(x) \to 0$ as $|x| \to \infty$.

(g_{11}) there is a $g_\infty \in C(\mathbb{R}^N \times \mathbb{R})$ which is T_i periodic in $x_i(1 \leq i \leq N)$ and satisfies (g_1)-(g_2)-(g_3) and

(∗) for all $x \in \mathbb{R}^N$ and $u \neq 0, s^{-1}g_\infty(x, su)u$ is nondecreasing in $s \in (0,\infty)$, such that $g_\infty(x,u) \leq g(x,u)$ and $g(x,u) - g_\infty(x,u) \to 0$ as $|x| \to \infty$ uniformly in $u \in (-R,R)$ for any $R > 0$.

We point out that the condition like (∗) was used earlier in [13] and some other papers. In [15] a similar assumption has been made for handling the existence of homoclinic orbits of a second order Hamiltonian system.

We will prove

Theorem 1.5. Let q satisfy (Q_3) and g satisfy (g_1) -(g_2)-(g_3) and (g_{11}). Then (E) has at least one nontrivial solution.

2. Preliminary results

To start with, we state some preliminary results concerning the variational setting associated with (E).

Suppose that q is a function defined on $\mathbb{R}^N$ satisfying

(Q) $q \in C(\mathbb{R}^N)$ and $\lim\limits_{|x|\to\infty} \inf q(x) = q_0 > 0$.

Then the Schrödinger operator $A \equiv -\Delta + q$ is defined as a selfadjoint operator with the domain $\mathcal{D}(A) \subset L^2(\mathbb{R}^N)$. Let $\sigma(A)$ (resp. $\sigma_p(A), \sigma_e(A)$) denote the spectrum (resp. discrete spectrum, essential spectrum) of A. Then $\sigma(A) = \sigma_p(A) \cup \sigma_e(A)$. Let $\{E(\nu); -\infty < \nu < \infty\}$ be the resolution of $A, U = 1 - E(0) - E(-0), |A|$ the absolute value of A, and $|A|^{1/2}$ the square root of $|A|$. Then U commutes with $A, |A|$ and $|A|^{1/2}$, and $A = |A|U$ is the polar decomposition of A (see [10]).

Let $W^{1,2}(\mathbb{R}^N)$ be the Sobolev space with the usual norm

$$\||u\||^2 = \int_{\mathbb{R}^N} (|\nabla u|^2 + u^2). \quad \forall u \in W^{1,2}.$$

Lemma 2.1. If q satisfies (Q), then $\sigma_e(A) \subset [\frac{q_0}{2}, \infty)$.

Proof. By (Q), there is $R > 0$ such that $q(x) > \frac{q_0}{2}$ for all $|x| \geq R$. Let $\chi \in C^\infty(\mathbb{R}_+, \mathbb{R}_+)$ be such that $0 \leq \chi(s) \leq 1, \chi(s) = 1$ if $s \geq R+1$, and $\chi(s) = 0$ if $s \leq R$. Define

$$\tilde{q}(x) = (1 - \chi(|x|))\frac{q_0}{2} + \chi(|x|)q(x), \quad \forall x \in \mathbb{R}^N,$$

and set $\tilde{A} \equiv -\Delta + \tilde{q}, B \equiv q - \tilde{q}$, both being regarded as selfafjoint operators acting in L^2. Then $A = \tilde{A} + B$. We claim that B is $\tilde{A}$-compact [10], i.e., for any sequence $\{u_n\} \subset \mathcal{D}(\tilde{A}) = \mathcal{D}(A)$ which is such that $\{\||u_n\||_{\tilde{A}}\}$ is bounded, $\{Bu_n\}$ contains a subsequence which is convergent in L^2, where

$$\|u_n\|_{\tilde{A}} = (\|u_n\|_{L^2}^2 + \|\tilde{A}u_n\|_{L^2}^2)^{1/2}$$

is the graph norm associated with $\tilde{A}$. Indeed, since

$$\||u_n\||^2 = \int_{\mathbb{R}^N} (|\nabla u_n|^2 + \frac{q_0}{2}u_n^2 + (1 - \frac{q_0}{2})u_n^2)$$

$$\leq (1 + \frac{q_0}{2})\|u_n\|_{\tilde{A}}^2 \leq const.,$$

by the Sobolev embedding theorem, there is a subsequence denoted again by $\{u_n\}$ such that

$$\int_{\mathbb{R}^N} (B(u_n - u_m))^2 = \int_{|x|\leq R+1} (B(u_n - u_m))^2 \to 0$$

as $n, m \to \infty$. Now by a Weyl's theorem [16], we see that

$$\sigma_e(A) = \sigma_e(\tilde{A} + B) = \sigma_e(\tilde{A}) \subset \sigma(\tilde{A}) \subset [\frac{q_0}{2}, \infty)$$

since $\tilde{A} \geq \frac{q_0}{2}$, thus proving the lemma.

As a consequence of Lemma 2.1, we have

Lemma 2.2. *If q satisfies (Q), then A has only a finite number of negative eigenvalues, and 0 is at most an isolated eigenvalue of A.*

Proof. This is because A is bounded from below, and by Lemma 2.1, the spectrum of A is discrete in $(-\infty, \frac{q_0}{2})$.

Taking into account Lemma 2.2, let n^- (resp. n^0) be the number of negative (resp. null) eigenvalues of A (counted in their multiplicities), and let $\bar{n} = n^- + n^0$. We then denote by $\lambda_1 \leq \cdots \leq \lambda_{n^-} < 0 = \lambda_{n^-+1} = \cdots = \lambda_{\bar{n}}$ the nonpositive eigenvalues, and by $e_1, \cdots, e_{\bar{n}}$ the corresponding orthonormal eigenfunctions ($Ae_i = \lambda_i e_i$). Moreover, it follows from Lemma 2.2 that

$$\tilde{\lambda} \equiv \inf\{|\lambda|; \lambda \in \sigma(A) \setminus \{0\}\} > 0. \tag{2.1}$$

Set

$$E^- = \mathrm{span}\{e_1, \cdots, e_{n^-}\},$$

$$E^0 = \mathrm{span}\{e_{n^-+1}, \cdots, e_{\bar{n}}\} = \ker A.$$

and $E = \mathcal{D}(|A|^{1/2})$, the domain of $|A|^{1/2}$. Clearly $C_0^\infty(\mathbb{R}^N)$ is dense in E, and E is a Hilbert space equipped with the inner product and norm:

$$(u, v) = (|A|^{1/2}u, |A|^{1/2}v)_{L^2} + (P^0 u, P^0 v)_{L^2},$$

$$\|u\| = (u, u)^{1/2},$$

where $(\cdot, \cdot)_{L^2}$ is the L^2 inner product, and $P^0 : L^2 \to E^0$ is the orthogonal projector. Let E^+ be the orthogonal complement of $(E^- \oplus E^0)$ in E. Then we have an orthogonal decomposition

$$E = E^- \oplus E^0 \oplus E^+.$$

By (2.1), it is easy to check that

$$\|u\|^2 \geq \min\{1, \tilde{\lambda}\} \|u\|_{L^2}^2, \quad \forall u \in E, \tag{2.2}$$

and that E is continuously embedded in $W^{1,2}$, and

$$\||u\||^2 \leq \|u\|^2 + (1 + |\inf q|) \|u\|_{L^2}^2. \tag{2.3}$$

Lemma 2.3. *If q satisfies (Q_3), then $\||\cdot\||$ and $\|\cdot\|$ are equivalent norms on E (and hence on $W^{1,2}$).*

Proof. This is clear since (Q$_3$) implies (Q), and (Q$_3$) also implies that $q(x)$ is bounded.

Lemma 2.4. If q satisfies (Q$_1$), then E is compactly embedded in L^p for $p \geq 2$ and $p < \frac{2N}{N-2}$ if $N \geq 3$, $p < \infty$ if $N = 2$, $p \leq \infty$ if $N = 1$.

Proof. It is easy to see from [16, Theorem XIII 64] that E is compactly embedded in L^2. For $N = 1$ and $p \in [2, \infty]$, refer to [8]. For $N \geq 2$ and $p > 2$ it follows from the inequality that

$$\|u\|_{L^p} \leq c\|u\|_{L^2}^{1-\theta}\||u|\|^{\theta}$$

where $\theta = \frac{(p-2)N}{2p}$ and c is independent of u.

Remark 2.1. If q satisfies (Q$_1$), by Lemma 2.4 (or 2.1), $\sigma(A) = \sigma_p(A)$ which can be arranged in $\lambda_1 \leq \lambda_2 \leq \cdots \to \infty$, and a corresponding system of eigenfunctions $\{e_n\}$ forms an orthonormal basis in L^2. Let n_λ denote the number of eigenvalues being less than or equivalent to λ. Then it is known by the Cwikel-Lieb-Rosenbljum inequality that

$$n_\lambda \sim (2^N \pi^{\frac{N}{2}} \Gamma(\frac{N}{2} + 1))^{-1} \int_{q(x) < \lambda} (\lambda - q(x))^{N/2} \tag{2.4}$$

as $\lambda \to \infty$ (see [16]) where $\Gamma(\cdot)$ is the Euler gamma function.

Lemma 2.5. If q satisfies (Q$_2$), then E is compactly embedded in L^p for $1 \leq p \in (\frac{2N}{2-\alpha+N}, 2)$.

Proof. First we assume $q(x) \geq 1$ for all $x \in \mathbb{R}^N$. Let $K \subset E$ be a bounded set, $\|u\| \leq M$ for all $u \in K$. By definition we will show that, for any $\varepsilon > 0$, K has a finite ε-net in L^p.

Let $k = \frac{2-\alpha}{2-p}$. Note that, for any $R > 0$ and $u \in E$, one has

$$\begin{aligned}
\int_{|x|>R} |u|^p &= \int_{\substack{|x|>R \\ |x|^k|u(x)|\leq 1}} |u|^p + \int_{\substack{|x|>R \\ |x|^k|u(x)|>1}} |u|^p \\
&\leq \int_{|x|>R} \frac{1}{|x|^{pk}} + \int_{\substack{|x|>R \\ |x|^k|u(x)|>1}} (|x|^k|u|)^p |x|^{-kp} \\
&\leq \int_{|x|>R} \frac{1}{|x|^{pk}} + \int_{|x|>R} |x|^{2-\alpha}|u|^2 \\
&\leq \int_{|x|>R} \frac{1}{|x|^{pk}} + \frac{1}{\beta(R)}\|u\|^2,
\end{aligned} \tag{2.5}$$

where $\beta(R) \equiv \inf_{|x|\geq R} q(x)|x|^{\alpha-2}$. Since $kp > N$ and $\beta(R) \to \infty$ as $R \to \infty$ by (Q$_2$), one can take R_0 large enough such that

$$\int_{|x|\geq R_0} \frac{1}{|x|^{pk}} + \frac{4M^2}{\beta(R_0)} < \frac{\varepsilon^2}{2},$$

and so from (2.5) it follows that

$$\int_{|x|\geq R_0} |u-v|^p < \frac{\varepsilon^2}{2}, \quad \forall u,v \in K. \tag{2.6}$$

On the other hand, by Sobolev compact embedding theorem, there are $u_1, \cdots, u_m \in K$ such that, for any $u \in K$, there exists u_i satisfying

$$\|u-u_i\|^2_{L^p(B(R_0))} < \frac{\varepsilon_2^2}{2} \tag{2.7}$$

where $B(R_0) = \{x \in \mathbb{R}^N; |x| < R_0\}$. (2.6) and (2.7) show that

$$\|u-u_i\|_{L^p(\mathbb{R}^N)} < \varepsilon,$$

i.e., K has a finite ε-net and so is precompact in L^p.

Next, in general, by (Q_2), $q(x)+a \geq 1$ for some $a > 0$ and all $x \in \mathbb{R}^N$. Notice that, as a set, $E = \mathcal{D}((A+a)^{\frac{1}{2}})$. One can introduce a norm on E by setting

$$\|u\|_a^2 = ((A+a)^{1/2}u, (A+a)^{1/2}u)_{L^2}.$$

By the above argument we know that $(E, \|\cdot\|_a)$ is compactly embedded in L^p for $1 \leq p \in (\frac{2N}{2-\alpha+N}, 2)$. Therefore in order to complete the proof it suffices to show that $\|\cdot\|_a$ and $\|\cdot\|$ are equivalent. In fact, for $u \in \mathcal{D}(A)$,

$$\||A|^{1/2}u\|_{L^2}^2 = (|A|u, u)_{L^2}$$

$$= (U(A+a)^{1/2}u, (A+a)^{1/2}u)_{L^2} - a(Uu, u)_{L^2}$$

$$\leq \|(A+a)^{1/2}u\|_{L^2}^2 + a\|u\|_{L^2}^2,$$

and on the other hand,

$$\|(A+a)^{1/2}u\|_{L^2}^2 = (Au, u)_{L^2} + a(u, u)_{L^2}$$

$$\leq \||A|^{1/2}u\|_{L^2}^2 + a\|u\|_{L^2}^2,$$

which, together with (2.2), shows

$$C_1\|u\|_a \leq \|u\| \leq C_2\|u\|_a, \quad \forall u \in \mathcal{D}(A). \tag{2.8}$$

Since $\mathcal{D}(A)$ is dense in E, (2.8) holds for all $u \in E$ by the continuity, thus proving the lemma.

Below, assume q satisfies (Q) and let

$$a(u,v) = (|A|^{1/2}Uu, |A|^{1/2}v)_{L^2}$$

be the bilinear form associated with A. Then for $u \in \mathcal{D}(A)$ and $v \in E$,

$$a(u, v) = \int_{\mathbb{R}^N} (\nabla u \nabla v + q(x) uv). \tag{2.9}$$

By the continuity, (2.9) holds for all $u, v \in E$. It is easy to see that E^-, E^0 and E^+ are orthogonal to each other with respect to $a(\cdot, \cdot)$, and moreover

$$a(u, v) = ((P^+ - P^-)u, v), \tag{2.10}$$
$$a(u, u) = \|u^+\|^2 - \|u^-\|^2 \tag{2.11}$$

where $P^\pm : E \to E^\pm$ are the orthogonal projectors.

3. Proofs of Theorem 1.1 and 1.2

We first prove Theorem 1.1. Let the assumptions of Theorem 1.1 be satisfied. Consider the variational functional corresponding to (E) defined by

$$f(u) = \frac{1}{2} \int_{\mathbb{R}^N} (|\nabla u|^2 + q(x) u^2) - \int_{\mathbb{R}^N} G(x, u). \tag{3.1}$$

By Lemma 2.4, (2.9) and (2.11), $f \in C^1(E, \mathbb{R})$,

$$f(u) = \frac{1}{2}(\|u^+\|^2 - \|u^-\|^2) - \int_{\mathbb{R}^N} G(x, u)$$

for all $u = u^- + u^0 + u^+ \in E^- \oplus E^0 \oplus E^+ = E$, and

$$f'(u)v = a(u, v) - \int_{\mathbb{R}^N} g(x, u)v, \quad \forall u, v \in E.$$

A standard argument shows that the critical points of f on E are solutions of (E). We are looking for the critical points of f.

Proof of Theorem 1.1. We argue step by step.

Step 1. f satisfies the Palais-Smale condition:
(PS) Suppose any sequence $\{u_n\} \subset E$ for which $f(u_n)$ is bounded and $\varepsilon_n \equiv \|f'(u_n)\| \to 0$ as $n \to \infty$ possesses a convergent subsequence.
Indeed, by (g_1),

$$
\begin{aligned}
f(u_n) - \tfrac{1}{2} f'(u_n) u_n &= \int_{\mathbb{R}^n} (\tfrac{1}{2} g(x, u_n) u_n - G(x, u_n)) \\
&\geq (\tfrac{1}{2} - \tfrac{1}{\mu}) \int_{\mathbb{R}^n} g(x, u_n) u_n \geq (\tfrac{\mu}{2} - 1) \int_{\mathbb{R}^N} G(x, u_n).
\end{aligned}
\tag{3.2}
$$

If $q(x) \geq 0$ for all $x \in \mathbb{R}^N$, it follows immediately that $\{\|u_n\|\}$ is bounded from (3.2) and

$$\|u_n\|^2 = \int_{\mathbb{R}^N} g(x, u_n)u_n + f'(u_n)u_n.$$

Assume $\dim(E^- \oplus E^0) > 0$ (so $q(x) < 0$ for some $x \in \mathbb{R}^N$). By (g_1)-(g_4),

$$G(x, u) \geq \underline{b}(|u|^\mu - |u|^2), \quad \forall x \in R^N \quad \text{and} \quad u \in \mathbb{R}. \tag{3.3}$$

By (3.2)– (3.3) and Lemma 2.4,

$$\|u_n\|_{L^\mu}^\mu \leq C(1 + \|u_n\|^2) \tag{3.4}$$

(here and after, C stands for a generic positive constant). Since $\dim (E^- \oplus E^0) < \infty$, by Hölder inequality, letting $\frac{1}{\mu'} + \frac{1}{\mu} = 1$,

$$\begin{aligned}
\|u_n^- + u_n^0\|_{L^2}^2 &= (u_n^- + u_n^0, u_n)_{L^2} \\
&\leq \|u_n^- + u_n^0\|_{L^{\mu'}}\|u_n\|_{L^\mu} \\
&\leq C\|u_n^- + u_n^0\|_{L^2}\|u_n\|_{L^\mu},
\end{aligned}$$

which, together with (3.4), yields

$$\|u_n^- + u_n^0\| \leq C(1 + \|u_n\|^{\frac{2}{\mu}}),$$

and consequently, by (3.2),

$$\begin{aligned}
\|u_n\|^2 &= \|u_n^- + u_n^0\|^2 + \|u_n^-\|^2 + 2f(u_n) + 2\int_{\mathbb{R}^N} G(x, u_n) \\
&\leq C(1 + \|u_n\| + \|u_n\|^{\frac{4}{\mu}}).
\end{aligned}$$

Hence $\|u_n\| \leq$ const. since $\mu > 2$.

Now using Lemma 2.4, a standard argument shows that $\{u_n\}$ has a convergent subsequence.

Step 2. There are $\rho, \sigma > 0$ such that

$$f|_{S_\rho} \geq \sigma \tag{3.5}$$

where $S_\rho = \{u \in E^+; \|u\| = \rho\}$. Indeed, by (g_2) and (g_3), for any $\varepsilon > 0$, there exists $C_\varepsilon > 0$ such that

$$G(x, u) \leq \varepsilon|u|^2 + C_\varepsilon|u|^\beta, \quad \forall x \in \mathbb{R}^N, \quad u \in \mathbb{R}, \tag{3.6}$$

and so by Lemma 2.4, for $u \in E^+$,

$$f(u) \geq \frac{1}{2}\|u\|^2 - C(\varepsilon\|u\|^2 + C_\varepsilon\|u\|^\beta).$$

Since $\beta > 2$, (3.5) follows by taking ε small enough.

Step 3. If $q(x) \geq 0$ for all $x \in \mathbb{R}^N$, for any $l \in \mathbb{N}$, we take $\{\varphi_i\}_{1 \leq i \leq l} \subset C_0^\infty$ with $\|\varphi_i\| = 1$ and supp $\varphi_i \subset B_i \backslash B_{i-1}$ where $B_i = \{x \in \mathbb{R}^N; |x| < i\}$, and set

$$E_l = \text{span}\{\varphi_1, \cdots, \varphi_l\}.$$

Then for any $u \in E_l$, supp $u \subset B_l$, and

$$f(u) = \tfrac{1}{2}\|u\|^2 - \int_{B_l} G(x, u)$$

$$\leq \tfrac{1}{2}\|u\|^2 + \tilde{a}_l - \tilde{b}_l\|u\|^\mu$$

with $\tilde{a}_l, \tilde{b}_l > 0$. Hence there is an $r_l > 0$ such that

$$f(u) \leq 0, \quad \forall u \in E_l \quad \text{with } \|u\| \geq r_l. \tag{3.7}$$

If dim $(E^- \oplus E^0) > 0$, taking $\varphi_i = e_{\bar{n}+i}$ (see Remark 2.1), $i = 1, \cdots, l$, setting

$$E_l = E^- \oplus E^0 \oplus \text{span}\{e_{\bar{n}+1}, \cdots, e_{\bar{n}+l}\},$$

and then using (3.3), one sees that the conclusion (3.7) still holds in this case.

Step 4. By Step 3, one can take $r > \rho$ large enough such that

$$f|_{\partial Q} \leq 0$$

where $Q = (B_r \cap (E^- \oplus E^0)) \oplus \{se_{\bar{n}+1}; 0 \leq s \leq r\}$. Let

$$c = \inf_{\gamma \in \Gamma} \max_{x \in Q} f(\gamma(x))$$

where $\Gamma = \{\gamma \in C(Q, E), \gamma|_{\partial Q} = id\}$. Now by the linking theorem (see [14]), $c \geq \sigma$ and is a critical value of f.

Suppose that $g(x, u)$ is also odd in u. Then f is even, $f(0) = 0$, and by Step 1-3, f satisfies all the assumptions of [14, Theorem 9, 12 and its proof]. Hence f has an unbounded sequence of critical values $c_l \to \infty$ as $l \to \infty$. Let $u_l \in E$ be such that $f(u_l) = c_l$ and $f'(u_l) = 0$. Then

$$c_l = \int_{\mathbb{R}^N} (\tfrac{1}{2} g(x, u_l) u_l - G(x, u_l)) \to \infty,$$

or

$$\int_{\mathbb{R}^N} (|\nabla u_l|^2 + q(x) u_l^2) \to \infty.$$

The proof is complete.

Remark 3.1. Let

$$g^+(x, u) = \begin{cases} g(x, u) & \text{if } u \geq 0 \\ 0 & \text{if } u < 0 \end{cases}$$

$$g^-(x, u) = \begin{cases} 0, & \text{if } u > 0 \\ g(x, u) & \text{if } u \leq 0. \end{cases}$$

$$G^\pm(x, u) = \int_o^u g^\pm(x, t)dt,$$

$$f^\pm(u) = \frac{1}{2} \int_{\mathbb{R}^N} (|\nabla u|^2 + q(x)u^2) - \int_{\mathbb{R}^N} G^\pm(x, u).$$

Then if $q(x) > 0$ for all x and satisfies (Q_1) and g satisfies $(g_1) - (g_2)- (g_3)$, along the line of the proof of Theorem 1.1, we see that $f^\pm$ has a nontrivial critical point $u^\pm$ such that, by the maximum principle, $u^+(x) > 0$ and $u^-(x) < 0$ for all $x \in \mathbb{R}^N$, i.e., (E) has at least two nontrivial solutions.

Now we give the following

Proof of Theorem 1.2. The proof will be completed by following [4, 11]. Hence we sketch it only.

Step 1. Since $q(x) > 0$ for all x, $E = E^+$. Let $S = \{u \in E; \|u\| = 1\}$. One introduces:

$$f(u) = \frac{1}{2}\|u\|^2 - \frac{1}{\mu} \int_{\mathbb{R}^N} p(x)|u|^\mu - \int_{\mathbb{R}^N} hu,$$

$$\overline{f}(u) = \frac{1}{2}\|u\|^2 - \frac{1}{\mu} \int_{\mathbb{R}^N} p(x)|u|^\mu$$

for $u \in E$, and

$$J(u) = \max_{\lambda \geq 0} f(\lambda u),$$

$$\overline{J}(u) = \max_{\lambda \geq 0} \overline{f}(\lambda u)$$

for $u \in S$. It is clear that $f \in C^2(E, \mathbb{R})$ and the critical points of f are solutions of (E). Let $J_a = \{u \in S; J(u) \geq a\}$ and $J^a = (u \in S; J(u) \leq a)$ for $a \in R$. We say that J satisfies $(PS)_a$ if for any $C \geq a$ and for any $\{u_n\} \subset S$ such that $a \leq J(u_n) \leq C$ and $\|J'(u_n)\| \to 0$ one can extract from $\{u_n\}$ a convergent subsequence.

One can verify that $J \in C(S, \mathbb{R})$, $J \geq 0$, $J \in C^2(J_a, \mathbb{R})$ for any $a > 0$. For any $u \in S$ with $J(u) > 0$, there exists a unique $\lambda = \lambda(u) > 0$ such that $J(u) = f(\lambda(u)u)$. Moreover J and f have the same positive critical values.

Using Lemma 2.4, it is easy to see that J satisfies $(PS)_a$ for any $a > 0$.

Step 2. There is a $\eta > 0$ such that for any $u \in S$ with $J(u) \geq 1$ and $\overline{J}(u) \geq 1$ one has

(i) if (p_1) occurs,

$$|J(u) - \overline{J}(u)| \leq \eta(J(u))^{1/2},$$

$$|J(u) - \overline{J}(u)| \leq \eta(\overline{J}(u))^{1/2};$$

(ii) if (p_2) occurs,

$$|J(u) - \overline{J}(u)| \leq \eta(J(u))^{\frac{1}{\mu}},$$

$$|J(u) - \overline{J}(u)| \leq \eta(\overline{J}(u))^{\frac{1}{\mu}}.$$

Step 3. Let S^k be the k-dimensional euclidean sphere, and set

$$M_k = \{A \subset S : A = \psi(S^k), \psi \text{ is odd and continuous}\}.$$

$$c_k = \inf_{A \in M_k} \max_{u \in A} \overline{J}(u),$$

$$\Gamma_k = \{A \subset S; A \text{ is symmetric and closed}, \gamma(A) \geq k\},$$

$$b_k = \inf_{A \in \Gamma_k} \max_{u \in A} \overline{J}(u),$$

where $\gamma(A) = \inf\{k;$ there is $\psi \in C(A, S^k), \psi$ being odd$\}$. Then $\{c_k\}$ and $\{b_k\}$ are critical values of $\overline{f}$ and $\overline{J}$, and by Borsuk-Ulam theorem, $b_k \leq c_k$. Moreover, if $0 < c_k < b < a < c_{k+1}$, then $\overline{J}^b$ is not contractible to a point in $\overline{J}^a$ (see [11]).

Let u be a solution of

$$-\Delta u + q(x)u - p(x)|u|^{\mu-2}u = 0, \quad x \in \mathbb{R}^N. \tag{3.8}$$

The generalized Morse index of $\overline{f}$ at $u, \sigma(u)$, is defined to be the dimension of the negative and null eigenspace of the Schrödinger operator

$$\overline{A}(u) \equiv -\Delta + (q(x) - (\mu - 1)p(x)|u|^{\mu-2})$$

acting on E.

By [4] we see that, for any $k \in \mathbb{N}$, there exists a critical point u_k of $\overline{f}$ such that $\overline{f}(u_k) = b_k$ and $\sigma(u_k) \geq k + 1$.

Step 4. There is $\overline{c} > 0$ such that for all $k \in \mathbb{N}$,

$$\overline{c}(k + 1)^\gamma \leq b_k \leq c_k \tag{3.9}$$

where $\gamma = \frac{2\mu}{N(\mu-2)}$. In fact, let u_k be as in Step 3 such that $\overline{J}'(u_k) = 0$, $\overline{J}(u_k) = b_k$ and $\sigma(u_k) \geq k + 1$. Since u_k satisfies (3.8), $u_k(x) \to 0$ and so $(q(x) - (\mu - 1)p(x)|u_k|^{\mu-2}) \to \infty$ as $|x| \to \infty$. Remark that $\sigma(u_k)$ is less than or equal to the number of negative and null eigenvalues of $\overline{A}(u_k)$ on L^2. Hence by (2.4) there is C_0 independent of k such that

$$\sigma(u_k) \leq C_0 \int_{q(x)<(\mu-1)p(x)|u_k|^{\mu-2}} |q(x) - (\mu-1)p(x)|u_k|^{\mu-2}|^{\frac{N}{2}}$$

$$\leq C_0 \|p\|_{L^{\frac{N}{\mu}}}^{\frac{N}{\mu}} \left(\int_{\mathbb{R}^N} (p(x)|u_k|^{\mu}) \right)^{\frac{N(\mu-2)}{2\mu}} \tag{3.10}$$

Since

$$b_k = \left(\frac{1}{2} - \frac{1}{\mu}\right) \int_{\mathbb{R}^N} p(x)|u_k|^{\mu},$$

we get immediately (3.9) by (3.10).

Finally, using Steps 1–4, along the line of the proof of [11, Thoerem 1], one sees that J has an unbounded sequence of critical values, thus proving the theorem.

4. Proofs of Theorem 1.3 and 1.4

In this section we deal with the sublinear cases.

Proof of Theorem 1.3. The proof will be divided into several steps.

Step 1. By (g_5)– (g_7), there are $0 < \underline{a} \leq \overline{a}$ and $0 < \underline{b} \leq \overline{b}$ such that

$$\underline{a}|u|^{\theta} \leq G(x,u) \leq \overline{a}|u|^{1+\nu}, \quad \forall x \in \mathbb{R}^N \quad \text{and} \quad |u| \leq 1, \tag{4.1}$$

$$\underline{b}|u|^{\overline{\theta}} \leq G(x,u) \leq \overline{b}|u|^{\theta}, \quad \forall x \in \mathbb{R}^N \quad \text{and} \quad |u| \geq 1. \tag{4.2}$$

Hence, by Lemma 2.4 and 2.5, the variational functional

$$f(u) = \int_{\mathbb{R}^N} G(x,u) - \frac{1}{2}\|u^+\|^2 + \frac{1}{2}\|u^-\|^2$$

for $u = u^- + u^0 + u^+ \in E$ is well-defined. Moreover, $f \in C^1(E, \mathbb{R})$, and by (2.10),

$$f'(u)v = \int_{\mathbb{R}^N} g(x,u)v - (u^+ - u^-, v), \quad \forall u, v \in E, \tag{4.3}$$

and I' is a compact mapping on E where $I(u) = \int_{\mathbb{R}^N} G(x,u)$. To see this, it suffices to show that I' is weakly continuous. Let $u_n \to u$ weakly in E. Note that by (g_6) and (4.1), $\frac{2N}{2-\alpha+N} < 1+\nu \leq \theta$. By Lemmas 2.4 and 2.5, one can assume that $u_n \to u$ strongly in $L^{1+\nu}$ and L^2 without loss of generality. By definition,

$$\|I'(u_n) - I'(u)\| = \sup_{\|\varphi\|=1} \left| \int_{\mathbb{R}^N} (g(x,u_n) - g(x,u))\varphi \right|.$$

(g_6) and (g_7) imply

$$|g(x,u)| \leq C(|u|^{\nu} + |u|), \quad \forall x \in \mathbb{R}^N \quad \text{and} \quad u \in \mathbb{R}. \tag{4.4}$$

For any $R > 0$, by (4.4) and Hölder inequality,

$$\left|\int_{|x|\geq R}(g(x,u_n)-g(x,u))\varphi\right| \leq C\int_{|x|\geq R}(|u_n|^\nu+|u|^\nu+|u_n|+|u|)|\varphi|$$

$$\leq C\left[\left(\int_{|x|\geq R}|u_n|^{1+\nu}+|u|^{1+\nu}\right)^{\frac{\nu}{1+\nu}}+\left(\int_{|x|\geq R}|u_n|^2+|u|^2\right)^{\frac{1}{2}}\right]. \tag{4.5}$$

For any $\varepsilon > 0$, one can take R_0 large enough such that

$$\left|\int_{|x|\geq R_0}(g(x,u_n)-g(x,u))\varphi\right| < \frac{\varepsilon}{2}, \quad \forall\|\varphi\|=1 \quad \text{and} \quad n\in\mathbb{N}$$

by (4.5). It is known that since $u_n\to u$ in L^2,

$$\|g(\cdot,u_n)-g(\cdot,u)\|_{L^2(B_{R_0})}\to 0.$$

There is n_0 such that

$$\left|\int_{|x|<R_0}(g(x,u_n)-g(x,u))\varphi\right| < \frac{\varepsilon}{2}$$

for all $\|\varphi\|=1$ and $n\geq n_0$. Hence $I'(u_n)\to I'(u)$.

Now by (4.3), the critical points of f on E are solutions of (E).

Step 2. f satisfies (PS). Let $\{u_n\}\subset E$ be such that $|f(u_n)|\leq$ const. and $\|f'(u_n)\|\to 0$. By (g_5),

$$f(u_n)-\frac{1}{2}f'(u_n)u_n \geq \left(1-\frac{\theta}{2}\right)\int_{\mathbb{R}^N}G(x,u_n). \tag{4.6}$$

For $u\in E$, let

$$u^1(x) = \begin{cases} u(x) & \text{if } |u(x)|<1, \\ 0 & \text{if } |u(x)|\geq 1 \end{cases}$$

$$u^2(x) = \begin{cases} 0 & \text{if } |u(x)|<1 \\ u(x) & \text{if } |u(x)|\geq 1. \end{cases}$$

Then (4.1), (4.2) and (4.6) imply

$$C(1+\|u_n\|) \geq \underline{a}\|u_n^1\|_{L^\theta}^\theta + \underline{b}\|u_n^2\|_{L^{\bar{\theta}}}^{\bar{\theta}}. \tag{4.7}$$

Since $\dim(E^-\oplus E^0)<\infty$, we have

$$\|u_n^-+u_n^0\|_{L^2}^2 \leq C\|u_n^-+u_n^0\|_{L^2}(\|u_n^1\|_{L^\theta}+\|u_n^2\|_{L^{\bar{\theta}}}),$$

which, together with (4.7), yields

$$\|u_n^-+u_n^0\| \leq C(1+\|u_n\|^{\frac{1}{\theta}}+\|u_n\|^{\frac{1}{\bar{\theta}}}).$$

Therefore

$$\|u_n\|^2 = \|u_n^- + u_n^0\|^2 + \|u_n^-\|^2 + 2\int_{\mathbb{R}^N} G(x, u_n) - 2f(u_n)$$

$$\leq C(1 + \|u_n\| + \|u_n\|^{\frac{2}{\theta}}),$$

i.e., $\|u_n\| \leq$ const., since $\overline{\theta} > 1$. Now since I' is compact, a standard argument shows that $\{u_n\}$ has a convergent subsequence.

Step 3. $f(u) \to -\infty$ for $u \in E^+$, as $\|u\| \to \infty$. In fact. by (4.1)– (4.2), for $u \in E^+$, as $\|u\| \to \infty$,

$$f(u) = \int_{\mathbb{R}^N} G(x, u) - \frac{1}{2}\|u\|^2 \leq C(\|u\|^{1+\nu} + \|u\|^{\theta}) - \frac{1}{2}\|u\|^2 \to -\infty,$$

since $1 + \nu \leq \theta < 2$. Hence there is $r > 0$ such that

$$f(u) \leq 0, \forall u \in E \quad \text{with} \quad \|u\| \geq r.$$

Let $Q = \{u \in E^+; \|u\| \leq r\}$.

Step 4. For any $l \in N, E_l = E^- \oplus E^0 \oplus \text{span}\{e_{\bar{n}+1}, \cdots, e_{\bar{n}+l}\}$, there are $\rho = \rho(l)$ and $\sigma = \sigma(l) > 0$ such that

$$f|_{\partial B_\rho \cap E_l} \geq \sigma. \tag{4.8}$$

Indeed, for $u \in E_l$ with $\|u\|_{L^\infty} \leq 1$, by (4.1),

$$f(u) \geq \underline{a}\|u\|_{L^\theta}^\theta - \frac{1}{2}\|u\|^2 \geq (a' - \frac{1}{2}\|u\|^{2-\theta})\|u\|^\theta$$

with some $a' > 0$, since $\dim E_l < \infty$.

Step 5. If $\dim(E^- \oplus E^0) = 0$, the above arguments show that f has a maximum (> 0) which yields a nontrivial solution of (E).

Assume $E^- \oplus E^0 \neq \{0\}$. Take $e_0 \in E^+$ with $\|e_0\| = 1$. Since $\dim (E^- \oplus E^0) < \infty$, for any $u = u^- + u^0 + se_0$ with $u^- \in E^-, u^0 \in E^0$ and $0 < s < 1$, by (4.1) –(4.2),

$$f(u) = \int_{\mathbb{R}} G(x, u) - \frac{1}{2}s^2 + \frac{1}{2}\|u^-\|^2$$

$$\geq \underline{a}\|u\|_{L^\theta}^\theta + \underline{b}\|u\|_{L^{\bar{\theta}}}^{\bar{\theta}} - \frac{1}{2}s^2 + \frac{1}{2}\|u^-\|^2$$

$$\geq a'(\|u\|^\theta + \|u\|^{\bar{\theta}}) - \frac{1}{2}s^2 + \frac{1}{2}\|u^-\|^2$$

$$\geq a's^\theta - \frac{1}{2}s^2 + \frac{1}{2}\|u^-\|^2.$$

Taking $s > 0$ small enough and setting $e = se_0$, we get a $\sigma > 0$ such that

$$f(u) \geq \sigma, \quad \forall u \in S \equiv E^- \oplus E^0 + e.$$

Since S and ∂Q link, by the Benci-Rabinowitz linking theorem [14], f has a critical value $c \geq \sigma$ which gives a nontrivial solution of (E).

Step 6. Finally, suppose that g is also odd in u. Then f is even and $f(0) = 0$. Let γ denote the $\mathbb{Z}_2$-genus and let, for any $1 \leq l \leq, \in \mathbb{N}$,

$$\Sigma_l = \{A \subset E \backslash \{0\}; A \text{ is closed and symmetric w.r.t. } 0, \gamma(A) \geq \bar{n} + l\}$$

$$c_l = \sup_{A \in \Sigma_l} \inf_{u \in A} f(u), \quad 1 \leq l \leq m.$$

Since f satisfies (PS), c_l is a critical value of f [14]. Since $\gamma(E_m \cap \partial B_\rho) = \bar{n} + m$, by (4.8) and Step 3,

$$0 < \sigma \leq \inf_{E_m \cap \partial B_\rho} f \leq c_m \leq c_{m-1} \leq \cdots \leq c_1 \leq \sup_{E^+} f < \infty.$$

Hence, f has m pairs of nontrivial critical points by the $\mathbb{Z}_2$-genus theory. Letting $m \to \infty$ we get the last conclusion of Theorem 1.3, and the proof is complete.

Proos of Theorem 1.4. Since the proof is quite similar to but simplier than that of Theorem 1.3, the details are omitted.

Remark 4.1. If $\dim E^- = n^- > 0$, we can show one more result. Indeed, we have

Theorem 4.1. Suppose that q satisfies (Q$_2$) with $\alpha < 2 - N$ and $n^- > 0$. Let g satisfy (g$_8$), (g$_{10}$) and

   (g$_{12}$) $G(x, u) \leq 0$ and $G(x, u) \to -\infty$ as $|u| \to \infty$ uniformly in $x \in \mathbb{R}^N$;
   (g$_{13}$) there are $a > 0, \omega > 0$ and $\theta \in (1, 2)$ such that

$$-G(x, u) \geq a|u|^\theta, \quad \forall x \in \mathbb{R}^N \quad \text{and} \quad |u| \leq \omega.$$

Then (E) has at least one nontrivial solution. Moreover, if $g(x, u)$ is odd in $u \in \mathbb{R}$, then (E) has n^- pairs of solutions.

Proof. We consider the functional

$$f(u) = \frac{1}{2}\|u^+\|^2 - \frac{1}{2}\|u^-\|^2 - \int_{\mathbb{R}^N} G(x, u)$$

for $u = u^- + u^0 + u^+ \in E$. Then $f \in C^1(E, \mathbb{R})$ and I' is compact on E where $I(u) = \int_{\mathbb{R}^N} G(x, u)$ as before. Consequently, the critical points of f on E are solutions of (E).

We verify that f satisfies (PS). Let $\{u_n\}$ be a (PS) sequence. By (g$_8$) and Lemma 2.5,

$$\|u_n^+\|^2 = f'(u_n)u_n^+ + \int_{\mathbb{R}^N} g(x, u_n)u_n^+ \leq C\|u_n^+\|.$$

Hence $\|u_n^+\| \leq$ const. Similarly, $\|u_n^-\| \leq$ const. If $\|u_n^0\| \to \infty$, then

$$v_n \equiv \frac{u_n^0}{\|u_n^0\|} \to v \in \partial B_1 \cap E^0.$$

Let $\Omega = \{x \in \mathbb{R}^N; v(x) \neq 0\}$. Then $|\Omega| > 0$ and

$$|u_n^0(x)| = \|u_n^0\| |v_n(x)| \to \infty, \quad \forall x \in \Omega.$$

By the Fatou's theorem and (g_{12}),

$$\liminf_{n \to \infty} \int_{\mathbb{R}^N} -G(x, u_n^0(x)) \geq \int_{\mathbb{R}^N} \liminf_{n \to \infty} [-G(x, u_n^0(x))] = \infty,$$

and so by (g_8),

$$-\int_{\mathbb{R}^N} G(x, u_n) \geq -\int_{\mathbb{R}^N} G(x, u_n^0) - M\|u_n^- + u_n^+\|_{L^1} \to \infty,$$

contradicting that $|f(u_n)| \leq$ const. Therefore $\|u_n\| \leq$ const., and as before $\{u_n\}$ has a convergent subsequence.

Next, we claim that there are ρ and $\sigma > 0$ such that

$$f(u) \geq \sigma, \quad \forall \in \partial B_\rho. \tag{4.9}$$

Let C_∞ be the constant such that $\|u^- + u^0\|_{L^\infty} \leq C_\infty \|u^- + u^0\|$ for $u = u^- + u^0 \in E^- \oplus E^0$, and set $r = (2C_\infty)^{-1}\omega$,

$$\Omega_1 = \{x \in R^N; |u^+(x)| \leq \frac{\omega}{2}\}, \quad \Omega_2 = \mathbb{R}^N \backslash \Omega_1.$$

Let $u \in B_r$. On $\Omega_1, |u(x)| \leq \|u^- + u^0\|_{L^\infty} + \frac{\omega}{2} \leq \omega$, and by ($g_{13}$),

$$-\int_{\Omega_1} G(x, u) \geq a \int_{\Omega_1} |u|^\theta.$$

On $\Omega_2, |u(x)| \leq 2|u^+(x)|$. Hence, letting

$$\|u^- + u^0\|_{L^2}^2 \theta' = \frac{\theta}{\theta-1}, \text{ we have } = (u^- + u^0, u)_{L^2}$$

$$\leq C\|u^- + u^0\|_{L^2} (\int_{\Omega_1} |u|^\theta + \int_{\Omega_2} |u|^\theta)^{\frac{1}{\theta}},$$

and consequently,

$$\|u^-\|^\theta + \|u^0\|^\theta \leq C\left[-\int_{\Omega_1} G(x, u) + \int_{\Omega_2} |u^+|^\theta\right]. \tag{4.10}$$

For $p \in (2, \overline{N}), \overline{N} = \frac{2N}{N-2}$ if $N \geq 3, \overline{N} = \infty$ if $N \leq 2$,

$$\int_{\mathbb{R}^N} |u^+|^p \geq \int_{\Omega_2} |u^+|^p \geq (\frac{\omega}{2})^p |\Omega_2|. \tag{4.11}$$

Setting $s = \frac{p}{\theta}$ and $s' = \frac{s}{s-1} = \frac{p}{p-\theta}$, by (4.11),

$$\int_{\Omega_2} |u^+|^\theta \leq |\Omega_2|^{\frac{1}{s'}} (\int_{\Omega_2} |u^+|^{s\theta})^{\frac{1}{s}}$$

$$\leq (\frac{\omega}{2})^{\theta-p} \|u^+\|_{L^p}^p \leq C\|u^+\|^p. \tag{4.12}$$

Combining (4.10) with (4.12), one sees that there is $\overline{C} > 0$ such that for all $u \in B_r$,

$$- \int_{\Omega_1} G(x, u) \geq \overline{C}[\|u^-\|^\theta + \|u^0\|^\theta - \|u^+\|^p],$$

and

$$f(u) \geq \tfrac{1}{2}\|u^+\|^2 - \tfrac{1}{2}\|u^-\|^2 - \int_{\Omega_1} G(x, u)$$

$$\geq (\tfrac{1}{2} - \|u^+\|^{p-2})\|u^+\|^2 + (\overline{C} - \tfrac{1}{2}\|u^-\|^{2-\theta})\|u^-\|^\theta + \overline{C}\|u^0\|^\theta,$$

and (4.9) follows by taking $\rho \leq r$ small enough since $p > 2 > \theta > 1$.

Finally, by (g$_8$) and Lemma 2.5, for $u \in E^-$,

$$f(u) = -\tfrac{1}{2}\|u\|^2 - \int_{\mathbb{R}^N} G(x, u)$$

$$\leq -\tfrac{1}{2}\|u\|^2 + M \int_{\mathbb{R}^N} |u| \leq -\tfrac{1}{2}\|u\|^2 + C\|u\| \to -\infty$$

as $\|u\| \to \infty$. Now a standard argument similar to that in the Steps 5-6 of the proof of Theorem 1.3 (see also [2]) yields the conclusions of Theorem 4.1.

5. Proof of Theorem 1.5

In the following let the assumptions of Theorem 1.5 be satisfied. Consider the functional f defined by (3.1). It is easy to verify that, in this case, again $f \in C^1(E, \mathbb{R})$ and the critical points of f on E are solutions of (E). Let $K = \{u \in E; f'(u) = 0\}$. Clearly, $0 \in K$.

By Lemma 2.3, $E = W^{1,2}$. For $u \in E$, set

$$\|u\|_\infty^2 = \int_{\mathbb{R}^N} (|\nabla u|^2 + q_\infty(x)u^2).$$

Then $\| \cdot \|_\infty$ is a norm on E which is equivalent to $\| \cdot \|$ and $\|| \cdot |\|$. On E we introduce the functional

$$f_\infty(u) = \frac{1}{2}\|u\|_\infty^2 - \int_{\mathbb{R}^N} G_\infty(x, u)$$

where $G_\infty(x, u) = \int_0^u g_\infty(x, t)dt$. Then $f_\infty \in C^1(E, \mathbb{R})$. Let $K_\infty = \{u \in E; f'_\infty(u) = 0\}$.

For $j = (j_1, \cdots, j_N) \in \mathbb{Z}^N$ and $u \in E$, let

$$(\tau_j u)(x) = u(x_1 + j_1 T_1, \cdots, x_N + j_N T_N).$$

Then by the assumptions,

$$f_\infty(\tau_j u) = f_\infty(u),$$

i.e., f_∞ is $\mathbb{Z}^N$-invariant.

Along the line of the proof of [6, Proposition 2.31], it is not difficult to check that, under our conditions, one also has

Lemma 5.1. Let $\{u_n\} \subset E$ be such that $f_\infty(u_n) \to b > 0$ and $f'_\infty(u_n) \to 0$. Then there exist an $l \in \mathbb{N}$ (depending on b), $v_1, \cdots, v_l \in K_\infty \backslash \{0\}$, a subsequence of u_n and the corresponding $\{k_n^i\} \subset \mathbb{Z}^N$ such that

$$\left\| u_n - \sum_{i=1}^l \tau_{k_n^i} v_i \right\|_\infty \to 0,$$

$$|k_n^i - k_n^j| \to \infty \quad \text{if} \quad i \neq j,$$

$$\sum_{i=1}^l f_\infty(v_i) = b.$$

Similarly, we also have (see also [12])

Lemma 5.2. Let $\{u_n\} \subset E$ be such that $f(u_n) \to b > 0$ and $f'(u_n) \to 0$. Then there exist a $v_0 \in K$, an $l \in \mathbb{N}$ (depending on b), $v_1, \cdots, v_l \in K_\infty \backslash \{0\}$, a subsequence of u_n and the corresponding $\{k_n^i\} \subset \mathbb{Z}^N$ such that

$$\left\| \left\| u_n - v_0 - \sum_{i=1}^l \tau_{k_n^i} v_i \right\| \right\| \to 0,$$

$$|k_n^i| \to \infty \quad \text{and} \quad |k_n^i - k_n^j| \to \infty \quad \text{if} \quad i \neq j,$$

$$f(v_0) + \sum_{i=1}^l f_\infty(v_i) = b.$$

Moreover, we have

$$f_\infty(u) = \frac{1}{2}\|u\|_\infty^2 + o(\|u\|_\infty^2) \quad \text{as} \quad u \to 0.$$

Hence there are $\sigma, \rho > 0$ such that

$$f_\infty(u) \geq \sigma, \quad \forall u \in S_\rho = \{u \in E; \|u\|_\infty = \rho\}. \tag{5.1}$$

By (g3), if $u \neq 0$, then

$$f_\infty(su) \to -\infty \quad \text{as} \quad |s| \to \infty. \tag{5.2}$$

Let

$$c_\infty = \inf_{\gamma \in \Gamma_\infty} \max_{t \in [0,1]} f_\infty(\gamma(t))$$

where

$$\Gamma_\infty = \{\gamma \in C([0,1], E); \gamma(0) = 0, \gamma(1) \in f_\infty^0 \backslash \{0\}\}$$

in which $f_\infty^0 = \{u \in E; f_\infty(u) \leq 0\}$. By (5.1), $c_\infty \geq \sigma$. Now the Ekeland's variational priniciple implies that there is $\{u_n\} \subset E$ such that

$$f_\infty(u_n) \to c_\infty \quad \text{and} \quad f_\infty'(u_n) \to 0.$$

Consequently, by Lemma 5.1, there is $v \in K_\infty \backslash \{0\}$ with

$$f_\infty(v) \leq c_\infty. \tag{5.3}$$

Let $b_\infty = \inf\{f_\infty(v); v \in K_\infty \backslash \{0\}\}$. By (5.3),

$$b_\infty \leq c_\infty. \tag{5.4}$$

Lemma 5.3. $c_\infty = b_\infty$.

Proof. The proof is well-known, see e.g., [7, 15]. By (5.4), it remains to show that

$$c_\infty \leq b_\infty. \tag{5.5}$$

Note that, for any $v \neq 0$, by (5.2), there is $s_v > 0$ such that $f_\infty(s_v v) \leq 0$, and by (*), $f_\infty(sv)$ attains its maximum along the ray $l_v = \{sv; s \geq 0\}$ at a point $s_v' v$ with $0 < s_v' < s_v$ characterized by $f_\infty'(s_v' v)v = 0$. Define

$$\gamma_v(t) = t(s_v v), \quad \forall t \in [0,1].$$

Then $\gamma_v \in \Gamma_\infty$ and $f_\infty(\gamma_v(t)) \leq f_\infty(s_v' v)$ for $t \in [0,1]$. In particular, for $v = in K_\infty \backslash \{0\}$, we get

$$c_\infty \leq \max_{t \in [0,1]} f_\infty(\gamma_v(t)) \leq f_\infty(v).$$

(5.5) follows.

Combining Lemma 5.2 with Lemma 5.3 we obtain immediately the following

Lemma 5.4. f satisfies $(PS)_c$ condition for any $c \in (0, c_\infty)$, i.e., if $\{u_n\} \subset E$ is such that $f(u_n) \to c$ and $f'(u_n) \to 0$, then $\{u_n\}$ has a convergent subsequence.

Now we are ready to give the following

Proof of Theorem 1.5. Take $\varphi \in E^+$ with $\|\varphi\| = 1$, and $v \in K_\infty \backslash \{0\}$ satisfying $f_\infty(v) = c_\infty$. Since f_∞ is $\mathbb{Z}^N$-invariant, without loss of generality, one can assume

$$\int_\Omega |v|^2 = \underline{a} > 0. \tag{5.6}$$

Set $X = \text{span}\{E^-, E^0, \varphi, v\}$. By (g$_3$), there is $r > 0$ such that

$$f(u) \leq 0 \quad \text{and} \quad f_\infty(u) \leq 0, \quad \forall u \in X \quad \text{with } \|u\| \geq r.$$

Let

$$Q = \{u = u^- + u^0 + s\varphi \in E; \|u\| \leq 2r, s \geq 0\},$$

$$\Gamma = \{\eta \in C(Q, E); \quad \eta|_{\partial Q} = id\}$$

$$\underline{c} = \inf_{\eta \in \Gamma} \max_{x \in Q} f(\eta(x)).$$

An argument similar to that of Step 2 of the proof of Theorem 1.2 shows that there are $0 < \tilde\rho < r, 0 < \tilde\sigma$ such that

$$f(u) \geq \tilde\sigma, \quad \forall u \in S_{\tilde\rho} = \{u \in E^+; \|u\| = \tilde\rho\}.$$

Hence, since Q and $S_{\tilde\rho}$ link, $\underline{c} \geq \tilde\sigma$.

Let $e = r\varphi, \bar{v} = 2r\frac{v}{\|v\|}$, and take $w \in E^- \oplus E^0$ satisfying $\|w\| = 2r$. Notice that

$$Q = \{tu + (1-t)e; \quad t \in [0,1] \text{ and } u \in \partial Q\}.$$

Writting $x = (t, u) = tu + (1-t)e$ for all $t \in [0,1]$ and $u \in \partial Q$, we define $\tilde\eta : Q \to E$ by setting

$$\tilde\eta((t,u)) = \begin{cases} 3t\bar{v} & t \in [0, \frac{1}{3}], \\ \bar{v}\cos\frac{3\pi}{2}(t - \frac{1}{3}) + w\sin\frac{3\pi}{2}(t - \frac{1}{3}) & t \in [\frac{1}{3}, \frac{2}{3}], \\ w\cos\frac{3\pi}{2}(t - \frac{2}{3}) + u\sin\frac{3\pi}{2}(t - \frac{2}{3}) & t \in [\frac{2}{3}, 1]. \end{cases}$$

Then $\tilde\eta \in \Gamma, f(\tilde\eta(t,u)) \leq 0$ for $(t,u) \in [\frac{1}{3}, 1] \times \partial Q$, and $f(\tilde\eta(t,u)) = f(3t\bar{v})$ for $(t,u) \in [0, \frac{1}{3}] \times \partial Q$. Moreover, one can take $\bar{t} \in (0, \frac{1}{3})$ such that

$$f(3\bar{t}\bar{v}) = \max_{x \in Q} f(\tilde\eta(x)).$$

By (Q$_3$) and (g$_{11}$), for any $u \in E$,

$$f_\infty(u) - f(u) = \frac{1}{2}\int_{\mathbb{R}^N}(q_\infty - q)u^2 + \int_{\mathbb{R}^N}(G(x,u) - G_\infty(x,u))$$

$$\geq \frac{1}{2}\int_{\mathbb{R}^N}(q_\infty - q)u^2.$$

Hence, taking into account (5.6),

$$f_\infty(3\bar{t}\bar{v}) - f(3\bar{t}\bar{v}) \geq \frac{18r^2\bar{t}^2}{\|v\|^2}\int_\Omega (q_\infty - q)v^2 > 0,$$

and consequently, as shown in the proof of Lemma 5.3,

$$f(3\bar{t}\bar{v}) < f_\infty(3\bar{t}\bar{v}) \leq f_\infty(v) \leq c_\infty.$$

Therefore, by definition,

$$\underline{c} \leq \max_{x \in Q} f(\tilde{\eta}(x)) < c_\infty,$$

i.e., $\underline{c} \in (0, c_\infty)$.

Now by Lemma 5.4 and the linking theorem (see [14]), $\underline{c}$ is a critical value of f. Hence f has a nontrivial critical point. The proof is thereby complete.

Remark 5.1. There are some existence results like Theorems 1.1 and 1.3 for a semilinear elliptic systems on $\mathbb{R}^N$, see a forthcoming paper [9].

References

[1] Alama S.- Li Y.Y., Existence of solutions for semilinear elliptic equations with indefinite lienar part, J. Diff. Eq. 96(1992), 89–115.

[2] Ambrosetti A. -Rabinowitz P., Dual variational methods in critical point theory and applications, J. Funct. Anal. 14(1973), 349–381.

[3] Ambrosetti A. -Struwe M., Existence of steady vortex rings in an ideal fluid, Arch. Rat. Mech. Anal. 108(1989), 97–109.

[4] Bahri A. -Lions P.L., Morse index of some nin-max critical points: I. application to multiplicity results, Comm. Pure Appl. Math. 41(1988), 1027–1037.

[5] Berestycki H. -Lions P. L., Nonlinear scalar field equations I, II, Arch. Rat. Mech. Anal. 82 (1983), 313-345, 347-375.

[6] Coti Zelati V. -Rabinowitz P., Homoclinic type solutions for a semilinear elliptic PDE on $\mathbb{R}^N$, SISSA (Trieste) preprint, 1991.

[7] Ding W. Y. -Ni W. M., On the existence of positive entire solutions of a semilinear elliptic equation, Arch. Rat. Mech. Anal. 91 (1986), 283-308.

[8] Ding Y. H., Existence and multiplicity results for homoclinic solutions to a class of Hamiltonian systems, preprint.

[9] Ding Y. H. -Li S. J., Existence of solutions for some superlinear or sublinear elliptic systems on $\mathbb{R}^N$, preprint.

[10] Kato T., " Perturbation Theory for Linear Operators", Springer-Verlag New York Inc., 1966.

[11] Li Y. Y., Existence of infinitely many critical values of some nonsymmetric functional, J. Diff. Eq., 95 (1992), 140-153.

[12] Lions P. L., On positive solutions of semilinear elliptic equations in unbounded domains, in "Nonlinear Diffusion Equations and their Equilibrium States II", W. M. Ni etc. ed. , Springer-Verlag, 1986.

[13] Nehari Z., On a class of nonlinear second order differential equations, Trans. Am. Math. Soc. 95(1960), 101-123.

[14] Rabinowitz P., Minimax methods in critical point theory with applications to differential equations, CBMS Reg. Conf. Ser. in Math. 65, A. M. S., Providence, 1986.

[15] Rabinowitz P. -Tanaka K., Some results on connecting orbits for a class of Hmailtonian systems, Math. Z. 206 (1991), 473-499.

[16] Reed M. -Simon B., " Methods of Modern Mathematical Physics, IV. Analysis of Operators." Academic Press, New York, 1978.

All Liminal AF-Algebras are Stable Isomorphic to Finite AF-Algebras

Zhaobo Huang

Institute of Mathematics, Fudan University, Shanghai

1. Introduction

In [5,6], we have already studied the relationships between the classification of AF-algebras defined by J. Cuntz and G. K. Pedersen [2] and their dimension groups. In this paper, we will continue to study the relation between finite AF-algebras and the liminal AF-algebras [8].

Recall that in [5], we have proved that the dimension groups of liminal AF-algebras are archimedean. In this paper, we shall further prove that they have finite scales.

Acknowledgement: This paper is one part of the author's thesis. He would like to thank sincerely his supervisor Prof. Shaozhong Yan for guidance and encouragement, and also thank Prof. Xiaoman Chen for help.

2. Theorems and their proofs

Our main theorems are the following:

Theorem A. If A is a liminal AF-algebra, then the dimension group of A has a finite scale. In other words, there is a finite AF-algebra which is stable isomorphic to A.

For the sake of simplicity, we shall only give a detailed proof of this theorem for a special class of the liminal C^*-algebras, i.e. the AF-algebras with continuous trace [8].

Theorem B. If A is an AF-algebra with continuous trace, then the dimension group $(G(A), G(A)_+)$ of A has a finite scale. In other words, there is a finite AF-algebra B with continuous trace which is stable isomorphic to A.

Before we embark on proving these theorems, we need to explain some notations. Recall that the dimension group of an AF-algebra A is the ordered group $(K_0(A), K_0(A)_+)$, the

---

1991 Mathematics Subject Classification: 46L35

scale of the dimension $\Gamma(A)$ is the image of the set $Proj(A)$, i.e. the projections of A, in $K_0(A)$. We call a scale finite if for every non-zero element $a \in \Gamma(A)$, there exists a natural number n such that $na \notin \Gamma(A)$. By Theorem 3.3 in [5], we know that an AF-algebra is finite iff its scale is finite.

In order to prove these theorems, we need to recall some preliminaries on Bratteli Diagram of AF-algebras. Let A be an AF-algebra, $A = \overline{\bigcup_{n=1}^{\infty} A_n}$, where A_n are finite dimensional C^*-algebras and $A_n \subset A_{n+1}$. The Bratteli diagram corresponding to $\{A_n\}_{n=1}^{\infty}$ is denoted by $(D, d, \mathfrak{U})$, where $D_n = \{(i,n)\}_{i=1}^{r(n)}, D = \cup_{n=1}^{\infty} D_n, \mathfrak{U} = \{u_n\}_{n=1}^{\infty}, D_n$ is called the n-th generation of D [1,7], $d: D \to N$ is a mapping from the set D to the natural number set N which satisfies the following condition

$$d((j, n+1)) \geq \sum_{i=1}^{r(n)} \beta_{ji}^{(n)} d((i,n)), \qquad \forall 1 \leq j \leq r(n+1) \tag{1}$$

where $\beta_{ji}^{(n)}$ are the elements of the matrix $u_n, u_n = (\beta_{ji}^{(n)})$.

Obviously, we can consider the n-th generation $(D_n, d|_{D_n})$ as the scaled dimension group $(Z^{r(n)}, Z_+^{r(n)}, \{d(i,n)\}_{i=1}^{r(n)})$ of the matrix algebra $M_{d(1,n)} \times M_{d(2,n)} \times \cdots \times M_{d(r(n),n)}$ and u_n as the map from $Z^{r(n)}$ to $Z^{r(n+1)}$, cf.[4]. We say that two Bratteli diagrams are equivalent if they represent the same AF-algebras (up to an isomorphism). From the corresponding relations between AF-algebras and their dimension groups[4], we can easily obtain the following two properties.

(A) If $\{m_n\}_{n=1}^{\infty}$ is a subsequence of N, $D_{m_n} = \{(i, m_n)\}_{i=1}^{r(m_n)}, D' = \cup_{n=1}^{\infty} D_{m_n}$, $d' = d|_{D'}$, and $u'_n = u_{m_n}$ if $m_{n+1} = m_n + 1, u'_n = u_{m_n} u_{m_n+1} \cdots u_{m_{n+1}-1}$ if $m_{n+1} > m_n + 1$, then we get another Bratteli diagram which is equivalent to the old one.

(B) Let $D'_1 = D_1$, $D'_{n+1} = D_{n+1}\backslash$ the set of all the descendants (see the following definition) of the points in D_n. If we delete the points in D'_n, we get a new Bratteli diagram which is equivalent to the original one.

Definition 1. (1) Let $x_1 = (i, n), x_2 = (j, m) \in D$ and $n > m$. If the element $\beta_{ij} \geq 1$ of the matrix U_m when $n = m + 1$, or the (i,j)-element $q \geq 1$ of the matrix $U_{n-1} \cdots U_{m+1} U_m$ when $n > m + 1$, we call x_1 a descendant of x_2, q being multiplicity, or x_2 is an ancestor of x_1, denoted by $x_1 \succ x_2$ or $x_2 \prec x_1$;

(2) If $\{x_n\}_{n=1}^{\infty}$ is a sequence of elements in D with $x_{n+1} \succ x_n$, then this sequence is called a chain of D. If every x_{n+1} is just in the next generation of x_n, this sequence is called complete;

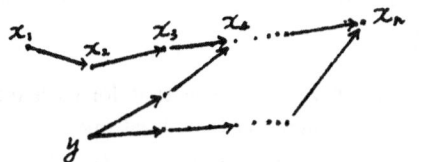

     Fig.1 Annihilating point   Fig.2 Circle point

(3) Let $x_1, x_2 \in D_n, x_3 \in D_m, (m > n)$. We call x_3 the annihilating point of x_1 and x_2, if $x_3 \succ x_1, x_2$ when $x_1 \neq x_2$, or if x_3 is a descendant of x_1 with multiplicity bigger than 1 when $x_1 = x_2$. If $\{x_n\}_{n=1}^{\infty}$ is a chain of D, $y \in D \setminus \{x_n\}_{n=1}^{\infty}$, and there exist descendants of y in $\{x_n\}_{n=1}^{\infty}$, then it contains the first descendant x_n of y, and we say that the point y annihilates at the chain $\{x_n\}_{n=1}^{\infty}$ with the annihilating point x_n. See Figure 1.

(4) If x_1 has a descendant x_2 with multiplicity bigger than 1, we say that it has a circle at x_1. See Figure 2.

To prove Theorem B, we need the following characterization for AF-algebras with continuous trace by their Bratteli diagrams [7].

Lemma 2. Let A be an AF-algebra, $(D, d, \mathfrak{U})$ be its Bratteli diagram. Then the following statements are equivalent:

(1) A is an AF-algebra with continuous trace.

(2) For every chain $\{x_n\}_{n=1}^{\infty}$ in D, there exists a positive integer m such that all the descendant points of x_m are with multiplicity one. In other words, the point x_m has not any circles.

By this theorem, it is easy to see that the AF-algebra represented by the Bratteli diagram in Figure 3(a) is not with continuous trace, for every point on the button line is with circles, but the AF-algebra represented by Figure 3(b) is with continuous trace by the above theorem.

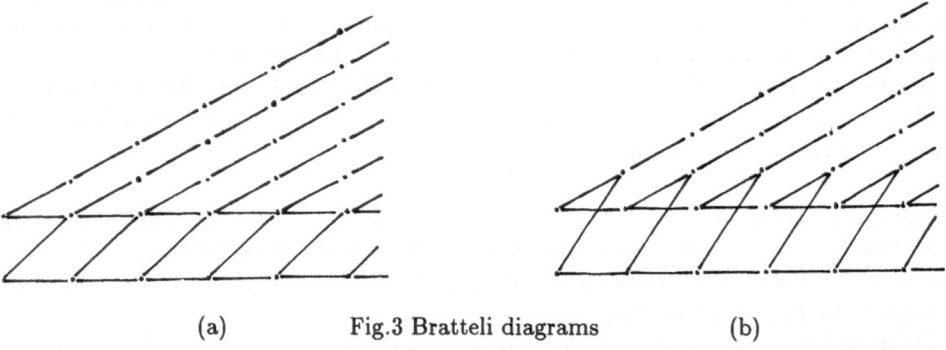

(a)   Fig.3 Bratteli diagrams   (b)

Lemma 3. Let A be an AF-algebra with continuous trace, $(D, d, \mathfrak{U})$ be one of its Bratteli diagrams. Then for each point $x_1 \in D$ there always exists a positive integer m such that all the descendants of x_1 in D_m have not any circles.

Proof. Suppose the statement is false, i.e. there is an $x_1 \in D$ such that for each integer $m > n_o$, where n_o is the generation number of $x_1 \in D$, there exists at least one descendant in D_m with circles. Without loss of generality, we may assume that $x_1 \in D_1$. Therefore, we can find an infinite number of elements $\{x_n\}_{n=2}^{\infty}$ in D with $x_n \in D_n$ and each x_n is a descendant of x_1 with circles. Since there are only finitely many descendants of x_1 in D_2

and $\{x_n\}_{n=3}^{\infty}$ are the descendants for one of these points, there is one descendant x_2' of x_1 in D_2 such that there are infinitely many descendants of x_2' in $\{x\}_{n=3}^{\infty}$; these points are still denoted by $\{x_n\}_{n=3}^{\infty}$. By this method, we can find a point x_3' in D_3 such that it is a descendant of x_2' and there are infinitely many descendants of x_3' in $\{x_n\}_{n=4}^{\infty}$. Proceeding inductively, we can construct a complete chain $\{x_n'\}_{n=1}^{\infty}, x_1' = x_1$ in D such that every x_n' has a circle, which contradicts the above lemma. Q.E.D.

By this lemma, we may simplify the Bratteli diagrams of AF-algebras with continuous trace as follows.

Lemma 4. Let A be an AF-algebra with continuous trace. Then there exists a Bratteli diagram $(D, d, \mathfrak{U})$ of A, such that there are not any circles for all of the elements in D.

We call such a Bratteli diagram simple.

Proof. Let A be an AF-algebra with continuous trace, $(D, d, \mathfrak{U})$ be a Bratteli diagram for it. By Lemma 3, there exists for every point x in D_1 a positive integer n_x such that all the descendants of x in D_{n_x} have not any circles. Since D_1 contains only finitely many points, there is a positive integer n_1 such that all the descendants in D_{n_1} for every point in D_1 have no circles; denote the set of all descendants of D_1 in D_{n_1} by $D_{n_1}(1)$, and denote the remaining points by $D_{n_1}(2)$ (Note: This set may be empty). We delete the generations between D_1 and D_{n_1}, rearrange the points in D_{n_1} such that the order numbers of the elements in $D_{n_1}(1)$ are smaller than those of $D_{n_1}(2)$. So we obtain a new Bratteli diagram which is equivalent to the old one, i. e. the AF-algebras represented by them are isomorphic; in fact, they are the same here.

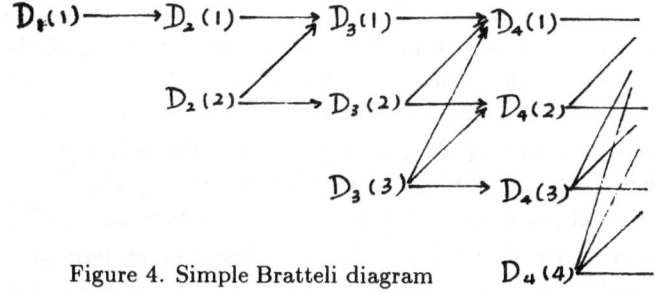

Figure 4. Simple Bratteli diagram

We may conduct the same operation for $D_{n_1}(2)$ to get a positive integer n_2 (Note: If $D_{n_1}(2)$ is empty, we may take any $n_2 > n_1$) such that all the descendants of $D_{n_1}(2)$ in D_{n_2} have no circles. Let $D_{n_2}(1)$ denote the set of descendants of D_1 in D_{n_2}, $D_{n_2}(2)$ denote the set of descendants of $D_{n_1}(2)$ in D_{n_2} except that of $D_{n_2}(1)$, and let the rest be denoted by $D_{n_2}(3)$. Delete the generations between D_{n_1} and D_{n_2}, and rearrange D_{n_2} such that the order numbers of $D_{n_2}(1), D_{n_2}(2), D_{n_2}(3)$ satisfy the relation that $D_{n_2}(1) < D_{n_2}(2) < D_{n_2}(3)$. We get a new Bratteli diagram which is equivalent to the old one. By the inductive method, we may construct a new Bratteli diagram $(D', d', \mathfrak{U}')$, $D_k' = D_{n_k}(1) \oplus D_{n_k}(2) \oplus \cdots \oplus D_{n_k}(k+1)$, such that the points in $D'_k \setminus D_{n_k}(k+1)$ have no circles. In this new Bratteli diagram, we remove D'_1 and $D_{n_k}(k+1)(k \geq 2)$ to obtain a Bratteli diagram $(D'', d'', \mathfrak{U}'')$ again, which no circles for every point in D''. See Figure 4.

From the above properties (A) and (B), it is easy to see that this Bratteli diagram is equivalent to the previous one. Q.E.D.

Proof of Theorem B. Let $(D, d, \mathfrak{U})$ be a simple Bratteli diagram for the AF-algebra A, shown as in Figure 4. Let $r(i, n)$ denote the number of elements in $D_n(i)$. Then $\sum_{i=1}^{n} r(i, n) = r(n)$ (where $r(i, n)$ may be equal to zero, in this case $D_n(i)$ and $D_m(i), m > n$ are empty sets). Starting from each point $(i, 1)(1 \leq i \leq r(1))$ of D_1, we construct a complete chain $\{x_n^{(i,1)}\}_{n=1}^{\infty}$ where $x_1^{(i,1)} = (i, 1)$. Let $J(i, 1)$ be the set of all descendants of $(i, 1)$, i. e. the ideal generated by $(i, 1)$. By Lemma 4, we have

(1) there are no points in $J(i, 1) \setminus \{x_n^{(i,1)}\}_{n=1}^{\infty}$ annihilating at the chain $\{x_n^{(i,1)}\}_{n=1}^{\infty}$.

(2) the points in $J_j(i, 1) = J(i, 1) \setminus \{x_n^{(j,1)}\}_{n=1}^{\infty}, i \neq j$ annihilate at most with one multiplicity at the chain $\{x_n^{(j,1)}\}_{n=1}^{\infty}$.

Remark 5. Since the points in $J_j(i, 1)$ are descendants of $(i, 1)$, there is at most one point in $\{J_j(i, 1) \cap D_n\} \subset D_n$ annihilating at $\{x_n^{(i,1)}\}_{n=1}^{\infty}$. Thus if there are points in $J_j(i, 1)$ annihilated at $\{x_n^{(j,1)}\}_{n=1}^{\infty}$, then there is a positive integer n_o such that there are no points in $J_j(i, 1) \cap D_n$ where $n > n_o$ annihilating at $\{x_n^{(j,1)}\}_{n=1}^{\infty}$, and in $J_j(i, 1) \cap D_{n_o}$ there is a unique point x'_{n_o} annihilating at $J_j(i, 1) \cap D_{n_o}$ with the annihilating point $x_{n_o+1}^{(j,1)}$. The points in $J_j(i, 1)$ which are annihilated at $\{x_n^{(j,1)}\}_{n=1}^{\infty}$ are ancestors of x'_{n_o}.

Since there are only finitely many elements in $D_1 = D_1(1)$, by Remark 5, there is a positive integer n_1 such that there are no points in $\cup_{n=n_1}^{\infty} D_n(1) \setminus \cup_{n=n_1}^{\infty} \{x_n^{(i,1)}\}$ which are annihilating at the chain $\{x_n^{(i,n)}\}_{n=1}^{\infty}$ constructed above for each $i : 1 \leq i \leq r(1)$. Let $D'_{n_1} = D_{n_1} \setminus \{x_{n_1}^{(i,1)}\}_{n=1}^{r(1)}$. By the choice of n_1, the chains starting from any points of $D'_{n_1}(1)$ do not intersect the chains $\{x_n^{(i,1)}\}_{n=1}^{\infty}, 1 \leq i \leq r(1)$. Proceeding inductively, we can construct a complete chain $\{x_n^{(i,n_1)}\}_{n=n_1}^{\infty}$ from each point in $D'_{n_1}, x_{n_1}^{(i,n_1)} = (i, n_1) \in D'_{n_1}$ and there is a positive integer n_2 such that there are no points in $\cup_{n=n_2}^{\infty} \cup_{k=1}^{n_1} D_n(k) \setminus \cup_{n=n_1}^{\infty} \{x_n^{(i,n)}\}$ annihilating at the chains $\{x_n^{(i,n_1)}\}_{n=n_1}^{\infty}$. Therefore, we have got a sequence of positive integers $\{n_k\}_{k=0}^{\infty}, n_o = 1$, chains $\{x_n^{(i,n_k)}\}_{n=n_k}^{\infty}, x_n^{(i,n_k)} = (i, n_k)$ and D'_{n_k} where $(i, n_k) \in D'_{n_k}$. Moreover, there are no points in $\cup_{n=n_{k+1}}^{\infty} \cup_{s=1}^{n_k} D_n(s)$ annihilating at all the chains constructed above starting from the points of $\cup_{j=0}^{k} D_{n_j}$. Here we make the convention that for the points in the chain itself, we say that these points does not annihilate at the chain.

If we delete the generations between n_k and n_{k+1}, we get a new Bratteli diagram which is equivalent to the old one. This new diagram has the following property:

There are complete chains from every point $D_k = D_{n_k}$ such that there are no points in $\cup_{j=k}^{\infty} \cup_{i=1}^{n_k} D_{n_j}(i)$ annihilating at these chains.

Now we are in a position to construct a finite scale for the dimension group $(G(A), G(A)_+)$, to construct a new function $d' : D \to N$ satisfying:

$$d'((j, n+1)) \geq \sum_{i=1}^{r(n)} \beta_{ji}^{(n)} d'((i, n)), \qquad \forall 1 \leq j \leq r(n+1) \tag{1'}$$

where $\beta_{ji}^{(n)}$ are the elements of the matrix u_n, $u_n = (\beta_{ji}^{(n)})$ such that the scale generated by d' is finite.

Set $d'(i, 1) = 1, 1 \leq i \leq r(1), d'(i, 2) = 1, 1 \leq i \leq r(2)$.

Set $d'(i, 3) = 1$, if $(i, 3) \in D_3(2) \cup D_3(3)$. From Figure 5, we know that the matrix u_2 can be written in the form $u_2 = \begin{bmatrix} u_{11}^2, u_{12}^2 \\ 0, u_{22}^2 \\ 0, 0 \end{bmatrix}$, where u_{ij}^2 is an $r(3, i) \times r(2, j)$-matrix.

Take an $r(2)$-vector $\alpha^2 = [a_1, a_2, \cdots, a_{r(2)}]$, when $1 \leq i \leq r(2, 1), a_i = d'(i, 2) = 1$, when $i > r(2, 1), a_i = 0$. Let the matrix u_2 act on α^2. We get an $r(3)$-vector b such that only the first $r(3, 1)$ elements are not equal to zero. Let $d'(i, 3) = b_i, 1 \leq i \leq r(3, 1)$. Suppose that $d'(i, n)$ has been constructed which satisfies $d'(i, n) = 1$ if $(i, n) \in D_n(n-1) \cup D_n(n)$. The matrix u_n is $r(n+1) \times r(n)$ and it can be decomposed as $u_n = (u_{ij}^n)$ where u_{ij}^n are $r(n+1, i) \times r(n, j)$-matrix, moreover, $u_{ij}^n = 0$ if $i > j, 1 \leq i \leq n+1, 1 \leq j \leq n$. Take an $r(n)-$ vector $\alpha^n = [a_1, a_2, \cdots, a_{r(n)}]$, where $a_i = 0$ if $i > \sum_{j=1}^{n-1} r(n, j), a_i = d'(i, n)$ if $1 \leq i \leq \sum_{j=1}^{n-1} r(n, j)$. Then $b_n = u_n \alpha^n$ is an $r(n+1)-$ vector. Let $d'(i, n+1) = b_i$ if $1 \leq i \leq \sum_{j=1}^{n-1} r(n+1, j)$; $d'(i, n+1) = 1$ if $i > \sum_{j=1}^{n-1} r(n+1, j)$. Thus we have constructed a new function $d' : D \to N$. If we delete the $D_n'(n)$ of the Bratteli diagram in Figure 5, we get a new Bratteli diagram D'. Restricting the function d' to D' which satisfies the equality (1'), it is easy to verify that d' restricted to the complete chains constructed in the preceding paragraph are bounded functions. In fact, the value of d' on the chain $\{x_{n_o}^{(i, n_o)}\}_{n=n_o}^{\infty}$ starting from (i, n_o) does not depend on n. For example, let $\{x_n^{(i,1)}\}_{n=1}^{\infty}$ be the chain starting from $(i, 1)$, where $d'(i, 1) = 1, d'(x_2^{(i,1)}) = 1$. By the construction of d', we know that the points which affect the value of $d'(x_{n+1}^{(i,1)})$ are in $D_n(i), 1 \leq i \leq n-1$. But the ideal generated by $D_n(i) \setminus \{x_n^{(i,1)}\}$ does not intersect the chain $\{x_n^{(i,1)}\}_{n=n_o}^{\infty}$, i.e. $\beta_{ji}^n = 0, (i, n) \in D_n(i) \setminus \{x_n^{(i,1)}\}, (j, n+1) = x_{n+1}^{(i,1)}$. Therefore, $d(x_{n+1}^{(i,1)}) = d(x_{n_o}^{(i,1)})$. Therefore, the scale generated by d' is finite which completes the proof of Theorem B.

Remark 6. We can prove Theorem A by a similar method with the help of the characterization by the Bratteli diagrams for the liminal AF-algebras given in [7]. We omit the details.

References

[1] O.Bratteli, Inductive limits of finite dimensional C^*-algebras, Trans. Amer. Math. Soc., 171(1972), 195–234.

[2] J.Cuntz and G.K.Pedersen, Equivalence and traces on C^*-algebras, J. Funct. Anal., 33(1979), 135–164.

[3] E.Effros, Dimensions and C^*-algebras, CBMS regional Conference Series in Math., AMS Vol.46, (1981).

[4] G.Elliott, On the classification of inductive limits of sequences ofsemisimple finite dimensional algebras, J.Algebra, 38(1976), 29–44.

[5] Z.B.Huang, Classification of AF-algebras and their dimension groups, Chin. Ann. of Math., 15B, 2(1994), 181–192.

[6] Z.B.Huang, Classification of AF-algebras and their dimension groups (II), to appear.

[7] A.J.Lazar and D.C.Taylar, Approximately finite dimensional C^*-algebras and Bratteli diagrams, Trans. AMS., 259(1980), 599–619.

[8] G.K.Pedersen, C^*-Algebras and their Automorphism Groups, Academy Press, London New York San Francisco, (1979).

The $(\mathcal{U} + \mathcal{K})$-Orbit of Essentially Normal Operators and Compact Perturbation of Strongly Irreducible Operators

Youqing Ji, Chunlan Jiang

Department of Mathematics, Jilin University, Changchun

Zhongyao Wang

Department of Mathematics, East China University of Science and Techlogy, Shanghai

1. Introduction

Let $\mathcal{H}$ be a complex, separable, infinite dimensional Hilbert space. $\mathcal{L}(\mathcal{H})$ denotes the set of all bounded linear operators acting on $\mathcal{H}$. We call

$$(\mathcal{U} + \mathcal{K})(T) = \{RTR^{-1}, \ R \in (\mathcal{U} + \mathcal{K})(\mathcal{H})\},$$

the $(\mathcal{U} + \mathcal{K})$ orbit of T, where $(\mathcal{U} + \mathcal{K})(\mathcal{H}) = \{R : R$ is invertible of the form unitary plus compact$\}$. $T \simeq A$ denotes $A \in (\mathcal{U} + \mathcal{K})(T)$. $\underset{U+K}{\simeq}$ is an equivalent relation. An operator T is strongly irreducible, i.e., $T \in (SI)$, if it does not commute with any nontrivial idempotent. An operator T is essentially normal if the self-commutor $[T, T^*] = T^*T - TT^*$ is compact. An operator T is called shift-like, if T can be written as a sum of a unilateral shift operator and a compact operator and $\sigma(T) = \bar{\mathcal{D}}$, dimker$(\lambda - T) = 0$, $\lambda \in \mathcal{D}$, where $\sigma(T)$ denotes the spectrum of T, $\mathcal{D}$ denotes the unit disk. P.S. Guinand and L. Marcoux [Mar] proved the following

Theorem G.M. Let T_1 and T_2 be shift-like and ϵ a positive number. Then there exist compact operators K_1 and K_2, $\|K_1\| < \epsilon$, $\|K_2\| < \epsilon$ such that

$$T_1 + K_1 \underset{\mathcal{U} + \mathcal{K}}{\simeq} T_2, \ T_2 + K_2 \underset{\mathcal{U} + \mathcal{K}}{\simeq} T_1.$$

In this paper, we shall strengthen the result of Guinand and Marcoux.

1991 Mathematics Subject Classification: 41A10, 47A15.

Our main results are the following theorems.

Theorem 1. Let Ω be an analytic Jordan region and

$$\mathcal{A}(\Omega, n) = \{T \in \ \mathcal{L}(\mathcal{H}) \text{ such that}$$

> 1. $\sigma(T) = \bar{\Omega}, \ \Omega \subset \rho_F(T)$,
>
> 2. $\text{ind}(T - \lambda) = -n, \text{dimker}(T - \lambda) = 0 \ \lambda \in \Omega$,
>
> 3. T is essentially normal.$\}$

where $\rho_F = \{\lambda; \ T - \lambda \text{ is Fredholm}\}$. Then for T_1 and T_2 in $\mathcal{A}(\Omega, n)$ and $\epsilon > 0$, there exist compact operators K_1 and K_2, $\|K_1\| < \epsilon$, $\|K_2\| < \epsilon$ satisfying that $T_1 + K_1 \simeq_{\mathcal{U}+\mathcal{K}} T_2$, $T_2 + K_2 \simeq_{\mathcal{U}+\mathcal{K}} T_1$.

Theorem 2. Let $T \in \mathcal{A}(\Omega, n)$. Then

$$\overline{(\mathcal{U} + \mathcal{K})(T)} = \{T \in \ \mathcal{L}(\mathcal{H}) \text{ such that}$$

> 1. A is essentially normal,
>
> 2. $\sigma(A) = \bar{\Omega}, \ \Omega \subset \rho_F(A)$,
>
> 3. $\text{ind}(A - \lambda) = -n, \ \lambda \in \Omega\}$.

Theorem 3. Let $T \in \mathcal{A}(\Omega, n)$ and ϵ be a positive number. Then there exists a compact K, $\|K\| < \epsilon$ such that $T + K \in (SI)$.

Theorem 3 answers partially an interesting question posed by D.A. Herrero: For an essential operator T with connected spectrum, is there a compact K such that $T + K \in (SI)$?

2. Proofs of Theorem 1 and Theorem 2

Let Ω be an analytic Jordan region. Then the Riemann Mapping Theorem asserts: Given z_0 in Ω, there exists an analytic isomorphism ϕ of Ω with unit disk $\mathcal{D}$, such that $\phi(0) = z_0$. By Schwarz Reflection Theorem, it is easily seen that ϕ has an analytic continuation to $\overline{\mathcal{D}}$. In the following statement, ϕ is always an analytic isomorphism of $\bar{\Omega}$ with $\bar{\mathcal{D}}$. It is obvious that $\phi \in \mathcal{H}^{(\infty)}(\mathcal{D})$. Let T_ϕ be a Toeplitz operator with symbol ϕ. Then $T_\phi^* \in B_1(\Omega_1^*)$ and $\sigma(T_\phi^*) = \bar{\Omega}^*$, where $\Omega^* = \{\lambda; \ \bar{\lambda} \in \Omega\}$; $B_n(\Omega)$ denotes the set of Cowen-Douglas operators of index n, i.e., if $B \in B_n(\Omega)$, then B satisfies
i) $\sigma(B) \supset \Omega$,
ii) $\text{dimker}(\lambda - B) = \text{ind}(\lambda - B) = n; \ \lambda \in \Omega$,

iii) $\vee\{\ker(\lambda - B); \ \lambda \in \Omega\} = \mathcal{H}$. And iii) can be replaced by (iii') $\vee\{\ker(\lambda_0 - B)^k; \ \lambda_0 \in \Omega, \ k = 1, 2, ...\} = \mathcal{H}$ (see [Cow-Dog]).

To prove Theorem 1 and Theorem 2, we shall need the following auxiliary results.

Lemma 2.1. Let ϕ be a bounded univalent analytic function on $\mathcal{D}$ and $\mathcal{M} = (\ker(T_\phi - \lambda)^*)^\perp$, $\lambda \in \Omega$. Then $T_\phi|_\mathcal{M}$ is unitary equivalent to T_ϕ ($T_\phi|_\mathcal{M} \simeq T_\phi$), where $T_\phi|_\mathcal{M}$ denotes a restriction of T_ϕ on $\mathcal{M}$.

Proof. Since $\mathcal{M}$ is a hyperinvariant subspace of T_ϕ and $\{T_\phi\}'$, the commutant of T_ϕ, is $H^\infty(\partial\mathcal{D})$, then $\mathcal{M}$ is an invariant subspace of T_z. By Berurling Theorem [Dog], we can find an inner function g such that $\mathcal{M} = T_g H^2(\partial\mathcal{D})$. Since $T_\phi^* \in B_1(\Omega^*)$, $g = \frac{z - a}{1 - \bar{a}z}$, $0 \le |a| < 1$. Thus T_g is a unitary operator from $H^2(\partial\mathcal{D})$ to $\mathcal{M}$. For $f \in H^2(\partial\mathcal{D})$,

$$T_g^*(T_\phi|_\mathcal{M})T_g f = T_\phi f.$$

This shows that $T_g^*(T_\phi|_\mathcal{M})T_g = T_\phi$.

Lemma 2.2. Let ϕ be a bounded univalent analytic function on $\mathcal{D}$ and $\Omega = \text{essran}\phi$. For $\lambda \in \Omega$, set

$$T = \begin{pmatrix} \lambda & 0 \\ D & T_\phi \end{pmatrix}_{H^2(\partial\mathcal{D})}^{\mathcal{C}}.$$

Then $T \underset{\mathcal{U}+\mathcal{K}}{\simeq} T_\phi$, where $D = \alpha e_0 \otimes 1$ is a rank one operator and $e_0 \in \ker(T_\phi - \lambda)^*$, $\|e_0\| = 1$.

Proof. Since $e_0 \in \ker(T_\phi - \lambda)^* = \text{ran}(T_\phi - \lambda)^\perp$, by Lemma 2.1,

$$T_\phi = \begin{pmatrix} \lambda & 0 \\ g_1 & T_\phi|_\mathcal{M} \end{pmatrix}_{\mathcal{M}=\ker(T_\phi-\lambda)^{*\perp}}^{\ker(T_\phi-\lambda)^*=\mathcal{C}} \cong \begin{pmatrix} \lambda & 0 \\ g_2 & T_\phi \end{pmatrix},$$

where g_1 and g_2 are rank one operators, $g_2 = f \otimes 1$, $f \in H^2(\partial\mathcal{D})$.

Since $e_0 \notin \text{ran}(T_\phi - \lambda)$ and T_ϕ is strongly irreducible, we can assert that $f = \beta e_0 + g_3$, where $g_3 \in \text{ran}(T_\phi - \lambda)$; $\beta \ne 0$. Thus we can find g_4 in $H^2(\partial\mathcal{D})$ such that $g_3 = (T_\phi - \lambda)g_4$. Setting $\alpha_1 = g_4 \otimes 1$, $X = \begin{pmatrix} 1 & 0 \\ D & T_\phi \end{pmatrix}$, then $X \in (\mathcal{U} + \mathcal{K})(\mathcal{H})$ and

$$X \begin{pmatrix} \lambda & 0 \\ g_2 & T_\phi \end{pmatrix} X^{-1} = \begin{pmatrix} \lambda & 0 \\ \beta e_0 \otimes 1 & T_\phi \end{pmatrix} \underset{\mathcal{U}+\mathcal{K}}{\simeq} \begin{pmatrix} \lambda & 0 \\ D & T_\phi \end{pmatrix}.$$

By the same argument, we have

Lemma 2.3. Assume that Ω and ϕ are as given in Lemma 2.2 and $g \notin \text{ran}(T_\phi - \lambda)$. Then

$$\begin{pmatrix} F & 0 \\ g \otimes 1 & T_\phi \end{pmatrix} \underset{\mathcal{U}+\mathcal{K}}{\simeq} T_\phi; \ \lambda \in \Omega.$$

Lemma 2.4. Let ϕ and Ω be given by Lemma 2.2. Then for a positive number ϵ, there exists a compact operator K_ϵ, $\|K_\epsilon\| < \epsilon$ such that

$$\begin{pmatrix} \lambda & 0 \\ C_1 & T_\phi \end{pmatrix}_{H^2(\partial D)}^{C^n} + K_\epsilon \overset{\simeq}{\underset{U+K}{\longrightarrow}} T_\phi,$$

where F is an $n \times n$ matrix and $\sigma(F) \subset \Omega$, $C \in \mathcal{L}(C^n, H^2(\partial D))$.

Proof. We shall proceed by induction on the dimension of the space on which F acts. Thus we begin by considering the case $n = 1$. We then have

$$T = \begin{pmatrix} \lambda & 0 \\ C_1 & T_\phi \end{pmatrix}_{H^2(\partial D)}^{C^n},$$

where $C_1 = f \otimes \alpha$ is a rank one operator. Choose a g in $H^2(\partial D)$ so that $f + g \notin \mathrm{ran}(T_\phi - \lambda)$ and $\|g\| < \epsilon$. Setting $K = \begin{pmatrix} 0 & 0 \\ g \otimes \alpha & 0 \end{pmatrix}$, then $\|K\| < \epsilon$. From Lemma 3.3, $T + K \overset{\simeq}{\underset{U+K}{\longrightarrow}} T_\phi$. We now assume that the result is true when F is an $(n-1) \times (n-1)$ matrix whose eigenvalues are in Ω. We shall prove the lemma to be true when F is an $n \times n$ matrix. It is obvious that there exists a unitary operator U such that

$$UTU^* = \begin{pmatrix} \lambda_1 & 0 & 0 \\ * & F_{n-1} & 0 \\ * & C_{n-1} & T_\phi \end{pmatrix},$$

where $F_{n-1} \in \mathcal{L}(C^{n-1})$, $\lambda_1 \in \sigma_p(F)$. According to our induction step, we can find a compact K_1 with $\|K_1\| < \epsilon/2$ such that

$$X_1(UTU^* + K_1)X_1^{-1} = \begin{pmatrix} \lambda_1 & 0 \\ h & T_\phi \end{pmatrix},$$

where $X_1 \in (\mathcal{U} + \mathcal{K})(\mathcal{H})$. Thus we can find a compact K_2 with $\|K_2\| < \dfrac{\epsilon}{2\|X_2\|\|X_2^{-1}\|}$ such that

$$X_2\left(\begin{pmatrix} \lambda_1 & 0 \\ h & T_\phi \end{pmatrix} + K_2\right) X_2^{-1} = T_\phi.$$

This complete the proof of Lemma 2.4.

Lemma 2.5. Let A and B be in $\mathcal{L}(\mathcal{H})$, $\tau_{A,B}$ denotes Rosenblum operator, i.e., $\tau_{AB}(X) = AX - XB$, $X \in \mathcal{L}(\mathcal{H})$. Set $\tau = \tau_{A,B}|_{\mathcal{K}(\mathcal{H})}$. Then $\tau^* = -\tau_{B,A}|_{\mathcal{C}_1}$, where τ^* is the adjoint operator of τ, $\mathcal{K}(\mathcal{H})$ and $\mathcal{C}_1$ denote the ideals of compact operator and trace class operator respectively.

Proof. Set $\mathcal{M} = \mathcal{K}(\mathcal{H})$. Then $\mathcal{M}$ is a Banach space and $\mathcal{M}^*$, the dual space of $\mathcal{M}$, is $\mathcal{C}_1$. For X_1 in $\mathcal{K}(\mathcal{H})$ and X_2 in $\mathcal{C}_1$, a computation shows that

$$
\begin{aligned}
(X_1, \tau^* X_2) &= \text{tr}(\tau X_1 \cdot X_2) \\
&= \text{tr}((AX_1 - X_1 B) X_2) = \text{tr}(AX_1 X_2) - \text{tr}(X_1 B X_2) \\
&= \text{tr}(X_2 A X_1) - \text{tr}(B X_2 X_1) \\
&= \text{tr}((X_2 A - B X_2) X_1).
\end{aligned}
$$

This show $\tau^* = -\tau_{B,A}|_{\mathcal{C}_1}$.

Lemma 2.6 Let ϕ be a bounded univalent analytic function on $\mathcal{D}$. Then $\text{ran}(\tau_{T_\phi, T_\phi}|_{\mathcal{K}(\mathcal{H})})$ is dense in $\mathcal{K}(\mathcal{H})$.

Proof. It suffices to show that $\ker \tau^* = \{0\}$. If $X \in \mathcal{C}_1 \cap \ker \tau^*$, this shows that $X \in \{T_\phi\}' = \{T_z\}'$. Note that there are no non-zero compact operators in $\{T_\phi\}'$, hence $X = 0$.

Lemma 2.7. Let $\mathcal{A}(n, \Omega)$ be given by Theorem 1 and ϕ be a bounded univalent analytic function on $\mathcal{D}$ such that $\Omega =$ essranϕ. Then for T in $\mathcal{A}(n, \Omega)$ and positive number ϵ, there exist a compact K with $||K|| < \epsilon$ such that $T + K_1 \underset{U+K}{\tilde{\sim}} \oplus_1^n T_\phi$.

Proof. Note that $\{e^{in\theta}\}_{n=0}^\infty$ is an ONB of $H^2(\partial \mathcal{D})$ and T_ϕ has a lower triangular matrix representation with respect to $\{e^{in\theta}\}_{n=0}^\infty$. Set $\mathcal{M}_m = \vee\{e^{in\theta}, \ n = 0, 1, ..., n\}$, and let P_m be the orthoghonal projection on $\oplus_1^n \mathcal{M}_m$. By BDF Theorem [Brow-Dog-Fill], we can find a unitary U and a compact K such that $UTU^* = \oplus_1^n T_\phi + K_0$. Pick a natural number m such that $||P_m K_0 P_m - K_0|| < \epsilon/8$. Setting $K_1 = P_m K_0 P_m - K_0$ and using Lemma 2.1, we can find a unitary U_1 such that

$$
U_1 (UTU^* + K_1) U_1^* = \begin{pmatrix} F & 0 & \cdots & \cdots & 0 \\ C_1 & T_\phi & & & \\ C_2 & & \ddots & & \\ \vdots & & & \ddots & \\ C_n & 0 & \cdots & \cdots & T_\phi \end{pmatrix},
$$

where F is an $nm \times nm$ matrix and C_k is a rank finite operator. By the upper semicontinuity of the spectrum, $\sigma(F) \subset \bar{\Omega}_{\epsilon/8} := \{\lambda \in \mathcal{C}; \text{dist}(\lambda, \bar{\Omega}) < \epsilon/8\}$. Thus we can perturb F by at most $\epsilon/4$ to obtain a new matrix F' satisfying $\sigma(F') \subset \Omega$. Set $C = F' - F$; then $||C|| < \epsilon/4$.

Thus we can find a compact K_2, $||K_2|| < \epsilon/4$ such that

$$U_1(UTU^* + K_1)U_1^* + K_2 = \begin{pmatrix} F' & 0 & \cdots & \cdots & 0 \\ C_1 & T_\phi & & & 0 \\ C_2 & & \ddots & & \\ \vdots & & & \ddots & \\ C_n & 0 & \cdots & \cdots & T_\phi \end{pmatrix} = A_1.$$

By Lemma 2.4, there exists X_1' in $(\mathcal{U} + \mathcal{K})(\mathcal{H})$ such that $X_1' \begin{pmatrix} F' & 0 \\ C_1 & T_\phi \end{pmatrix} X_1'^{-1} = T_\phi$. Thus we can find a X_1 in $(\mathcal{U} + \mathcal{K})(\mathcal{H})$ such that

$$X_1 A_1 X_1^{-1} = \begin{pmatrix} T_\phi & 0 & 0 & \cdots & 0 \\ C_1' & T_\phi & & & 0 \\ \vdots & & & \ddots & \\ C_{n-1}' & 0 & \cdots & \cdots & T_\phi \end{pmatrix} = A_2,$$

where C_k' is a rank finite operator. Applying Lemma 2.6 to each C_k', we can find compact operators D_k and E_k such that

$$T_\phi E_k - E_k T_\phi = B_k + D_k,$$

and

$$||D_k|| \leq \frac{\epsilon}{||X_1||||X_1^{-1}||8^k}.$$

Set

$$K_3 = \begin{pmatrix} 0 & 0 & \cdots & 0 \\ D_1 & 0 & \cdots & 0 \\ \vdots & \vdots & \ddots & \vdots \\ D_{n-1} & 0 & \cdots & 0 \end{pmatrix},$$

$$K_2 = \begin{pmatrix} I & & & \\ E_1 & I & & \\ \vdots & & \ddots & \\ E_{n-1} & \cdots & \cdots & I \end{pmatrix}.$$

Then K_3 is compact with $||K_3|| < \dfrac{\epsilon}{||X_1||||X_1^{-1}||8}$ and $X_3 \in (\mathcal{U} + \mathcal{K})(\mathcal{H})$. A computation shows that

$$X_3(A_2 + K_3)X_3^{-1} = \oplus_1^n T_\phi.$$

This complete the proof of the lemma.

We are now in a position to prove Theorem 1.

For an analytic Jordan region Ω, we can find a univalent analytic function ϕ on $\bar{\mathcal{D}}$ such that $\phi(\bar{\mathcal{D}}) = \bar{\Omega}$. Furthermore, we can find a univalent analytic function φ and positive numbers δ_1 and δ_2 such that $\varphi = \phi^{-1}$ and $\varphi(\Omega_{\delta_1}) = \mathcal{D}_{\delta_2}$, where $\mathcal{D}_{\delta_2} = \{z; \text{dist}(\lambda, \mathcal{D}) < \delta_2\}$. If $T \in \mathcal{A}(\Omega, n)$, then $\varphi(T)$ is n-shift like.

An operator A is called n-shift like, if A satisfies the following properties.

a) A is essentially normal,

b) $\sigma(A) = \bar{\mathcal{D}}$, $\rho_F(A) \supset \mathcal{D}$,

c) $\text{ind}(\lambda - A) = -\text{dimker}(\lambda - A) = -n$, $\lambda \in \mathcal{D}$.

For T in $\mathcal{A}(\Omega, n)$, set $A = \varphi(T)$, $B = \oplus_1^n T_z$ and set $m_1 = \sup\{|\phi(z)|; |z| = 1 + z_2\}$; $m_2 = \sup\{\|(z - B)^{-1}\|; |z| = 1 + z_2\}$. From Theorem 2 of [Jia-Cao-Sun], there exist compact K and X in $(\mathcal{U} + \mathcal{K})(\mathcal{H})$ such that

$$X(B + K)X^{-1} = A$$

and $\|K_1\| < \min\{\frac{1}{(m+1)M_2}, \delta_2\}$, where m is a positive number with $\frac{m_1 m_2(1 + \delta_2)}{m} < \epsilon$. This shows that $\sigma(B + K_1) = \bar{\mathcal{D}}$. Thus

$$\phi(X(B + K_1)X^{-1}) = X\phi(B + K)X^{-1} = \phi(A) = \phi\varphi(T) = T.$$

$$K' = \phi(B + K) - \oplus_1^n T_\phi = \phi(B + K) - \phi(B)$$

$$= \frac{1}{2\pi i} \int_{|z|=1+\delta_2} \phi(z)((z - B - K)^{-1} - (z - A)^{-1})dz$$

$$= \frac{1}{2\pi i} \int_{|z|=1+\delta_2} \phi(z) \sum_{n=1}^{\infty} ((z - B)^{-1}K)^n (z - A)^{-1})dz,$$

$$\|K'\| \leq \frac{2\pi m_1(1 + \delta_2)}{2\pi} m_2 \sum_{n=1}^{\infty} \|(z - B)^{-1}\|^n \|K\|^n$$

$$= m_1 m_2 (1 + \delta_2) \sum_{n=1}^{\infty} \left(\frac{1}{m+1}\right)^n$$

$$= \frac{m_1 m_2 (1 + \delta_2)}{m} < \epsilon.$$

Since K is compact, K' is compact. This shows that $\oplus_1^n T_\phi + K' \underset{\mathcal{U}+\mathcal{K}}{\simeq} T$.

By the above argument and Lemma 2.7, for T_1 and T_2 in $\mathcal{A}(\Omega, n)$, we can find a compact operator K_1 with $\|K_1\| < \epsilon/2$ such that $X_1(T + K_1)X_1^{-1} = \oplus_1^n T_\phi$, where $X_1 \in (\mathcal{U} + \mathcal{K})(\mathcal{H})$, and we can find a compact operator K_2 with $\|K_2\| \leq \epsilon/2\|X_1\|\|X_1^{-1}\|$ such that $\oplus_1^n T_\phi + K_2 \simeq T_2$. Therefore $T_1 + K_1 + X_1^{-1}K_2 X_1 \underset{\mathcal{U}+\mathcal{K}}{\simeq} T_2$ and $\|K_1 + X_1^{-1}K_2 X_1\| < \epsilon$. Similarily, there exists a compact K' with $\|K'\| < \epsilon$ such that $T_2 + K' \underset{\mathcal{U}+\mathcal{K}}{\simeq} T_1$. This complete the proof of Theorem 1.

For T in $\mathcal{L}(\mathcal{H})$, setting $\text{minind}T = \min(\text{dimker}T, \text{dimker}T^*)$, we have

Lemma 3.8 ([Her, Lemma 3.46]). Let $\sigma(T)$ be a perfect set and ϵ a positive number. Then there exists a compact K with $\|K\| < \epsilon$ such that $\mathrm{minind}(T + K - \lambda) = 0$, $\lambda \in \rho_F(T)$.

By Lemma 3.8 and Theorem 1, we obtain immediately Theorem 2.

3. Proof of Theorem 3

Lemma 3.1 ([Jzz-Sun] Theorem 2.8]). Let A be a strongly irreducible operator with $\|A\| \leq 1$ and ϕ be a univalent analytic function on $\bar{\mathcal{D}}$. Then $\phi(T)$ is strongly irreducible.

Lemma 3.2 [Jia-Cao-Sun]. Let ϵ be a positive number. Then there is a compact K with $\|K\| < \epsilon$ such that $T_{z^n} + K \in (SI)$.

By Lemma 3.1 and Lemma 3.2, we have

Lemma 3.3. Let Ω be an analytic connected Cauchy region and n a natural number. Then there exists a strongly irreducible operator A in $\mathcal{A}(\Omega, n)$.

Now, we begin the proof of Theorem 3. Let $T \in \mathcal{A}(\Omega, n)$. By Lemma 3.3, there exists a strongly irreducible A in $\mathcal{A}(\Omega, n)$. From Theorem 1, we can find a compact K with $\|K\| < \epsilon$ such that $T + K \underset{\mathcal{U}+\mathcal{K}}{\tilde{\approx}} A$. This complete the proof of Theorem 3.

References

[Bro-Dog-Fil] L.G. Brown, P.G. Douglas and P.F. Fillmore, Unitary equivalence modelo the compact operators and extensions of C^*-algebras, Probleedings of conference on operator theory, Halifax, Nova scotia 1973 58-128.

[Dou] R.G. Douglas, Banach algebras techniques in operator theory. (New York and London: Academic Press, 1972).

[Her] D.A. Herrero, Approximation of Hilbert space operators, Vol 1 and ed., Pitman Res. Notes Mayh. 224 (1990).

[Jia-Cao-Sun] C.L. Jiang, G.F. Cao and S.L. Sun, A completely irreducible compact perturbation of a class of essentially normal operator, Northeast Math. J. 10(4), 1994, 427-433.

[JZJ-Sun] Z.J. Jiang and S.L. Sun, On completely irreducible operator, Acta Scientiaum Naturalium University, Jilinonsis 4(1992) 20-29.

[Mar] L. Marcoux, The closure of the $(\mathcal{U}+\mathcal{K})$-orbit of shift-like operators, Indiana University Math. J. 41(4)(1992).

The Complete Irreducibility of Analytic Toeplitz Operators

Chunlan Jiang, Guangfu Cao and Shanli Sun

Department of Mathematics, Jilin University, Changchun

1. Introduction

Let $\mathcal{L}(H)$ denote the algebra of all bounded linear operators acting on a complex, separable, infinite dimensional Hilbert space H, $\mathcal{D}$ denote the unit disc in the complex plane $\mathcal{C}$, $\mathcal{T}$ be the boundary of $\mathcal{D}$, i.e., $\mathcal{T} = \{z \in \mathcal{C}, |z| = 1\}$, $H^2(\mathcal{T})$ the Hardy space on $\mathcal{T}$, $H^\infty(\mathcal{D})$ the space of bounded analytic functions on $\mathcal{D}$ and $\mathcal{A}(\mathcal{D})$ the disc algebra. For T in $\mathcal{L}(H)$, $\sigma(T)$ denotes the spectrum of T and $\sigma_e(T)$ the essential spectrum of T, $\rho_F(T) = \mathcal{C} \setminus \sigma_e(T)$, $\sigma_r(T) = \{\lambda \in \mathcal{C}; (\lambda - T)$ is not right invertible$\}$, $\sigma_l(T) = \{\lambda \in \mathcal{C}; (\lambda - T)$ is not left invertible$\}$, $\rho_r(T) = \mathcal{C} \setminus \sigma_r(T)$, $\rho_l(T) = \mathcal{C} \setminus \sigma_l(T)$. For ϕ in $H^\infty(\mathcal{D})$, define $T_\phi f = \phi f$, $\forall f \in H^2(\mathcal{T})$. T_ϕ is called an analytic Toeplitz operator with the symbol ϕ. $\{T_\phi\}'$ denotes the commutant of T_ϕ.

An operator is said to be completely irreducible, if it does not commute with any non-trivial idempotent operator (see [JZJ], [Gil]).

Z.J. Jiang raised the following question.

Question 1. Which in analytic Toeplitz operators are completely irreducible?

In Section 2, we obtain a characterization of the complete irreducibility of some analytic Toeplitz operators.

P.R. Halmos proved that each operator can be written as a sum of an irreducible operator and a compact operator (see [Hal]).

A natural question is posed by D.A. Herrero.

Question 2. Can each operator with connected spectrum be written as a sum of a completely irreducible operator and a compact operator?

1991 Mathematics Subject Classification: 41A10, 47A55

For essentially normal operators with connected spectrum, C.L. Jiang, S.H. Sun and Z.Y. Wang [Jia-Sun-Wan] proved that the answer to Question 2 is "yes".

In Section 3, we obtain the following result.

Theorem JCS. Let $\phi \in \mathcal{A}(\mathcal{D})$ and ϵ be a positive number. Then there exists a compact operator K with $\|K\| < \epsilon$ such that $T_\phi + K$ is completely irreducible.

2. Complete irreducibility of analytic Toeplitz operators

Theorem 2.1. Let $\phi \in \mathcal{A}(\mathcal{D})$ and $\mathrm{clos}\phi(\mathcal{D}) \neq \phi(\mathcal{T})$. Then T_ϕ is completely irreducible if and only if $\{T_\phi\}' = \{T_z\}'$.

Proof. The "if" part is obvious. It is sufficient to prove the "only if" part. If T_ϕ is completely irreducible, since $\mathrm{clos}\phi(\mathcal{D}) \neq \phi(\mathcal{T})$, $\sigma(T_\phi) \setminus \sigma_e(T_\phi) \neq \emptyset$. Without loss of generality, we can assume that $0 \in \sigma(T_\phi) \setminus \sigma_e(T_\phi)$, i.e., T_ϕ is a Fredholm operator but T_ϕ is not invertible. Thus, $H^2(\mathcal{T}) \ominus \phi H^2(\mathcal{T})$ is finite dimensional. ¿From Beurling's Theorem [Dou], we can find a finite Blaschke product χ and an outer function F such that $\phi = \chi \cdot F$. From Theorem 5 of [Cow], there exists a finite Blaschke product g such that $\phi(z) = f(g(z))$. Using the Main Theorem of [Tho], we have

$$\{T_\phi\}' = \{T_\chi\}' \cap \{T_F\}' = \{T_g\}'.$$

Since T_ϕ is completely irreducible, $\{T_\phi\}'$ and, furthermore, $\{T_g\}'$ contain no nontrivial idempotent operators. This shows that T_g is completely irreducible. Since g is a finite Blaschke product, $\{T_g\}' = \{T_z\}'$.

Theorem 2.2. Let ϕ be in $H^\infty(\mathcal{D})$, χ and F be the inner and outer factors of ϕ, respectively, i.e., $\phi = \chi \cdot F$. If $\chi(z) = z^n$, $n \geq 1$, and $F(z) = a_0 + a_{p_1} z^{p_1} + ... + a_{p_k} z^{p_k} + ...$, $a_{p_k} \neq 0, 1, 2, ...$, then T_ϕ is completely irreducible if and only if there exists a natural number k_0 such that

$$g \cdot c \cdot d(n, p_1, p_2, ..., p_{k_0}) = 1,$$

where $g \cdot c \cdot d(n, p_1, p_2, ..., p_{k_0})$ denotes the greatest common divisor of $(n, p_1, p_2, ..., p_{k_0})$.

Proof. "$\Rightarrow$". Assume that T_ϕ is completely irreducible. If $g \cdot c \cdot d(n, p_1, p_2, ..., p_k) \geq 2$, for each natural number k, then there exists a natural number s, $s > 1$ such that $s|n$ and $s|p_k$, $k = 1, 2, ...$. Hence, ϕ is a function of z^s and $\{T_{z^s}\}' \subset \{T_\phi\}'$. Since $s > 1$, T_{z^s} is completely reducible, which contradicts that T_ϕ is completely irreducible.

   "$\Leftarrow$". It is obvious.

Theorem 2.3. Let $f(z) = a_{p_1} z^{p_1} + ... + a_{p_k} z^{p_k} + ...$ be an entire function. Then T_f is completely irreducible if and only if there exists a natural number k such that $g \cdot c \cdot d(p_1, p_2, ..., p_k) = 1$.

Proof. By Theorem 1 of [Bak-Ded-Ull], we can find a natural number s such that $\{T_f\}' = \{T_{z^s}\}'$. Thus if T_f is completely irreducible, then T_{z^s} is completely irreducible, and $s = 1$. Hence, there exists a natural number k_0 such that

$$g \cdot c \cdot d(p_1, p_2, ..., p_{k_0}) = 1.$$

If there exists a natural number k_0 such that $g \cdot c \cdot d(p_1, p_2, ..., p_{k_0}) = 1$, since $\{T_f\}' = \{T_{z^s}\}'$, it is sufficent to prove that $s = 1$. If $s > 1$, then there exists a natural number p_m such that $s \nmid p_m$. Setting $p_{m_0} = \min\{p_m : s \nmid p_m\}$,

$$P_{m_0}(z) = \sum_{k=0}^{m_0-1} a_{p_k} z^{p_k},$$

then $P_{m_0}(z)$ is a function of z^s and $\{T_{z^s}\}' = \{T_{f-P_{m_0}}\}'$. Since $f(z) - P_{m_0}(z) = \sum_{k=m_0}^{\infty} a_{p_k} z^{p_k}$, it follows from Lemma 4 of [Ped-Won] that $s | p_{m_0}$. This contradicts the assumption of p_{m_0}.

3. Proof of Theorem JCS

Firstly, we need the following lemmas.

Lemma 3.1. Let $B \in \mathcal{L}(H)$ and $\mathcal{M} = (\ker(B - \lambda)^{*n})^{\perp}$, where $\lambda \in \rho_l(B)$ and n is a natural number. Then $B|_{\mathcal{M}}$ is similar to B and $P_{\mathcal{M}} B^*|_{\mathcal{M}}$ similar to B^*, where $P_{\mathcal{M}}$ is the orthogonal projection onto $\mathcal{M}$.

Proof. Since $\mathcal{M} = \mathrm{ran}(\lambda - B)^n$ and $\ker(\lambda - B)^n = \{0\}$, $(\lambda - B)^n$ is invertible in $\mathcal{L}(H, \mathcal{M})$. Setting $A_1 = (\lambda - B)^n$, $B_1 = B|_{\mathcal{M}}$, then $A_1^{-1} B_1 A_1 = B$, i.e., $B|_{\mathcal{M}}$ is similar to B. The second conclusion is a direct consequence of the first one.

Lemma 3.2. Let $A \in \mathcal{L}(H)$. If there exists a λ in $\rho_F(A)$ and $e_0 \notin \mathrm{ran}(\lambda - A)^*$ satisfying that $\dim \ker(\lambda - A) = \mathrm{ind}(\lambda - A) = 1$, then

$$A \sim \begin{pmatrix} \lambda & 1 \otimes e_0 \\ 0 & A \end{pmatrix} \quad \text{on } \mathcal{C} \oplus H,$$

where, as well as in the following $A \sim B$ denotes that A is similar to B.

Proof. Denoting $\mathcal{M} = \ker(\lambda - A)^\perp$, then

$$A = \begin{pmatrix} \lambda & c \\ 0 & (A^*|_{\mathcal{M}})^* \end{pmatrix} \begin{matrix} \ker(\lambda - A) \\ \mathcal{M} \end{matrix},$$

where c is a rank one operator and $A^*|_{\mathcal{M}} \sim A^*$ by Lemma 3.1. Thus

$$A \sim \begin{pmatrix} \lambda & c_1 \\ 0 & A \end{pmatrix} \begin{matrix} C \\ H \end{matrix}$$

where c_1 is a rank one operator, i.e., $c_1 = 1 \otimes f$, for some f in H. Since $e_0 \notin \mathrm{ran}(\lambda - A)^*$, $f = \alpha e_0 + (A - \lambda)^* f_1$, $f_1 \in H$. If $\alpha = 0$, then $f = (A - \lambda)^* f_1$, so $c_1 = 1 \otimes (A - \lambda)^* f_1$. A simple computation indicates that

$$A - \lambda \sim \begin{pmatrix} 0 & 0 \\ 0 & A - \lambda \end{pmatrix}.$$

This contradicts the condition $\dim\ker(\lambda - A) = 1$, thus $\alpha \neq 0$. Set

$$X = \begin{pmatrix} \alpha^{-1} & -\alpha^{-1} \otimes f_1 \\ 0 & I \end{pmatrix};$$

then

$$XAX^{-1} \sim X \begin{pmatrix} \lambda & c_1 \\ 0 & A \end{pmatrix} X^{-1} = \begin{pmatrix} \lambda & 1 \otimes e_0 \\ 0 & A \end{pmatrix}.$$

Lemma 3.3. For a given operator $A \in \mathcal{L}(H)$, let $T = \begin{pmatrix} F & c \\ 0 & A \end{pmatrix}$ on $C^n \oplus H$, where F is an $n \times n$ matrix with $\sigma(F) \subset \sigma(A) \cap \rho_F(A)$ and $\dim\ker(\lambda - A) = \mathrm{ind}(\lambda - A) = 1$ for $\lambda \in \sigma(F)$. Then for $\epsilon > 0$, there exists a compact K with $\|K\| < \epsilon$ such that

$$T + K \sim A.$$

Proof. When $n = 1$, $T = \begin{pmatrix} \lambda & c \\ 0 & A \end{pmatrix}$, where $c = \alpha \otimes f, \alpha \in C$ and f in H. If $f \in \mathrm{ran}(\lambda - A)^*$, choosing a unit vector e_0 in $\ker(\lambda - A)$, then $f + e_0 \notin \mathrm{ran}(\lambda - A)^*$. Set

$$K = \begin{pmatrix} 0 & \epsilon(\alpha \otimes e_0) \\ 0 & 0 \end{pmatrix}.$$

By Lemma 3.2, $T + K \sim A$. It is obvious that $\|K\| < \epsilon$.

If $f \notin \mathrm{ran}(A - \lambda)^*$, by Lemma 3.2, $T \sim A$.

Assuming that the conclusion of the lemma is true for $n \leq k - 1$, we shall prove that the lemma is true for $n = k$. Let $\lambda_0 \in \sigma(F)$. Then there exists a unitary operator U such that

$$U \begin{pmatrix} F & c \\ 0 & A \end{pmatrix} U^* = \begin{pmatrix} \lambda_0 & c_1 & c_2 \\ 0 & F_1 & c_3 \\ 0 & 0 & A \end{pmatrix} \begin{matrix} C \\ C^{n-1} \\ H \end{matrix} = A_1,$$

where F_1 is a $(k-1) \times (k-1)$ matrix.

By the induction assumption, there exist a compact K_1 with $||K_1|| < \epsilon/2$ and an invertible X_1 such that

$$X_1(A_1 + K_1)X_1^{-1} = \begin{pmatrix} \lambda_0 & c \\ 0 & A \end{pmatrix}.$$

Repeating the proof of the "$n = 1$" part, we can obtain the lemma.

An operator A is called a Cowen-Douglas operator of index 1, if there exists a connected open set Ω such that a) $\sigma(A) \supset \Omega$, b) $\Omega \subset \rho_F(A)$, c) $\cup\{\ker(\lambda - A), \ \lambda \in \Omega\} = H$, d) $\dim\ker(\lambda - A) = \text{ind}(\lambda - A) = 1$, $\lambda \in \Omega$. In brief, we write $A \in B_1(\Omega)$.

Now, we begin the proof of Theorem JCS.

For $\phi \in \mathcal{A}(\mathcal{D})$, it is well known that T_ϕ is essentially normal and $\text{ind}(T_\phi - \lambda) < 0$ for $\lambda \in \rho_F(T_\phi) \cap \sigma(T_\phi)$.

Let $\{\Omega_i\}_{i=1}^l$ be the components of $\rho_F(T_\phi^*) \cap \sigma(T_\phi^*)$ and $\sigma = \sigma_e(T_\phi^*)$. It is clear that σ is a connected set. Assuming that $\text{ind}(\lambda - T_\phi^*) = n_i$, $\lambda \in \Omega_i$, then by the Main Theorem of [Jia-Wan], we can find an operator E and a sequence of operators $\{B_i\}_{i=1}^l$ such that

1. $B_i \in B_1(\Omega_i)$, $i = 1, ..., l$;
2. $E = \text{diag}\{\lambda_1, \lambda_2, ...\}$, where $\{\lambda_i\}_{i=1}^\infty$ is a dense subset of σ and $\text{card}\{n, \ \lambda_n = \lambda_k\} = \infty$;
3. $T = (\oplus_1^l \oplus_1^{n_i} B_i) \oplus E$ is essentially normal;
4. The spectral picture of T is the same as that of T_ϕ^*,
5. For a positive number ϵ, there exists a compact K with $||K|| < \epsilon$ such that $T + K$ is completely irreducible.

Thus, it is suffcient to prove that for any positive number δ, there exists a compact K_1 with $||K_1|| \leq \delta$ such that $T_\phi^* + K_1$ is similar to T.

Without loss of generality, we assume that $l < +\infty$. By BDF Theorem [Bro-Dou-Fil], there exists a compact K such that $T + K = T_\phi^*$, where

$$T = (\oplus_1^l \oplus_1^{n_i} B_i) \oplus E \text{ on } (\oplus_1^l \oplus_1^{n_i} H_i) \oplus H_E.$$

Pick μ_i in Ω_i and let $\mathcal{M}_{n_i} = \oplus_1^{n_i} \ker(\mu_i - B_i)^m$, $\mathcal{M}_1 = \oplus_1^l \mathcal{M}_{n_i}$. Set $\{e_k^E\}_{k=1}^\infty$ to be an ONB of H_E such that $Ee_k^E = \lambda_k e_k^E$ and set $\mathcal{M}_2 = \cup\{e_k^E, \ k = 1, 2, ..., m\}$.

Let P_{m_1} and P_{m_2} be orthogonal projection onto $\mathcal{M}_1$ and $\mathcal{M}_2$, respectively. Set $P_m = P_{m_1} + P_{m_2}$. Then $P_m \xrightarrow{\text{SOT}} I$, $m \to \infty$. Thus there is a natural number m_0 such that

$$||P_{m_0} K P_{m_0} - K|| < \delta/8$$

and

$$T + K - P_{m_0} K P_{m_0} \simeq \begin{pmatrix} L_{11} & L_{12} & 0 \\ 0 & P_{m_01}^\perp(\oplus_1^l \oplus_1^{n_i} B_i)P_{m_01}^\perp & 0 \\ 0 & 0 & P_{m_02}^\perp E P_{m_02}^\perp \end{pmatrix},$$

where "$\cong$" denotes "unitary equivalent", L_{11} and L_{12} are rank finite operators.

By the upper continuity of the spectrum, $\sigma(L_{11}) \subset \sigma(T)_{\delta/8} = \{z; \operatorname{dist}\{z, \sigma(T)\} < \delta/8\}$. Then we can find an operator L with $\|L\| < \delta/4$ such that

$$\sigma(L_{11} + L) \subset \bigcup_{i=1}^{l} \Omega_i \cup \left(\sigma(T) \setminus \bigcup_{i=1}^{l} \partial\Omega_i \right).$$

Setting $K_1 = \begin{pmatrix} L & 0 & 0 \\ 0 & 0 & 0 \\ 0 & 0 & 0 \end{pmatrix}$, then $\|K_1\| < \delta/4$. Applying Lemma 3.1 and noting that $\dim\ker(\lambda_k - F) = \infty$, we can find an invertible X_1 such that

$$X_1(T + K - P_{m_0} K P_{m_0} + K_1) X_1^{-1} = \begin{pmatrix} L_{11} + L & L_{12} & 0 \\ 0 & \oplus_1^l \oplus_1^{n_i} B_i & 0 \\ 0 & 0 & E \end{pmatrix}.$$

Let $L_{11} + L = G$. Then G is a matrix on the finite dimensional space. Since $\sigma(G) \subset \cup_{i=1}^{l} \Omega_i \cup (\sigma(T) \setminus \cup_{i=1}^{l} \partial\Omega_i)$, $\sigma_e(G) \cap \sigma_r(\oplus_1^l \oplus_1^{n_i} B_i) = \emptyset$. By Lemma 3.22 of [Her], there exists an invertible operator Y such that

$$Y \begin{pmatrix} G & L_{12} \\ 0 & \oplus_1^l \oplus_1^{n_i} B_i \end{pmatrix} Y^{-1} = \begin{pmatrix} G & 0 \\ 0 & \oplus_1^l \oplus_1^{n_i} B_i \end{pmatrix}.$$

Since G is a matrix on the finite dimensional space, G is similar to $\oplus_1^n G_k$, where $\sigma(G_k) \subset \Omega_i$ or $\sigma(G_k) \subset \sigma(E)$. Without loss of generality, we assume that $n = l+1$ and $\sigma(G_k) \subset \Omega_k$, $k = 1, 2, ..., l$, and $\sigma(G_{l+1}) \subset \sigma(E)$; G_{l+1} is a diagonal matrix. By the above argument, we can find an invertible operator X_2 such that

$$X_2 X_1 (T + K - P_{m_0} K P_{m_1} + K_1) X_1^{-1} X_2^{-1}$$

$$= \left(\oplus_{i=1}^{l} \begin{pmatrix} G_i & 0 \\ 0 & \oplus_1^{n_i} B_i \end{pmatrix} \right) \oplus \begin{pmatrix} G_{l+1} & 0 \\ 0 & E \end{pmatrix} = A.$$

Applying Lemma 3.3 to $\begin{pmatrix} G_i & 0 \\ 0 & \oplus_1^{n_i} B_i \end{pmatrix}$ and noting that $\dim\ker(\lambda_k - E) = \infty$, we can find a compact K_2, $\|K_2\| < \delta/4\|X_1\|\|X_1^{-1}\|\|X_2\|\|X_2^{-1}\|$ such that $A + K_2 \sim (\oplus_{i=1}^{l} \oplus_{i=2}^{n_i} B_i) \oplus E = T$. This shows that there exists a compact K_3 with $\|K_3\| < \epsilon$ such that $T_\phi^* + K_3 \sim T$. This complete the proof of the theorem.

References

[Bak-Ded-Ull] I.N. Baker, J.A. Deddens and J.L. Ullman, A theorem on entire functions with applications to Toeplitz operators, Duke Math. J. 41 (1974), 703-745.

[Cow] C.C. Cowen, The Commutant of an analytic Toeplitz operator, Trans. Amer. Math. Soc. 239 (1978), 1-31.

[Ded-Won] J.A. Deddens and T.K. Wong, The commutant of analytic Toeplitz operators, Trans. Amer. Math. Soc. 184 (1973), 261-273.

[Dou] R.G. Douglas, Banach algebra techniques in operator theory, Academic Press, 1972.

[Gil] F. Gilfeather, Strongly irreducibility of operators, Indiana Univ. Math. J. 22(1972), 393-397.

[Hal] P.R. Halmos, Irreducible operators, Mich. Math. J. 15(1968), 234-245.

[Her] D.A. Herrero, Approximation of Hilbert space operators, vol 1, 2nd. ed. Pitman Res. Notes Math., 224(1990).

[Jia-Sun-Wan] C.L. Jiang, S.H. Sun and Z.Y. Wang, Essentially normal operator + compact operator = strongly irreducible, to appear Int. Equ. and Operator Theory.

[Jia-Wan] C.L. Jiang and Z.Y. Wang, The spectral picture and the close of the similarity orbit of strongly irreducible operator, to appear J. of Int. Equ. and operator Theory.

[JZJ] Z.J. Jiang, A lecture on operator theory, the report at the seminar of functional analysis in Jilin Univ., Changchun, 1979.

[Tho] J.E. Thomson, Intersection of commutants of analytic Toeplitz operators, Proc. Amer. Math. Soc. 52(1975) 1305-1310.

Classification of Real Von Neumann Algebras (I)

Bingren Li

Institute of Mathematics, Academia Sinica, Beijing

Introduction

Let M be a real Von Neumann (VN simply) algebra on a real Hilbert space H, and p, q be two projections of M. Similarly to the complex case ([1]), $p \sim q$ if there exists $v \in M$ such that $p = v^*v$ and $q = vv^*$; $p \preceq q$ if there exists a projection $q_1 \in M$ such that $p \sim q_1$ and $q_1 \leq q$. Then we have the definitions of finiteness, infiniteness and others for real VN algebras and their projections. And we have the unique decomposition:

$$M = M_1 \bigoplus M_2 \bigoplus M_3,$$

where M_1, M_2, M_3 are finite, semi-finite and properly infinite, purely infinite real VN algebras respectively.

In the following, we shall study the finite, properly infinite, semi-finite, pruely infinite real VN algebras separately. We can find that most of the results on (complex) VN algebras of these types hold in the real case. In particular, a real VN algebra M is finite, properly infinite, semi-finite or purely infinite, if and only if so is the (complex) VN algebra $M_c = M \dot{+} iM$. Therefore, the first classification of real VN algebras can be reduced to that of the complex case.

However, in the second classification of real VN algebras there are some differences. We shall discuss it in another paper.

1991 Mathematics Subject Classification:46L10

Partially supported by NSF of China.

1. Finite real VN algebras

Definition 1.1 Let M be a real VN algebra.

A (real) linear functional φ on M is said to be positve, denoted by $\varphi \geq 0$, if $\varphi(a^*a) \geq 0$, and $\varphi(a^*) = \varphi(a), \forall a \in M$.

A positive functional φ on M is said to be tracial, if

$$\varphi(a^*a) = \varphi(aa^*), \forall a \in M.$$

Then it is easy to see that $\varphi(a) = \varphi(u^*au), \forall a \in M$, $u \in U(M)$, where $U(M)$ is the set of all unitary elements of M. Moreover, if φ is also normal, then by the σ-density of the (real) linear span $[U(M)]$ of $U(M)$ in M ([2]) we have $\varphi(ab) = \varphi(ba), \forall a, b \in M$.

Similar to [1, Th.6.3.8], we have

Theorem 1.2. Let M be a finite real VN algebra. Then $\#K(a) = 1, \forall a^* = a \in M$, where

$$K(a) = \overline{Co\{u^*au \mid u \in U(M)\}} \cap Z,$$

" $\llcorner$ " is the norm closure, and $Z = Z(M)$ is the center of M.

Now we get a map $T : M_h \longrightarrow Z_h$, i.e., $\{T(a)\} = K(a), \forall a \in M_h$, where M_h, Z_h are the self-adjoint parts of M, Z respectively. Clearly, T is (real) linear and

$$T(M_+) \subset Z_+, T(z) = z, T(a) = T(u^*au),$$

$\forall z \in Z_h, a \in M_h, u \in U(M)$, where M_+, Z_+ are the positive parts of M, Z respectively.

Further, similarly to [1, Th.6.3.10] the following theorem is obvious.

Theorem 1.3. Let M be a real VN algebra. Then M is finite, if and only if there exists a faithful family of normal tracial real states on M.

Now we point out

Theorem 1.4. Let M be a real VN algebra on a real Hilbert space H. Then M is finite, if and only if the (complex) VN algebra $M_c = M \dotplus iM$ (on complex Hilbert space $H_c = H \dotplus iH$) is finite.

Proof. The sufficiency is obvious. Now let M be finite. By Theorem 1.3, there is a faithful family F of normal tracial real states on M. We shall show that $F_c = \{\varphi_c \mid \varphi \in F\}$ is also faithful for M_c, where $\varphi_c(a + ib) = \varphi(a) + i\varphi(b), \forall \varphi \in F, a, b \in M$.

Let $(a + ib) \in M_{c+}$, and $a, b \in M$. Since

$$< a\xi, \xi > = < (a + ib)\xi, \xi > \geq 0, \forall \xi \in H,$$

it follows that $a \in M_+$. If $\varphi_c(a+ib) = 0, \forall \varphi \in F$, then $\varphi(a) = 0, \forall \varphi \in F$, and $a = 0$. Further, by $b^* = -b$ and $ib = (a + ib) \geq 0$ we have $b = 0$. Therefore, F_c is also faithful for M_c, and the proof is completed.

Now we discuss the (real) linear map $T : M_h \longrightarrow Z_h$, where M is a finite real VN algebra, and $\{T(a)\} = K(a), \forall a \in M_h$. Since M_c is also finite and $U(M) \subset U(M_c)$, it follows that T is the restriction of T_c on M_h, where T_c is the central valued trace from M_c to Z_c (the center of M_c), i.e.,

$$\{T_c(x)\} = \overline{Co\{u^*xu \mid u \in U(M_c)\}} \cap Z_c, \forall x \in M_c.$$

Lemma 1.5. Let M be a finite real VN algebra. Then

$$T_c(\bar{x}) = \overline{T_c(x)}, \forall x \in M_c,$$

where $\bar{x} = a - ib$ if $x = a + ib \in M_c$ and $a, b \in M$. Consequently, $T_c(M) \subset Z, T_c(M_k) \subset Z_k$, where M_k, Z_k are the skew self- adjoint parts of M, Z respectively.

Proof. Let $x \in M_c$. Then there is a sequence

$$\{x_n\} \subset Co\{u^*xu \mid u \in U(M_c)\}$$

such that $x_n \xrightarrow{\|\ \|} T_c(x) \in Z_c$. Since $\| \bar{y} \|=\| y \|, \forall y \in M_c$, it follows that $\bar{x}_n \xrightarrow{\|\ \|} \overline{T_c(x)} \in Z_c$. By

$$\overline{u^*xu} = \bar{u}^*\bar{x}\bar{u}, \overline{u} \in U(M_c), \forall u \in U(M_c),$$

we have $\bar{x}_n \in Co\{u^*\bar{x}u \mid u \in U(M_c)\}, \forall n$, and

$$\overline{T_c(x)} \in \overline{Co\{u^*\bar{x}u \mid u \in U(M_c)\}} \cap Z_c = \{T_c(\bar{x})\}.$$

Thus $\overline{T_c(x)} = T_c(\bar{x}), \forall x \in M_c$. In particular, $T_c(M) \subset Z$. Moreover, $T_c(M_k) \subset Z_k$ since $T_c(x^*) = T_c(x)^*, \forall x \in M_c$. The proof is completed.

Definition 1.6. Let M be a finite real VN algebra, and T_c be the central valued trace from M_c to Z_c. Then $T = T_c \mid M$ is called the central valued trace on M, i.e.,

$$\{T(a)\} = \overline{Co\{u^*au \mid u \in U(M_c)\}} \cap Z, \forall a \in M.$$

Clearly, $T : M \longrightarrow Z$ is (real) linear, $T(M_+) = Z_+, T(M_h) \subset Z_h, T(M_k) \subset Z_k, T(a^*) = T(a)^*, \forall a \in M$, and

$$\{T(a)\} = \overline{Co\{u^*au \mid u \in U(M)\}} \cap Z, \forall a \in M_h.$$

Moreover, Proposition 6.3.14 of [1] also holds for our T, and the proof is the same.

2. Properly infinite real VN algebras

Similar to [1, Th.6.4.4], we have

Theorem 2.1. Let M be a real VN algebra. Then the following statements are equivalent:
1) M is properly infinite;
2) there exists an orthogonal sequence $\{p_n\}$ of projections in M such that

$$\sum_n p_n = 1, p_n \sim 1, \forall n;$$

3) there exists a projection p of M such that $p \sim (1-p) \sim 1$.

Now we point out

Theorem 2.2. Let M be a real VN algebra. Then M is properly infinite, if and only if $M_c = M \dot{+} iM$ is properly infinite.

Proof. Let M_c be properly infinite. If z is a finite central projection of M , then Mz is a finite real VN algebra. By Theorem 1.4, $(Mz)_c = M_c z$ is finite. Thus, z is also a finite central projection of M_c, and z must be zero, i.e., M is also properly infinite.

Conversely, suppose that M is properly infinite. By Theorem 2.1, there is a projection p of M such that $p \sim (1-p) \sim 1$. This relation also holds in M_c. Therefore, M_c is also properly infinite. The proof is completed.

3. Semi-finite real VN algebras

Definition 3.1. Let M be a real VN algebra. $\varphi : M_+ \longrightarrow [0, +\infty]$ is called a trace on M_+ if

$$\varphi(a + b) = \varphi(a) + \varphi(b), \varphi(\lambda a) = \lambda \varphi(a),$$

$\forall a, b \in M_+$ with $\lambda \geq 0$ (where we define $o \cdot \infty = 0$), and $\varphi(a^*a) = \varphi(aa^*), \forall a \in M$.

Now let φ be a trace on M_+. Denote

$$\mathcal{N} = \{a \in M \mid \varphi(a^*a) < \infty\}$$

and $\mathcal{M} = \mathcal{N}^2$, the (real) linear span of the subset $\{ab \mid a, b \in \mathcal{N}\}$. Clearly, $\mathcal{M}$ and $\mathcal{N}$ are two * two-sided ideals of M, and $\mathcal{M} \subset \mathcal{N}$. Moreover, $\mathcal{M}$ is called the definition ideal of φ.

If $a \in \mathcal{M}_+$ and $\varphi(a) < \infty$, then $a^{\frac{1}{2}} \in \mathcal{N}$, $a = a^{\frac{1}{2}} \cdot a^{\frac{1}{2}} \in \mathcal{M}_+ = \mathcal{M} \cap \mathcal{M}_+$. Conversely, let

$$a = \sum_j x_j^* y_j \in \mathcal{M}_+ = \mathcal{M} \cap \mathcal{M}_+,$$

where $x_j, y_j \in \mathcal{N}$. Since

$$x^* y + y^* x = \frac{1}{2}\{(x+y)^*(x+y) - (x-y)^*(x-y)\}.$$

$\forall x, y \in \mathcal{M}$, it follows that

$$
\begin{aligned}
a &= \frac{a+a^*}{2} = \sum_j \frac{1}{2}(x_j^* y_j + y_j^* x_j) \\
&= \frac{1}{4}\sum_j \{(x_j + y_j)^*(x_j + y_j) - (x_j - y_j)^*(x_j - y_j)\} \quad (1) \\
&\le \frac{1}{4}\sum_j (x_j + y_j)^*(x_j + y_j).
\end{aligned}
$$

By $x_j, y_j \in \mathcal{N}, \forall j$, we have $\varphi(a) < \infty$. Hence,

$$\mathcal{M} \cap \mathcal{M}_+ = \mathcal{M}_+ = \{a \in M_+ \mid \varphi(a) < \infty\}.$$

Moreover, from (1) we can see that

$$\mathcal{M}_h = [\mathcal{M}_+] = \mathcal{M}_+ - \mathcal{M}_+,$$

where $\mathcal{M}_h$ is the self-adjoint part of $\mathcal{M}$, and $[\mathcal{M}_+]$ is the (real) linear span of $\mathcal{M}_+$.

Let $x \in \mathcal{M}$, and $x = uh$ be the polar decomposition of x. Then

$$h = u^* x \in \mathcal{M} \cap \mathcal{M}_+ = \mathcal{M}_+, (\overset{\infty}{\in} \in \mathcal{N},$$

and $x = uh^{\frac{1}{2}} \cdot h^{\frac{1}{2}}$. Thus, $\mathcal{M} = \{ab \mid a, b \in \mathcal{N}\}$.

Now $(\varphi \mid \mathcal{M}_+)$ can be extended to a (real) linear functional on $\mathcal{M}$ (still denoted by φ) if we define

$$
\begin{cases}
\varphi(a - b) &= \varphi(a) - \varphi(b), \quad \forall a, b \in \mathcal{M}_+, \\
\varphi \mid \mathcal{M}_k &= 0,
\end{cases}
$$

where $\mathcal{M}_k$ is the skew self-adjoint part of $\mathcal{M}$, and $\mathcal{M} = \mathcal{M}_h \dotplus \mathcal{M}_k = (\mathcal{M}_+ - \mathcal{M}_+) \dotplus \mathcal{M}_k$. Further, we get a (complex) linear functional

$$\varphi_c : \mathcal{M}_c = \mathcal{M} \dotplus i\mathcal{M} \longrightarrow \mathbb{C},$$

i.e.,

$$\varphi_c(a + ib) = \varphi(a) + i\varphi(b),$$

$\forall a, b \in \mathcal{M}$. If $a, b \in \mathcal{M}$ are such that $(a + ib) \in \mathcal{M}_{c+} = \mathcal{M}_c \cap \mathcal{M}_{c+}$ (where $\mathcal{M}_c = \mathcal{M} \dotplus i\mathcal{M}$), then it is easy to see that $a \in \mathcal{M}_+$ and $b \in \mathcal{M}_k$. Hence

$$\varphi_c(a + ib) = \varphi(a), \forall(a + ib) \in \mathcal{M}_{c+} \text{ with } a, b \in \mathcal{M}.$$

Now we define a function $\psi : M_{c+} \longrightarrow [0, +\infty]$ as follows:

$$\psi(a + ib) = \begin{cases} +\infty, & \text{if } (a + ib) \in M_{c+} \backslash \mathcal{M}_{c+}, \\ \varphi_c(a + ib) = \varphi(a), & \text{if } (a + ib) \in \mathcal{M}_{c+}. \end{cases}$$

We claim that ψ is a trace on $M_{c+}, \psi \mid M_+ = \varphi$, and $\psi(\bar{x}) = \psi(x), \forall x \in M_{c+}$, where $\bar{x} = a - ib$ if $x = a + ib \in M_c$ and $a, b \in M$ (see Lemma 1.5).

In fact, clearly we have

$$\psi(x + y) = \psi(x) + \psi(y), \psi(\lambda x) = \lambda \psi(x),$$

$\forall x, y \in M_{c+}$ with $\lambda \geq 0$. Further, if $a, b \in M$ are such that $(a + ib)^*(a + ib) \in \mathcal{M}_{c+}$, then

$$(a^*a + b^*b) \in \mathcal{M}_+, \text{ and } (a^*b - b^*a) \in \mathcal{M}_k,$$

and

$$\psi((a + ib)^*(a + ib)) = \varphi(a^*a) + \varphi(b^*b) < \infty.$$

Hence, $a, b \in \mathcal{N}$, $(aa^* + bb^*) \in \mathcal{M}_+$, $(ba^* - ab^*) \in \mathcal{M}_k$, and

$$\begin{aligned} \psi((a + ib)(a + ib)^*) &= \varphi(aa^*) + \varphi(bb^*) \\ &= \varphi(a^*a) + \varphi(b^*b) \\ &= \psi((a + ib)^*(a + ib)). \end{aligned}$$

Similarly, if $a, b \in M$ are such that $(a + ib)(a + ib)^* \in \mathcal{M}_{c+}$, then we also have $a, b \in \mathcal{N}$ and

$$\psi((a + ib)(a + ib)^*) = \psi((a + ib)^*(a + ib)).$$

Therefore,

$$(a + ib)^*(a + ib) \in \mathcal{M}_{c+} \iff (a + ib)(a + ib)^* \in \mathcal{M}_{c+}$$
$$\iff a, b \in \mathcal{N}$$

and ψ is a trace on M_{c+}. Clearly, $\psi \mid M_+ = \varphi$. Moreover, it is easy to see that $\bar{x} \in M_{c+}$ if $x \in M_{c+}$. Thus, $x \in \mathcal{M}_{c+} \iff \bar{x} \in \mathcal{M}_{c+}$. By the definition of ψ, we have $\psi(\bar{x}) = \psi(x), \forall x \in M_{c+}$.

Now we study the definition ideal of ψ. Clearly, $\{x \in M_{c+} \mid \psi(x) < \infty\} = \mathcal{M}_{c+}$. By [1], the definition ideal of ψ is $[\mathcal{M}_{c+}]$ (complex linear span of $\mathcal{M}_{c+}$). Moreover, from the above discussion we have

$$\{x \in M_c \mid \psi(x^*x) < \infty\} = \mathcal{N} \dot{+} i\mathcal{N} = \mathcal{N}_c.$$

and

$$[\mathcal{M}_{c+}] = \mathcal{N}_c^2 = \mathcal{N}^2 + i\mathcal{N}^2 = \mathcal{M} + i\mathcal{M} = \mathcal{M}_c.$$

Therefore, the definition ideal of ψ is $\mathcal{M}_c = \mathcal{M} \dot{+} i\mathcal{M}$.

By [1], ψ can be extended to a (complex) linear functional on $\mathcal{M}_c$, still denoted by ψ. We point out that $\psi \mid \mathcal{M} = \varphi$. Since $\psi \mid \mathcal{M}_+ = \varphi$, it suffices to prove $\psi(k) = 0$, $\forall k^* = -k \in \mathcal{M}$. Let $k \in \mathcal{M}_k$ and write

$$k = (a_1 + ib_1) - (a_2 + ib_2) + i(a_3 + ib_3) - i(a_4 + ib_4),$$

where $(a_j + ib_j) \in \mathcal{M}_{c+}$ (so $a_j \in \mathcal{M}_+, b_j \in \mathcal{M}_k$), $1 \le j \le 4$. Since $k^* = -k$, it follows that

$$k = i(a_3 + ib_3) - i(a_4 + ib_4).$$

By $k \in \mathcal{M}$, we can see $a_3 = a_4$. Therefore,

$$\begin{aligned} \psi(k) &= i\psi(a_3 + ib_3) - i\psi(a_4 + ib_4) \\ &= i\varphi(a_3) - i\varphi(a_4) = 0. \end{aligned}$$

Now by [1, Proposition 6.5.2], we have

$$\varphi(ab) = \psi(ab) = \psi(ba) = \varphi(ba),$$

$\forall a \in \mathcal{M}, b \in \mathcal{M}$, or $a, b \in \mathcal{N}$.

Furthermore, we say that the above ψ is unique, i.e., if $\psi' : \mathcal{M}_{c+} \longrightarrow [0, +\infty]$ is a trace, and $\psi' \mid \mathcal{M}_+ = \varphi, \psi'(\bar{x}) = \psi'(x), \forall x \in \mathcal{M}_{c+}$, then $\psi' = \psi$.

Indeed, denote the definition ideal of ψ' by $\mathcal{M}'_c$. Clearly, $x \in \mathcal{M}'_c \Longleftrightarrow \bar{x} \in \mathcal{M}'_c$. So we can write

$$\mathcal{M}'_c = \mathcal{M}' \dot{+} i\mathcal{M}',$$

where $\mathcal{M}' \subset \mathcal{M}$. Clearly,

$$\mathcal{M}'_+ = \mathcal{M}'_c \cap \mathcal{M}_+ = \{a \in \mathcal{M}_+ \mid \varphi(a) = \psi'(a) < \infty\} = \mathcal{M}_+.$$

Let $\mathcal{N}'_c = \{x \in \mathcal{M}_c \mid \psi'(x^*x) < \infty\}$. If $(a + ib) \in \mathcal{N}'_c$ and $a, b \in \mathcal{M}$. Then

$$(a + ib)^*(a + ib) = (a^*a + b^*b) + i(a^*b - b^*a) \in \mathcal{M}'_c = \mathcal{M}' + i\mathcal{M}'.$$

Thus, $(a^*a + b^*b) \in \mathcal{M}' \cap \mathcal{M}_+ = \mathcal{M}'_+ = \mathcal{M}_+$, $\varphi(a^*a + b^*b) < \infty$, and $a, b \in \mathcal{N}$. Conversely, if $a, b \in \mathcal{N}$, then by $\psi' \mid \mathcal{M}_+ = \varphi$, $\mathcal{N} \subset \mathcal{N}'_c$ we have $(a + ib) \in \mathcal{N}'_c$. Therefore,

$$\mathcal{N}'_c = \mathcal{N} \dot{+} i\mathcal{N},$$

$$\mathcal{M}'_c = \mathcal{N}'^2_c = \mathcal{N}^2 \dot{+} i\mathcal{N}^2 = \mathcal{M} \dot{+} i\mathcal{M} = \mathcal{M}_c$$

and $\mathcal{M}' = \mathcal{M}$. Now ψ' can be extend to a (complex) linear functional on $\mathcal{M}'_c = \mathcal{M}_c$, still denoted by ψ'.

If $(a + ib) \in \mathcal{M}'_{c+} = \mathcal{M}_{c+}$ and $a, b \in \mathcal{M}$, then $a \in \mathcal{M}_+$ and $b^* = -b \in \mathcal{M}_k$. Since $(a - ib) \in \mathcal{M}'_{c+} = \mathcal{M}_{c+}$, and

$$\varphi(a) + i\psi'(b) = \psi'(a + ib) = \psi'(a - ib) = \varphi(a) - i\psi'(b),$$

it follows that $\psi'(a + ib) = \varphi(a), \psi'(b) = 0$. Therefore,

$$\psi'(a + ib) = \begin{cases} +\infty, & \text{if } (a + ib) \in \mathcal{M}_{c+} \backslash \mathcal{M}_{c+}, \\ \varphi(a), & \text{if } (a + ib) \in \mathcal{M}_{c+}, \end{cases}$$

i.e., $\psi' = \psi$.

By the above discussion, we have

Proposition 3.2. Let M be a real VN algebra, and φ be a trace on M_+. Then there is a unique trace ψ on M_{c+} such that

$$\psi \mid M_+ = \varphi, \ \psi(\bar{x}) = \psi(x), \forall x \in \mathcal{M}_{c+},$$

where $M_c = M \dot{+} iM$. Moreover, let $\mathcal{N} = \{a \in M \mid \varphi(a^*a) < \infty\}$, $\mathcal{N}_\psi = \{x \in M_c \mid \psi(x^*x) < \infty\}$, $\mathcal{M} = \mathcal{N}^2$ (the definition ideal of φ), and $\mathcal{M}_\psi = \mathcal{N}_\psi^2$ (the definition ideal of ψ). Then

$$\mathcal{M} = \{ab \mid a, b \in \mathcal{N}\}, \mathcal{M}_h = \mathcal{M}_+ - \mathcal{M}_+, \mathcal{M}_+ = \{a \in M_+ \mid \varphi(a) < \infty\},$$

$$\mathcal{M}_\psi = \mathcal{M} \dot{+} i\mathcal{M} = \mathcal{M}_c, \mathcal{N}_\psi = \mathcal{N} \dot{+} i\mathcal{N} = \mathcal{N}_c,$$

and

$$\psi(a + ib) = \begin{cases} +\infty, & \text{if } (a + ib) \in \mathcal{M}_{c+} \backslash \mathcal{M}_{c+} \\ \varphi(a), & \text{if } (a + ib) \in \mathcal{M}_{c+}. \end{cases}$$

Still denote the natural extensions of φ, ψ on $\mathcal{M}, \mathcal{M}_c$ by φ, ψ respectively. Then

$$\varphi \mid \mathcal{M}_k = 0, \ \psi \mid \mathcal{M} = \varphi,$$

and

$$\varphi(ab) = \varphi(ba), \forall a \in \mathcal{M}, b \in M, \text{or } a, b \in \mathcal{N}.$$

Definition 3.3. ψ is called the trace (on M_{c+}) determined by the trace φ (on M_+).

Proposition 3.4. Let M be a real VN algebra, φ be a trace on M_+, and ψ be the trace on M_{c+} determined by φ, where $M_c = M \dot{+} iM$.

1) φ is semi-finite (i.e., for any $a \in M_+$ and $a \neq 0$, there exists $b \in M_+$ and $b \neq 0$ such that $b \leq a$ and $\varphi(b) < \infty$), if and only if the definition ideal $\mathcal{M}$ of φ is σ-dense in M. Consequently, φ is semi-finite, if and only if ψ is semi-finite.

Moreover, if φ is semi-finite, then for any projection p of M we have

$$p = \sup\{q \mid q \text{ is a projection of } M, q \leq p, \text{and } \varphi(q) < \infty\}.$$

2) φ is normal (i.e., for any bounded increasing net $\{a_\ell\} \subset M_+$ we have $\varphi(\sup_\ell a_\ell) = \sup_\ell \varphi(a_\ell)$), if and only if ψ is normal.

3) If φ is normal, then $(1 - z) = \{p \mid p \text{ is a projection of } M \text{ and } \varphi(p) = 0\}$ is a central projection of M, $M(1 - z) = \{a \in M \mid \varphi(a^*a) = 0\}$, and $\varphi \mid (Mz)_+$ is faithful. Denote $z = s(\varphi)$, and $s(\varphi)$ is called the support of φ.

4) Let φ be normal, and $s(\varphi) = z$. Then $s(\psi) = z$. Consequently, φ is faithful if and only if ψ is faithful.

Proof. 1) Similarly to [1, Prop. 6.5.4.]. Moreover, if $\mathcal{M}$ is σ-dense in M, then $\mathcal{M}_c = \mathcal{M} \dotplus i\mathcal{M}$ (the definition ideal of ψ) is obviously σ-dense in M_c. Now if ψ is semi-finite, then for any $0 \neq a \in M_+ \subset M_{c+}$ there are $b, c \in M$ such that $0 \neq (b + ic) \in \mathcal{M}_{c+}$, $(b + ic) \leq a$, and $\psi(b + ic) < \infty$. By Proposition 3.2, $\psi(b + ic) = \varphi(b) < \infty$. Since $a \geq (b + ic) \geq 0$, it follows that $a \geq b \geq 0$. If $b = 0$, then $ic = b + ic \geq 0$, and $c = 0$. It is impossible, since $b + ic \neq 0$. Thus, $0 \neq b \in M_+$, $b \leq a$, and $\varphi(b) < \infty$, i.e., φ is also semi-finite.

2) Clearly, φ is normal if ψ is normal. Now let φ be normal, $\{(a_\ell + ib_\ell)\}$ be a bounded increasing net in M_{c+}, and $(a + ib) = \sup_\ell(a_\ell + ib_\ell)$. It is easy to see that $\{a_\ell\}$ will be a bounded increasing net in M_+, $a = \sup_\ell a_\ell$ and $b_\ell \xrightarrow{s} b$.

Notice the following fact. If $(a + ib) \in M_{c+}$ (clearly $a \in M_+$, $b \in M_k$), then $2a \geq (a + ib)$. In particular, if $(a + ib) \in M_{c+}$ and $a \in \mathcal{M}_+$, then $(a + ib) \in \mathcal{M}_{c+}$, and $b \in \mathcal{M}$. In fact, by $(a - ib) \in M_{c+}$ we can see that

$$2a - (a + ib) = a - ib \geq 0, \ 2a \geq (a + ib).$$

Moreover, if $a \in \mathcal{M}_+ \subset \mathcal{M}_{c+}$, then

$$\psi(a + ib) \leq \psi(2a) = 2\psi(a) < \infty.$$

Thus, $(a + ib) \in \mathcal{M}_c = \mathcal{M} \dotplus i\mathcal{M}$, and $b \in \mathcal{M}$.

If there is an index ℓ_o such that $\psi(a_{\ell_o} + ib_{\ell_o}) = +\infty$, then $\psi(a_\ell + ib_\ell) = +\infty$, $(a_\ell + ib_\ell) \in M_{c+} \backslash \mathcal{M}_{c+}$, $\forall \ell \geq \ell_o$. By the above fact, we have $a_\ell \notin \mathcal{M}_+$, $\forall \ell \geq \ell_o$. Then $\varphi(a_\ell) = +\infty$, $\forall \ell \geq \ell_o$, and

$$\varphi(a) = \sup_\ell \varphi(a_\ell) = +\infty.$$

Hence, $a \notin \mathcal{M}_+$, $(a + ib) \in M_{c+} \backslash \mathcal{M}_{c+}$, and

$$\psi(a + ib) = +\infty = \sup_\ell \psi(a_\ell + ib_\ell).$$

If $\psi(a_\ell + ib_\ell) < \infty, \forall \ell$, then

$$a_\ell \in \mathcal{M}_+, b_\ell \in \mathcal{M}_k, \forall \ell,$$

and $\psi(a_\ell + ib_\ell) = \varphi(a_\ell) \nearrow \varphi(a)$. Further, if $\varphi(a) < \infty$, i.e., $a \in \mathcal{M}_+$, then by the above fact we have $(a + ib) \in \mathcal{M}_{c+}$. So $\psi(a + ib) = \varphi(a) = \sup_\ell \varphi(a_\ell) = \sup_\ell \psi(a_\ell + ib_\ell)$. If $\varphi(a) = +\infty$, then $a \notin \mathcal{M}_+$, $(a + ib) \notin \mathcal{M}_{c+}$, and

$$\psi(a_\ell + ib_\ell) = \varphi(a_\ell) \nearrow \varphi(a) = +\infty = \psi(a + ib).$$

Therefore, ψ is also normal.

3) Similarly to [1, Prop.6.5.5] (noticing that the (real) linear span $[U(M)]$ of $U(M)$ is σ-dense in M).

4) By $\psi(1 - z) = \varphi(1 - z) = 0$, we can see that $\psi \mid (M_c(1 - z))_+ = 0$. Now it suffices to show that ψ is faithful on $(M_c z)_+$. We may assume $z = 1$.

Let $(a + ib) \in M_{c+}$ and $a, b \in M$, and $\psi(a + ib) = 0$. Of course, $(a + ib) \in M_{c+}$, $a \in M_+$, $b \in M_k$. So $\varphi(a) = \psi(a + ib) = 0$. Since φ is faithful, it follows that $a = 0$. Further, $b = 0$ since $ib = (a + ib) \geq 0$. Therefore, ψ is also faithful.

The proof is completed.

Theorem 3.5. Let M be a real VN algebra, and $M_c = M \dotplus iM$. Then the following statements are equivalent:

1) M is semi-finite;

2) there exists a faithful semi-finite normal trace on M_+;

3) M_c is semi-finite.

Proof. 1) $\Longleftrightarrow$ 2), similarly to [1, 6.5.7 and 6.5.8].

2) $\Longrightarrow$ 3) is immediate from Proposition 3.4 and [1,Th.6.5.8].

Now let M_c be semi-finite. Then there exists a faithful semi-finite normal trace ψ' on M_{c+}. Let

$$\psi(x) = \frac{1}{2}\{\psi'(x) + \psi'(\bar{x})\}, \forall x \in M_{c+}.$$

Clearly, ψ is a faithful normal trace on M_{c+}, and $\psi(\bar{x}) = \psi(x), \forall x \in M_{c+}$. We claim that ψ is also semi-finite. In fact, if x is a non-zero element of M_{c+}, then there is $0 \neq y \in M_{c+}$ such that $y \leq x$ and $\psi'(y) < \infty$. Of course, $0 \neq \bar{y} \in M_{c+}$. Then there is $0 \neq \bar{z} \in M_{c+}$ such that

$$\bar{z} \leq \bar{y}, \text{and } \psi'(\bar{z}) < \infty.$$

Clearly, $0 \leq z \leq y \leq x, z \neq 0$, and $\psi'(z) \leq \psi'(y) < \infty$. Hence, $\psi(z) = \frac{1}{2}\{\psi'(z) + \psi(\bar{z})\} < \infty$, and ψ is semi-finite.

Further, let $\varphi = \psi \mid M_+$. Then φ is a trace on M_+, and ψ is the trace determined by φ. By Proposition 3.4, φ is also faithful semi-finite normal. Therefore, M is semi-finite.

The proof is completed.

Corollary 3.6. If M is a semi-finite real VN algebra, then M' is also semi-finite.

4. Purely infinite real VN algebras

Propsition 4.1. Let M be a real VN algebra. Then M is purely infinite, if and only if there is not any non-zero semi-finite normal trace on M_+.

The proof runs similarly to [1, Prop.6.6.2].

Theorem 4.2. Let M be a real VN algebra, and $M_c = M \dotplus iM$. Then M is purely infinite, if and only if M_c is purely infinite.

Proof. Let M_c be purely infinite. If M is not purely infinite, then by Propsition 4.1 there is a non-zero semi-finite normal trace φ on M_+. Let ψ be the trace on M_{c+} determined by φ. By Proposition 3.4, ψ is also semi-finite normal. Of course, ψ is non-zero. This is a contradiction, since M_c is purely infinite. Hence, M must be purely infinite.

Now let M be purely infinite. If M_c is not purely infinite, then there is a non-zero semi-finite normal trace ψ' on M_{c+}. Let

$$\psi(x) = \frac{1}{2}\{\psi'(x) + \psi'(\bar{x})\}, \forall x \in M_{c+}.$$

Clearly, ψ is also a non-zero semi-finite normal trace on M_{c+} and $\psi(\bar{x}) = \psi(x), \forall x \in M_{c+}$. Let

$$\varphi = \psi \mid M_+.$$

Then φ is a semi-finite normal trace on M_+, and ψ is a the trace determined by φ (see Propositions 3.2 and 3.4). By the uniqueness (see Proposition 3.2), φ must be non-zero. This is a contradiction, since M is purely infinite. Hence M_c must be purely infnite.

The proof is completed.

Corollary 4.3. If M is a purely infinte real VN algebra, then M' is also purely infinite.

References

[1] Li Bingren, Introduction to operator algebras, World Sci., Singapore, 1992.
[2] Li Bingren, Real finite dimensional C^*-algebras, to appear.

C-Cosine Operator Functions and the Applications to the Second Order Abstract Cauchy Problem

Haiyan Wang

Department of Mathematics, Southeast University, Nanjing

Shengwang Wang

Department of Mathematics, Nanjing University, Nanjing

1. Introduction

Many physical problems may be modelled as a second order abstract Cauchy problem

$$\begin{cases} \frac{d^2}{dt^2}u(t,x,y) = Au(t,x,y), & t \in R \\ u(0,x,y) = x, & u'(0,x,y) = y, \end{cases} \tag{1.1}$$

where A is a linear operator on a Banach space X, and $t \to u(t,x,y) \in C(R,X)$. From [6], the well-posedness of (1.1) corresponds to A generating a strongly continuons cosine operator function. When (1.1) is not well-posed, at least in its original formulation, a useful concept for dealing with it is a C-cosine operator function (Definition 2.1). In [9], it has been proved that if A generates a C-cosine operator function then (1.1) is C-well-posed, i.e., (1.1) has a unique mild solution for every initial value in the image of C and a unique solution for every initial value in C(D(A)).

However, in order that (1.1) has all these solutions, it is not necessary that A itself generates a C-cosine operator function, even if A is closed and C commutes with A. This is in contrast to the strongly continuous case, that is, when $C = I$. When A is closed, $\rho(A) \neq \phi$ and (1.1) has a unique solution, for all x,y in $D(A)$; it follows automatically that A generates an exponentially bounded strongly continuous cosine operator function (see[6, Theorem 2. 8. 2]).

1991 Mathematics subject Classification: 47D05, 34G10, 35G10

Project supported by the National Science Foundation of China

Here is a simple example. Let $G \equiv \frac{d}{ds}$, on $X = L^\infty(R)$, with maximal domain. Then G is closed and $\rho(G) \supset \{\lambda : Re\lambda \neq 0\}$. Let $H = G^2$. Then H is closed and $\rho(H) \supset \{\lambda : \arg \lambda \neq \pi\}$. Let B be the restriction of H to $D(H^2)$, the domain of H^2, that is,

$$D(B) \equiv D(H^2), \quad Bx = Hx, \quad \forall x \in D(B).$$

Let $A \equiv \overline{B}$, the closure of B. Let $C = (1 - H)^{-2}$, and define a C-cosine operator function $\{C(t)\}_{t\in R}$ by

$$(C(t)f)(s) = \frac{1}{2}[(Cf)(s+t) + (Cf)(s-t)].$$

It is not hard to show that H is the generator of $\{C(t)\}_{t\in R}$. The domain of A equals the graph closure of the domain of H^2, which may be shown to equal $(1 - H)^{-1}\overline{D(H)}$, which does not equal $D(H) = (1 - H)^{-1}(L^\infty(R))$. Therefore A does not generate a C-cosine operaror function. But (1.1) has a unique bounded mild solution for every x in the image of C.

In this paper, we will write down exactly what conditions on A are equivalent to (1.1) having a unique mild solution, for every initial value in the image of C, where C commutes with A. We shall see that this is equivalent to (1.1) having a unique solution for every initial value in $C(D(A))$ if the resolvent set of A is nonempty (Theorem 3.3). In the exponentially bounded case, the condition that the resolvent set of A is nonempty can be replaced by the condition that the C-resolvent of A is nonempty (Theorem 3.7).

We introduce a C-cosine operator function for A (Definition 2.5). We will also say that A has a C-cosine operator function. When C is injective and A is closed, having a C-cosine operator function is equivalent to (1.1) having a unique mild solution, for every initial value in the image of C. Thus, from the point of view of the abstract Cauchy problem (1.1), there is no difference between A having a C-cosine operator function and A generating a C-cosine operator function, than it to show that A itself is the generator.

When A generates a C-cosine operator function $\{C(t)\}_{t\in R}$, then $\{C(t)\}_{t\in R}$ is a C-cosine operator function for A. However, it is sometimes sufficient to have an extension of A be the generator. This is desirable, because it is sometimes difficult to determine or describe the exact domain of the generator. The most natural choice of a domain may produce an operator A, that has a C-cosine operator function, but is not its generator.

In Section 2 we introduce a C-cosine operator function for A, and the generator of a C-cosine operator function; C is bounded and commutes with A, but may not be injective. We give some basic properties of the generator, $\tilde{A}$, its relationship with A.

In Section 3 we give the main results. These include numerous equivaient conditions for A to have a C-cosine operator function, when A is closed and C is injective. In particular, it is equivalent to (1.1) having a unique mild solution, for every x, y in the image of C. When the resolvent set of A is nonempty, it is equivalent to (1.1) having a unique solution, for every x, y in $C(D(A))$. When the image of C is dense, it is equivalent to $C^{-1}AC$ generating the C-cosine operator function. In terms of a Laplace transfom, there are also numerous equivalent conditions for A to have an exponentially bounded C-cosine operator function, when A is closed and C is injective. In this case it suffices to assume that the C-resolvent set of A is nonempty.

In dealing with the first order abstract Cauchy problem, the concepts of C-semigroup generated by A and C-semigroup for A have been introduced in [4,5], respectively.

All operators are linear on a Banach space X. We will write $D(A)$, Im (A) for the domain and image of the operator A, respectively. We will write $[D(A)]$ for the normed vector space with the graph norm $||x||_{[D(A)]} \equiv ||x|| + ||Ax||$. We will write $B(X)$ for the space of bounded linear operators from X into itself.

2. Definitions and basic properties

Definitions 2.1. The strongly continuous family of operators $C(t) : R \to B(X)$ is a C-cosine operator function if

(1) $C(0) = C$;
(2) $C(t+s)C + C(t-s)C = 2C(s)C(t)$, $\forall t, \ s \in R$.

$\{C(t)\}_{t \in R}$ is nondegerate if $C(t)x \equiv 0$, for all $t \in R$, only when $x = 0$.

Obviously, if $\{C(t)\}_{t \in R}$ is a C-cosine operator function, then $C-(-t)C = C(t)C, C(t)C(s) = C(t)C(s) = C(s)C(t)$, $\forall t, \ s \in R$.

Proposition 2.2. A C-cosine operator function $\{C(t)\}_{t \in R}$ is nondegenerate if and only if C is in jective.

Proof. Suppose C is injective. It is obvious by Definition 2.1 (1) that $\{C(t)\}_{t \in R}$ is nondegenerate. Conversely, assume that $Cx = 0$ for some $x \in X$. From Definition 2.1 (2), we have $C(s)C(t)x = 0$, $\forall s, \ t \in R$. This implies $C(t)x \in \cap\{\ker C(s) : s \in R\} = \{0\}$ for all $t \in R$. By the nondegeneration of $C(t), \forall t \in R, x = 0$. Here C is injective.

In this paper, we always assume that $\{C(t)\}_{t \in R}$ is nondegenerate.

Definition 2.3 [9]. Suppose C is injective. The generator $\tilde{A}$, of a C-cosineoperator function $\{C(t)\}_{t \in R}$, is defined by

$$\tilde{A}x = C^{-1} \left(\lim_{t \to 0} \frac{1}{t^2} (C(t)x + C(-t)x - 2C) \right),$$

with $D(\tilde{A})$ defined to be the set of all x such that the limit exists and is in the image of C.

Proposition 2.4 [9, Theorem 2.1]. Suppose $\tilde{A}$ is the generator of a C-cosine operator function. Then

(1) $\tilde{A}$ is closed, and Im $(c) \subseteq \overline{D(A)}$.

(2) For every $x \in X, t \in R$, $\int_0^t (t-s)C(s)x\,ds \in D(\tilde{A})$ and

$$\tilde{A} \int_0^t (t-s)C(s)x\,ds = C(t)x - Cx.$$

(3) If $x \in D(\tilde{A})$, then $C(t)x \in D(\tilde{A})$ and

$$\frac{d^2}{dt^2}C(t)x = C(t)\tilde{A}x = \tilde{A}C(t)x, \ \forall t \in R.$$

(4) $\tilde{A} = C^{-1}\tilde{A}C$.

Proof. (1)-(3) can be found in [9, Theorem 2.1]. We now prove (4). Letting $t = 0$ in (3) of this proposition, we obtain $\tilde{A} \subseteq C^{-1}\tilde{A}C$. Next, assume $x \in D(C^{-1}\tilde{A}C)$. Then $Cx \in D(\tilde{A})$ and

$$C(C(t)x - Cx) = C(t)Cx - C^2x$$
$$= C \int_0^t (t-s)C(s)C^{-1}\tilde{A}Cx\,ds.$$

By the injectivity of C,

$$\lim_{t\to 0} \frac{1}{t^2}(C(t)x + C(-1)x - 2Cx)$$
$$= \lim_{t\to 0} \frac{1}{t^2}\left(\int_0^t (t-s)C(s)C^{-1}\tilde{A}Cx\,ds + \int_0^{-t} (-t-s)C(s)C^{-1}\tilde{A}Cx\,ds \right)$$
$$= \tilde{A}Cx \in Im(C).$$

Hence $x \in D(\tilde{A})$ and $\tilde{A}x = C^{-1}\tilde{A}Cx$. This implies $C^{-1}\tilde{A}C \subseteq \tilde{A}$.

Definition 2.5. Suppose A is closable. We say that the C-cosine operator function $\{C(t)\}_{t\in R}$ is a C-cosine operator function for A if

(1) $C(t)A \subseteq AC(t), \quad t \in R$;

(2) for $x \in X, t \in R$, we have $\int_0^t (t-s)C(s)x\,ds \in D(A)$ and

$$A \int_0^t (t-s)C(s)x\,dx = C(t)x - Cx.$$

We will also say that A has a C-cosine operator function or has the C-cosine operator function $\{C(t)\}_{t\in R}$.

Theorem 2.6. Suppose C is injective, A is closed, and $\{C(t)\}_{t\in R}$ is a C-cosine operator function for A. Then

(1) $\{C(t)\}_{t\in R}$ is unique;

(2) $Im(C) \subseteq \overline{D(A)}$;

(3) $C^{-1}AC$ is the generator of $\{C(t)\}_{t\in R}$, i.e., $C^{-1}AC = \tilde{A}$, where $\tilde{A}$ is Defined in Definition 2.3.

Proof. (1) Assume that both $\{C_1(t)\}_{t\in R}, \{C_2(t)\}_{t\in R}$ are C-cosine operator functions for A. For every $x \in X$, we have

$$\frac{d}{ds}[C_1(t-s)\int_0^s (s-r)C_2(r)x dr]$$

$$= -\int_0^{t-s} C_1(u)\left(A\int_0^s (s-r)C_2(r)x dr\right) du + C_1(t-s)\int_0^s C_2(r)x dr$$

$$= -\int_0^{t-s} C_1(r)C_2(r)x dr)du + C_1(t-s)\int_0^s C_2(r)x dr + \int_0^{t-s} C_1(r)Cx dr,$$

and hence, by integrating with respect to s from 0 to t,

$$C\int_0^t (t-r)C_2(r)x dr$$

$$= -\int_0^t \int_0^{t-s} C_1(r)C_2(s)x drds + \int_0^t C_1(t-s)\int_0^s C_2(r)x drds$$

$$+ \int_0^t \int_0^{(t-s)} C_1(r)x drds.$$

Integrating the last two integrals by parts to obtain

$$C\int_0^t (t-r)C_2(r)x drds = \int_0^t C_1(t-s)Cx drds = C\int_0^t (t-r)C_1(r)x dr.$$

C being injective, one obtains, by differentiating twice the integrals, that $C_2(t)x = C_1(t)x$, $\forall x \in X$, $t \in R$.

(2) For any $x \in X, n \in N$, define

$$x_n = n^2 \int_0^{\frac{1}{n}} \left(\frac{1}{n} - s\right) C(s)x ds.$$

Then $x_n \in D(A)$, for all $n\mathbb{N}$, and $x_n \to \frac{1}{2}Cx$, as $n \to \infty$, thus $Cx \in \overline{D(A)}$.

(3) For any $x \in D(\tilde{A})$,

$$C(t)x - Cx = \int_0^t (t-s)C(s)\tilde{A}x ds.$$

Definition 2.5 implies

$$A\int_0^t (t-s)C(s)x ds = \int_0^t (t-s)C(s)\tilde{A}x ds.$$

A being closed. Differentiate with respec to t to obtain $C(t)x \in D(A)$, and $AC(t)x = C(t)\tilde{A}x$. Setting $t = 0$, one obtains $ACx = C\tilde{A}x$, and hence $\tilde{A} \subseteq C^{-1}AC$. Conversely, for any $x \in D(A)$,

$$\lim_{t \to 0} \frac{1}{t^2} [C(t)x + C(-t)x - 2Cx]$$
$$= \lim_{t \to 0} \frac{1}{t^2} \left[\int_0^t (t-s)C(s)Axds + \int_0^{-t} (-t-s)C(s)Axds \right]$$
$$= CAx.$$

This implies $C\tilde{A}x = CAx$. From the injectivity of C, $A \subseteq \tilde{A}$ and hence $C^{-1}AC \subseteq C^{-1}\tilde{A}C = \tilde{A}$ by Proposition 2.4 (4).

Corollary 2.7. If A is closed and there exists a strongly continuous cosine oper ator function for A, then A is the generator.

Corollary 2.8. Suppose $\{C(t)\}_{t \in R}$ is a C-cosine operator function generated by $\tilde{A}$. Then $\{C(t)\}_{t \in R}$ is the C-cosine operator function for $\tilde{A}$.

Theorem 2.9. Suppose $\{C(t)\}_{t \in R}$ is a C-cosine operator function generated by an extension of A, A is closed and densely defined and $C(t)$ leaves $D(A)$ invariant, for all $t \in R$. Then $\{C(t)\}_{t \in R}$ is a C-cosine operator function for A.

Proof. Let $\tilde{A}$ be the generator of $\{C(t)\}_{t \in R}$. From Corollary 2.8, $\{C(t)\}_{t \in R}$ is the C-cosine operator function for $\tilde{A}$. This, together with the fact that $C(t)$ leaves $D(A)$ invariant, asserts that

$$C(t)Ax = C(t)\tilde{A}x = \tilde{A}C(t)x = AC(t)x,$$

for every $x \in D(A), t \in R$. Hence

$$A\left(\int_0^t (t-s)C(s)xds \right) = \int_0^t (t-s)C(s)Axds = \int_0^t (t-s)C(s)\tilde{A}xds = C(t)x - Cx.$$

Since $D(A)$ is dense, and A is closed, the same is true for all $x \in X$.

Theorem 2.10. Suppose C is injective, A is closed. If $C^{-1}AC$ is the generator of the C-cosine operator function $\{C(t)\}_{t \in R}$ and $CA \subseteq AC, \overline{Im(C)} = X$, then $\{C(t)\}_{t \in R}$ is a C-cosine operator function for A.

Proof. Let $x \in X, C^{-1}AC$ be the generator of $\{C(t)\}_{t \in R}$. Then

$$C(t)x - Cx = C^{-1}AC \int_0^t (t-s)C(s)xds.$$

Therefore

$$C(t)Cx - C^2x = A \int_0^t (t-s)C(s)xds.$$

Since $\overline{Im(C)} = X$, A is closed, then for any $y \in X$,

$$C(t)y - Cy = A \int_0^t (t-s)C(s)yds.$$

To prove $C(t)A \subseteq AC(t), t \in R$, let $x \in X(A) \subseteq D(C^{-1}AC)$, then

$$C(t)x - Cx = \int_0^t (t-s)C(s)C^{-1}ACxds = \int_0^t (t-s)C(s)Axds.$$

Therefore

$$A \int_0^t (t-s)C(s)xds = \int_0^t (t-s)C(s)Axds, \quad x \in D(A).$$

Differentiating twice with respect to t we have $C(t)x \in D(A)$, and

$$AC(t) = C(t)Ax, \quad x \in D(A).$$

Then $\{C(t)\}_{t \in R}$ is a C-cosine operator function for A.

3. Main results

Definition 3.1. A solution of (1.1) is $u(t,x,y)$ such that $t \to u(t,x,y) \in C(R,[D(A)]) \cap C^2(R,X)$, satisfying (1.1). A mild solution of (1.1) is $u(t,x,y)$ such that $t \to u(t,x,y) \in C(R,X)$, and for all $t \in R$, $\int_0^t (t-r)u(r,x,y)dr \in D(A)$, with

$$u(t,x,y) = A \int_0^t (t-r)u(r,x,y)dr + x + ty.$$

Proposition 3.2 ([7, **Proposition** 1.4]). Let $\lambda \in R$ and let A be a closed linear operator satisfying the conditions
  (a) $\lambda - A$ is injective,
  (b) $D((\lambda - A)^{-1}) \supseteq Im(C)$,
  (c) for $x \in D(A), Cx \in D(A)$ and $ACx = CAx$.
Then we have
  (i) $C(D(A)) \subseteq C(D(C^{-1}AC)) \subseteq (\lambda - A)^{-1}C(X)$,
  (ii) $C(D(A)) = (\lambda - A)^{-1}C(X)$ if and only if $\lambda \in \rho(A)$,
  (iii) $C^{-1}AC = A$ if $\rho(A) \neq \phi$.

To prove the equivalence of Theorem 3.4 (i) and (v), we need the following

Theorem 3.3. Suppose $\lambda \in \rho(A)$. Then (1.1) has a unique mild solution for all $x,y \in Im(C)$ if and only if (1.1) has a unique solution for all $x,y \in (\lambda - A)^{-1}[Im(C)]$.

Proof. "⇒" Let $x, y \in (\lambda - A)^{-1}[Im(C)]$, Then there exist $x_1, y_1 \in Im(C)$, such that $x = (\lambda - A)^{-1}x_1, y = (\lambda - A)^{-1}y_1$. Suppose $u(t, x_1, y_1)$ is the mild solution of (1.1), let $\tilde{u}(t, x, y) \equiv (\lambda - A)^{-1}u(t, x_1, y_1)$. Then

$$
\begin{aligned}
\tilde{u}(t, x, y) &= (\lambda - A)^{-1}A \int_0^t (t - r)u(r, x_1, y_1)dr. \\
&\quad + (\lambda - A)^{-1}x_1 + t(\lambda - A)^{-1}y_1 \\
&= \int_0^t (t - r)A(\lambda - A)^{-1}u(r, x_1, y_1)dr + x + ty.
\end{aligned}
$$

Then $\tilde{u}(t, x, y) \in C(R, [D(A)]) \cap C^2(R, X)$, and

$$
\frac{d^2}{dt^2}\tilde{u}(t, x, y) = A(\lambda - A)^{-1}u(t, x_1, y_1) = A\tilde{u}(t, x, y)
$$

$$
\tilde{u}(0, x, y) = x, \quad \tilde{u}'(0, x, y) = y.
$$

Therefore $\tilde{u}(t, x, y)$ is the solution of (1.1).

For any $x, y \in (\lambda - A)^{-1}[Im(C)]$, let $\tilde{u}_1(t, t, y), \tilde{u}(t, x, y)$ be the solution of (1.1), denote $u_0(t, x_1, y_1) = (\lambda - A)[\tilde{u}_1(t, x, y) - \tilde{u}_1(t, x, y)]$, then

$$
u_0(t, x_1, y_1) = A \int_0^t (t - r)u_0(r, x_1, y_1)dr.
$$

From the uniqueness of the mild solution of (1.1), we have $u_0(t, x_1, y_1) \equiv 0, t \in R$. Thus $\tilde{u}_1(t, x, y) = \tilde{u}_2(t, x, y), t \in R$.

"⇐" Let $x, y \in Im(C)$. Then $x_1 = (\lambda - A)^{-1}x, y_1 = (\lambda - A)^{-1}y \in (\lambda - A)^{-1}[Im(C)]$. Suppose $\tilde{u}(t, x_1, y_1)$ is the solution of (1.1). Then

$$
\frac{d^2}{dt^2}\tilde{u}(t, x_1, y_1) = A\tilde{u}(t, x_1, y_1),
$$

$$
\tilde{u}(0, x_1, y_1) = x_1, \quad \tilde{u}'(0, x_1, y_1) = y_1.
$$

Integrating twice, one obtains

$$
\tilde{u}(t, x_1, y_1) = A \int_0^t (t - r)\tilde{u}(r, x_1, y_1)dr + x_1 + ty.
$$

Since $\tilde{u}(t, x_1, y_1) \in D(A)$, so we let

$$
\begin{aligned}
u(t, x, y) &\equiv (\lambda - A)\tilde{u}(t, x_1, y_1) \\
&= A \int_0^t (t - r)(\lambda - A)\tilde{u}(r, x_1, y_1)dr + (\lambda - A)x_1 + t(\lambda - A)y_1 \\
&= A \int_0^t (t - r)u(r, x, y)dr + x + ty.
\end{aligned}
$$

Therefore $u(t, x, y)$ is the mild solution fo (1.1).

For any $x, y \in Im(C)$, let $u_1(t, x, y), u_2(t, x, y)$ be the mild solutin of (1.1), denote $\tilde{u}_0(t, x_1, y_1) \equiv (\lambda - A)^{-1}[u_1(t, x_1, y_1) - u_2(t, x, y)]$. Then

$$\frac{d^2}{dt^2} \tilde{u}_0(t, x_1, y_1) = A\tilde{u}_0(t, x_1, y_1),$$

$$\tilde{u}_0(0, x_1, y_1) = 0, \quad \tilde{u}_0'(0, x_1, y_1) = 0.$$

From the uniqueness of the solution of (1.1), we have $\tilde{u}_0(t, x_1, y_1) \equiv u_2(t, x, y), \quad t \in R$.

Theorem 3.4. Suppose C is injective. Then the following are equivalent:

(i) $CA \subseteq AC$, (1.1) has a unique mild solution for all $x, y \in Im(C)$.

(ii) There exists a strongly continuous family of operators $\{C(t)\}_{t\in R}$, such that

(a) $C(t)A \subseteq AC(t), \quad t \in R$;

(b) for any $x \in X, \quad t \in R, \int_0^t (t - s)C(s)x\,ds \in D(A)$, and

$$A \int_0^t (t - s)C(s)x\,ds = C(t)x - Cx.$$

(iii) $CA \subset AC$, (1.1) has a unique mild solution for every $x, y \in Im(C)$ and there exists a strongly continuous family of operators $\{C(t)\}_{t\in R}$ satisfying (ii, b).

(iv) There exists a C-cosine operator function $\{C(t)\}_{t\in R}$ for A. If $\rho(A) \neq \phi$, then (i)-(iv) are equivalent to the following.

(v) $CA \subseteq AC$, (1.1) has a unique solution for every $x, y \in C(D(A))$.

(vi) There exists a C-cosine operator function $\{C(t)\}_{t\in R}$ generated by A. If $\overline{In(C)} = X$, then (i)-(iv) are equivalent to the following.

(vii) $CA \subseteq AC$, there exists a C-cosine operator function $\{C(t)\}_{t\in R}$ generated by $C^{-1}AC$.

Proof. (i)$\Rightarrow$(ii) let $x, y \in Im(C)$, (1.1) has a unique mild solution $u(t, x, y)$. For any $t \in R$, define a family of operators $\{C(t)\}_{t\in R}$ on X:

$$C(t)x \equiv u(t, Cx, 0). \tag{3.1}$$

First we prove $C(t) \in B(X)$ for any $t \in R$. Since

$$u(t, Cx, 0) = A \int_0^t (t - r)u(r, Cx, 0)dr + Cx, \quad t \in R, x \in X,$$

let $x_0 \to x, C(t)_{X_n} \to \phi(t)$, uniformly on compact subsets of $(-\infty, \infty)$. Then

$$C(t)_{X_n} = A \int_0^t (t - r)C(r)_{X_n}dr + C_{X_n}.$$

Thus, letting $n \to \infty$, and using the fact that A is closed and the convergence is uniform, we find that $\int_0^t (t-r)\phi(r)dr \in D(A)$, and

$$\phi(t) = A \int_0^t (t-r)\phi(r)dr + Cx, \quad t \in R.$$

From the uniqueness of the mild solution fo (1.1), we have

$$\phi(r) = u(t, Cx, 0) = C(t)x.$$

Then $C(t)$ is closed on X for any $t \in R$, by the closed graph theorem, $C(t) \in B(X)$ for any $t \in R$.

Let $x \in D(A)$, denote

$$w(t) \equiv \int_0^t (t-r) \int_0^r (r-s)u(s, CAx, 0)dsdr + \frac{1}{2}t^2 Cx.$$

Then

$$\begin{aligned} Aw(t) &= \int_0^t (t-r)A \int_0^r (r-s)u(s, CAx, 0)dsdr + \frac{1}{2}t^2 ACx \\ &= \int_0^t (t-r)[u(r, CAx, 0) - CAx]dr + \frac{1}{2}t^2 ACx \\ &= \int_0^t (t-r)u(r, CAx, 0)dr = \frac{d^2}{dt^2}w(t) - Cx. \end{aligned}$$

From the uniqueness of the mild solution of (1.1), we have

$$w(t) = \int_0^t (t-r)u(r, CAx, 0)dr.$$

By differentiating twice with respect to t, we obtain

$$u(r, CAx, 0) = \int_0^t (t-r)u(r, CAx, 0)dr + Cx \in D(A).$$

Then

$$\begin{aligned} AC(t)x &= Au(t, Cx, 0) = A \int_0^t (t-r)u(r, CAx, 0)dr + ACx \\ &= u(r, CAx, 0) = C(t)Ax. \end{aligned}$$

This proves (a). (b) is obvious since $u(t, x, y)$ is the mild solution of (1.1) and $C(t)$ is defined by (3.1).

(ii)$\to$(iii) It is obvious that $CA \subseteq AC$. Suppose $v(\cdot) \in C((-\infty, \infty), X)$ and satisfies

$$v(t) = A \int_0^t (t-r)v(r)dr.$$

For $t, s \in R$, using (a), (b), we have

$$\frac{d}{ds}[C(t-s)\int_0^s (s-r)v(r)dr + \int_0^{t-s} C(\tau)\int_0^s v(r)dr d\tau]$$
$$= -\int_0^{t-s} C(\tau)\left(A\int_0^s (s-r)v(r)dr\right)d\tau + C(t-s)\int_0^s v(r)dr - C(t-s)\int_0^s v(r)dr$$
$$+ \int_0^{t-s} C(\tau)v(s)dr\tau = 0.$$

By integrating with respect to s from 0 to t, we have

$$C\int_0^t (t-r)v(r)dr = 0.$$

Since C is injective, we have

$$\int_0^t (t-r)v(r)dr = 0.$$

By differentating twice, we have $v(t) = 0$. This implies the mild solution of (1.1) is unique.

(iii) $\Rightarrow$ (iv) For $x \in D(A)$, define

$$C_1(t)x \equiv \int_0^t (t-s)C(s)Axds + Cx.$$

Since $\int_0^t (t-s)C(s)Axds \in D(A), CA \subseteq AC$, we have

$$A\int_0^t (t-s)C_1(s)xds = \int_0^t (t-s)AC_1(s)xds$$
$$= \int_0^t (t-s)A\int_0^s (s-r)C(r)Axdrds + \int_0^t (t-s)ACxds$$
$$= \int_0^t (t-s)C(s)Axds = C_1(t)x - Cx.$$

By the uniqueness of mild solutions of (1.1), it follows that

$$C(t)x = C_1(t)x = \int_0^t (t-s)C(s)Axds + Cx.$$

This implies that

$$A\int_0^t (t-s)C(s)xds = \int_0^t (t-s)C(s)Axds. \tag{3.2}$$

We may differentiate twice both sides of (3.2) to conclude that $C(t)x \in D(A)$, with $AC(t)x = C(t)Ax$.

We now prove that $\{C(t)\}_{t \in R}$ is a C-cosine operator function. By the uniqueness of mild solutions of (1.1) and (ii,b), it is obvious that $C(t)C(s) = C(s)C(t), C(t)C = CC(t), C(-t) = C(t), t, s \in R$. For $x \in X, s \in R$, by (ii,b),

$$C(t)C(s)x = A\int_0^t (t-r)C(r)C(s)xdr + CC(s)x,$$

$$C(t \pm s)Cx = A \int_0^{t \pm s} (t \pm s - r)C(r)Cx dr + C^2 x.$$

Therefore

$$
\begin{aligned}
C(t+s)Cx + C(t-s)Cx &= A \int_s^{t+s} (t+s-r)C(r)Cx dr \\
&\quad +A \int_0^s (t+s-r)C(r)Cx dr + C^2 x \\
&\quad +A \int_{-s}^{t-s} (t-s-r)C(r)Cx dr + A \int_0^{-s} (t-s-r)C(r)Cx dr \\
&= A \int_0^{t-r} (t-r)[C(r+s)Cx + C(r-s)Cx] dr \\
&\quad +A \int_0^s [(t+s-r)-(t-s+r)]C(r)Cx dr + 2C^2 x \\
&= A \int_0^t (t-r)[(t+s)Cx + C(r-s)Cx] dr \\
&\quad +2 \left[A \int_0^s (s-r)C(r)Cx dr \right] \\
&= A \int_0^t (t-r)[C(r+s)Cx + C(r-s)Cx] dr + 2C(s)Cx.
\end{aligned}
$$

By the uniqueness of mild solutions of (1.1), we have

$$C(t+s)Cx + C(t-s)Cx = 2C(t)C(s)x = 2C(s)C(t)x.$$

This implies that $\{C(t)\}_{t \in R}$ is a C-cosine operator function for A.

(iv)$\Rightarrow$ (i) It is obvious that $CA \subseteq AC$. Let $x, y \in Im(C)$, denote $u(t, x, y) = C(t)C^{-1}x + \int_0^t C(s)C^{-1}y ds$. Then $\int_0^t u(s, x, y)ds \in D(A)$, and

$$
\begin{aligned}
u(t, x, y) &= C(t)C^{-1}x + \int_0^t C(s)C^{-1}y ds \\
&= x + A \int_0^t (t-s)C(s)C^{-1}x ds + ty + A \int_0^t (t-s) \int_0^s C(r)C^{-1}y dr ds \\
&= A \int_0^t (t-s)u(s, x, y)ds + x + ty.
\end{aligned}
$$

This implies that $u(t, x, y)$ is the mild solution of (1.1). The proof of the uniqueness is the same as in (ii)$\Rightarrow$(iii).

If $\rho(A) \neq \rho$, by Theorem 3.3, Proposition 3.2, we have (i)$\Leftrightarrow$(v); by Proposition 3.2 (iii), we have (iv)$\Leftrightarrow$(vi).

If $\overline{Im(C)} = X$, by theorem 2.10, we (iv)$\Leftrightarrow$(vii).

Corollary 3.5. The following are equivalent:

(i) (1.1) has a unique mild solution for all $x, y \in X$.

(ii) There exists a strongly continuous family of operators $\{C(t)\}_{t \in R}$, such that

(a) $C(t)A \subseteq AC(t)$, $t \in R$,

(b) for any $x \in X$, $t \in R$, $\int_0^t (t - s)C(s)x\,ds \in D(A)$, and

$$A \int_0^t (t - s)C(s)x\,ds = C(t)x - x.$$

(iii) (1.1) has a unique mild solution for every $x, y \in X$ and there exists a strongly contnuous family of operators $\{C(t)\}_{t \in R}$ satisfying (ii,b).

(iv) There exists a strongly continuous cosine operator function generated by A. If $\rho(A) \neq \rho$, then (i)-(iv) are equivalent to the following:

(v) (1.1) has a unique solution for all $x, y \in D(A)$.

Definition 3.6. The complex number λ is in $\rho_c(A)$, the C-resolvent set of A, if $\lambda - A$ is injective and $Im(C) \subseteq Im(\lambda - A)$.

In the exponentially bounded case the condition $\rho(A) \neq \phi$ in Theorem 3.3 can be replaced by the condition that $\rho_c(A)$ contains those $\lambda's$ with λ sufficiently large.

Theorem 3.7. Supose C is injective, $CA \subseteq AC$ and there exists $\omega > 0$ such that $\lambda^2 \in \rho_c(A)$, for $\lambda > \omega$. Then (1.1) has a uniqure mild sdution $u(t, x, y)$ which is $O(e^{\omega|t|})$ for all $x, y \in Im(C)$ if and only if (1.1) has a unique solution $\tilde{u}(t, x, y)$ such that $\tilde{u}(t, x, y), \tilde{u}''(t, x, y)$ are $O(e^{\omega|t|})$ for all $x, y \in (\lambda^2 - A)^{-1}C(X)$.

Proof. "$\Rightarrow$" Let $x_0, y_0 \in X, u(t, Cx_0, Cy_0)$ be the unique mild solution of (1.1). Then by Theorem 3.4 there exists a C-cosine operator function $\{C(t)\}_{t \in R}$ for A, and $u(t, Cx_0, Cy_0) = C(t)x_0 + \int_0^t C(s)y_0\,ds$, hence

$$\begin{aligned}
u(s, Cu(t, Cx_0, Cy_0), 0) &= C(s)u(t, Cx_0, Cy_0) \\
&= C(s)C(t)x_0 + \int_0^t C(s)C(r)y_0\,dr \\
&= \frac{1}{2}[C(t + s)Cx_0 + C(t - s)Cx_0 + \int_0^t (C(r + s)Cy_0 + C(r - s)Cy_0)dr] \\
&= \frac{1}{2}C[C(t + s)Cx_0 + \int_0^{t+s} C(r)y_0\,dr + C(t - s)x_0 + \int_0^{t-s} C(r)y_0)dr \\
&\quad + \int_s^0 C(r)y_0\,dr + \int_{-s}^0 C(r)y_0\,dr] \\
&= \frac{1}{2}C[(u(t + s), Cx_0, Cy_0) + u(t - s, Cx_0, Cy_0)].
\end{aligned}$$

(3.3)

Let $x, y \in (\lambda^2 - A)^{-1}C(X)$. Then there exist $x_1, y_1 \in X$, such that $x = (\lambda^2 - A)^{-1}Cx_1, y = (\lambda^2 - A)^{-1}Cy_1$. Suppose $u(t, Cx_1, Cy_1)$ is the $O(e^{\omega|t|})$ mild solution of (1.1).

Then for $Re\lambda > \omega$,

$$\int_0^\infty e^{-\lambda s} u(s, Cx_1, Cy_1) ds = \lambda^2 \int_0^\infty e^{-\lambda s} A \int_0^s (s-r) u(r, Cx_1, Cy_1) dr ds \in D(A)$$

and

$$A \int_0^\infty e^{-\lambda s} u(s, Cx_1, Cy_1) ds = \lambda^2 \int_0^\infty e^{-\lambda s} A \int_0^\infty e^{-\lambda s} A \int_0^s (s-r) u(r, Cx_1, Cy_1) dr ds$$
$$= \lambda^2 \int_0^\infty e^{-\lambda s} [u(s, Cx_1, Cy_1) - Cx_1 - sCy_1] ds$$
$$= \lambda^2 \int_0^\infty e^{-\lambda s} u(s, Cx_1, Cy_1) ds - \lambda Cx_1 - Cy_1.$$

Hence

$$(\lambda^2 - A) \int_0^\infty e^{-\lambda s} u(s, Cx_1, Cy_1) ds = \lambda Cx_1 + Cy_1.$$

Substituting 0 for y_1, $u(t, Cx_1, Cy_1)$ for x_1, from (3.3) we have

$$Cu(t, Cx_1, Cy_1) = \frac{1}{\lambda}(\lambda^2 - A) \int_0^\infty e^{-\lambda s} u(s, Cu(t, Cx_1, Cy_1, 0) ds$$
$$= \frac{1}{2\lambda} C(\lambda^2 - A) \int_0^\infty e^{-\lambda s} [u(t+s), Cx_1, Cy_1) + u(t-s, Cx_1, Cy_1)] ds.$$

Since C is injective, then

$$u(t, Cx_1, Cy_1) = \frac{1}{2\lambda}(\lambda^2 - A) \int_0^\infty e^{-\lambda s} [u(t+s), Cx_1, Cy_1) \qquad (3.4)$$
$$+ u(t-s, Cx_1, Cy_1)] ds \in Im(\lambda^2 - A).$$

Therefore, we can define

$$\tilde{u}(t, x, y) \equiv (\lambda^2 - A)^{-1} u(t, Cx_1, Cy_1) \qquad (3.5)$$
$$= \frac{1}{2\lambda} \int_0^\infty e^{\lambda s} [u(t+s, Cx_1, Cy_1) + u(t-s, Cx_1, Cy_1)] ds.$$

As in the proof of Theorem 3.3, we know that $\tilde{u}(t, x, y)$ is the solution of (1.1). Since $u(t, Cx_1, Cy_1)$ is $O(e^{\omega|t|})$, by (3.5), $\tilde{u}''(t, x, y)$ is also $O(e^{\omega|t|})$.

"$\Rightarrow$" Let $x, y \in Im(C)$. Then $x_1 = (\lambda^2 - A)^{-1} x$, $y_1 = (\lambda^2 - A)^{-1} y \in (\lambda^2 - A)^{-1} C(X)$. Let $\tilde{u}(t, x_1, y_1)$ be the solution of (1.1). As in the proof of Theorem 3.3, we know $u(t, x, y) \equiv (\lambda^2 - A) \tilde{u}(t, x_1, y_1)$ is the mild solution of (1.1). From $u(t, x, y) = \lambda^2 \tilde{u}(t, x_1, y_1) - A\tilde{u}(t, x_1, y_1) = \lambda^2 \tilde{u}(t, x_1, y_1) - \tilde{u}''(t, x_1, y_1)$, $u(t, x, y)$ is $O(e^{\omega|t|})$.

The uniqueness has been proved in Theorem 3.3.

Theorem 3.8. Suppose C is injective, $\omega > 0$. Then the following are equivalent.

(i) $CA \subseteq AC$, (1.1) has a unique mild solution $u(t, x, y)$ which is $O(e^{\omega|t|})$ for all $x, y \in Im(C)$.

(ii) There exists a strongly continuous family of operators $\{C(t)\}_{t \in R}$ which is $O(e^{\omega|t|})$, such that

(a) $C(t)A \subseteq AC(t), \quad t \in R$;

(b) for any $x \in X, \quad t \in R, \int_0^t (t - s)C(s)x ds \in D(A)$, and $A \int_0^t (t - s)C(s)x ds = C(t)x - Cx$.

(iii) There exists a C-cosine operator function $\{C(t)\}_{t \in R}$ for A which is $O(e^{\omega|t|})$.

(iv) $CA \subseteq AC$, for any $\lambda > \omega, \lambda^2 \in P_c(A)$ and there exists a strongly continuous family of operators $\{C(t)\}_{t \in R}$ which is $O(e^{\omega|t|}), C(t) = C(-t), t \in R$, and

$$\lambda(\lambda^2 - A)^{-1}Cx = \int_0^\infty e^{-lt}C(t)x dt, \quad x \in X. \tag{3.6}$$

(v) For any $\lambda > \omega, \lambda^2 \in P_c(A), CA \subseteq AC$ and there exists a C-cosine operator function $\{C(t)\}_{t \in R}$ generated by $C^{-1}AC$ which is $O(e^{\omega|t|})$.

In addition, if there exists $\lambda > \omega$, such that $\lambda^2 \in \rho_c(A)$, then (i)-(v) are equivalent to

(vi) $CA \subseteq AC$, (1.1) has a unique solution $\tilde{u}(t, x, y)$ that both $\tilde{u}(t, x, y), \tilde{u}''(t, x, y)$ are $O(e^{\omega|t|})$.

Furthermore, if there exists $\lambda > \omega$, such that $\lambda^2 \in \rho(A)$, then (i)-(vi) are equivalent to

(vii) There exists a C-cosine operator function $\{C(t)\}_{t \in R}$ generated by A which is $O(e^{\omega|t|})$.

Proof. (i)$\Rightarrow$(ii). As in the proof of theorem 3.4, $C(t)x = u(t, Cx, 0)$, using Banach-Steinhaus theorem, $C(t)$ is $O(e^{\omega|t|})$.

(ii)$\Rightarrow$(iii), (iii)$\Rightarrow$(i) are obvious.

(iii)$\Rightarrow$(iv). It is obvious that $C(t) = C(-t), t \in R$. For any $\lambda > \omega, x \in X$.

$$
\begin{aligned}
(\lambda^2 - A) &\int_0^\infty e^{-\lambda t}C(t)x dt \\
&= \lambda^2 \int_0^\infty e^{-\lambda t}C(t)x dt - \lambda^2 \int_0^\infty e^{-\lambda t} A \int_0^t (t - s)C(s)x ds dt \\
&= \lambda^2 \int_0^\infty e^{-\lambda t}C(t)x dt - \lambda^2 \int_0^\infty e^{-\lambda t}[C(t)x - Cx] dt \\
&= \lambda Cx.
\end{aligned}
$$

Since $C(t)A \subseteq AC(t), t \in R$, then for any $\lambda > \omega, x \in X(A)$,

$$\int_0^\infty e^{-\lambda t}C(t)(\lambda^2 - A)x dt = (\lambda^2 - A) \int_0^\infty e^{-\lambda t}C(t)x dt = Cx.$$

Therefore $\lambda^2 \in \rho_c(A)$, and

$$\lambda(\lambda^2 - A)^{-1}Cx = \int_0^\infty e^{-\lambda t}C(t)x dt, \quad x \in X.$$

(iv)$\Rightarrow$(ii). Since $CA \subseteq AC$, the uniqueness of the Laplace transform and the fact that $C(t)$ is continuous imply that, for all $t \in R, C(t)$ commutes with all C-resolvents. This implies that, for all $x \in D(A), C(t)x \in D(A)$, with $C(t)Ax = AC(t)x$.

Without loss of generality (by translating A if necessary), suppose $\omega < 0$. Define, for any $t \geq 0$,

$$\tilde{C}(t)x \equiv \int_0^t (t-s)C(s)x\,ds, \quad x \in X.$$

By (iv),

$$\frac{1}{\lambda}(\lambda^2 - A)^{-1}Cx = \int_0^\infty e^{\lambda t}\tilde{C}(t)x\,dt, \quad \forall x \in X, \quad \lambda > 0. \tag{3.7}$$

(3.7) implies that

$$\left\| \left(\frac{d}{d\lambda}\right)^k \left(\frac{1}{\lambda}(\lambda^2 - A)^{-1}Cx\right) \right\| \leq Mk\lambda^{-(k+1)}\|x\|, \quad \lambda > 0, \quad k \in N, \tag{3.8}$$

while (3.6) implies that

$$\left\| \left(\frac{d}{d\lambda}\right)^k (\lambda(\lambda^2 - A)^{-1}Cx) \right\| \leq Mk\lambda^{-(k+1)}\|x\|, \quad \lambda > 0, \quad k \in N \tag{3.9}$$

for some constant M. Assertions (3.8) and (3.9), along with the identity

$$A\left(\frac{1}{\lambda}(\lambda^2 - A)^{-1}Cx\right) = \lambda(\lambda^2 - A)^{-1}Cx - \frac{1}{\lambda}Cx$$

imply that

$$\left\| \left(\frac{d}{d\lambda}\right)^k \left(\frac{1}{\lambda}(\lambda^2 - A)^{-1}C\right) \right\|_{[D(A)]} \leq M_1 k\lambda^{-(k+1)}\|x\|, \quad \lambda > 0, \quad k \in N$$

for some constant M_1. By [1, Theorem 1.1], there exists $C_x : [0,\infty) \to [D(A)]$ such that $C_x(0) = 0$ and

$$\frac{1}{\lambda}(\lambda^2 - A)^{-1}C_x = \lambda \int_0^\infty e^{-\lambda t}C_x(t)\,dt.$$

Comparing this with (3.7) tells us that $\int_0^t \tilde{C}(s)x\,ds = C_x(t) \in D(A)$, for any $x \in X$. Assertion (3.7) now implies that

$$\frac{1}{2}t^2 Cx = \tilde{C}(t)x - A\int_0^t (t-s)\tilde{C}(s)x\,ds.$$

Since A is closed, we may differentiate this twice to conclude that $\int_0^t (t-s)C(s)x\,ds = \tilde{C}(t)x \in D(A)$, with

$$C(t)x - Cx = A\int_0^t (t-s)C(s)x\,ds, \quad x \in X, t \geq 0 \tag{3.10}$$

Since $C(t) = C(-t)$, (3.10) also holds for $t \in R$.

(iii)$\Rightarrow$(v). It is obvious by Theorem 2.6(3) and (iv).

(v)$\Rightarrow$(iv). By [9, Theorem 2.4], for any $\lambda > \omega, \lambda^2 - C^{-1}AC$ is injective, $Im(\lambda^2 - C^{-1}AC) \supseteq Im(C)$ and for $x \in X$.

$$\lambda(\lambda^2 - C^{-1}AC)^{-1}Cx = \int_0^\infty e^{-\lambda t}C(t)x dt.$$

Since $\lambda^2 - A \subseteq \lambda^2 - C^{-1}AC, \lambda^2 - A$ is injective. This, togather with $Im(\lambda^2 - A) \supseteq Im(C)$, gives

$$(\lambda^2 - A)^{-1}C = C^{-1}C(\lambda^2 - A)^{-1}C = [C^{-1}(\lambda^2 - A)^{-1}C]C = (\lambda^2 - C^{-1}AC)^{-1}C.$$

Hence

$$\lambda(\lambda^2 - A)^{-1}Cx = \int_0^\infty e^{-\lambda t}C(t)x dt.$$

And $C(t) = C(-t), t \in R$ is obvious.

In addition, if there exists $\lambda > \omega$, such that $\lambda^2 \in \rho_c(A)$, by Theorem 3.7, (i)$\Leftrightarrow$(vi). Furthermore, if there exists $\lambda > \omega$, such that $\lambda^2 \in \rho(A)$, then by Theorem 3.4, (i)-(vi)$\Leftrightarrow$(vii).

Corollary 3.9. Let $\omega > 0$. The following are equivalent:

(i) (1.1) has a unique mild solution $u(t, x, y)$ which is $O(e^{\omega|t|})$ for all $x, y \in X$.

(ii) There exists a strongly continuous family of operators $\{C(t)\}_{t \in R}$ which is $O(e^{\omega|t|})$, such that

(a) $C(t)A \subseteq AC(t), t \in R$.

(b) for any $x \in X, t \in R, \int_0^t (t-s)C(s)x ds \in D(A)$, and $A \int_0^t (t-s)C(s)x ds = C(t)x - x$.

(iii) There exists a strongly continuous cosine operator function $\{C(t)\}_{t \in R}$ generated by A which is $O(e^{\omega|t|})$.

(iv) For any $\lambda > \omega, \lambda^2 \in \rho(A)$ and there exists a strongly continuous operator family $\{C(t)\}_{t \in R}$ which is $O(e^{\omega|t|}), C(t) = C(-t).t \in R$, and

$$\lambda(\lambda^2 - A)^{-1}x = \int_0^\infty e^{-\lambda t}C(t)x dt.$$

In addition, if there exists $\lambda > \omega$, such that $\lambda^2 \in \rho(A)$, then (i)-(iv) are equivalent to

(v) (1.1) has a unique solution $\tilde{u}(t, x, y), \tilde{u}''(t, x, y)$ are $O(e^{\omega|t|})$ for all $x, y \in D(A)$.

References

[1] W. Arendt, Resolvent positive operators, Proc. London Math. Soc. 54(1987), 321-349.

[2] W. Arendt, Vector-valued Laplace transforms and Cauchy problems, Israel J. Math. 59(1987), 327-352.

[3] R. deLaubenfels, C-semigroups and the Cauchy problem, J. Funct. Anal., 111(1993), 44-61.

[4] R. deLaubenfels, Existence Families, Functional Calculi and Evolution Equations, Lecture Notes is Math., Spring Verlag, to appear.

[5] R. deLaubenfels, G.Z. Sun and S.W. Wang, Regularized semigroups, existence families and the abstract Cauchy problem, to submit.

[6] J.A. Goldstein, Semigroups of Linear Operators and Applications, Oxford, New York, 1985.

[7] N. Tahaki and I. Miyadera, C-semigroup and the abstract Cauchy problem, J. Math. Anal. Appl., 170(1992), 196-206.

[8] H.Y. Wang, On the spectral mapping theorem for C-cosine operator functions, J. Southeast Univ., 24(1994) 5, 82-86.

[9] H.Y. Wang, C-cosine operator functions and the second order abstract Cauchy problem, Northeast Math. J., 11(1995)1, 1-10.

Integrated Semigroups, Cosine Families
and Higher Order Abstract
Cauchy Problems

Tijun Xiao

Department of Mathematics, Yunnan Teachers' University, Kunming

Jin Liang

Kunming Institute of Technology, Kunming

Abstract. By a direct investigation on the higher order abstract Cauchy problem: $u^{(n)}(t) + \sum_{i=0}^{n-1} A_i u^{(i)}(t) = 0, (t \geq 0); u^{(k)}(0) = u_k, (0 \leq k \leq n-1)$, some new results concerning the existence, uniqueness and continuous dependence (in some sense) on the initial data of its solutions are obtained; in the case of $n = 2$, we improve the corresponding results by Neubrander [5], where the second order problem is reduced to a first order system, and techniques from the theory of "integrated semigroups" are employed. In addition, we carry out a further study on the special case when $-A_{n-2}$ generates an "integrated cosine family " and $A_{n-1} = 0, D(A_i) \supset D(A_{n-2}), (0 \leq i \leq n-2)$.

1. General case

This paper is concerned with the higher order abstract Cauchy problem

$$(ACP_n) \qquad \begin{cases} u^{(n)}(t) + \sum_{i=0}^{n-1} A_i u^{(i)}(t) = 0, (t \geq 0), \\ u^{(k)}(0) = u_k, (0 \leq k \leq n-1), \end{cases}$$

where $n \geq 2, A_0, \cdots, A_{n-1}$ are closed linear operators on a Banach space E. A function $u(\cdot) \in C^n(R^+, E)$ is said to be a solution of (ACP_n) if for $0 \leq i \leq n-1, u^{(i)}(t) \in D(A_i)(t \geq 0), A_i u^{(i)}(\cdot) \in C(R^+, E)$ and (ACP_n) is satisfied everywhere.

1991 Mathematics Subject Classification: 47D05, 34G10

Both authors are supported by the National NSF of China and the AFSF of Yunnan Province.

Throughout this paper, r is a nonnegative integer. The characteristic polynomial of the equation is written as

$$P_\lambda = \lambda^n + \sum_{i=0}^{n-1} \lambda^i A_i$$

with

$$\text{domain } D(P_\lambda) = \bigcap_{i=0}^{n-1} D(A_i);$$

by R_λ we denote P_λ^{-1}, if the inverse exists. $L(X,Y)$ will indicate the Banach space of all bounded linear operators from X to Y (where X, Y are Banach spaces), $L(X) = L(X, X)$. By $\mathcal{L}$, we denote the Laplace transform

$$\mathcal{L}(f(t))(\lambda) := \int_0^\infty e^{-\lambda t} f(t) dt, \quad \text{(for } \lambda \text{ sufficiently large)},$$

where $f : R^+ \to X$ is continuous and exponentially bounded. If A is a closed linear operator with domain $D(A)$ in a Banach space $(X, \|\cdot\|)$, then

$$\|x\|_m^A := \|x\| + \|Ax\| + \cdots + \|A^m x\|, \quad (x \in D(A^m)).$$

For $f \in C(R^+, X)$,

$$J^m f(t) = \begin{cases} f(t) & \text{if } m = 0 \\ \int_0^t \dfrac{(t-s)^{m-1}}{(m-1)!} f(s) ds & \text{if } m = 1, 2, 3, \cdots. \end{cases}$$

By M_n, we will denote the operator matrix

$$\begin{pmatrix} -A_{n-1} & I & 0 & \bullet & \bullet & 0 \\ -A_{n-2} & 0 & I & \bullet & \bullet & 0 \\ \bullet & & \bullet & \bullet & \bullet & \bullet \\ \bullet & & \bullet & \bullet & \bullet & I \\ -A_0 & 0 & 0 & \bullet & \bullet & 0 \end{pmatrix}$$

with domain $D(M_n) = (D(A_0) \cap \cdots \cap D(A_{n-1})) \times E^{n-1} \subset E^n$.

Definition 1.1. Let X, Y be Banach spaces, $\omega > 0$. A function $H : (\omega, \infty) \to L(X, Y)$ (resp. $h : (\omega, \infty) \to X$) is in the class $\mathcal{L} - L(X, Y)$ (resp. $\mathcal{L} - X$) if there exists a strongly continuous (resp. continuous), exponentially bounded $F : R^+ \to L(X, Y)$ (resp. $f : R^+ \to X$) such that $\mathcal{L}(F(t)x)(\lambda) = H(\lambda)x$ for $x \in X, \lambda > \omega$ (resp. $\mathcal{L}(f(t))(\lambda) = h(\lambda)$ for $\lambda > \omega$). F (resp. f) is called the determining function of H (resp. h).

Definition 1.2. Let $\omega > 0$. An infinite differentiable function $H : (\omega, \infty) \to L(E)$ is said to be in the class ψ, if

$$\sup\{\|(\lambda - \omega)^{m+1} H^{(m)}(\lambda)/m!\|; \quad \lambda > \omega, m = 0, 1, 2, \cdots, \} < \infty.$$

Definition 1.3. A closed linear operator (on E) B is called the generator of an r-times integrated semigroup, if there is $\omega > 0$ such that $(\omega, \infty) \subset \rho(B)$ (the resolvent) and $\lambda^{-r}(\lambda - B)^{-1} \in \mathcal{L} - L(E)$.

In [5], Neubrander considered (ACP_2). By reducing (ACP_2) to the first order system

$$w'(t) = M_2 w(t), (t \geq 0); w(0) = (u_0, u_1 + A_1 u_0),$$

and looking for conditions ensuring that M_2 generate an r-times integrated semigroup, the following results are given.

Theorem I[5,§5]. Assume that R_λ exists for λ sufficiently large. If either

    (a) $\lambda^{1-r} R_\lambda, \lambda^{-r} A_0 R_\lambda \in \mathcal{L} - L(E)$ or

    (b) $\lambda^{2-r} R_\lambda, \lambda^{1-r} A_0 R_\lambda \in \psi$ or

    (c) $\|\lambda^{3-r} R_\lambda\|, \|\lambda^{2-r} A_0 R_\lambda\|$ are bounded on a right half plane   or

    (d) $D(A_0) \cap D(A_1)$ is dense in E, and $\lambda^{1-r} R_\lambda, \lambda^{-r} A_0 R_\lambda \in \psi$, then for every initial value (u_0, u_1) with $(u_0, u_1 + A_1 u_0) \in D(M_2^{r+2})$, (ACP_2) has a unique solution $u(\cdot)$ satisfying $\|u(t)\| \leq Ce^{\omega t}\|(u_0, u_1 + A_1 u_0)\|_r^{M_2}$ $(t \geq 0)$ for some constants $C, \omega > 0$.

In this paper, succeeding [7] and [8], we continue our study on (ACP_n) by direct consideration of them. We obtain

Theorem 1.4. Suppose that R_λ exist for λ sufficiently large. If either

    (i) For each $0 \leq i \leq n-1, \lambda^{i-r-1} A_i R_\lambda \in \mathcal{L} - L(E)$, or

    (ii) For each $0 \leq i \leq n-1, \lambda^{i-r} A_i R_\lambda \in \psi$, or

    (iii) For each $0 \leq i \leq n-1, \|\lambda^{i-r+1} A_i R_\lambda\|$ is bounded on a right half plane, or

    (iv) $D(A_0) \cap \cdots \cap D(A_{n-1})$ is dense in E, and for each $0 \leq i \leq n-1, \lambda^{i-r-1} A_i R_\lambda \in \psi$, then for each initial value $(u_0, u_1, \cdots, u_{n-1})$ with

$$(1.1) \qquad \left(\sum_{k=0}^{n-1} A_k u_k, \sum_{k=1}^{n-1} A_{k-1} u_k, \cdots, A_0 u_{n-2} + A_1 u_{n-1}, A_0 u_{n-1} \right) \in D(M_n^r),$$

(ACP_n) has a unique solution $u(\cdot)$ satisfying that for $t \geq 0$,

$$(1.2) \qquad \|u(t)\| \leq Ce^{\omega t} \cdot \begin{cases} \displaystyle\sum_{k=0}^{n-1} \|u_k\| + \sum_{k=0}^{n-2} \sum_{i=0}^{k} \|A_i u_k\| & \text{if } r \leq n-1 \\[3mm] \displaystyle\sum_{k=0}^{n-1} \left(\|u_k\| + \sum_{i=0}^{k} \|A_i u_k\| \right) & \text{if } r = n \\[3mm] \displaystyle\sum_{k=0}^{n-1} \left(\|u_k\| + \sum_{m=0}^{r-n} \|x_{k,m}^{(1)}\| + \sum_{i=2}^{n} \|x_{k,r-n}^{(i)}\| \right) & \text{if } r \geq n+1, \end{cases}$$

for some constants $C, \omega > 0$, where for each $1 \le i \le n, 0 \le m \le r, x_{k,m}^{(i)}$ denotes the ith component of $M_n^m x_k$,

$$(1.3) \qquad x_k := (A_k u_k, A_{k-1} u_k, \cdots, A_0 u_k, 0, \cdots, 0), \quad (0 \le k \le n-1).$$

Remark. For $n = 2$, Theorem 1.4 improves Theorem I by replacing $\lambda^{-r} A_0 R_\lambda \in \mathcal{L} - L(E)$ in (a), $\lambda^{1-r} A_0 R_\lambda \in \psi$ in (b), the boundedness of $\|\lambda^{2-r} A_0 R_\lambda\|$ in (c), $\lambda^{-r} A_0 R_\lambda \in \psi$ in (d), respectively by weaker conditions: $\lambda^{-1-r} A_0 R_\lambda \in \mathcal{L} - L(E)$ in (i), $\lambda^{-r} A_0 R_\lambda \in \psi$ in (ii), the boundedness of $\|\lambda^{1-r} A_0 R_\lambda\|$ in (iii), $\lambda^{-1-r} A_0 R_\lambda \in \psi$ in (iv), thus arriving at a (slightly) stronger conclusion. In fact, (i) is equivalent to

$$\lambda^{1-r} R_\lambda, \quad \lambda^{-1-r} A_0 R_\lambda \in \mathcal{L} - L(E),$$

by using the identity

$$(1.4) \qquad \lambda R_\lambda u + A_1 R_\lambda u + \lambda^{-1} A_0 R_\lambda u = \lambda^{-1} u, \quad (u \in E).$$

For (ii), (iii) or (iv), we have a similar equivalence. Moreover, we note

$$(u_0, u_1 + A_1 u_0) \in D(M_2^{r+2})$$
$$\Longleftrightarrow \quad u_0 \in D(A_1) \text{ and } (A_0 u_0 + A_1 u_1, A_0 u_1) \in D(M_2^r).$$

Hereafter, the following basic facts will be used freely: For $u \in E, m \in N, \lambda^{-m} u \in \mathcal{L} - E$; if $H(\lambda) \in \mathcal{L} - L(E)$ (resp. $h(\lambda) \in \mathcal{L} - E$), so is $\lambda^{-1} H(\lambda)$ (resp. $\lambda^{-1} h(\lambda)$).

Proof of Theorem 1.4. Let hypothesis (i) hold, $(u_0, u_1, \cdots, u_{n-1})$ satisfy (1.1), and x_k be as in (1.3).

It can be verified easily that $\lambda \in \rho(M_n)$ for λ sufficiently large and

$$(\lambda - M_n)^{-1} = \begin{pmatrix} \lambda^{n-1} R_\lambda & \lambda^{n-2} R_\lambda & \bullet\bullet\bullet & \lambda R_\lambda & R_\lambda \\ & & * & & \end{pmatrix}$$

Then we see that for $0 \le k, j \le n-1, 1 \le m \le r$,

$$\lambda^{j-n}(\lambda - M_n)^{-1} x_k = \lambda^{j-n-1} x_k + \lambda^{j-n-2} M_n x_k + \cdots + \lambda^{j-n-m} M_n^{m-1} x_k$$

$$+ \lambda^{j-n-m}(\lambda - M_n)^{-1} M_n^m x_k.$$

Taking the first components on the two sides of the above equality yields that for $0 \le k, j \le n-1, 1 \le m \le r$,

$$(1.5) \qquad \sum_{i=0}^{k} \lambda^{i-k+j-1} R_\lambda A_i u_k = \lambda^{j-n-1} A_k u_k + \cdots + \lambda^{j-n-m} x_{k,m-1}^{(1)}$$

$$+ \sum_{i=1}^{n} \lambda^{1-i} \lambda^{j-m-1} R_\lambda x_{k,m}^{(i)}.$$

Letting $m = r$ in (1.5), we see by hypothesis (i) that

(1.6) $$A_j \sum_{i=0}^{k} \lambda^{i-k+j-1} R_\lambda A_i u_k \in \mathcal{L} - E, \quad (0 \le k, j \le n-1).$$

Set, for λ sufficiently large,

$$U(\lambda; j) = \sum_{k=j}^{n-1} \lambda^{j-k-1} u_k - \sum_{k=0}^{n-1} \sum_{i=0}^{k} \lambda^{i-k+j-1} R_\lambda A_i u_k, \quad (0 \le j \le n-1),$$

$$U(\lambda; n) = -\sum_{k=0}^{n-1} \sum_{i=0}^{k} \lambda^{i-k+n-1} R_\lambda A_i u_k.$$

Then by virtue of (1.6), we have

(1.7) $$A_j U(\lambda; j) \in \mathcal{L} - E, \quad (0 \le j \le n-1).$$

Observe

$$-U(\lambda; n)$$

$$= \sum_{k=0}^{n-1} \left[-\sum_{i=0}^{k} \lambda^{i-k-1} (P_\lambda - \lambda^n)(R_\lambda A_i u_k) + \sum_{i=0}^{k} \lambda^{i-k-1} A_i u_k \right]$$

$$= \sum_{k=0}^{n-1} \left[\sum_{j=0}^{k} \lambda^{j-k-1} A_j u_k - \sum_{j=0}^{n-1} \lambda^j A_j \sum_{i=0}^{k} \lambda^{i-k-1} R_\lambda A_i u_k \right]$$

$$= \sum_{j=0}^{n-1} A_j U(\lambda; j); \text{ namely}$$

(1.8) $$U(\lambda; n) + \sum_{j=0}^{n-1} A_j U(\lambda; j) = 0, \text{ (for } \lambda \text{ sufficiently large).}$$

This together with (1.7) implies that $U(\lambda; n) \in \mathcal{L} - E$, and therefore

$$U(\lambda; j) \in \mathcal{L} - E, \quad (0 \le j \le n).$$

Let

(1.9) $$U(\lambda; 0) = \mathcal{L}(u(t))(\lambda).$$

We observe

$$U(\lambda; j+1) = \lambda U(\lambda; j) - u_j, \quad (0 \le j \le n-1).$$

Therefore, making use of the fact that

(1.10) $$\begin{cases} \text{If } \mathcal{L}(f_2(t))(\lambda) = \lambda \mathcal{L}(f_1(t))(\lambda) - \tilde{u}, \text{ where } \tilde{u} \in E, \\ f_1, f_2 \text{ are } E\text{-valued functions, then} \\ f_1(t) \text{ is differentiable and } f_1'(t) = f_2(t)(t \ge 0), f_1(0) = \tilde{u}, \end{cases}$$

we obtain that $u(t)$ is n-times differentiable for $t \geq 0$,

(1.11) $$u^{(j)}(0) = u_j, \quad (0 \leq j \leq n-1),$$

and

(1.12) $$U(\lambda; j) = \mathcal{L}(u^{(j)}(t))(\lambda), \quad (0 \leq j \leq n).$$

This combined with (1.7) implies that $u^{(j)}(t) \in D(A_j), (t \geq 0, 0 \leq j \leq n-1)$, and

(1.13) $$A_j U(\lambda; j) = \mathcal{L}(A_j u^{(j)}(t))(\lambda), \quad (0 \leq j \leq n-1),$$

due to the fact that

(1.14) $$\begin{cases} \text{If } h(\lambda) = \mathcal{L}(f(t))(\lambda), Ah(\lambda) = \mathcal{L}(g(t))(\lambda), \\ \text{then } f(t) \in D(A), (t \geq 0), \text{ and } Af(t) = g(t). \end{cases}$$

Thus, (1.12) and (1.13) together show that for λ sufficiently large,

$$\mathcal{L}\left(u^{(n)}(t) + \sum_{j=0}^{n-1} A_j u^{(j)}(t) \right)(\lambda) = U(\lambda; n) + \sum_{j=0}^{n-1} A_j U(\lambda; j) = 0,$$

by noting (1.8). Now, the uniqueness of Laplace transform implies

$$u^{(n)}(t) + \sum_{j=0}^{n-1} A_j u^{(j)}(t) = 0, \quad (t \geq 0).$$

This together with (1.11) indicates that $u(\cdot)$ is a solution as desired.

From the identity

(1.15) $$\lambda^{n-1} R_\lambda = \lambda^{-1} - \sum_{i=0}^{n-1} \lambda^{i-1} A_i R_\lambda,$$

we see that $\lambda^{n-r-1} R_\lambda \in \mathcal{L}-L(E)$, since by hypothesis $\lambda^{i-r-1} A_i R_\lambda \in \mathcal{L}-L(E) (0 \leq i \leq n-1)$
Let $F_n(t)$ be the determining function of $\lambda^{n-r-1} R_\lambda$, i.e.,

$$\lambda^{n-r-1} R_\lambda u = \mathcal{L}(F_n(t)u)(\lambda), \quad (u \in E).$$

From (1.9), we know

$$\mathcal{L}(u(t))(\lambda) = \sum_{k=0}^{n-1} (\lambda^{-k-1} u_k - \sum_{i=0}^{k} \lambda^{i-k-1} R_\lambda A_i u_k).$$

Therefore, if $r = n$,

$$\mathcal{L}(u(t))(\lambda) = \mathcal{L}\left(\sum_{k=0}^{n-1} \left(\frac{t^k}{k!} u_k - \sum_{i=0}^{k} J^{n-r-i+k} F_n(t) A_i u_k \right) \right)(\lambda);$$

if $r \leq n - 1$, using (1.15) again, one obtains

$$\mathcal{L}(u(t))(\lambda) = \sum_{k=0}^{n-2}\left(\lambda^{-k-1}u_k - \sum_{i=0}^{k}\lambda^{i-k-1}R_\lambda A_i u_k\right) + R_\lambda u_{n-1}$$

$$= \mathcal{L}\left(\sum_{k=0}^{n-2}\left(\frac{t^k}{k!}u_k - \sum_{i=0}^{k}J^{n-r-i+k}F_n(t)A_i u_k\right) + J^{n-r-1}F_n(t)u_{n-1}\right)(\lambda);$$

if $r \geq n + 1$, we have by virtue of (1.3) with $m = r - n$ that

$$\mathcal{L}(u(t))(\lambda)$$

$$= \mathcal{L}\left(\sum_{k=0}^{n-1}\left(\frac{t^k}{k!}u_k - \frac{t^n}{n!}A_k u_k - \cdots - \frac{t^{r-1}}{(r-1)!}x^{(1)}_{k,r-n-1} - \sum_{i=1}^{n}J^{i-1}F_n(t)x^{(i)}_{k,r-n}\right)\right)(\lambda).$$

From these observations, (1.2) follows immediately, by using the uniqueness of Laplace transform. Finally, by hypothesis, for each $0 \leq k \leq n - 1, u \in \bigcap_{i=k+1}^{n-1}D(A_i)$,

$$\lambda^{-k-r-1}\sum_{i=k+1}^{n}\lambda^i R_\lambda A_i u \in \mathcal{L} - E, \quad (A_n := I).$$

Let $G_k(t)u$ be the determining function of $\lambda^{-k-r-1}\sum_{i=k+1}^{n}\lambda^i R_\lambda A_i u$. Then it can be easily verified (by (1.15)) that for $u \in D(A_0) \cap \cdots \cap D(A_{n-1})$,

$$\mathcal{L}\left(G_0(t)u + \int_0^t G_{n-1}(s)A_0 u \, ds - \frac{t^r}{r!}u\right)(\lambda) = 0,$$

and for $u \in \bigcap_{i=k}^{n-1}D(A_i), (1 \leq k \leq n - 1)$,

$$\mathcal{L}\left(G_k(t)u + \int_0^t(G_{n-1}(s)A_k u - G_{k-1}(s)u)ds\right)(\lambda) = 0.$$

It follows that for $t \geq 0, 1 \leq k \leq n - 1$,

$$(1.16) \qquad G_0(t)u = \frac{t^r}{r!}u - \int_0^t G_{n-1}(s)A_0 u \, ds, \quad \left(u \in \bigcap_{i=0}^{n-1}D(A_i)\right),$$

$$(1.17) \qquad G_k(t)u = \int_0^t(G_{k-1}(s)u - G_{n-1}(s)A_k u)ds, \quad \left(u \in \bigcap_{i=k}^{n-1}D(A_i)\right).$$

Let now $v(\cdot)$ be an arbitrary solution of (ACP_n) with $v^{(k)}(0) \in \bigcap\limits_{i=k+1}^{n-1} D(A_k)(0 \leq k \leq n-1)$.

Then $v^{(k)}(t) \in \bigcap\limits_{i=k}^{n-1} D(A_i)$ for each $t \geq 0, 0 \leq k \leq n-1$. Accordingly, we have by (1.16) and (1.17) that for $t \geq s \geq 0$,

$$\frac{d}{ds}\left[\sum_{k=0}^{n-1} G_k(t-s)v^{(k)}(s)\right]$$

$$= -\frac{(t-s)^{r-1}}{(r-1)!}v(s) + G_{n-1}(t-s)A_0 v(s) - \sum_{k=1}^{n-1}[G_{k-1}(t-s)v^{(k)}(s)$$

$$-G_{n-1}(t-s)A_k v^{(k)}(s)] + \sum_{k=0}^{n-1} G_k(t-s)v^{(k+1)}(s)$$

$$= -\frac{(t-s)^{r-1}}{(r-1)!}v(s),$$

which gives (noting $G_k(0)v = 0$) that

$$\int_0^t \frac{(t-s)^{r-1}}{(r-1)!}v(s)ds = \sum_{k=0}^{n-1} G_k(t)v^{(k)}(0).$$

The uniqueness of solutions follows.

Arguing as in [5], we deduce that either (ii) or (iii) or (iv) implies (i). Thus, the proof is complete.

Corollary 1.5. Let the hypothesis in Theorem 1.4 hold. Then (ACP_n) has a unique solution for each $(u_0, u_1, \cdots, u_{n-1})$ satisfying

$$u_k \in D_k := \begin{cases} \bigcap\limits_{i=0}^{k} \Omega_{k,i} & \text{if } 0 \leq k \leq r-1 \\ \bigcap\limits_{i=0}^{k-r} D(A_i) \bigcap\limits_{i=k-r+1}^{k} \Omega_{k,i} & \text{if } k \geq r, \end{cases}$$

where

$$\Omega_{k,i} = \bigcap_{(i_1,\cdots,i_p)\in\Lambda_{k,i}} D(A_{n-i_p}\cdots A_{n-i_1}A_i),$$

$$\Lambda_{k,i} = \left\{(i_1,\cdots,i_p) : \sum_{m=1}^{p} i_m \geq i-k+r > \sum_{m=1}^{p-1} i_m, 1 \leq i_1,\cdots,i_p \leq n, p \in \mathbb{N}\right\}$$

Example. Consider the Cauchy problem

(1.18) $$\begin{cases} u''(t) - Au'(t) - (aA^2 + bA + cI)u(t) = 0, t \geq 0, \\ u(0) = u_0, u'(0) = u_1, \end{cases}$$

where $a, b, c \in \mathbb{C}$ (the set of complex numbers), $a \neq 0$. In this case,

$$R_\lambda = (\lambda^2 - \lambda A - aA^2 - bA - cI)^{-1}.$$

Set, as in [4],

$$f_{1,2}(\lambda) = \tfrac{1}{2a}(-\lambda - b \pm [(1 + 4a)\lambda^2 + 2b\lambda + b^2 - 4ac]^{1/2}),$$

$$g(\lambda) = [(1 + 4a)\lambda^2 + 2b\lambda + b^2 - 4ac]^{-1/2},$$

$$H_\omega = \{f_{1,2}(\lambda); \lambda \in \mathbb{C}, \operatorname{Re}\lambda > \omega\}, \quad (\omega > 0).$$

Suppose that there exist constants $C, \omega > 0, q \in \{-1, 0, 1, 2, \cdots\}$ such that for each $\lambda \in H_\omega, \lambda \in \rho(A)$ and

$$\|R(\lambda, A)\| \leq C|\lambda|^q.$$

Then there is a constant C_0 such that for $\operatorname{Re}\lambda \in \omega$,

$$\|\lambda^{-q+1} R_\lambda\| \leq \quad |\lambda|^{-q+1}\|g(\lambda)[R(f_1(\lambda), A) - R(f_2(\lambda), A)]\|$$

$$\leq \quad C_0, \quad \text{noting } [4, (8.5)];$$

$$\|\lambda^{-q} A R_\lambda\| \leq \quad |\lambda|^{-q}\|g(\lambda)[f_1(\lambda)R(f_1(\lambda), A)$$

$$- f_2(\lambda)R(f_2(\lambda), A)]\| \leq C_0.$$

Further, making use of (1.4) gives that there is $C_1 > 0$ such that

$$\|\lambda^{-q-1}(aA^2 + bA + cI)R_\lambda\| \leq C_1, \quad (\operatorname{Re}\lambda > \omega).$$

Now applying Corollary 1.5 with $n = 2, r = q + 2, A_1 = -A, A_0 = -aA^2 - bA - cI$, we conclude that (1.18) has a unique solution for each $(u_0, u_1) \in D(A^{q+5}) \times D(A^{q+4})$. Here

$$D_0 = \Omega_{0,0} = D(A^{r+3}) = D(A^{q+5});$$

if $r = 1$, i.e., $q = -1$,

$$D_1 = D(A_0) \cap \Omega_{1,1} = D(A^2) \cap D(A^{r+2}) = D(A^{r+2}) = D(A^{q+4}),$$

if $r \geq 2$, i.e., $q \geq 0$,

$$D_1 = \Omega_{1,0} \cap \Omega_{1,1} = D(A^{r+2}) \cap D(A^{r+2}) = D(A^{q+4}).$$

Definition 1.6. If hypothesis (i) in Theorem 1.4 holds, then we say $P_\lambda \in \Phi_l(n, r)$.

Theorem 1.7 (Perturbation case). Let $n \geq r+1$, $\lambda^n + \sum_{i=0}^{n-1} \lambda^i A_i \in \Phi_l(n,r)$. Let $B_0, \cdots, B_{n-r-1}$ be closed linear operators satisfying that for each $0 \leq m \leq n-r-1$, $A_m + B_m$ is closed; there is i_m with $m+r+1 \leq i_m \leq n$ such that $D(B_m) \supset D(A_{i_m})$, $\rho(A_{i_m}) \neq \phi$. Then

$$\lambda^n + \sum_{i=0}^{n-1} \lambda^i A_i + \sum_{m=0}^{n-r-1} \lambda^m B_m \in \Phi_l(n,r).$$

Proof. For each $0 \leq m \leq n-r-1$, take $\lambda_{i_m} \in \rho(A_{i_m})$. Obviously,

$$B_m(\lambda_{i_m} - A_{i_m})^{-1} \in L(E), \quad (0 \leq m \leq n-r-1),$$

since $D(B_m) \supset D(A_{i_m})$. By hypothesis, we have

$$\lambda^{n-r-1} R_\lambda, \ \lambda^{i_m - r - 1} A_{i_m} R_\lambda \in \mathcal{L} - L(E), \quad (0 \leq m \leq n-r-1).$$

It follows that

$$\lambda^m R_\lambda, \lambda^m A_{i_m} R_\lambda \in \mathcal{L} - L(E), \quad (0 \leq m \leq n-r-1),$$

by noting $m \leq i_m - r - 1$. From these observations, we deduce that for $0 \leq m \leq n-r-1$,

$$
\begin{aligned}
(1.19) \qquad \lambda^m B_m R_\lambda = \ &\lambda_{i_m}[B_m(\lambda_{i_m} - A_{i_m})^{-1}]\lambda^m R_\lambda \\
&- [B_m(\lambda_{i_m} - A_{i_m})^{-1}]\lambda^m A_{i_m} R_\lambda \in \mathcal{L} - L(E).
\end{aligned}
$$

Now, for λ sufficiently large, set

$$H_0(\lambda) = \sum_{m=0}^{n-r-1} \lambda^m B_m R_\lambda,$$

$$K_m(\lambda) = \begin{cases} \lambda^{m-r-1} A_m R_\lambda & \text{if } n-r \leq m \leq n \\ \lambda^{m-r-1}(A_m + B_m)R_\lambda & \text{if } 0 \leq m \leq n-r-1. \end{cases}$$

Then the hypothesis and (1.19) together imply that for $0 \leq m \leq n$,

$$H_0(\lambda), K_m(\lambda) \in \mathcal{L} - L(E).$$

Let, for λ sufficiently large,

$$(1.20) \qquad \begin{cases} H_0(\lambda)u = \mathcal{L}(S(t)u)(\lambda), (u \in E) \\ K_m(\lambda)u = \mathcal{L}(S_m(t)u)(\lambda), (u \in E, 0 \leq m \leq n). \end{cases}$$

Then for $u \in E, 0 \leq m \leq n, \lambda$ sufficiently large,

$$
\begin{aligned}
(1.21) \qquad K_m(\lambda)[I + H_0(\lambda)]^{-1}u = \ &K_m(\lambda)u + K_m(\lambda)\{[I + H_0(\lambda)]^{-1}u - u\} \\
= \ &\mathcal{L}(S_m(t)u + S_m(t) * \tilde{S}(t)u)(\lambda),
\end{aligned}
$$

where

$$\tilde{S}(t) = \sum_{m=1}^{\infty} (-S(t))^{*m}, \quad (t \geq 0),$$

$*m$ indicating the mth convolution power. On the other hand,

$$\sum_{i=0}^{n} \lambda^i A_i + \sum_{m=0}^{n-r-1} \lambda^m B_m = R_\lambda \left[I + \sum_{m=0}^{n-r-1} \lambda^i B_m R_\lambda \right].$$

This combined with (1.21) shows the result as asserted.

2. Special case

This section deals with the following Cauchy problem:

(2.1)
$$\begin{cases} u^{(n+2)}(t) = Au^{(n)}(t) + \sum_{i=0}^{n-1} B_i u^{(i)}(t), (t \geq 0) \\ u^{(j)}(0) = u_j, (0 \leq j \leq n+1), \end{cases}$$

where $B_0, \cdots, B_{n-1}$ are closed linear operators on E and A is the generator of an r-times integrated cosine family of bounded linear operators on E. Integrated cosine families were introduced by Arendt and Kellermann [1].

Definition 2.1. Let B be a closed linear operator on E. B is called the generator of an r-times integrated cosine family if there is $\omega > 0$ such that $(\omega, \infty) \subset \rho(B)$, and $\lambda^{1-r}(\lambda^2 - B)^{-1} \in \mathcal{L} - L(E)$.

An important sufficient condition for an operator B to be a generator of an integrated cosine family is given by the following result, which is an immediate consequence of [1, Proposition 3.1].

Proposition 2.2. Let B be a closed linear operator on E satisfying that there exist constants $C, \omega > 0$ such that for $\text{Re}\lambda > \omega, \lambda^2 \in \rho(B)$ and $\|(\lambda^2 - B)^{-1}\| \leq C(1+|\lambda|)^{r_0}$ for some $r_0 \in \{-2, -1, 0, 1, 2, \cdots\}$. Then B generates an $(r_0 + 3)$-times integrated cosine family.

By E_1, we denote the Banach space $D(A)$ endowed with the graph norm $\|u\|_{E_1} = \|u\| + \|Au\|$; for a real number b, $[b]$ will be the least integer $> b - 1$ if $b \geq 0$, and $[b] = 0$ if $b < 0$.

Theorem 2.3. Let $A, B_0, \cdots, B_{n-1}$ be as above. Assume that for each $0 \leq i \leq n - 1, D(A) \subset D(B_i)$ such that $B_i D(A) \subset D(A^{[\frac{1}{2}(i-n+r+2)]})$ for $n - r \leq i \leq n - 1$. Then (2.1) has a unique solution $u(\cdot)$ for $u_n \in D(A^{[\frac{1}{2}(r+3)]})$, $u_{n+1} \in D(A^{[\frac{1}{2}(r+2)]})$, $u_j \in \bigcap_{i=0}^{j} D(A^{[\frac{1}{2}(r-i+1)]}$ $B_{j-i})$, $(0 \le j \le n-1)$. Moreover, there exist constants $C, \omega > 0$ such that for $t \ge 0$,

$$
(2.2) \quad \|u(t)\| + \|Au(t)\| \le Ce^{\omega t}\left\{ \sum_{l=0}^{[\frac{1}{2}(r+3)]} \|A^l u_n\| + \sum_{i=0}^{[\frac{1}{2}(r+2)]} \|A^l u_{n+1}\| \right.
$$

$$
\left. + \sum_{j=0}^{n-1}\sum_{i=0}^{j}\sum_{l=0}^{[\frac{1}{2}(r-i+1)]} \|A^l B_{j-i} u_j\| + \sum_{i=0}^{n-1}(\|u_i\| + \|Au_i\|)\right\}.
$$

Proof. First, we have

$$
(2.3) \qquad \lambda^{1-r}(\lambda^2 - A)^{-1} \in \mathcal{L} - L(E),
$$

since A is the generator of an r-times integrated cosine family. Take $a \in \rho(A)$. Observe that for k being any positive integer, $u \in D(A^k)$,

$$
(\lambda^2 - A)^{-1}u = \{\lambda^{-2} + \lambda^{-2}A(\lambda^2 - A)^{-1}\}u = \cdots
$$

$$
= \{\lambda^{-2} + \cdots + \lambda^{-2k}A^{k-1} + \lambda^{-2k}A^k(\lambda^2 - A)^{-1}\}u.
$$

It follows from (2.3) that for k being any integer with $0 \le k \le \frac{1}{2}r$,

$$
(2.4) \qquad \lambda^{1-r}\lambda^{2k}(\lambda^2 - A)^{-1}(a - A)^{-k} \in \mathcal{L} - L(E),
$$

by noting $A^l(a - A)^{-k} \in L(E)$, $(l = 1, \cdots, k)$. Therefore for k being any integer with $0 \le k \le \frac{1}{2}(r+2)$,

$$
(2.5) \qquad \lambda^{1-r}\lambda^{2k-2}A(\lambda^2 - A)^{-1}(a - A)^{-k} \in \mathcal{L} - L(E),
$$

because of the following equalities: for $1 \le k \le \frac{1}{2}(r+2)$,

$$
\lambda^{1-r}\lambda^{2k-2}A(\lambda^2 - A)^{-1}(a - A)^{-k}
$$

$$
= \lambda^{1-r}\lambda^{2(k-1)}(\lambda^2 - A)^{-1}(a - A)^{-(k-1)} \cdot A(a - A)^{-1},
$$

$$
\lambda^{1-r}\lambda^{-2}A(\lambda^2 - A)^{-1} = \lambda^{1-r}(\lambda^2 - A)^{-1} - \lambda^{1-r}\lambda^{-2}.
$$

From (2.4) and (2.5), it follows that for $0 \le i \le n-1$, $m = 0, 1$,

$$
\lambda^{i-n}A^m(\lambda^2 - A)^{-1}(a - A)^{-[\frac{1}{2}(i-n+r+2)]} \in \mathcal{L} - L(E),
$$

by noting $2 \cdot [\frac{1}{2}(i - n + r + 2)] \ge i - n + r + 1$. Consequently, for $0 \le i \le n-1$,

$$
(2.6) \qquad \lambda^{i-n}(\lambda^2 - A)^{-1}(a - A)^{-[\frac{1}{2}(i-n+r+2)]} \in \mathcal{L} - L(E, E_1),
$$

by noting (1.14). By hypothesis,

$$(a - A)^{[\frac{1}{2}(i-n+r+2)]} B_i \in L(E_1, E), \quad (0 \le i \le n - 1),$$

which implies, by (2.6), that

$$\lambda^{i-n}(\lambda^2 - A)^{-1} B_i \in \mathcal{L} - L(E_1), \quad (0 \le i \le n - 1).$$

Now let

(2.7)
$$\sum_{i=0}^{n-1} \lambda^{i-n}(\lambda^2 - A)^{-1} B_i u = \mathcal{L}(x(t)u)(\lambda),$$

for $u \in D(A), \lambda$ sufficiently large, where $\{x(t)\}_{t \ge 0}$ is a strongly continuous, exponentially bounded family of bounded linear operators on E_1. Set

$$V_\lambda = \left\{ I - \sum_{i=0}^{n-1} \lambda^{i-n}(\lambda^2 - A)^{-1} B_i \right\}^{-1}$$

Then for $u \in D(A), \lambda$ sufficiently large,

(2.8)
$$(V_\lambda - I)u = \mathcal{L}\left\{ \sum_{i=0}^{\infty} (x(t))^{*i} u \right\}(\lambda).$$

Using (2.4) and (2.5) again, we obtain that for $p = -2, -1, 0, \cdots, n - 1$,

(2.9)
$$\lambda^{-p-1}(a - A)^{q(p)}(\lambda^2 - A)^{-1} \in \mathcal{L} - L(E),$$

where

(2.10)
$$q(p) = \begin{cases} -[\frac{1}{2}(r - p - 1)] & \text{if } p \le r - 1 \\ 1 & \text{if } p \ge r. \end{cases}$$

Now for $u \in E, \lambda$ sufficiently large, $p, q(q)$ as above, let

(2.11)
$$\lambda^{-p-1}(a - A)^{q(p)}(\lambda^2 - A)^{-1} u = \mathcal{L}(y_p(t)u)(\lambda).$$

We see easily that

$$y_p(t)(a - A)^{-1} = (a - A)^{-1} y_p(t), \quad (p = -2, -1, 0, 1, \cdots, n - 1),$$

which implies

$$Ay_p(t)u = y_p(t)Au, \quad (u \in D(A), t \ge 0, -2 \le p \le n - 1).$$

So $\{y_p(t)\}_{t \ge 0}$ is also a strongly continuous, exponentially bounded family of bounded linear operators on E_1. It follows from (2.8) and (2.11) that for $u \in D(A), \lambda$ sufficiently large, $-2 \le p \le n - 1, q(p)$ as in (2.10),

(2.12)
$$\lambda^{-p-1} V_\lambda (a - A)^{q(p)}(\lambda^2 - A)u = \mathcal{L}(z_p(t)u)(\lambda),$$

where

$$z_p(t) = y_p(t) + y_p(t) * \sum_{m=1}^{\infty} (x(t))^{*m}.$$

Setting, for λ sufficiently large

$$R_\lambda = \left(\lambda^{n+2} - \lambda^n A - \sum_{i=0}^{n-1} \lambda^i B_i \right)^{-1},$$

then $\lambda^n R_\lambda = V_\lambda (\lambda^2 - A)^{-1}$. Consequently, we have by (2.12) that

$$(2.13) \qquad \lambda^n R_\lambda \left(u_{n+1} + \lambda u_n + \sum_{j=0}^{n-1} \sum_{p=0}^{j} \lambda^{-p-1} B_{j-p} u_j \right) = \mathcal{L}(v(t))(\lambda),$$

where

$$v(t) = z_{-1}(t)(a - A)^{[\frac{1}{2}r]} u_{n+1} + z_{-2}(t)(a - A)^{[\frac{1}{2}(r+1)]} u_n$$

$$+ \sum_{j=0}^{n-1} \sum_{p=0}^{j} z_p(t)(a - A)^{q(p)} B_{j-p} u_j \in C(R^+, E_1),$$

by noting $(a - A)^{[\frac{1}{2}r]} u_{n+1}, (a - A)^{[\frac{1}{2}(r+1)]} u_n \in D(A)$ and

$$(a - A)^{q(p)} B_{j-p} u_j \in D(A), \quad (0 \le p \le j, 0 \le j \le n-1),$$

by hypothesis. From (2.13) and the identity

$$\lambda^n R_\lambda = \lambda^{-2} \lambda^n A R_\lambda + \lambda^{-2} \sum_{i=0}^{n-1} \lambda^i B_i R_\lambda + \lambda^{-2},$$

we deduce by the uniqueness of Laplace transform that for $t \ge 0$,

$$v(t) = \int_0^t (t-s) A v(s) ds + \int_0^t (t-s) \sum_{i=0}^{n-1} B_i J^{n-i} v(s) ds$$

$$+ t u_{n+1} + u_n + \int_0^t (t-s) I_0(s) ds,$$

where for $s \ge 0$,

$$I_0(s) = \sum_{j=0}^{n-1} \sum_{p=0}^{j} \frac{s^p}{p!} B_{j-p} u_j.$$

Hence $v''(t) = Av(t) + \sum_{i=0}^{n-1} B_i J^{n-i} v(t) + I_0(t), (t \ge 0); v(0) = u_n, v'(0) = u_{n+1}$. Accordingly,

$u(t) = \sum_{j=0}^{n-1} \frac{t^j}{j!} u_j + J^n v(t), (t \ge 0)$ is a solution of (2.1). (2.2) follows from (2.14) and (2.15).

For the uniqueness, we observe from (2.4) and (2.5) that $\lambda^{-r-1}(\lambda^2 - A)^{-1} \in \mathcal{L}-L(E, E_1)$. So $\lambda^{-r-1}\lambda^n R_\lambda = \lambda^{-r-1}V_\lambda(\lambda^2 - A)^{-1} \in \mathcal{L} - L(E, E_1)$ due to (2.8). Thus R_λ is polynomially bounded, which completes the proof by arguing as in the proof of [6, Theorem 4.1.2].

For higher order abstract Cauchy problem, associated with strongly continuous semigroups, cosine families or integrated semigroups, see [2, 3, 7, 8].

References

[1] W. Arendt and H. Kellermann, Integrated solutions of Volterra integro-differential equations and applications, Pitman Res. Notes in Math., 190(1989), 21–51.

[2] R. deLaubenfels, Integrated semigroups and integrodifferential equations, Math. Z., 204(1990), 501–514.

[3] F. Neubrander, Wellposedness of higher order abstract Cauchy problems, Trans. Amer. Math. Soc., 295(1986), 257–290.

[4] F. Neubrander, Integrated semigroups and their applications to the abstract Cauchy problem, Pacific J. Math., 135(1988), 111–157.

[5] F. Neubrander, Integrated semigroups and their application to complete second order Cauchy problems, Semigroup Forum, 38(1989), 233–251.

[6] A. Pazy, Semigroups of linear operators and applications to partial differential equations, Springer-Verlag, New York and Berlin, 1983.

[7] T.J. Xio (Xiao) and J. Liang, On complete second order linear differential equations in Banach space, Pacific J. Math., 142(1990), 175–195.

[8] T.J. Xio (Xiao) and J. Liang, The Cauchy problem for higher order abstract differential equations, Chin. Ann. of Math., A14: 5(1993), 23–35.

A Class of Toeplitz Operators on the Hardy Space

Kehe Zhu

Department of Mathematics, State University of New York, Albany, USA

1. Introduction

Let $\mathbb{D}$ be the open unit disk in the complex plane. The Hardy space H^2 consists of analytic functions

$$f(z) = \sum_{n=0}^{\infty} a_n z^n, \quad z \in \mathbb{D},$$

such that

$$\|f\| = \left[\sum_{n=0}^{\infty} |a_n|^2 \right]^{\frac{1}{2}} < +\infty.$$

It will be convenient to identify the Hardy space H^2 with the closed subspace of $L^2(\partial\mathbb{D}, d\theta)$ consisting of functions f whose Fourier series is of the form

$$f(t) = \sum_{n=0}^{\infty} a_n e^{\text{int}}.$$

We let $A(\mathbb{D})$ denote the disk algebra. Recall that a function is in $A(\mathbb{D})$ if it is continuous on the closed disk $\overline{\mathbb{D}}$ and analytic in $\mathbb{D}$.

In this paper we introduce a class of Toeplitz operators on the Hardy space H^2 which generalizes the classical Toeplitz operators on H^2. Thus for a finite complex Borel measure μ on the closed disk $\overline{\mathbb{D}}$ we define a Toeplitz type operator T_μ as follows:

$$T_\mu f(z) = \int_{\overline{\mathbb{D}}} \frac{f(w)d\mu(w)}{1 - z\overline{w}}, \quad z \in \mathbb{D}, f \in A(\mathbb{D}).$$

It is obvious that T_μ maps every function in $A(\mathbb{D})$ to an analytic function in $\mathbb{D}$.

When μ is positive, we shall determine exactly when T_μ induces a bounded operator on H^2. In this case, we shall also determine when T_μ induces a compact operator on H^2. The

1991 Mathematics Subject Classification: 47B35 and 30D55.

Research supported by the National Science Foundation

question of when T_μ belongs to the trace class will be discussed as well. As a by-product of our analyis we shall obtain several characterizations of Carleson measures on $\mathbb{D}$.

2. Boundedness

We begin with some examples of measures on $\overline{\mathbb{D}}$.

First, if $\mu = d\theta$, the normalized arc-length measure on $\partial\mathbb{D}$, then T_μ is the identity operator on H^2. More generally, if $\mu = \varphi d\theta$, where φ is a bounded function on $\partial\mathbb{D}$, then T_μ becomes the classical Toeplitz operator T_φ on H^2. See [1] and [4] for information about classical Toeplitz operators on H^2.

Second, if $\mu = dA$, the normalized area measure on $\mathbb{D}$, then T_μ is a compact operator on H^2. Actually, in this case, $T_\mu f$ behaves like an anti-derivative of f. More interestingly, if $d\mu(z) = dA(z)/\overline{z}$, then $T_\mu f$ is indeed an anti-derivative of f.

Recall that a positive Borel measure μ on $\mathbb{D}$ is called a Carleson measure if there exists a constant $C > 0$ such that

$$\int_{\mathbb{D}} |f(z)|^2 d\mu(z) \leq C\|f\|^2$$

for all $f \in H^2$, where $\|\ \|$ is the norm in H^2. Carleson measures can be characterized geometrically in terms of the so-called Carleson squares.

For any $z \in \mathbb{D}$ we define a set S_z as follows. If $z = 0$, we let $S_z = \mathbb{D}$. If $z = re^{it} \neq 0$, then we let

$$S_z = \{se^{i\theta} : r \leq s \leq 1, |\theta - t| \leq \pi(1 - r)\}.$$

The set S_z will be called the Carleson square at z. It is well known that a positive Borel measure μ on $\mathbb{D}$ is a Carleson measure if and only if

$$\mu(S_z) \leq C(1 - |z|), \quad z \in \mathbb{D},$$

for some constant $C > 0$; see [3].

For a finite complex Borel measure μ on $\overline{\mathbb{D}}$ we define two functions $\tilde{\mu}$ and $\hat{\mu}$ on $\mathbb{D}$ as follows:

$$\tilde{\mu}(z) = \int_{\mathbb{D}} \frac{1 - |z|^2}{|1 - z\overline{w}|^2} d\mu(w), \quad z \in \mathbb{D},$$

and

$$\hat{\mu}(z) = \frac{\mu(S_z)}{1 - |z|^2}, \quad z \in \mathbb{D}.$$

These two transforms of μ will play an important role in our analysis of the Toeplitz type operator T_μ.

Theorem 1. Suppose μ is a finite positive Borel measure on $\overline{\mathbb{D}}$. Then T_μ induces a bounded operator on H^2 if and only if $\mu = \lambda + \varphi d\theta$, where λ is a Carleson measure on $\mathbb{D}$ and φ is a bounded and nonnegative function on $\partial\mathbb{D}$.

Proof. Since $A(\mathbb{D})$ is dense in H^2, T_μ induces a bounded inear operator on H^2 if and only if there exists a constant $C > 0$ such that

$$\|T_\mu f\| \le C\|f\|, \quad f \in A(\mathbb{D}),$$

where $\|\ \ \|$ is the norm in H^2. Let $\langle\ ,\ \rangle$ be the inner product in H^2. Then it is easy to see that

$$\langle T_\mu f, g \rangle = \int_{\mathbb{D}} f(z)\overline{g(z)}d\mu(z)$$

and so

$$\langle T_\mu f, f \rangle = \int_{\mathbb{D}} |f(z)|^2 d\mu(z)$$

for all f and g in $A(\mathbb{D})$. If $\mu = \lambda + \varphi d\theta$, then the first identity above along with the Cauchy-Schwarz inequality shows that T_μ is bounded on H^2.

To prove the other part of the theorem we decompose μ as follows:

$$\mu = \lambda + \varphi d\theta + \mu_0,$$

where λ is the restriction of μ to $\mathbb{D}$, $\varphi d\theta$ is the absolutely continuous (with respect to $d\theta$) part of the restriction of μ to $\partial\mathbb{D}$, and μ_0 is remaining singular (with respect to $d\theta$) measure on $\partial\mathbb{D}$. Since μ is positive, each of its three parts above is positive.

Now assume T_μ induces a bounded linear operator on H^2. Then clearly T_λ, T_φ, and T_{μ_0} all induce bounded operators on H^2. The boundedness of T_λ, together with the second identity in the first paragraph of the proof, implies that λ is a Carleson measure. The boundedness of T_φ implies that φ is bounded; this is a well-known fact in the classical theory of Toeplitz operators; see [4]. It remains to show that the boundedness of T_{μ_0} on H^2 forces μ_0 to be the zero measure.

For every $z \in \mathbb{D}$ let k_z be the normalized reproducing kernel of H^2 at z. Thus

$$k_z(w) = \frac{\sqrt{1-|z|^2}}{1-\overline{z}w}, \quad w \in \mathbb{D}.$$

Since each k_z is a unit vector in H^2, and since

$$\tilde{\mu}_0(z) = \langle T_{\mu_0} k_z, k_z \rangle, \quad z \in \mathbb{D},$$

the boundedness of T_{μ_0} implies that $\tilde{\mu}_0$ is bounded in $\mathbb{D}$. Since μ_0 is supported on $\partial\mathbb{D}$, it follows that

$$\tilde{\mu}_0(z) = \int_0^{2\pi} \frac{1-|z|^2}{|1-ze^{-it}|^2}d\mu_0(t)$$

is the Poisson integral of the measure μ_0. In order words, if S is the singular inner function corresponding to μ_0, then

$$|S(z)| = \exp[-\tilde{\mu}_0(z)], \quad z \in \mathbb{D}.$$

The boundedness of $\tilde{\mu}_0$ then shows that $|S|$ is bounded below on $\mathbb{D}$, which is impossible unless $\mu_0 = 0$. This completes the proof of the theorem.

Corollary 2. Suppose μ is finite positive Borel measure on $\overline{\mathbb{D}}$. Then the following conditions are equivalent.

(1) T_μ is bounded on H^2.
(2) $\tilde{\mu}$ is bounded on $\mathbb{D}$.
(3) $\hat{\mu}$ is bounded on $\mathbb{D}$.

Proof. If μ puts no mass on $\partial\mathbb{D}$, then the equivalences are standard facts about Carleson measures. See [4] or [3] for example. The general case follows from decomposing μ into its interior part and boundary part.

3. Compactness

To describe the compactness of the Toeplitz operators T_μ on H^2 we need the notion of vanishing Carleson measures on $\mathbb{D}$. Recall that a positive Borel measure μ on $\mathbb{D}$ is called a vanishing Carleson measure if for every $\varepsilon > 0$ there exists $\delta \in (0,1)$ such that

$$\mu(S_z) < \varepsilon(1 - |z|), \quad \delta < |z| < 1.$$

See [4] for more information about vanishing Carleson measures.

Theorem 3. Suppose μ is a finite positive Borel measure on $\overline{\mathbb{D}}$. Then T_μ induces a compact operator on H^2 if and only if μ puts no mass on $\partial\mathbb{D}$ and μ is a vanishing Carleson measure on $\mathbb{D}$.

Proof. First assume T_μ is compact. Apply Theorem 1 to write $\mu = \lambda + \varphi d\theta$. The compactness of T_μ clearly implies the compactness of T_λ and T_φ. It is well known that T_φ is compact on H^2 if and only if $\varphi = 0$; see [4] for example. The compactness of T_λ implies that

$$\tilde{\lambda}(z) = \langle T_\lambda k_z, k_z \rangle \to 0$$

as $|z| \to 1^-$, since $k_z \to 0$ weakly in H^2 as $|z| \to 1^-$. By Theorem 8.2.5 of [4] the measure λ is a vanishing Carleson measure on $\mathbb{D}$.

Next assume that μ puts no mass on $\partial\mathbb{D}$ and that μ is a vanishing Carleson measure on $\mathbb{D}$. We use an approximation argument to show that T_μ must be compact. For any $r \in (0,1)$ let μ_r be the restriction of μ to

$$\mathbb{D}_r = \{z : |z| < r\}.$$

Carefully examining the proof of Theorem 1, one finds that $\|T_\mu\|$ is comparable to

$$\|\mu\| = \sup\{\hat{\mu}(z) : z \in \mathbb{D}\}.$$

It follows that

$$\|T_\mu - T_{\mu_r}\| \to 0, \quad r \to 1^-,$$

if μ is a vanishing Carleson measure. It is easy to see that each T_{μ_r} is compact on H^2. Thus T_μ is compact whenever μ is a vanishing Carleson measure on $\mathbb{D}$.

Corollary 4. Suppose μ is a finite positive Borel measure on $\overline{\mathbb{D}}$. Then the following conditions are equivalent.

(1) T_μ induces a compact operator on H^2.

(2) $\tilde{\mu}(z) \to 0$ as $|z| \to 1^-$.

(3) $\hat{\mu}(z) \to 0$ as $|z| \to 1^-$.

Proof. By well-known results about Carleson measures it is easy to see that each of conditions (2) and (3) is equivalent to the conditions that μ puts no mass on $\partial\mathbb{D}$ and that μ is a vanishing Carleson measure on $\mathbb{D}$.

4. Trace estimates

In this section we present some partial results about trace estimates for Toeplitz operators T_μ on H^2. Since the compactness of T_μ implies that μ puts no mass on $\partial\mathbb{D}$, at least in the case of positive measures, it suffices to consider measures on the open disk when we consider trace estimates for Toeplitz operators.

Proposition 5. Suppose μ is a vanishing Carleson measure on $\mathbb{D}$. Then T_μ belongs to the trace class on H^2 if and only if the function $(1 - |z|)^{-1}$ is μ-integrable. Moreover, we have the following trace formula

$$Tr(T_\mu) = \int_{\mathbb{D}} \frac{d\mu(z)}{1 - |z|^2}.$$

Proof. Recall that

$$\langle T_\mu f, f \rangle = \int_{\mathbb{D}} |f(z)|^2 d\mu(z)$$

whenever f and g are in the disk algebra $A(\mathbb{D})$. Let $\{e_n\}$ be the canonical orthonormal basis for H^2. Thus for $n \geq 0$ we have $e_n(z) = z^n$, $z \in \mathbb{D}$. It follows that

$$\sum_{n=0}^{\infty} \langle T_\mu e_n, e_n \rangle = \int_{\mathbb{D}} \frac{d\mu(z)}{1 - |z|^2}.$$

This clearly gives the desired result.

Corollary 6. If μ is a vanishing Carleson measure on $\mathbb{D}$, then T_μ belongs to the trace class on H^2 if and only if

$$\int_0^{2\pi} \tilde{\mu}(re^{it})dt = O(1 - r)$$

as $r \to 1^-$.

Proof. It is easy to see that

$$\frac{1}{2\pi} \int_0^{2\pi} \tilde{\mu}(re^{it})dt = \int_{\mathbb{D}} \frac{(1 - r^2)d\mu(w)}{1 - r^2|w|^2}$$

for all $r \in (0, 1)$. Since

$$\int_{\mathbb{D}} \frac{d\mu(w)}{1 - |w|^2} < +\infty$$

if and only if

$$\sup_{0 < r < 1} \int_{\mathbb{D}} \frac{d\mu(w)}{1 - r^2|w|^2} < +\infty,$$

the desired result is then clear.

Note that for any finite positive Borel measure μ on $\mathbb{D}$ the function $\tilde{\mu}(z)/(1 - |z|^2)$ is subharmonic, so that its Laplacian is nonnegative on $\mathbb{D}$.

Corollary 7. If μ is a vanishing Carleson measure on $\mathbb{D}$, then T_μ is a trace class operator on H^2 if and only if

$$\int_{\mathbb{D}} (1 - |z|^2) \Delta \left[\frac{\tilde{\mu}(z)}{1 - |z|^2} \right] dA(z) < +\infty,$$

where Δ is the Laplacian on $\mathbb{D}$.

Proof. Note that

$$\frac{\partial^2}{\partial z \partial \bar{z}} \left[\frac{\tilde{\mu}(z)}{1 - |z|^2} \right] = \int_{\mathbb{D}} \frac{|w|^2 d\mu(w)}{|1 - z\bar{w}|^4}.$$

The desired result is then a consequence of Theorem 5, after an application of Fubini's theorem and Lemma 4.2.2 of [4].

We have only considered trace class Toeplitz operators in this section. It is natural to ask when the Toeplitz operator T_μ belongs to a Schatten class S_p. We do not even know the answer for $p = 2$, the Hilbert-Schmidt case.

5. A moment problem

Given a finite complex Borel measure μ on $\overline{\mathbb{D}}$, the sequence $\{\mu_n\}$ defined by

$$\mu_n = \int_{\overline{\mathbb{D}}} z^n d\mu(z), n \geq 0,$$

is called a moment sequence of μ. In this section we consider a certain generalized moment sequence $\{\mu_{n,m}\}$ defined by

$$\mu_{n,m} = \int_{\overline{\mathbb{D}}} z^n \overline{z}^m d\mu(z), \quad n, m \geq 0.$$

Let M_μ be the infinite matrix whose entry at the nth row and mth column is $\mu_{n,m}$.

Proposition 8. Let μ be a finite positive Borel measure on $\mathbb{D}$. Then μ is a Carleson measure if and only if M_μ is a bounded operator on l^2. And μ is a vanishing Carleson measure if and only if M_μ is a compact operator on l^2.

Proof. This is a consequence of the previous results on boundedness and compactness of T_μ and the easily checked fact that T_μ as an operator on H^2 is unitarily equivalent to M_μ as an operator on l^2, provided one of them is bounded.

Proposition 9. Suppose μ is a finite positive Borel measure on $\mathbb{D}$. For f in $A(\mathbb{D})$ we let $\{f_n\}$ be the sequence defined by

$$f_n = \int_{\mathbb{D}} f(z)\overline{z}^n d\mu(z), \quad n \geq 0.$$

Then μ is a Carleson measure if and only if the mapping $f \mapsto \{f_n\}$ extends to a bounded linear operator from H^2 into l^2. And μ is a vanishing Carleson measure if and only if $f \mapsto \{f_n\}$ extends to a compact linear operator form H^2 into l^2.

Proof. This is again a consequence of our earlier results on boundedness and compactness of Toeplitz operators T_μ on H^2. We omit the routine details.

References

[1] R. Douglas, Banach Algebra Techniques in Operator Theory, Academic Press, New York, 1972.

[2] P. Duren, Theory of H^p Spaces, Academic Press, New York, 1970.

[3] J. Garnett, Bounded Analytic Functions, Academic Press, New York, 1981.

[4] K. Zhu, Operator Theory in Function Spaces, Marcel Dekker, New York, 1990.

Other *Mathematics and Its Applications* titles of interest:

A.M. Samoilenko: *Elements of the Mathematical Theory of Multi-Frequency Oscillations.* 1991, 314 pp.   ISBN 0-7923-1438-7

Yu.L. Dalecky and S.V. Fomin: *Measures and Differential Equations in Infinite-Dimensional Space.* 1991, 338 pp.   ISBN 0-7923-1517-0

W. Mlak: *Hilbert Space and Operator Theory.* 1991, 296 pp. ISBN 0-7923-1042-X

N.Ja. Vilenkin and A.U. Klimyk: *Representation of Lie Groups and Special Functions. Volume 1: Simplest Lie Groups, Special Functions, and Integral Transforms.* 1991, 608 pp.   ISBN 0-7923-1466-2

N.Ja. Vilenkin and A.U. Klimyk: *Representation of Lie Groups and Special Functions. Volume 2: Class I Representations, Special Functions, and Integral Transforms.* 1992, 630 pp.   ISBN 0-7923-1492-1

N.Ja. Vilenkin and A.U. Klimyk: *Representation of Lie Groups and Special Functions. Volume 3: Classical and Quantum Groups and Special Functions.* 1992, 650 pp.   ISBN 0-7923-1493-X

(Set ISBN for Vols. 1, 2 and 3: 0-7923-1494-8)

K. Gopalsamy: *Stability and Oscillations in Delay Differential Equations of Population Dynamics.* 1992, 502 pp.   ISBN 0-7923-1594-4

N.M. Korobov: *Exponential Sums and their Applications.* 1992, 210 pp.
     ISBN 0-7923-1647-9

Chuang-Gan Hu and Chung-Chun Yang: *Vector-Valued Functions and their Applications.* 1991, 172 pp.   ISBN 0-7923-1605-3

Z. Szmydt and B. Ziemian: *The Mellin Transformation and Fuchsian Type Partial Differential Equations.* 1992, 224 pp.   ISBN 0-7923-1683-5

L.I. Ronkin: *Functions of Completely Regular Growth.* 1992, 394 pp.
     ISBN 0-7923-1677-0

R. Delanghe, F. Sommen and V. Soucek: *Clifford Algebra and Spinor-valued Functions. A Function Theory of the Dirac Operator.* 1992, 486 pp.
     ISBN 0-7923-0229-X

A. Tempelman: *Ergodic Theorems for Group Actions.* 1992, 400 pp.
     ISBN 0-7923-1717-3

D. Bainov and P. Simenov: *Integral Inequalities and Applications.* 1992, 426 pp.
     ISBN 0-7923-1714-9

I. Imai: *Applied Hyperfunction Theory.* 1992, 460 pp.   ISBN 0-7923-1507-3

Yu.I. Neimark and P.S. Landa: *Stochastic and Chaotic Oscillations.* 1992, 502 pp.
     ISBN 0-7923-1530-8

H.M. Srivastava and R.G. Buschman: *Theory and Applications of Convolution Integral Equations.* 1992, 240 pp.   ISBN 0-7923-1891-9

Other *Mathematics and Its Applications* titles of interest: